Development and Applications in Solubility

Development and Applications in Solubility

Edited by

T.M. Letcher
University of Kwazulu-Natal, Durban, South Africa

RSC Publishing

ISBN-10: 0-85404-372-1
ISBN-13: 978-0-85404-372-9

A catalogue record for this book is available from the British Library

Published by The Royal Society of Chemistry,
Thomas Graham House, Science Park, Milton Road,
Cambridge CB4 0WF, UK

Registered Charity Number 207890

For further information see our web site at www.rsc.org

Typeset by Macmillan India Ltd, Bangalore, India
Printed by Henry Ling Ltd, Dorchester, Dorset, UK

Foreword

In 2002, the International Association of Chemical Thermodynamics (IACT) was established, and since then it has enthusiastically and effectively continued the role of former International Union of Pure and Applied Chemistry (IUPAC) Commission I.2 on chemical thermodynamics in promoting world-wide activities in the field of thermodynamics. In particular, the importance of thermodynamics in basic research, industrial application and teaching has been highlighted by a series of key publications initiated by IUPAC and continued by IACT: "Chemical Thermodynamics" in 1999, "Chemical Thermodynamics for Industry" in 2004, and "Developments and Applications in Solubility", scheduled to appear in 2006. Contributions originating from IACT are not routine activities: they are fuelled by our intent to show the scientific and engineering community the major role played by thermodynamics in the applied sciences.

Solubility is an extraordinarily wide field, which, therefore, needs a large variety of tools for scientific investigation. This area of physical chemistry encompasses experimental measuring techniques, theory, modelling and simulation, and (industrial) application. With the solubility of solids, liquids and gases in other media themselves being solid, liquid or gaseous, a large range of specialized experimental equipments/methods can be found; most of them are reported in the present book. In general, instrumentation has benefited from the tremendous progress in electronics, micromechanics, informatics *etc.*, which has yielded new sophisticated sensors, detectors and controllers. On the other hand, modelling and *ab initio* calculations are also expanding rapidly. While these computational predictive methods tend to substitute measurements, one must realize that they can never fully replace them.

Science without recognizable application is becoming less important these days. With solubility, this danger does not exist: solubility phenomena are abundant in ordinary, everyday life, in small or large-scale industrial processes, and in the biological field. Key topics are, for instance, the solubility of gases in aqueous systems including biological fluids, the use of supercritical fluids for extraction or separation, or its use as a medium for chemical reactions.

This new book entitled "Developments and Applications of Solubility" assembles 24 chapters authored by renowned specialists. It continues the tradition of its two predecessors by offering high-quality contributions covering great parts of the topic indicated in the title. As President of IACT, I have the pleasure and honour to thank the editor, Professor Trevor M. Letcher, for his strong involvement in this enterprise, and the authors, whose liberally contributed expertise made it possible and will guarantee success.

Jean-Pierre E. Grolier
President of IACT

Preface

Solubility is one of the most basic and important of thermodynamic properties, and a property which underlies most industrial processes. This book is a collection of 24 chapters involving recent research works, all related to *Solubility*. The objective is to bring together research from disparate disciplines which have a bearing on *Solubility*. Links between these chapters, we believe, could lead to new ways of solving problems and looking at new and also old *Solubility*-related issues. Underlying this philosophy is our inherent belief that a book is still an important vehicle for the dissemination of knowledge.

Our book, *Developments and Applications in Solubility*, has its origins in committee meetings of the International Association of Chemical Thermodynamics (IACT). It is a project produced under the auspices of the International Union of Pure and Applied Chemistry (IUPAC). In true IUPAC image, the authors, which represent some of the most important names in their respective fields, come from many countries around the world, including: Australia, Austria, Finland, France, Germany, Ireland, The Netherlands, New Zealand, Portugal, Slovenia, South Africa, Switzerland, Poland, United Kingdom and the United States of America.

The book highlights the theory, techniques, interesting and new results, modeling and simulation, and industrial applications related to *Solubility*. It includes chapters on:

- the fundamentals of solubility in terms of thermodynamics,
- data banks,
- solubility of gases in ionic liquids, polymers, molten salts, water and in sea water,
- solubility phenomena related to "green" chemicals,
- isotope effects,
- inorganic solids in industry,
- organic solids in industry,
- modelling, predictions and simulation techniques including COSMO-RS, and industrial processes including:
- hydrometallurgical leaching,
- impurities in cryogenic liquids,
- BTEX and acid gases,

- reaction design,
- supercritical systems,
- pharmaceutical and cosmetic industries,
- carbon dioxide in chemical processes
- and solubility issues related to the oil industry.

I wish to record my special thanks to Professor Glenn Hefter, Professor Rubin Battino and Dr Justin Salminen who were part of the task team, to the 46 authors and to the publishers, the Royal Society of Chemistry, who have all helped in producing this useful and informative book on the importance and applications of solubility in our chemical industry.

T.M. Letcher
Stratton on the Fosse
Somerset

Contents

Theory, Techniques and Results

Theory, Techniques and Results

CHAPTER 1

Thermodynamics of Nonelectrolyte Solubility

EMMERICH WILHELM

Institute of Physical Chemistry, University of Wien, Währinger Straße 42, Wien (Vienna) A-1090, Austria

> *Magic means rather different things to different people.*
> Brigadier Donald Ffellowes in "*The Kings of the Sea*", by S.E. Lanier, *The Magazine of Fantasy & Science Fiction*, November 1968.

1.1 Introduction

The liquid state is one of the three principal states of matter. The majority of chemical synthesis reactions are carried out in the liquid state, and separation processes usually involve liquid/fluid states, *i.e.* solutions. Thus, not surprisingly, for a century and a half experimental investigations of physical properties of solutions and of phase equilibria involving solutions (vapour–liquid equilibrium: VLE; liquid–liquid equilibrium: LLE; solid–liquid equilibrium: SLE; solid–vapour equilibrium: SVE) have held a prominent position in physical chemistry. The scientific insights gained in these studies can hardly be overrated, and have been of immense value for the development of the highly formalized, general discipline of mixture thermodynamics, for instance by providing idealized solution models, such as the one based on the Lewis–Randall (LR) rule, or the one based on Henry's law (HL). In addition to its profound theoretical interest, this topic includes many important practical, industrial applications in chemical process design, in the environmental sciences, in geochemistry, in biomedical technology and so forth. Water is the most abundant liquid on the earth, and because it sustains life as we know it, it is also the most important liquid solvent. The preponderance of scientific papers dealing with aqueous solutions is thus not surprising. We note that the study of the solubility in water of the rare gases and of simple hydrocarbons have provided fundamental information on hydrophobic effects that are thought to be of pivotal importance for the formation and stability of higher order structures of biological substances, such as proteins, nucleic acids, and cell membranes.

Evidently, this short review cannot possibly be comprehensive, and I shall focus on just a few topics which reflect my current research interests and idiosyncrasies. For instance, VLE with supercritical solutes, that is the solubility of gases in liquids, will be discussed in some detail, and so will the van't Hoff type analysis of high-precision solubility data. SLE and SVE will not be considered at all. Almost inevitably, pride of place will be given to the Henry fugacity,[1] or Henry's law constant, which is one of most misunderstood thermodynamic quantities. The goal is to clarify some points often overlooked, and to dispel misconceptions frequently encountered in the literature.

1.2 Thermodynamics

In this section I will present a brief overview of classical thermodynamics applicable to nonelectrolyte solutions in general,[2] and to solutions of gases in liquids in particular.[1,3–6] When discussing solutions, one is either interested in *single-phase properties*, such as partial molar volumes, or in quantities which characterize the equilibrium *solubility* itself, for instance the amount of substance i, the solute, dissolved in a given amount of solvent j in the presence of *both coexisting phases*. The equations governing VLE and LLE will be considered first. For details see refs. 1 and 2.

A general criterion for phase equilibrium at temperature T and pressure P is the equality of the chemical potential μ_i^π of each constituent component i in all coexisting phases π, or equivalently, the equality of the fugacity f_i^π of each component in all coexisting phases. Thus, for the specific case of VLE (π = V or L),

$$f_i^{\mathrm{V}}(T, P, \{x_i^{\mathrm{V}}\}) = f_i^{\mathrm{L}}(T, P, \{x_i^{\mathrm{L}}\}), \quad i = 1, 2, \ldots, N \tag{1}$$

where N is the number of components present, each with mole fraction x_i^{V} in the vapour phase and x_i^{L} in the liquid phase. Similarly, for LLE (π = L' or L")

$$f_i^{\mathrm{L}'}\left(T, P, \left\{x_i^{\mathrm{L}'}\right\}\right) = f_i^{\mathrm{L}''}\left(T, P, \left\{x_i^{\mathrm{L}''}\right\}\right), \quad i = 1, 2, \ldots N \tag{2}$$

From now on, however, I shall confine attention to *binary* systems, where $i = 1$ or 2.

Two entirely *equivalent* formal procedures are commonly used to establish the link with experimental reality:

(I) When using the fugacity coefficient of component i in solution in phase π, which quantity is defined by

$$\phi_i^\pi(T, P, \{x_i^\pi\}) = f_i^\pi(T, P, \{x_i^\pi\})/x_i^\pi P \tag{3}$$

and adopting the convenient notation $x_i^{\mathrm{V}} = y_i$, $x_i^{\mathrm{L}} = x_i$, and dropping the superscript L where unambiguously permissible, the condition for thermodynamic equilibrium (VLE) may be expressed as

$$y_i \phi_i^{\mathrm{V}}(T, P, y_i) = x_i \phi_i^{\mathrm{L}}(T, P, x_i) \tag{4}$$

and for LLE as

$$x'_i\phi_i^{L'}(T,P,x'_i) = x''_i\phi_i^{L''}(T,P,x''_i) \tag{5}$$

This approach is called, for obvious reasons, the (ϕ, ϕ) method.

(II) In the second procedure, the component fugacities in the vapour phase are again expressed in terms of fugacity coefficients, whereas the liquid-phase fugacities of the components are expressed in terms of appropriately normalized liquid-phase activity coefficients.

When based on the LR rule the convention is called *symmetric*, and the corresponding activity coefficient is given by

$$\gamma_i^{LR}(T,P,x_i) = f_i^{L}(T,P,x_i)/x_i f_i^{L*}(T,P) \tag{6}$$

where the superscript asterisk denotes, as always, a pure substance property: $f_i^{L*}(T,P) = P\phi_i^{L*}(T,P)$ is the *fugacity of pure component i* in either a real or a hypothetical liquid state at (T,P) of the liquid solution, and $\phi_i^{L^*}(T,P)$ is its fugacity coefficient.

When based on HL the convention is called *unsymmetric*, and leads to

$$\gamma_i^{HL}(T,P,x_i) = f_i^{L}(T,P,x_i)/x_i h_{i,j}(T,P) \tag{7}$$

where $h_{i,j}(T,P)$ is the *Henry fugacity of i dissolved in liquid j* at (T,P) of the liquid solution.[1] This quantity is also known as Henry's law constant. It is defined for any phase π (L or V) by

$$\lim_{x_i^\pi \to 0}\left(f_i^\pi/x_i^\pi\right) = \left(df_i^\pi/dx_i^\pi\right)_{x_i^\pi=0} = h_{i,j}^\pi(T,P) \tag{8}$$

where all the operations are at *constant T and P*.

The VLE conditions may thus be recast into

$$\phi_i^{V}(T,P,y_i)y_i P = \gamma_i^{LR}(T,P,x_i)x_i f_i^{L*}(T,P) \tag{9}$$

or, equivalently, into

$$\phi_i^{V}(T,P,y_i)y_i P = \gamma_i^{HL}(T,P,x_i)x_i h_{i,j}(T,P) \tag{10}$$

where the superscript π = L of the Henry fugacity has been dropped for convenience. This approach is called the (ϕ, γ) method.

For LLE we may write either

$$x'_i\gamma_i^{LR}(T,P,x'_i) = x''_i\gamma_i^{LR}(T,P,x''_i) \tag{11}$$

or equivalently

$$x'_i\gamma_i^{HL}(T,P,x'_i) = x''_i\gamma_i^{HL}(T,P,x''_i) \tag{12}$$

For VLE, there exists in principle a third procedure in which the component fugacities in the liquid phase *as well as* in the vapour phase are expressed in terms of activity coefficients ($\gamma_i^{L,LR}$, $\gamma_i^{V,LR}$, $\gamma_i^{L,HL}$, $\gamma_i^{V,HL}$). However, to the best of my knowledge, it has never been utilized.

By definition, for component i in solution in *any* phase π, Equation (3) applies, hence according to Equation (8) the important, *generally valid* relation

$$\begin{aligned}\phi_i^{\pi\infty}(T,P) &= \lim_{x_i^\pi \to 0} \phi_i^\pi(T,P,x_i^\pi) \\ &= \frac{1}{P} \lim_{x_i^\pi \to 0} \left[f_i^\pi(T,P,x_i^\pi)/x_i^\pi \right] \\ &= h_{i,j}^\pi(T,P)/P \end{aligned} \tag{13}$$

is obtained,[1,4–6] where $\phi_i^{\pi\infty}(T,P)$ is the fugacity coefficient of i at infinite dilution in the phase π.

As this juncture, several points should be emphasized. While

$$f_i^{\mathrm{L}*}(T,P)/P = \phi_i^{\mathrm{L}*}(T,P) = \exp\left(G_i^{\mathrm{R,L}*}(T,P)/RT\right) \tag{14}$$

is a property, at (T,P), of *pure liquid i* ($G_i^{\mathrm{R,L}*}$ denotes the residual molar Gibbs energy), the Henry fugacity defined by Equation (8) for the liquid solution phase (π = L) is a *liquid-phase* property which depends on (T,P) and the chemical identity of *both* solute i and solvent j (hence the double subscript!):[1,4–6]

$$h_{i,j}(T,P)/P = \phi_i^{\mathrm{L}\infty}(T,P) = \exp\left(\mu_i^{\mathrm{R,L}\infty}(T,P)/RT\right) \tag{15}$$

Here, $\phi_i^{\mathrm{L}\infty}$ is the fugacity coefficient of component i at infinite dilution in the liquid solvent j, and $\mu_i^{\mathrm{R,L}\infty}$ is the corresponding residual chemical potential.

The various quantities corresponding to the two conventions introduced above are, of course, related. For instance,

$$\gamma_i^{\mathrm{LR}} = \phi_i^{\mathrm{L}}/\phi_i^{\mathrm{L}*} \tag{16}$$

$$\gamma_i^{\mathrm{HL}} = \phi_i^{\mathrm{L}}/\phi_i^{\mathrm{L}\infty} \tag{17}$$

$$\gamma_i^{\mathrm{LR}\infty} = h_{i,j}/f_i^{\mathrm{L}*} \tag{18}$$

where $\gamma_i^{\mathrm{LR}\infty}$ denotes the activity coefficient at infinite dilution. For details see ref. 1 and the literature cited therein.

Equations (4), (9), and (10) may each serve as a rigorous thermodynamic basis for the treatment of VLE. The decision as to which approach should be adopted for solving actual problems is by and large a matter of taste and/or convenience, yet is subject to important practical constraints.

VLE involving fairly simple fluids may conveniently be treated in terms of the (ϕ, ϕ) approach, Equation (4), because the use of a single equation of state (EOS) valid for *both* phases V and L has some computational advantage and a certain aesthetic appeal. However, since no generally satisfactory EOS for dense fluids of practical, that is technical, importance has as yet been developed, this approach is rather limited. The situation is further aggravated by the sensitivity of results on the so-called mixing rules and combining rules,[7,8] which have always an empirical flavour.

At low to moderate pressures, data reduction and VLE calculations are preferably based upon the classical (ϕ, γ) formalism expressed by Equations (9) and (10). Here, an EOS is required only for the low-density vapour phase for which satisfactory models based on virial coefficients are available, while for the liquid phase a suitable activity coefficient model is introduced.

For LLE, similar comments apply: in the majority of cases the (γ, γ) method is used.

Gas solubilities are usually measured at isothermal conditions. Since the equilibrium composition varies with total pressure, for each composition the quantities ϕ_i^{V}, ϕ_i^{L}, γ_i^{LR}, γ_i^{HL}, $f_i^{\mathrm{L}*}$, and $h_{2,1}$ refer to a *different* pressure. For the reduction and correlation of solubility data it is customary and advantageous to select for each temperature the vapour pressure $P_{\mathrm{s},1}(T)$ of the solvent as reference pressure (the subscript s always indicates saturation condition). For temperatures well below the critical temperature of the solvent, the respective correction terms, known as *Poynting integrals*, are usually quite small.[1–3,7,9,10] If so desired, conversion to any other reference pressure is, in principle, straightforward.

According to Equation (8), the Henry fugacity of solute 2 dissolved in liquid solvent 1 is defined by

$$h_{2,1} = \lim_{x_2 \to 0} (f_2^{\mathrm{L}}/x_2) \tag{19}$$

For VLE, because of the phase equilibrium criterion (1), f_2^{L} may be set equal to the fugacity of the solute in the coexisting vapour phase, that is

$$f_2^{\mathrm{L}} = f_2^{\mathrm{V}} = \phi_2^{\mathrm{V}} y_2 P \tag{20}$$

At the vapour pressure $P_{\mathrm{s},1}$, the Henry fugacity pertaining to the *liquid phase* is thus rigorously accessible from *isothermal* VLE measurements at decreasing total pressure $P \to P_{\mathrm{s},1}$ according to

$$h_{2,1}(T, P_{\mathrm{s},1}) = \lim_{x_2 \to 0} \left[\phi_2^{\mathrm{V}}(T, P, y_2) y_2 P/x_2\right] \tag{21}$$

Entirely equivalent expressions relating the Henry fugacity to limiting slopes, (see Equation (8)), may be derived. We note that from the VLE measurements at $P > P_{\mathrm{s},1}$ the liquid-phase activity coefficient γ_2^{HL} may be extracted, though frequently experimental imprecision precludes obtaining reliable results.

Another versatile and widely used measure of the solubility of a gas in a liquid is the *Ostwald coefficient*.[1,3,9–11] It is defined by

$$L_{2,1}(T, P) = \left(\rho_2^{\mathrm{L}}/\rho_2^{\mathrm{V}}\right)_{\mathrm{equil}} \tag{22}$$

where $\rho_2 = n_2/v = x_2/V = x_2\rho$, with the appropriate superscript L or V, is the amount-of-substance concentration of solute 2 in either the liquid-phase solution or in the *coexisting* vapour-phase solution at T and equilibrium pressure P.

The amounts of solvent 1 and solute 2 are denoted by n_1 and n_2, respectively, $v = (n_1 + n_2)V$, $V = \rho^{-1}$ is the molar volume of the solution (L or V), and ρ is the (total) molar density of the solution. Thus in contradistinction to the Henry fugacity, the Ostwald coefficient is a *distribution coefficient* pertaining to the solute dissolved in the coexisting phases L and V. It therefore *always* refers to T and P of the *actual* VLE. After some algebraic manipulation one can show that[12]

$$\begin{aligned} L_{2,1}^{\infty}(T, P_{s,1}) &= \lim_{P \to P_{s,1}} L_{2,1}(T, P) \\ &= \frac{RT}{h_{2,1}(T, P_{s,1}) V_{s,1}^{L*}} Z_{s,1}^{V*} \phi_2^{V\infty}(T, P_{s,1}) \end{aligned} \tag{23}$$

where $Z_{s,1}^{V*} = P_{s,1} V_{s,1}^{V*}/RT$ is the compression factor of pure saturated solvent vapour, $V_{s,1}^{V*}$ is the molar volume of pure saturated solvent vapour, $V_{s,1}^{L*}$ is the molar volume of pure saturated liquid solvent, and $\phi_2^{V\infty}$ is the fugacity coefficient of the solute in the vapour phase at infinite dilution. When correlating solubility data over wide temperature ranges up to the critical point, it might be advantageous to use $L_{2,1}^{\infty}$ instead of $h_{2,1}$.[13,14]

The most important application of VLE relations is in the design of zseparation processes. A frequently used measure of the tendency of a given component to distribute itself in one or the other equilibrium phase is the *vapour-liquid distribution* coefficient or *K-value* of solute 2 in solvent 1, $K_{2,1}(T,P) = (y_2/x_2)_{equil}$. Using Equation (4) the *general* expression

$$K_{2,1}(T, P) = \phi_2^{L}(T, P, x_2)/\phi_2^{V}(T, P, y_2) \tag{24}$$

is obtained, which establishes the link with EOS calculations. The infinite-dilution limit of this quantity may thus also be expressed as[1,6,13]

$$K_{2,1}^{\infty}(T, P_{s,1}) = h_{2,1}(T, P_{s,1})/P_{s,1}\phi_2^{V\infty}(T, P_{s,1}) \tag{25}$$

$$K_{2,1}^{\infty}(T, P_{s,1}) = \gamma_2^{LR\infty}(T, P_{s,1})\phi_2^{L*}(T, P_{s,.1})/\phi_2^{V\infty}(T, P_{s,1}) \tag{26}$$

or

$$K_{2,1}^{\infty}(T, P_{s,1}) = V_{s,1}^{V*}/V_{s,1}^{L*} L_{2,1}^{\infty}(T, P_{s,1}) \tag{27}$$

Infinite-dilution quantities are usually used for selecting selective solvents for extractive distillation or extraction ($\gamma_i^{LR\infty}$ is needed) or gas absorption (h_{ij} is needed) (see, for instance, ref. 15).

1.3 Subtleties of Approximation

Taking into account the pressure dependence of $h_{ij}(T,P)$ and $\gamma_2^{HL}(T, P, x_2)$,[1,3] the equilibrium criterion Equation (10) may be recast into the *key equation* for *isothermal* VLE data treatment (data reduction and correlation) within the *unsymmetric* convention:

$$\ln \gamma_2^{\mathrm{HL}}(T, P_{\mathrm{s},1}, x_2) = \ln\left(\frac{\phi_2^{\mathrm{V}}(T, P, y_2)y_2 P}{x_2 h_{2,1}(T, P_{\mathrm{s},1})}\right) - \int_{P_{\mathrm{s},1}}^{P} \frac{V_2^{\mathrm{L}}(T, P, x_2)}{RT}\,\mathrm{d}P \tag{28}$$

This equation provides the rigorous basis for the determination of the activity coefficients γ_2^{HL} from *isothermal* solubility data measured at various total pressures P. The argument of the logarithmic term on the right-hand side of Equation (28) is a dimensionless group containing the experimental data, the Henry fugacity already extracted therefrom *via* Equation (21), and the vapour-phase fugacity coefficient of the solute which must be either known from independent experiments or calculated from a suitable EOS, say, the virial equation. In order to evaluate the second term on the right-hand side, *i.e.* the Poynting integral, information is needed on the pressure dependence as well as the composition dependence of the partial molar volume V_2^{L} of the solute in the liquid phase. Each data point thus yields a constant-temperature, constant-pressure activity coefficient $\gamma_2^{\mathrm{HL}}(T,\ P_{\mathrm{s},1},\ x_2)$, which may be represented as a function of composition by any appropriate correlating equation compatible with the number and the precision of the experimental results. This is, then, the reward for exacting and tedious experimental work on the solubility of a gas in a liquid: the *Henry fugacity* $h_{2,1}(T,P_{\mathrm{s},1})$ and a *correlating equation for* $\gamma_2^{\mathrm{HL}}(T,\ P_{\mathrm{s},1},\ x_2)$. This classical sequential approach is almost universally adopted in this field and simply reflects the focusing of interest on the solute in a composition range close to pure solvent.

In the key relation (28), the influence of composition on the liquid-phase fugacity has been separated formally from the influence of pressure. However, rigorous evaluation of the Poynting integral would require detailed knowledge of the pressure dependence *and* the composition dependence of the partial molar volume at each temperature of interest. Such comprehensive information is rarely available, whence for the great majority of solutions approximations at various levels of sophistication must be introduced to make the problem tractable.[1,3] The situation becomes particularly unsatisfactory at high pressures and/or when the critical region is approached, where Poynting corrections become significant. In fact, theoretical models predict that the partial molar volume of the solute is proportional to the compressibility of the solvent near its critical point,[16] with the effects of this divergence being already felt relatively far from the critical point.[17] The pioneering experiments of Wood and collaborators[18,19] have fully confirmed these expectations.

With few exceptions, typical gas-solubility measurements do not cover large composition ranges, while at the same time experimental scatter often tends to obscure the composition dependence of any derived activity coefficient. Thus, practicality usually dictates very simple correlating equations for γ_2^{HL} containing rarely more than one adjustable parameter. Using a two-suffix *Margules* equation and approximating $V_2^{\mathrm{L}}(T,\ P,\ x_2)$ by a pressure-independent partial molar volume at infinite dilution $V_2^{\mathrm{L}\infty}(T, P_{\mathrm{s},1})$, the *Krichevsky–Ilinskaya* equation[20] is obtained:

$$\ln\left(\frac{\phi_2^{\mathrm{V}}(T,P,y_2)y_2P}{x_2h_{2,1}(T,P_{\mathrm{s},1})}\right) - \frac{(P-P_{\mathrm{s},1})V_2^{\mathrm{L}\infty}(T,P_{\mathrm{s},1})}{RT} = A\left(x_1^2-1\right) \qquad (29)$$

where $A = A(T, P_{\mathrm{s},1})$ is a system-specific parameter. The error introduced by assuming $V_2^{\mathrm{L}\infty}$ to be pressure-independent may be estimated, for instance, *via* a *modified Tait* equation.[3,21] If one now assumes $\gamma_2^{\mathrm{HL}} = 1$, independent of composition, the *Krichevsky–Kasarnovsky* equation[22] is obtained:

$$\ln\left(\frac{\phi_2^{\mathrm{V}}(T,P,y_2)y_2P}{x_2h_{2,1}(T,P_{\mathrm{s},1})}\right) = \frac{(P-P_{\mathrm{s},1})\,V_2^{\mathrm{L}\infty}(T,P_{\mathrm{s},1})}{RT} \qquad (30)$$

It has frequently been used for the determination of $V_2^{\mathrm{L}\infty}$ from gas-solubility measurements at elevated pressures. However, the solubility may then be already appreciable and hence the underlying assumptions too severe. Values of $V_2^{\mathrm{L}\infty}$ obtained in this way should always be regarded with caution and may be unreliable.

The next popular simplification neglects the Poynting term, which leads to

$$\phi_2^{\mathrm{V}}(T,P,y_2)y_2P = x_2h_{2,1}(T,P_{\mathrm{s},1}) \qquad (31)$$

And finally, with the assumption $\phi_2^{\mathrm{V}} = 1$, that is the vapour phase is regarded as a perfect-gas mixture, the simplest and most familiar version of HL,

$$P_2 = x_2h_{2,1}(T,\ P_{\mathrm{s},1}) \qquad (32)$$

is obtained, where $P_2 = y_2P$ is the partial pressure of the gaseous solute.

Evidently, the partial molar volume of the solute in the liquid solution is of importance in the reduction and correlation of accurate gas-solubility measurements. The preferred experimental methods for its determination are either precision dilatometry or precision densimetry.[18,19,23–28] For a survey of estimation methods see refs. 1, 3, and 4. Of special note is the capability of semi-empirical versions of scaled particle theory[29,30] to predict $V_2^{\mathrm{L}\infty}(T, P_{\mathrm{s},1})$ reasonably well, even for aqueous solutions, where the minima found experimentally for argon and oxygen dissolved in water,[18,25,26] respectively, are semi-quantitatively reproduced:[10,31]

$$V_2^{\mathrm{L}\infty}(T,P_{\mathrm{s},1}) = V_{\mathrm{cav}} + \kappa_{T,\mathrm{s},1}^{\mathrm{L}*}(\mu_{\mathrm{int}} + RT) \qquad (33)$$

Here, V_{cav} is the partial molar volume associated with cavity formation, $\kappa_{T,\mathrm{s},1}^{\mathrm{L}*}$ the isothermal compressibility of the pure liquid solvent at saturation, and μ_{int} the partial molar Gibbs energy of interaction. For many solvents, a self-consistent set of *effective* Lennard–Jones (6,12) parameters has been given by Wilhelm and Battino.[32] The correlational and predictive powers of this method can be substantially improved by introducing the concept of *temperature-dependent* size parameters.[33,34]

As pointed out above, when using the classical sequential approach exemplified by Equations (21) and (28), a vapour-phase EOS is required for calculating $\phi_2^{\mathrm{V}}(T, P, y_2)$. The majority of gas-solubility measurements lie in the low to moderate pressure domain, whence the virial EOS, either explicit in pressure or in molar volume, is most convenient. The computational convenience associated with a volume-explicit EOS leads to the widely used approximation for the mixture compression factor at fairly *low pressures*,

$$\begin{aligned} Z^{\mathrm{V}}(T, P, y_2) &= PV^{\mathrm{V}}/RT \\ &= 1 + (RT)^{-1}P(y_1B_{11} + y_2B_{22} + y_1y_2\delta_{12}) \end{aligned} \tag{34}$$

where the second virial coefficients with identical subscripts refer to pure components 1 and 2, respectively, and

$$\delta_{12} = 2B_{12} - (B_{11} + B_{22}) \tag{35}$$

B_{12} designates a composition-independent interaction virial coefficient (cross-coefficient). The corresponding expression for the vapour-phase fugacity coefficient is

$$\ln \phi_i^{\mathrm{V}} = \frac{P}{RT}\left(B_{ii} + y_j^2\delta_{12}\right), \quad i,j = 1,2, \quad i \neq j \tag{36}$$

The fugacity coefficient of the solute at infinite dilution in the vapour phase is thus given by

$$\ln \phi_2^{\mathrm{V}\infty} = \frac{P}{RT}(2B_{12} - B_{11}) \tag{37}$$

and the fugacity coefficient of pure component i by

$$\ln \phi_i^{\mathrm{V}*} = PB_{ii}/RT \tag{38}$$

The quite popular rule of thumb, $\phi_2^{\mathrm{V}}(T, P, y_2) = \phi_2^{\mathrm{V}*}(T, P)$, is in general inapplicable for the evaluation of $\phi_2^{\mathrm{V}\infty}$ since it requires $B_{12} = (B_{11} + B_{22})/2$.

Frequently, experimental results on second virial coefficients and/or second interaction virial coefficients[35] are not available. In particular this is the case at low reduced temperatures, where adsorption is significant. Even for water vapour, perhaps the best investigated fluid, the situation below about 400 K is not entirely satisfactory and subject to intensive research.[36,37] Important contributions come from flow calorimetric measurements of the isothermal Joule–Thomson coefficient, which have the advantage that adsorption errors are avoided, and measurements can be made at considerably lower pressures and temperatures than in conventional (P,V,T) methods.[38] Thus one has to rely quite heavily on semi-empirical correlation methods, which are almost all based on the *extended corresponding states theorem*. One of the most popular and reliable methods is that originally due to Tsonopoulos,[39] which since its inception in 1974 has been revised and extended several times:[40]

$$\begin{aligned} B_{ii,\mathrm{r}}(T_\mathrm{r}) \equiv B_{ii}P_{\mathrm{c},i}/RT_{\mathrm{c},i} &= B^{(0)}(T_\mathrm{r}) + \omega_i B^{(1)}(T_\mathrm{r}) \\ &\quad + \hat{a}_i B^{(2)}(T_\mathrm{r}) + \hat{b}_i B^{(3)}(T_\mathrm{r}) \end{aligned} \tag{39}$$

Here, $B_{ii,\mathrm{r}}(T_\mathrm{r})$ is the reduced second virial coefficient of pure substance i at a reduced temperature $T_\mathrm{r} = T/T_{\mathrm{c},i}$, $T_{\mathrm{c},i}$ is the critical temperature, $P_{\mathrm{c},i}$ is the critical pressure, and ω_i is the acentric factor. The $B^{(i)}(T_\mathrm{r})$ are the universal *Tsonopoulos functions*, and $\hat{a}_i$ and $\hat{b}_i$ are quantities for specific compound classes, such as ketones, alkylhalides, 1-alkanols, *etc.* For hydrogen-bonded substances both parameters $\hat{a}_i$ and $\hat{b}_i$ must be used. For instance, for the normal 1-alkanols (except methanol) $\hat{a}_i = 0.0878$ and $\hat{b}_i$ appears to be a function of the reduced dipole moment $\mu_{i,\mathrm{r}}$ defined by[41]

$$\mu_{i,\mathrm{r}} = \left(N_\mathrm{L}\mu_i^2/4\pi\varepsilon_0 V_{\mathrm{c},i} k_\mathrm{B} T_{\mathrm{c},i}\right)^{1/2} \tag{40}$$

Here, N_L is Avogadro's constant, μ_i is the numerical value of the permanent molecular dipole moment of substance i, ε_0 is the permittivity of vacuum, $V_{\mathrm{c},i}$ is the critical molar volume, and k_B is Boltzmann's constant.

If one wishes to use Equation (39) to calculate the reduced second virial cross-coefficient $B_{ij,\mathrm{r}}(T_\mathrm{r}) = B_{ij}P_{\mathrm{c},ij}/RT_{\mathrm{c},ij}$ at a reduced temperature $T_\mathrm{r} = T/T_{\mathrm{c},ij}$, appropriate combination rules have to be devised to obtain the characteristic interaction parameters $T_{\mathrm{c},ij}$, $P_{\mathrm{c},ij}$, $V_{\mathrm{c},ij}$, ω_{ij}, $\hat{a}_{ij}$, and $\hat{b}_{ij}$ to replace the corresponding pure-substance quantities. For details see the original literature and refs. 1, 7, and 8.

Evidently, property estimation methods and correlation methods based on generalized corresponding states approaches require reliable data on critical properties and acentric factors. Since Henry fugacities and related quantities of interest are usually referred to orthobaric conditions, reliable vapour pressure data are indispensable and must be judiciously selected. A valuable source for all these quantities is the book by Poling *et al.*[42] For the most important solvent, water, the International Association for the Properties of Water and Steam (IAPWS) recommends[43] $T_\mathrm{c} = 647.096$ K, $P_\mathrm{c} = 22.064$ MPa, and $\rho_\mathrm{c} = 322\ \mathrm{kg\cdot m^{-3}}$. The molar mass of the international standard water with respect to isotopic composition (Vienna Standard Mean Ocean Water, VSMOW) is $18.015268 \times 10^{-3}\ \mathrm{kg\cdot mol^{-1}}$. An equation representing the vapour pressure of liquid water at most temperatures within current experimental uncertainty (*ca.* $\pm$ 0.025%) has been given by Wagner and Pruss[44] in the form of a six-constant Wagner-type vapour pressure equation.

Once experimental Henry fugacities for a specific solvent-solute system have been collected over a certain temperature range, the question arises as to their most satisfactory mathematical representation as a function of temperature. In the absence of theoretically well-founded models of general validity, essentially empirical fitting equations have to be used, subject however, to some important thermodynamic constraints. Depending on the choice of variables, that is T or T^{-1}, for expanding the enthalpy of solution, either the *Clarke–Glew* equation[45]

$$\begin{aligned}\ln\left[h_{2,1}(T, P_{\mathrm{s},1})/Pa\right] = {} & A_0 + A_1(T/\mathrm{K})^{-1} \\ & + A_2\ln(T/\mathrm{K}) + \sum_{i=3}^{n} A_i(T/\mathrm{K})^{i-2}\end{aligned} \tag{41}$$

or the *Benson–Krause* (BK) equation[46,47]

$$\ln\left[h_{2,1}(T, P_{s,1})/\mathrm{Pa}\right] = \sum_{i=0}^{m} a_i (T/\mathrm{K})^{-i} \tag{42}$$

is obtained. Note that the three-term version of Equation (41) is the well-known *Valentiner* equation.[48] On the basis of the ability to fit accurate $h_{2,1}$ data over reasonably large temperature ranges, and of simplicity, the BK power series in T^{-1} appears to be superior.

In some (elementary) chemistry textbooks there appears to be some confusion concerning the qualitative dependence of solubility on temperature. In fact, the sweeping claim that "the solubility of a gas in a liquid decreases with increasing temperature" is misleading/incorrect when the entire liquid range between the triple point ($T_{t,1}$) and the critical point of the solvent is considered. For many systems, the following behaviour is well documented: at low temperatures near $T_{t,1}$, the solubility expressed as, say, mole fraction solubility x_2 of gas dissolved at a convenient low partial pressure (traditionally, $P_2 = 1$ atm $= 101.325$ kPa), first *decreases* with increasing temperature, then passes through a *minimum* to *increase* steeply when the solvent critical temperature is approached. Such a behaviour is, of course, also reflected by the temperature dependence of the Henry fugacity, that is to say, $h_{2,1}(T,P_{s,1})$ first *increases* with increasing temperature, then goes through a *maximum* to *decrease* steeply, when $T_{c,1}$ is approached, as found , respectively, for argon, krypton, oxygen, methane, *etc.* dissolved in water.[1,9,49]

Any correlation for $h_{2,1}(T,\ P_{s,1})$ extending up to the critical region *must* incorporate the thermodynamically correct *limiting behaviour* of the Henry fugacity for $T \rightarrow T_{c,1}$ and $P_{s,1} \rightarrow P_{c,1}$:[1,4–6,13]

$$\lim_{T \rightarrow T_{c,1}} h_{2,1}(T, P_{s,1}) = P_{c,1}\phi_2^{V\infty}(T_{c,1}, P_{c,1}) \tag{43}$$

This *exact* limiting value follows directly from the generally valid Equation (13) and the equilibrium condition prevailing at the critical point, that is

$$\phi_2^{V\infty}(T_{c,1}, P_{c,1}) = \phi_2^{L\infty}(T_{c,1}, P_{c,1}) \tag{44}$$

No elaborate derivation is necessary.[50] Equation (43) also shows that Hayduk's assertion[51] that the solubilities of gases in a given solvent tend to coincide at a temperature near the solvent's critical is not true.

When the critical point of the solvent is approached along the coexistence curve, for *volatile* solutes the limiting temperature derivative of the Henry fugacity is $-\infty$.[52,53]

During the last 15 years or so, a number of equations for presenting the temperature dependence of $h_{2,1}(T,\ P_{s,1})$ between the triple point temperature and the critical temperature of the solvent were developed to incorporate the thermodynamically correct limiting behaviour indicated above. For details I refer to refs. 1 and 13, and the original literature.[54–57]

Until recently, precision measurements of Henry fugacities over temperature ranges sufficiently large to permit *van't Hoff analysis* of the solubility data, constituted the only reliable source of information on partial molar enthalpy changes on solution, $\Delta H_2^{\infty}(T, P_{s,1}) = H_2^{L\infty} - H_2^{pg^*}$, and *a fortiori* on partial molar heat capacity changes on solution, $\Delta C_{P,2}^{\infty}(T, P_{s,1}) = C_{P,2}^{L\infty} - C_{P,2}^{pg^*}$, of sparingly soluble gases in liquids. Here, $H_2^{L\infty}$ is the partial molar enthalpy of the solute at infinite dilution in the liquid solvent, $C_{P,2}^{L\infty}$ is the partial molar heat capacity at constant pressure of the solute at infinite dilution in the liquid solvent, and $H_2^{pg^*}$ and $C_{P,2}^{pg^*}$ are, respectively, the molar enthalpy and the molar heat capacity at constant pressure of the pure solute in the perfect-gas state (pg). If the BK equation is selected for the correlation of experimental Henry fugacities with temperature,

$$\frac{\Delta H_2^{\infty}(T, P_{s,1})}{RT} = \sum_{i=1}^{m} i a_i (T/\mathrm{K})^{-i} + \frac{V_2^{L\infty}}{R} \frac{\mathrm{d}P_{s,1}}{\mathrm{d}T} \tag{45}$$

and

$$\begin{aligned} \frac{\Delta C_{P,2}^{\infty}(T, P_{s,1})}{R} = & -\sum_{i=2}^{m} i(i-1) a_i (T/\mathrm{K})^{-i} + 2\frac{T}{R}\frac{\mathrm{d}V_2^{L\infty}}{\mathrm{d}T}\frac{\mathrm{d}P_{s,1}}{\mathrm{d}T} \\ & -\frac{T}{R}\left(\frac{\partial V_2^{L\infty}}{\partial P}\right)_T \left(\frac{\mathrm{d}P_{s,1}}{\mathrm{d}T}\right)^2 + \frac{TV_2^{L\infty}}{R}\frac{\mathrm{d}^2 P_{s,1}}{\mathrm{d}T^2} \end{aligned} \tag{46}$$

are obtained.[1,58] Until recently, the supplemental terms in Equations (45) and (46) containing the slope $(\mathrm{d}P_{s,1}/\mathrm{d}T)$ and the curvature $(\mathrm{d}^2P_{s,1}/\mathrm{d}T^2)$ of the orthobaric curve – now referred to in the literature[53,54] as *Wilhelm terms* – have been overlooked. Their contributions increase rapidly with increasing temperature. In fact, $V_2^{L\infty}$ of a gas at infinite dilution in a liquid solvent *diverges* to $+\infty$ at the critical point of the solvent, and the partial molar enthalpy at infinite dilution, $H_2^{L\infty}$ will diverge in exactly the same manner. Since $C_{P,2}^{L\infty} = (\partial H_2^{L\infty}/\partial T)_P$, the partial molar heat capacity at constant pressure at infinite dilution will diverge as $(\partial \kappa_{T,s,1}^{L^*}/\partial T)_P$, *i.e.* $C_{P,2}^{L\infty}$ will tend to $+\infty$ as $T_{c,1}$ is *approached from lower temperatures*, and to $-\infty$, as $T_{c,1}$ is *approached from higher temperatures* (at $P = P_{c,1}$). The important experiments of Wood *et al.* confirm these expectations.[18,19,59,60]

In ref. 1, I have presented a comprehensive comparison of enthalpy changes on solution, $\Delta H_2^{L\infty}$, and heat capacity changes on solution $\Delta C_{P,2}^{L\infty}$, obtained from van't Hoff analysis of *high-precision solubility data* with *calorimetrically* determined values. Besides our own results[9,10,31,61] on Ar, O_2, CH_4, C_2H_6, and C_3H_8 dissolved in liquid water, those of Krause and Benson[54] on the rare gases He through Xe have been included. The calorimetrically determined enthalpy changes on solution were obtained either at the Thermochemistry Laboratory in Lund, Sweden, or in the Chemistry Department of the University of Colorado in Boulder, USA.[62–69] With the exception of one set of *direct* heat capacity measurements on argon dissolved in water,[59] all heat capacity changes on solution were obtained from the temperature dependence of the enthalpy changes on solution, *i.e.* $\Delta C_{P,2}^{\infty} = (\partial \Delta H_2^{\infty}/\partial T)_P$.

Evidently, comparing van't Hoff derived enthalpy changes (*one* differentiation level) and heat capacity changes (*two* differentiation levels) with directly obtained high-quality calorimetric results is a severe test of solubility data. In general, the agreement was found to be completely satisfactory, that is it was usually within the combined experimental error. What better tribute to both experimental ingenuity and state-of-the-art data treatment can one wish for?

1.4 Concluding Remarks

Chemical thermodynamics of solutions continues to be a developing field. The major impetus comes from continuing advances in instrumentation leading to increased precision, accuracy and speed of measurements, and from increasing ranges of application (higher temperatures, higher pressures, smaller concentrations).[70] This is paralleled by advances in the statistical-mechanical treatment of solutions, and by increasingly sophisticated computer simulations which provide new insights and stimulating connections at a microscopic level. In this review, I have concisely presented the thermodynamic formalism relevant to the study of dilute solutions of nonelectrolytes. Two intimately related topics have been dealt with prominently: (a) adequate discussion of solution behaviour in terms of the *Henry fugacity* and related *activity coefficients*, and (b) *reconciliation* of results for caloric quantities derived from solubility measurements, that is *via van't Hoff analysis*, with those measured directly with *calorimeters*. Though outside the scope of this article, I would like to point out the increasing number of solubility studies with a strong biophysical and/or biomedical flavour. While my own perception of their importance may not be shared by all, it appears safe to state that they will greatly stimulate applied research in the coming decade: cross-fertilization is becoming increasingly important.

References

1. E. Wilhelm, in *Measurement of the Thermodynamic Properties of Multiple Phases*, Experimental Thermodynamics, R.D. Weir and Th.W. de Loos (eds), Vol VII, 2005, Elsevier, Amsterdam, 137–176. (Equations (7.101) and (7.102) contain the same *printing error*: in the first term on the right hand side of both equations, *i.e.* in the summation term, the power of (T/K) should be $-i$, instead of -1, that is $(T/\mathrm{K})^{-i}$ should be used).
2. H.C. Van Ness and M.M. Abbott, *Classical Thermodynamics of Nonelectrolyte Solutions*, McGraw-Hill, New York, 1982.
3. E. Wilhelm, *CRC Crit. Rev. Analyt. Chem.*, 1985, **16**, 129.
4. E. Wilhelm, *Fluid Phase Equil.*, 1986, **27**, 233.
5. E. Wilhelm, *Thermochim. Acta*, 1987, **119**, 17.

6. E. Wilhelm, *Thermochim. Acta*, 1990, **162**, 43.
7. J.M. Prausnitz, R.N. Lichtenthaler and E.G. Azevedo, *Molecular Thermodynamics of Fluid-Phase Equilibria*, 3rd edn, Prentice-Hall PTR, Upper-Saddle River, NJ, 1999.
8. F. Kohler, J. Fischer and E. Wilhelm, *J. Mol. Struct.*, 1982, **84**, 245.
9. T.R. Rettich, Y.P. Handa, R. Battino and E. Wilhelm, *J. Phys. Chem.*, 1981, **85**, 3230.
10. T.R. Rettich, R. Battino and E. Wilhelm, *J. Chem. Thermodyn.*, 2000, **32**, 1145.
11. R. Battino, *Fluid Phase Equil.*, 1984, **15**, 231.
12. E. Wilhelm, in *Solubility Data Series (IUPAC)*, R. Battino (ed.), Vol 10. Pergamon Press, Oxford, 1982, pp. XX-XXVIII.
13. E. Wilhelm, in *Molecular Liquids: New Perspectives in Physics and Chemistry*, NATO ASI Series, Series C: Mathematical and Physical Sciences, J.J.C. Teixeira-Dias (ed.), Vol 379. Kluwer, Dordrecht, 1992, 175–206.
14. A.A. Chialvo, Y.V. Kalyuzhnyi and P.T. Cummings, *AIChE J.*, 1996, **42**, 571.
15. J. Gmehling and C. Möllmann, *Ind. Eng. Chem. Res.*, 1998, **37**, 3112.
16. J.C. Wheeler, *Ber. Bunsenges. Phys. Chem.*, 1972, **76**, 308.
17. J.M.H. Levelt Sengers, in: *Supercritical Fluid Technology: Reviews in Modern Theory and Applications*, T.J. Bruno and J.F. Ely (eds.), CRC Press, Boca Raton, 1991, 1–56.
18. D.R. Biggerstaff and R.H. Wood, *J. Phys. Chem.*, 1988, **92**, 1988.
19. L. Hnedkovsky, R.H. Wood and V. Majer, *J. Chem. Thermodyn.*, 1996, **28**, 125.
20. I.R. Krichevsky and A.A. Ilinskaya, *Acta Physicochim. URSS*, 1945, **20**, 327.
21. E. Wilhelm, *J. Chem. Phys.*, 1975, **63**, 3379.
22. I.R. Krichevsky and J.S. Kasarnovsky, *J. Am. Chem. Soc.*, 1935, **57**, 2168.
23. J.C. Moore, R. Battino, T.R. Rettich, Y.P. Handa and E. Wilhelm, *J. Chem. Eng. Data*, 1982, **27**, 22.
24. Y.P. Handa, P. D'Arcy and G.C. Benson, *Fluid Phase Equil.*, 1982, **8**, 181.
25. N. Bignell, *J. Phys. Chem.*, 1984, **88**, 5409.
26. N. Bignell, *J. Phys. Chem.*, 1987, **91**, 1687.
27. P. Izak, I. Cibulka and A. Heintz, *Fluid Phase Equil.*, 1995, **107**, 235.
28. T. Zhou and R. Battino, *J. Chem. Eng. Data*, 2001, **46**, 331.
29. H. Reiss, *Adv. Chem. Phys.*, 1965, **9**, 1.
30. R.A. Pierotti, *Chem. Rev.*, 1976, **76**, 717.
31. T.R. Rettich, R. Battino and E. Wihelm, *J. Solution Chem.*, 1992, **21**, 987.
32. E. Wilhelm and R. Battino, *J. Chem. Phys.*, 1971, **55**, 4012.
33. E. Wilhelm, *J. Chem. Phys.*, 1973, **58**, 3558.
34. G. Schulze and J.M. Prausnitz, *Ind. Eng. Chem. Fundam.*, 1981, **20**, 175.
35. Landolt-Börnstein, *Numerical Data and Functional Relationships in Science and Technology, New Series*, Group IV: Physical Chemistry, Vol 21: *Virial Coefficients of Pure Gases and Mixtures*; Subvolume A: *Virial Coefficients*

of Pure Gases, Subvolume B: *Virial Coefficients of Mixtures*, M. Frenkel and K.N. Marsh (eds.), Springer, Berlin, 2002 and 2003.
36. L. Haar, J.S. Gallagher and G.S. Kell, *NBS/NRC Steam Tables*, Hemisphere Publishing Corporation, New York, 1984.
37. A.H. Harvey and E.W. Lemmon, *J. Phys. Chem. Ref. Data*, 2004, **33**, 369.
38. M.L. McGlashan and C.J. Wormald, *J. Chem. Thermodyn.*, 2000, **32**, 1489.
39. C. Tsonopoulos, *AIChE J.*, 1974, **20**, 263.
40. C. Tsonopoulos and J.H. Dymond, *Fluid Phase Equil.*, 1997, **133**, 11.
41. S. Bo, R. Battino and E. Wilhelm, *J. Chem. Eng. Data*, 1993, **38**, 611. Correction: *ibid.*, 1996, **41**, 644.
42. B.E. Poling, J.M. Prausnitz and J.P. O'Connell, *The Properties of Gases and Liquids*, 5th edn, McGraw-Hill, New York, 2001.
43. IAPWS 1992, reprinted in *Physical Chemistry of Aqueous Systems: Meeting the Needs of Industry*, H.J. White, Jr., J.V. Sengers, D.B. Neumann and J.C. Bellows (eds.), Begell House, New York, 1995, A101–A102.
44. W. Wagner and A. Pruss, *J. Phys. Chem. Ref. Data*, 1993, **22**, 783.
45. E.C.W. Clarke and D.N. Glew, *Trans. Faraday Soc.*, 1966, **62**, 539.
46. B.B. Benson and D. Krause Jr., *J. Chem. Phys.*, 1976, **64**, 689.
47. B.B. Benson, D. Krause Jr. and M.A. Peterson, *J. Solution Chem.*, 1979, **8**, 655.
48. S. Valentiner, *Z. Phys.*, 1927, **42**, 253.
49. R. Crovetto, R. Fernandez-Prini and M.L. Japas, *J. Chem. Phys.*, 1982, **76**, 1077.
50. D. Beutier and H. Renon, *AIChE J.*, 1978, **24**, 1122.
51. W. Hayduk and H. Laudie, *AIChE J.*, 1973, **19**, 1233.
52. W. Schotte, *AIChE J.*, 1985, **31**, 154.
53. M.L. Japas and J.M.H. Levelt Sengers, *AIChE J.*, 1989, **35**, 705.
54. D. Krause Jr. and B.B. Benson, *J. Solution Chem.*, 1989, **18**, 823.
55. A.H. Harvey and J.M.H. Levelt Sengers, *AIChE J.*, 1990, **36**, 539.
56. A.H. Harvey, *AIChE J.*, 1996, **42**, 1491.
57. A.H. Harvey, R. Crovetto and J.M.H. Levelt Sengers, *AIChE J.*, 1990, **36**, 1901.
58. E. Wilhelm, *Thermochim. Acta*, 1997, **300**, 159. (Equation (12) contains a *printing error*: the *minus sign* of the first term on the right hand side is missing, *i.e.* the summation term should be $-\Sigma i(i-1)A_i(T/\mathrm{K})^{-i}$).
59. D.R. Biggerstaff, D.E. White and R.H. Wood, *J. Phys. Chem.*, 1985, **89**, 4378.
60. D.R. Biggerstaff and R.H. Wood, *J. Phys. Chem.*, 1988, **92**, 1994.
61. T.R. Rettich, R. Battino and E. Wilhelm, in preparation. A preliminary version will be presented at the THERMO International 2006 in Boulder, Colorado, USA, 30 July – 4 August 2006.
62. S.J. Gill and I. Wadsö, *J. Chem. Thermodyn.*, 1982, **14**, 905.
63. S.F. Dec and S.J. Gill, *Rev. Sci. Instrum.*, 1984, **55**, 765.
64. S. F. Dec and S.J. Gill, *J. Solution Chem.*, 1984, **13**, 27.
65. G. Olofsson, E. Oshodj, E. Qvarnström and I. Wadsö, *J. Chem. Thermodyn.*, 1984, **16**, 1041.

66. S.F. Dec and S.J. Gill, *J. Solution Chem.*, 1985, **14**, 417.
67. S.F. Dec and S.J. Gill, *J. Solution Chem.*, 1985, **14**, 827.
68. H. Naghibi, S.F. Dec and S.J. Gill, *J. Phys. Chem.*, 1986, **90**, 4621.
69. H. Naghibi, S.F. Dec and S.J. Gill, *J. Phys. Chem.*, 1987, **91**, 245.
70. H.L. Clever and R. Battino, in *The Experimental Determination of Solubilites*, G.T. Hefter and R.P.T. Tomkins (eds.), Wiley, New York, 2003, 101–150.

CHAPTER 2

Thermodynamics of Electrolyte Solubility

EARLE WAGHORNE

School of Chemistry and Chemical Biology, University College Dublin, Belfield, Dublin 4, Ireland

2.1 Introduction

The precipitation of electrolytes has importance in applications ranging from the formation of geological deposits, the fouling of industrial machinery to the purification of fine chemicals. Sometimes we need to avoid precipitation, sometimes to induce it and sometimes to understand what is going to happen under the prevailing conditions. The solubility of an electrolyte is influenced by a wide range of factors, including ion association, variation in ionic activity coefficients, complexation and temperature.

The complexity of these influences, even for a relatively simple system,[1] can be illustrated by consideration of the solubility of $CaSO_4$ in aqueous media containing $NaClO_4$ and Na_2SO_4. Figure 1 shows the variation in $CaSO_4$ solubility as a function of $NaClO_4$ concentration in the simple three component systems, $CaSO_4$, $NaClO_4$ and H_2O. In this case, the solubility passes through a maximum at around 1.5–2 mol kg^{-1} $NaClO_4$. Additionally there is a marked temperature dependence, with the solubility increasing between 273.5 and 298 K and then decreasing substantially at 423 K.

The situation becomes more complex when Na_2SO_4 is added to the system.[1] The results shown in Figure 2 are for the solubility of $CaSO_4$ in solutions having a Na_2SO_4 and $NaClO_4$ at a constant ionic strength. In these cases, the solubility decreases sharply with the concentration of Na_2SO_4. At all concentrations the solubility is highest at the intermediate ionic strength, 1.95 mol dm^{-3}.

Solubility is an equilibrium property and as such is amenable to thermodynamic analysis provided that sufficient information is available. The purpose of this chapter then is to describe the thermodynamic framework within which electrolyte solubilities are described.

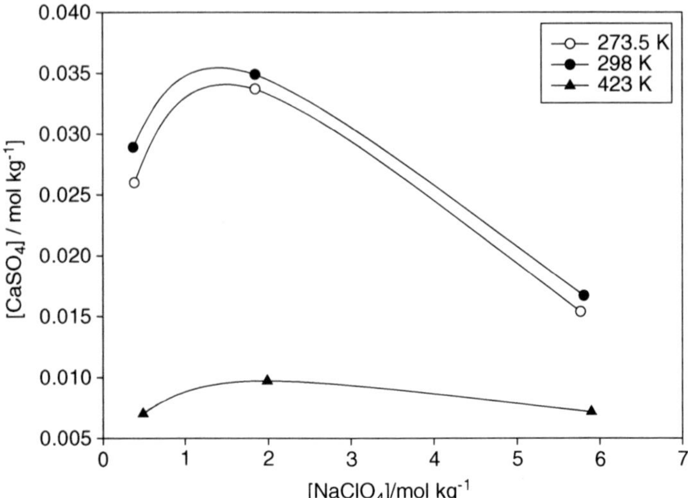

Figure 1 *The solubility of $CaSO_4$ as a function of the concentration of $NaClO_4$ in aqueous solution at 273.5, 298 and 423 K.*

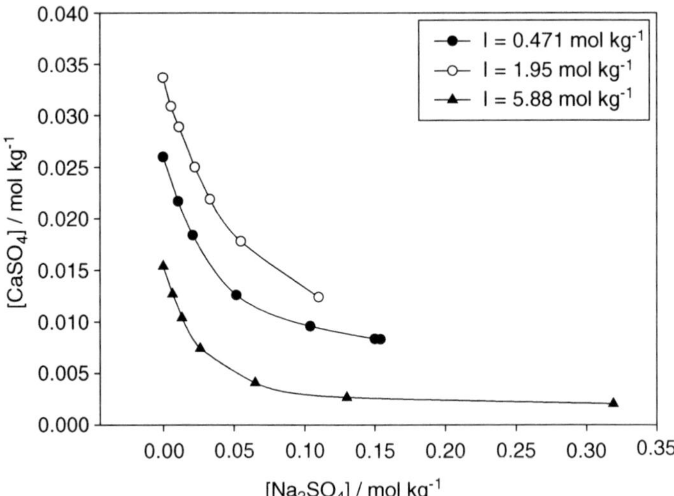

Figure 2 *The solubility of $CaSO_4$ as a function of the concentration of Na_2SO_4 in aqueous solution at ionic strengths 0.471, 1.95 and 5.88 mol kg^{-1}; the ionic strength is maintained by the addition of $NaClO_4$.*

2.2 The Solubility Product

The simplest case to consider is the dissolution of a simple electrolyte in a solvent; dissolution can be represented by the equilibria

$$(A_mB_n)_s \rightleftharpoons mA^{n+} + nB^{m-} \tag{1}$$

$$(A_mB_n)_s \rightleftharpoons (A_mB_n)_{sln} \tag{2}$$

where $(A_mB_n)_s$ represents the solid electrolyte and $(A_mB_n)_{sln}$ the undissociated electrolyte in solution. It is convenient to treat the undissociated electrolyte as an ion pair (see below) in which case the equilibria are written in the alternative, but equivalent form as

$$(A_mB_n)_s \rightleftharpoons mA^{n+} + nB^{m-} \tag{3}$$

$$mA^{n+} + nB^{m-} \rightleftharpoons (A_mB_n)_{sln} \tag{4}$$

Equilibrium (3) is characterized by its equilibrium constant:

$$K_{s0} = \frac{\{A^{m+}\}^n\{B^{n-}\}^m}{\{(A_nB_m)_s\}} \tag{5}$$

where $\{A\}$ represents the activity of A. The standard state for solids is taken as the pure solid at the temperature and pressure of the system, and so the activity of $(A_mB_n)_s$ is unity and Equation (5) becomes the familiar expression for the solubility product of the electrolyte.

$$K_{s0} = \{A^{m+}\}^n\{B^{n-}\}^m \tag{6}$$

The solubility product is connected to the main thermodynamic parameters through the standard state free energy as

$$\Delta_{sln}G^0 = -RT\ln K_{s0} = \Delta_{sln}H^0 - T\Delta_{sln}S^0 \tag{7}$$

where $\Delta_{sln}H^0$ and $\Delta_{sln}S^0$ are the standard state enthalpy and entropy of solution. The temperature effect on K_{s0} is

$$\left(\frac{\partial \ln K_{s0}}{\partial T}\right) = \left(\frac{\Delta_{sln}H^0}{RT^2}\right) \tag{8}$$

The $\Delta_{sln}H^0$ values of electrolytes composed of the larger monovalent ions tend to be positive, resulting in the common decrease in solubility with decreasing temperature, while those of polyvalent ions tend to be negative (exothermic).[2] Thus, the solubilities of simple electrolytes can increase or decrease with increasing temperature.

Systems with more than one electrolyte present a more complex situation. Several solid phases are possible in such cases as, for example, in the case of mixtures of $AgNO_3$ and NaCl in water, where AgCl precipitates at relatively low concentrations. In more complex systems any species for which the solubility product is exceeded will precipitate. In principle for a system with p different cations and q different anions, the number of possible binary electrolyte solid phases is the product of p and q.

2.3 Ion Pairing

Ion pairing can occur in dilute solutions of many electrolytes, particularly those with multivalent ions, and for all electrolytes in sufficiently concentrated solutions. Ion pairing is generally more pronounced in non-aqueous solvents, which have lower dielectric constants than water. For a symmetric electrolyte the uncharged ion pair is prevalent although triple ions and higher aggregates are possible at high concentrations:

$$A^{n+} + B^{n-} \rightleftharpoons (AB)_{sln} \tag{9}$$

Thermodynamically, this is represented by the association constant

$$K_a = \frac{\{AB\}_{sln}}{\{A\}\{B\}} \tag{10}$$

In effect, the ion pairs represent a reservoir of electrolyte in the solution and increase the solubility; that is, for the simple 1:1 ion pair in a binary system, the solubility, s, is given by

$$s = [A] + [AB] = [B] + [AB] \tag{11}$$

It is obvious that the complexity of the system increases for unsymmetric electrolytes or for mixed electrolyte systems. Thus, for example, in a simple AB_2 electrolyte, at least two equilibria need to be considered

$$A^{2+} + 2B^- \rightleftharpoons AB^+ + B^- \rightleftharpoons AB_2 \tag{12}$$

each with its associated association constant, and the total solubility, is

$$s = [A^{2+}] + [AB^+] + [AB_2] \tag{13}$$

Again, at higher concentrations larger aggregates are possible.

In mixed electrolyte solutions, the situation becomes increasingly complex with each added electrolyte. Thus in a simple mixture of two 1:1 electrolytes, AB and MX say, four simple neutral ion pairs are possible, AB, AX, MB and MX. Without explicitly writing the possibilities it is clear that mixed electrolytes containing unsymmetric electrolytes are correspondingly more complex and that the complexity grows with each additional electrolyte.

The ion pairing has been studied for a wide range of systems, including non-aqueous and mixed aqueous solvents, There is one review of ion pairing with limited application to solubility[3] but a relatively large base of experimental data.

Equations (9) and (12) are deceptively simple, and disguise how resistant ion pair formation has been to both theoretical and experimental treatment.

Bjerrum[4,5] proposed, reasonably, that the motion of ions would be coupled when the energy of attraction between them exceeded the thermal energy. For solely coulombic interactions his theory predicts a distance within which the electrostatic attraction between ions is greater than $2kT$, which will be sufficient to couple the motions of the ions; the resulting Bjerrum distance, q, then is

$$q = \frac{|z_1 z_2| e^2}{2\varepsilon kT} \tag{14}$$

where z_1 represents the charge on the ion, e and ε are the electron charge and solvent dielectric constant, respectively, k is Boltzmann's constant and T the

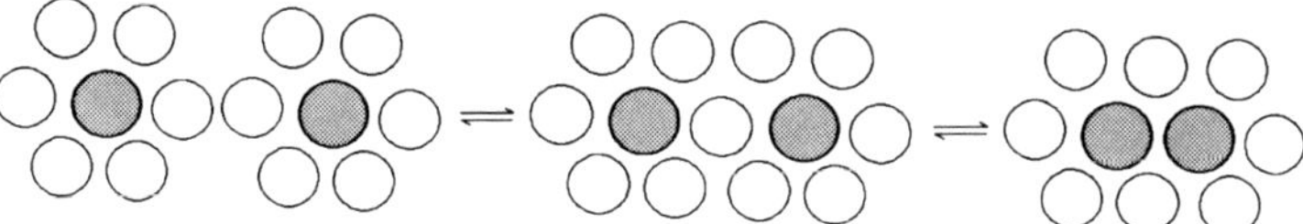

Figure 3 *The equilibria among ion pairs separated by 2 or 1 solvent molecules and contact-ion pairs.*

temperature. The treatment takes account of only electrostatic interactions and neglects the molecularity of the solvent. Nevertheless, for dilute solutions it provides a route to estimating ion pairing at low concentrations. The situation is, however, more complex and the strong interactions between ions and solvent molecules results in three ion pair configurations. Figure 3 illustrates the three commonly assumed structures, the first in which the ions retain their individual solvation shells and so are separated by two solvent molecules, the second in which the ions share some part of their solvation shells and so are separated by one solvent molecule and the third where the ions are in contact and share a common solvation shell. These are generally described as solvent separated and contact ion pairs. In solution it is likely that intermediate structures also occur.

The presence of species such as these creates an experimental difficulty, not; in that different techniques will have different sensitivities to the species present. Thus, for example, conductance will see only the dissociated ions and the presence of ion pairs is determined by difference from the experimental molar conductance and that expected for a strong electrolyte. In contrast spectroscopic techniques, including dielectric relaxation spectroscopy detect the ion pairs directly. However, they are likely to have variable sensitivity to the different types of ion pairs, resulting in technique specific values of the association constant. The resulting scatter in the reported values can be large; for example, the literature values for the ion pair association constant of $MgSO_4$ in water range from *ca.* 0 to 200, expressed on the molar concentration scale.[6–8]

2.4 Complexation

The formation of complexes provides a route to increased solubility. Several equivalent representations of the speciation in these systems have been used.[9] The formation of mononuclear complexes between silver ions and halide or pseudo-halide ions are typical and serve as an example. Thus the interaction of a solid 1:1 electrolyte, MX with excess anions, X^-, can be represented by equilibria with the general form

$$MX_s + (n-1)X^- \rightleftharpoons (MX_n)^{1-n} \tag{15}$$

for which the equilibrium constants have the form

$$K_{sn} = \frac{\{MX_n\}^{1-n}}{\{MX\}\{X^-\}^{n-1}} \tag{16}$$

and we have:

$$MX_s \rightleftharpoons M^+ + X^-, \quad K_{s0} \tag{6}$$

$$MX_s \rightleftharpoons MX, \quad K_{s1} \tag{17}$$

$$MX_s + X^- \rightleftharpoons (MX_2)^-, \quad K_{s2} \tag{18}$$

and so on.

Phenomenologically ion complexation is similar to ion pairing. Thus, we can write the series of equilibria as

$$M^+ + nX^- \rightleftharpoons (MX_n)^{1-n} \tag{19}$$

for which the equilibrium constants, β_n, are given by

$$\beta_n = \frac{\left\{(MX_n)^{1-n}\right\}}{\{M^+\}\{X^-\}^n} \tag{20}$$

or, completely equivalently, we can write the equilibria in the form

$$(MX_n)^{1-n} + X^- \rightleftharpoons (MX_{n+1})^{-n} \tag{21}$$

for which the equilibrium constants, K_n, are

$$K_n = \frac{\{(MX_{n+1})^{-n}\}}{\left\{(MX_n)^{1-n}\right\}\{X^-\}} \tag{22}$$

Clearly, these different representations are equivalent and interconvertible; the relationships among them are

$$\beta_n = K_1 K_2 \ldots K_n \tag{23}$$

$$K_{sn} = K_{s0}\beta_n \tag{24}$$

$$K_n = \frac{K_{sn}}{K_{s(n-1)}} \tag{25}$$

The formation of complexes becomes more important at high concentrations of the complexing ion and is likely to be more extensive in non-aqueous solvents, particularly in dipolar aprotic solvents, where the solvation of anions is weaker, leading to stronger complexation.

Polynuclear complexes can form at high concentrations and for the equilibria

$$mM^+ + nX^- \rightleftharpoons (M_mX_n)^{m-n} \tag{26}$$

the equilibrium constants, β_{mn}, are given by

$$\beta_{mn} = \frac{\{(M_mX_n)^{m-n}\}}{\{M^+\}^m\{X^-\}^n} \tag{27}$$

In mixed electrolyte solutions mixed ligand complexes may also form; again it is convenient to write these as

$$m\mathrm{M}^+ + n\mathrm{X}^- + p\mathrm{Y}^- \rightleftharpoons (\mathrm{M}_m\mathrm{X}_n\mathrm{Y}_p)^{m-n-p} \tag{28}$$

for which the constants, β_{mnp}, are

$$\beta_{mnp} = \frac{\{(\mathrm{M}_m\mathrm{X}_n\mathrm{Y}_p)^{m-n-p}\}}{\{\mathrm{M}^+\}^m\{\mathrm{X}^-\}^n\{\mathrm{Y}^-\}^p} \tag{29}$$

The above was written for complexation involving monovalent ions; however, extension to ions with higher charges or to complexation by uncharged molecules is straightforward. As an example, the data for AgBr in aqueous solutions containing ammonia and added bromide, are listed in Table 1.[10]

The consequences of complex formation in this system can be seen in Figure 4, where the solubility of AgBr is plotted as a function of NH_4Br concentration in

Table 1 *Equilibrium constants for the reactions of AgBr with NH_3 and Br^- in aqueous solutions*

Reaction	*Equilibrium constant*
$AgBr_{(s)} \rightleftharpoons Ag^+ + Br^-$	$K_{s0} = 1.2 \times 10^{-12}$
$AgBr_{(s)} + 2NH_3 \rightleftharpoons (Ag(NH_3)_2)^+ + Br^-$	$K_{s2,0} = 2.6 \times 10^{-5}$
$AgBr_{(s)} + 2NH_3 \rightleftharpoons Ag(NH_3)_2Br$	$K_{s2,1} = 5.3 \times 10^{-5}$
$AgBr_{(s)} + 2NH_3 + Br^- \rightleftharpoons (Ag(NH_3)_2Br_2)^-$	$K_{s2,2} = 7.0 \times 10^{-5}$
$Ag^+ + 2NH_3 \rightleftharpoons (Ag(NH_3)_2)^+$	$\beta_{2,0} = 1.4 \times 10^7$
$Ag^+ + 2NH_3 + Br^- \rightleftharpoons (Ag(NH_3)_2Br)$	$\beta_{2,1} = 4.4 \times 10^7$
$Ag^+ + 2NH_3 + 2Br^- \rightleftharpoons (Ag(NH_3)_2Br_2)^-$	$\beta_{2,2} = 5.8 \times 10^7$

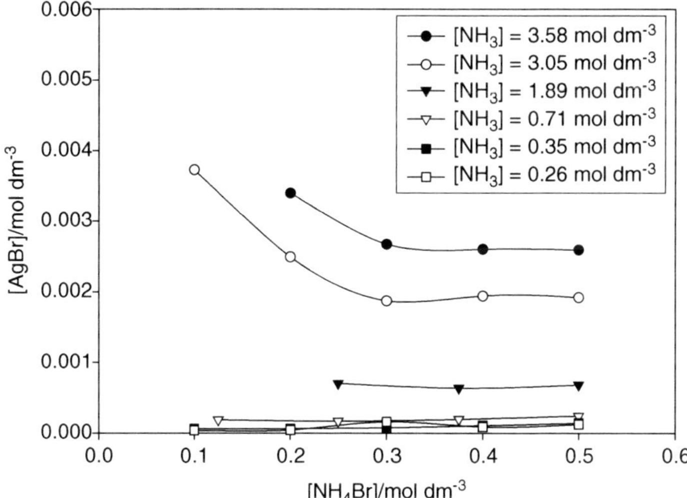

Figure 4 *Solubility of AgBr in solutions containing NH_3, NH_4Br and NH_4ClO_4. Total-added electrolyte concentration is 1.00 mol dm^{-3}.*

solutions containing varying concentrations of NH_3. The total concentration of NH_4^+, and hence of the ionic strength, were maintained at 1 mol dm^{-3} with added NH_4ClO_4. At the lowest concentration of NH_3, the solubility of AgBr increases slightly with increasing Br^- concentration, reflecting complexation by bromide ion. A much larger increase in concentration results from increasing the NH_3 concentration, with the solubility increasing by two orders of magnitude ($4 \times 10^{-5} - 350 \times 10^{-5}$ mol dm^{-3}) for a 10-fold increase in the ammonia concentration. Somewhat surprisingly, at the higher ammonia concentrations the solubility decreases with increasing bromide concentration.

2.5 Electrolyte Activities

The activity, $\{i\}$, of a species, i, is defined by

$$\mu_i = \mu_i^0 + RT\ln\{i\} \tag{30}$$

where μ_i and μ_i^0 are the chemical potential of the species in the system and its chemical potential in its standard state. The activity of the species is written as a function of its concentration, $[i]$, as

$$\{i\} = [i]\gamma_i \tag{31}$$

where γ_i the activity coefficient of the species. The activity coefficient contains all non-ideal contributions to the chemical potential and becomes unity as the composition of the solution approaches that of the standard, or reference, state. Standard states are arbitrarily chosen, typically as an ideal solution having unit concentration. In practice, this is a hypothetical solution in which there are no solute–solute interactions (as at infinite dilution) but the concentration is unity. This is the solution that would be formed if the solute obeyed Henry's law up to unit concentration.

Dissociation of electrolytes introduces two solutes, the anion and the cation into solution. In principle, these solutes have individual activity coefficients but, since there is no experiment that allows the measurement of thermodynamic properties of individual ions, all that can be measured is the mean ionic activity coefficient, $\gamma_\pm$, which is defined in terms of the cationic and anionic activity coefficients, γ_+ and γ_- as

$$\gamma_\pm = \sqrt[m+x]{\gamma_+^m \gamma_-^x} \tag{32}$$

where m and x are the numbers of cations and anions in the undissociated electrolyte.

A search of the literature shows that experimental values of activity coefficients are available for a range of systems, including mixtures of electrolytes but there appear to be no recent reviews of these.[11,12]

Prediction of electrolyte activity coefficients is one of the classical problems in physical chemistry and is elegantly outlined in Robinson and Stokes classic book.[5] The defining characteristic of ions is that they carry a net charge and so the principal interactions between ions, and the largest contribution to the activity coefficient are coulombic. Debye and Hückel solved the problem for a system of purely electrostatic interactions between point charges surrounded by a dielectric continuum. The resulting limiting law

$$\log \gamma_{\pm} = -A|z_+z_-|\sqrt{I} \tag{33}$$

where z_i is the charge on species i, I the ionic strength ($=1/2\ \Sigma[i]\ z_i^2$) and A is given by

$$A = \sqrt{\frac{2\pi N_A}{1000}}\frac{e^3}{2.303\ k^{3/2}}\frac{1}{(\varepsilon T)^{3/2}} \tag{34}$$

should apply for sufficiently dilute solutions and in fact does. In Equation (30), N_A is Avagadro's number, e and k are the electron charge and Boltzmann's constant and ε the dielectric constant (relative permittivity) of the solvent. This takes no account of the ionic volumes and so includes distributions in which the ionic charges lie within the sum of the ionic radii, this becoming increasingly important as the concentration increases. The extended Debye–Hückel equation, taking account of ion sizes is

$$\log \gamma_{\pm} = -\frac{A|z_+z_-|\sqrt{I}}{1 + Ba\sqrt{I}} \tag{35}$$

where a represents the distance of closest approach between the ions and B is given by

$$B = \sqrt{\frac{8\pi N_{\mathrm{A}} e^2}{1000k}}\frac{1}{\sqrt{\varepsilon T}} \tag{36}$$

In aqueous solutions, for which there is a substantial body of data, Equation (35), with a suitable value of a, represents the activity coefficients satisfactorily up to an ionic strength of about 0.1.

At still higher concentrations interactions other than the coulombic interactions between the ionic charges become important, these include ion solvent interactions, which further restrict the volumes accessible to the ionic charges and short-range interactions between the ions. These are assumed to make a contribution to the activity coefficient that is linear in the concentration, leading to

$$\log \gamma_{\pm} = -\frac{A|z_+z_-|\sqrt{I}}{1 + Ba\sqrt{I}} + bI \tag{37}$$

where b is a constant that will vary from electrolyte to electrolyte.

These approaches are compared in Figure 5, where the experimental data for NaCl in aqueous solution[5] are compared with the predictions of the three equations. The fact that Equation (37) fits the data up to around 3 mol dm^{-3}, with only two adjustable parameters, is striking.

While Equation (37) will account for activity coefficients up to relatively high-electrolyte concentrations it requires values for the a and b parameters, which are obtained by fitting the activity coefficients to the equation. Thus, it is very good for interpolating or, with less confidence, extrapolating the activity coefficients.

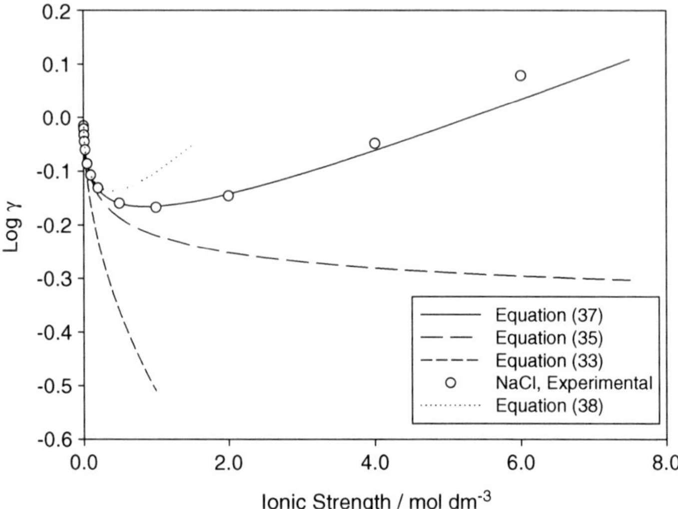

Figure 5 *Comparison of experimental activity coefficients for NaCl in aqueous solutions with those calculated via Equations (32), (34), (36) and (37). The values of the adjustable constants were $a = 4.00$ Å and $b = 0.0551$ mol dm^{-3} (Experimental data were taken from Robinson and Stokes.[5]).*

In the absence of experimental data an equation without adjustable parameters can be used to estimate the activity coefficients. Probably, the most common of these is that due to Davies[13]

$$\log \gamma_{\pm} = -A|z_{+}z_{-}|\left(\frac{\sqrt{I}}{1+\sqrt{I}} - 0.3I\right) \tag{38}$$

which gives reasonable estimates of activity coefficients up to an ionic strength of around 0.5 mol dm^{-3}. In fact, an earlier equation by Davies, in which the coefficient is set to 0.1 rather than 0.3, gives better agreement with activity coefficients of 1:1 electrolytes in water. The log $\gamma_{\pm}$ values calculated using the Davies equation, for a 1:1 electrolyte in water are included in Figure 5.

2.6 Pitzer Theory

In a series of papers, Pitzer and co-workers[14–18] developed an alternative approach, in which the activity coefficient is treated in terms of a virial expansion as

$$\frac{G^{\mathrm{E}}}{RT} = n_{\mathrm{w}}f + \frac{1}{n_{\mathrm{w}}}\sum_{ij}\lambda_{ij}n_i n_j + \frac{1}{n_{\mathrm{w}}^2}\sum_{ijk}\mu_{ijk}n_i n_j n_k \tag{39}$$

where n_{w} is the number of moles of water, n_i, n_j and n_k are the numbers of moles of the i, j and k ions; f is a modified Debye–Hückel term that is not ion specific, and the λ_{ij} and μ_{ijk} are the second and third virial coefficients and effectively take account of short-range interactions between ions, considered two and three at a time. The second virial coefficient is taken to be ionic strength dependent, which allows the treatment of systems where ion pairing occurs, without explicit consideration of the speciation present in the solution.

Differentiation of Equation (39) with respect to n_w or n_i leads to expressions for the osmotic or ionic activity coefficient, respectively; the resulting equations are expressed with a change in variable from numbers of moles to molal concentrations.

There is a considerable reservoir of parameters for aqueous electrolyte solutions so that the approach provides a route to the thermodynamics of electrolytes in complex or highly concentrated solutions where, commonly, there are not sufficient data to carry out a detailed analysis of the species present. The availability of the relevant parameters is much less in non-aqueous or mixed aqueous solvents.

2.7 Treatment of Non-Aqueous or Mixed Aqueous Solvents

There are far fewer data available in non-aqueous or mixed aqueous solvents. There is a considerable body of work for electrolytes at high dilution, which is commonly reported as free energies, enthalpies or entropies of transfer of the electrolyte from some reference solvent. Water is commonly the reference but conversion among reference solvents is, in any case, straightforward. The free energy of transfer, $\Delta_t G^0(R \rightarrow S)$, for example, can be considered to be the difference between the free energies of solution, $\Delta_t G^0(i)$, of the electrolyte in the solvent and reference solvent. That is

$$\Delta_t G^0(R \rightarrow S) = \Delta_{sln} G^0(S) - \Delta_{sln} G^0(R) \tag{40}$$

The enthalpies and entropies of transfer are similarly defined.

The thermodynamic solubility product can be calculated from its value in the reference solvent and the corresponding transfer-free energy as

$$\ln K_{s0}(S) = \ln K_{s0}(R) + \frac{\Delta_t G(R \rightarrow S)}{RT} \tag{41}$$

Equally, the transfer-free energies are calculable from the transfer enthalpies and entropies. Transfer enthalpies and entropies of electrolytes have been reviewed[2,19] and there are earlier reviews containing solution-free energies[20] and transfer-free energies.[21,22] There is also a recent review of transfer-free energies of cations, estimated using and extrathermodynamic assumption;[23] these data allow the calculation of the transfer-free energy of an electrolyte from that of another with a common anion, for example, for a 1:1 electrolyte as

$$\Delta_t G(\mathrm{MX}) = \Delta_t G(\mathrm{NX}) - \Delta_t G(N^+) + \Delta_t G(M^+) \tag{42}$$

References

1. R. Kalyanaraman, L.B. Yeats and W.L. Marshall, *J. Chem. Therm.*, 1973, **5**, 899.
2. G. Hefter, Yu. Marcus and W.E. Waghorne, *Chem. Rev.*, 2002, **102**, 2773.
3. A. Macchioni, *Chem. Rev.*, 2005, **105**, 2039.
4. N. Bjerrum, *Danske Vidensk. Selsk.*, 1926, **7**, 108.
5. R.A. Robinson and R.H. Stokes, *Electrolyte Solutions*, 2nd edn; Butterworths and Co., London, 1970.
6. R. Buchner, T. Chen and G. Hefter, *J. Phys. Chem.*, 2004, **108**, 2365.
7. J.A. Rard, *J. Chem. Thermodyn.*, 1997, **29**, 533.
8. F. Malatesta and R. Zamboni, *J. Solution Chem.*, 1997, **26**, 791.
9. M. Salomon, *IUPAC Solubility Data Series*, 1979, Vol **3**, p. ix, Pergamon Press.
10. I. Leiden and G. Persson, *Acta Chem. Scand.*, 1961, **15**, 1141.
11. W.J. Hamer and Y.-C. Wu, *J. Phys. Chem. Ref. Data*, 1972, **1**, 1047.
12. R.N. Goldberg, B.R. Staples, R.L. Nuttall and R. Arbuckle, *NBS Special Publication*, 1977, **485**.
13. C.W. Davies, *Ion Association*, Butterworths, London, 1962.
14. K.S. Pitzer, *J. Phys. Chem.*, 1973, **77**, 268.
15. K.S. Pitzer and G. Mayorga, *J. Phys. Chem.*, 1973, **77**, 2300.
16. K.S. Pitzer and G. Mayorga, *J. Solution Chem.*, 1974, **3**, 539.
17. K.S. Pitzer and J.J. Kim, *J. Am. Chem. Soc.*, 1974, **96**, 5701.
18. K.S. Pitzer, *J. Solution Chem.*, 1975, **4**, 249.
19. Y. Marcus, *Pure Appl. Chem.*, 1985, **57**, 1103.
20. H.P. Bennetto, *Ann. Rep., A, Phys. Inorg. Chem.*, 1973, **70**, 223.
21. B.G. Cox, *Ann. Rep., A, Phys. Inorg. Chem.*, 1973, **70**, 249.
22. B.G. Cox and W.E. Waghorne, *Chem. Soc. Rev.*, 1980, **9**, 381.
23. C. Kalidas, G.T. Hefter and Y. Marcus, *Chem. Rev.*, 2000, **100**, 819.

CHAPTER 3

Experimental, Calculated and Predicted Solubilities - Basis for the Synthesis and Design of Thermal Separation Processes

JÜRGEN GMEHLING[1] AND WILFRIED CORDES[2]

[1] *Carl von Ossietzky University Oldenburg, Institute for Pure and Applied Chemistry – Industrial Chemistry, D-26111, Oldenburg, Germany*

[2] *DDBST Software and Separation Technology GmbH, Marie-Curie-Straße 10, D-26129, Oldenburg, Germany*

3.1 Introduction

A prerequisite for the synthesis and design of thermal separation processes is a reliable knowledge of the phase equilibrium behavior of the system to be separated. Depending on the separation process, different phase equilibria are of importance. While for the design of distillation processes the vapor–liquid equilibrium (VLE) behavior is required, for absorption, extraction and crystallization processes the solubility of gases, liquids or solids in the liquid phase or the supercritical fluid has to be known. For a safe design of all these separation processes reliable experimental solubility data would be most desirable. Although a large amount of data have been published and stored in factual data banks, the required data are often missing. In these cases thermodynamic models (g^E-models, equations of state) can be applied, which allow the calculation of the phase equilibrium behavior from limited experimental information, *e.g.* binary data. But the number of binary data is restricted. To supplement the solubility database, powerful and reliable predictive models, in most cases group contribution methods, can be applied today.

To use the full potential of experimental, calculated and predicted data, the Dortmund Data Bank (DDB) and reliable thermodynamic models were integrated in a sophisticated software package for the straightforward synthesis and design of thermal separation processes.

3.2 Thermodynamic Fundamentals

In the last few years advanced thermodynamic models with a large range of applicability have been developed to calculate or predict the required solubilities.

To derive suitable expressions for the calculation of phase equilibria one always can start from the isofugacity condition introduced by Lewis:[1]

$$f_i^{\alpha} = f_i^{\beta} \tag{1}$$

To get a connection to the measurable quantities composition, temperature and pressure auxiliary quantities such as activity coefficients γ_i and fugacity coefficients φ_i are required, which account for the deviation from ideal behavior (Raoult's law respectively ideal gas behavior).

The required relations for the calculation of the solubility of liquids, gases and solids in liquids and supercritical gases are summarized below. Starting from the isofugacity condition the following two relations can be derived to calculate liquid–liquid equilibria (LLE) or distribution coefficients K_i:

$$\left(x_i\, \varphi_i^{\mathrm{L}}\right)^{\mathrm{E}} = \left(x_i\, \varphi_i^{\mathrm{L}}\right)^{\mathrm{R}}, \quad K_i = \frac{x_i^{\mathrm{E}}}{x_i^{\mathrm{R}}} = \frac{\left(\varphi_i^{\mathrm{L}}\right)^{\mathrm{R}}}{\left(\varphi_i^{\mathrm{L}}\right)^{\mathrm{E}}} \tag{2}$$

$$\left(x_i\, \gamma_i\right)^{\mathrm{E}} = \left(x_i\, \gamma_i\right)^{\mathrm{R}}, \quad K_i = \frac{x_i^{\mathrm{E}}}{x_i^{\mathrm{R}}} = \frac{\gamma_i^{\mathrm{R}}}{\gamma_i^{\mathrm{E}}} \tag{3}$$

where:

L = liquid phase
x_i = mole fraction of component i in the liquid phase
E = extract phase
R = raffinate phase

Also for the calculation of gas solubilities both auxiliary quantities can be applied.

The resulting equations are

$$x_i\, \varphi_i^{\mathrm{L}} = y_i\, \varphi_i^{\mathrm{V}} \tag{4}$$

$$x_i\, \gamma_i^{*}\, H_{\mathrm{i,j}} = y_i\, \varphi_i^{V}\, P \tag{5}$$

where:

$H_{\mathrm{i,j}}$ = Henry constant
V = vapor phase
y_i = mole fraction of component i in the vapor phase
γ_i^{*} = activity coefficient of component i (asymmetrical normalization)

By a thermodynamic cycle it can be shown that in the case of eutectic systems the solubility of solids in liquids can be calculated using the following slightly simplified equation, when besides the activity coefficients γ_i the phase transition enthalpies (Δh_{m}, Δh_{tr}) and temperatures (T_{m}, T_{tr}) are known. If no phase transition in the solid phase occurs, the second term on the right-hand side can be neglected.

$$x_i \gamma_i^* H_{i,j} = y_i \varphi_i^V P \qquad (6)$$

where:

T_m = melting temperature, K
T_{tr} = transition temperature, K
Δh_m = heat of fusion, J mol^{-1}
Δh_{tr} = heat of transition, J mol^{-1}
R = general gas constant, J mol^{-1} K
T = absolute temperature, K

For the calculation of salt solubilities in water, the equilibrium constant (solubility product) K of the "chemical reaction"

$$MX \cdot nH_2O \rightleftharpoons v_+ M^+_{aq} + v_- X^-_{aq} + nH_2O$$

together with a suitable electrolyte model can be used, whereby the required equilibrium constant K can be obtained directly from tabulated standard thermodynamic properties for the salt and the cations and anions in aqueous solution:

$$K = a_{M^+}^{\nu_+} a_{X^-}^{\nu_-} a_{\mathrm{H_2O}}^n = (m_\pm \gamma_\pm)^\nu a_{\mathrm{H_2O}}^n \qquad (7)$$

$$\gamma_\pm = \left(\gamma_+^{\nu_+} \gamma_-^{\nu_-}\right)^{1/\nu}, \quad m_\pm = \left(m_+^{\nu_+} m_-^{\nu_-}\right)^{1/\nu} \qquad (8)$$

with $\nu = \nu_+ + \nu_-$

where:

a = activity
$m_\pm$ = mean molality
$\gamma_\pm$ = mean activity coefficient
ν = stoichiometric factor
$+$ = cation
$-$ = anion

The solubility of high boiling liquids in supercritical fluids can be calculated in the same way as the VLE behavior using an equation of state.

$$x_i \varphi_i^{\mathrm{L}} = y_i \varphi_i^{\mathrm{V}} \qquad (9)$$

In the case of high boiling solids, the equation of state is only required to account for the fugacity coefficient φ_2^{V} in the vapor phase, while the standard fugacity f_2^{os} (vapor pressure P_2^{s}, Poynting factor Poy_2, saturation fugacity coefficient φ_2^{s}) of the solid 2 is calculated from other information:

$$f_2^{\mathrm{os}} = \varphi_2^{\mathrm{s}} P_2^{\mathrm{s}} \mathrm{Poy}_2 = \varphi_2^{\mathrm{s}} P_2^{\mathrm{s}} \exp \frac{v_2^{\mathrm{s}}(P - P_2^{\mathrm{s}})}{RT} = y_2 \varphi_2^{\mathrm{V}} P \qquad (10)$$

where:

v_2^{s} = molar volume of the solid
P = total pressure

In all cases for the calculation of solubilities as a function of temperature and pressure, either activity coefficients or fugacity coefficients are required. While for the calculation of fugacity coefficients an equation of state, *e.g.* the

Soave–Redlich–Kwong-[2] or the Peng–Robinson[3] equation of state with reliable mixing rules has to be applied, for the calculation of activity coefficients g^E-models, such as the Wilson,[4] NRTL[5] or UNIQUAC[6] model can be used. Both approaches allow the calculation of multicomponent systems using only binary information. When the required binary data are missing, today reliable group contribution methods, such as UNIFAC[7–9] or modified UNIFAC,[10–11] respectively group contribution equations of state, *e.g.* PSRK[12–13] or VTPR[14–15] with a large range of applicability can be applied to predict the required activity or fugacity coefficients.

For the calculation of salt solubilities mean activity coefficients are required. They can be calculated or predicted with the help of an electrolyte model, *e.g.* LIQUAC[16–17] or LIFAC.[17–18]

3.3 Available Solubility Data

As mentioned before, for the straightforward synthesis of absorption, extraction and crystallization processes reliable experimental solubility data under the desired operating conditions would be most desirable. A large number of solubility data have been published. The greatest part of the published data has been stored in factual data banks. The largest factual data bank for thermophysical properties is the Dortmund Data Bank (DDB), which was started in 1973 at the University of Dortmund with the idea of using the vast amount of available experimental VLE data for the development of a group contribution method for the prediction of VLE. While at the beginning only VLE data were stored for the development of original UNIFAC,[8–9] in the meantime all kind of thermophysical properties for pure components and mixtures are stored in this data bank. Since 1989 this data bank is continuously updated by DDBST GmbH,[19] located in Oldenburg/Germany.

With the DDB the engineer has access to nearly all world-wide published experimental data for pure compounds and mixtures. For example, more than 15,600 data sets are available for LLE, approximately 18,000 data sets with gas solubilities for non-electrolyte and electrolyte systems, nearly 12,000 data sets for salt solubilities, more than 16,000 data sets with solid–liquid equilibria (SLE) and an enormous number of solubility data of high boilers (liquids, solids) in supercritical fluids. In the meantime even more than 10,000 experimental data points for systems with polymers, ionic liquids, pharmaceuticals, *etc.* are available. The data stored in DDB are used in-house by a number of companies for the synthesis and design of chemical processes. An internet access to the data base is also available. Only a small part of the data (10–15%) has been published in unified form.[20–24]

3.4 Software Package (DDBSP)

In Figure 1 the rough structure of the software package is shown. It can be applied for various applications of industrial interest. It is particularly useful

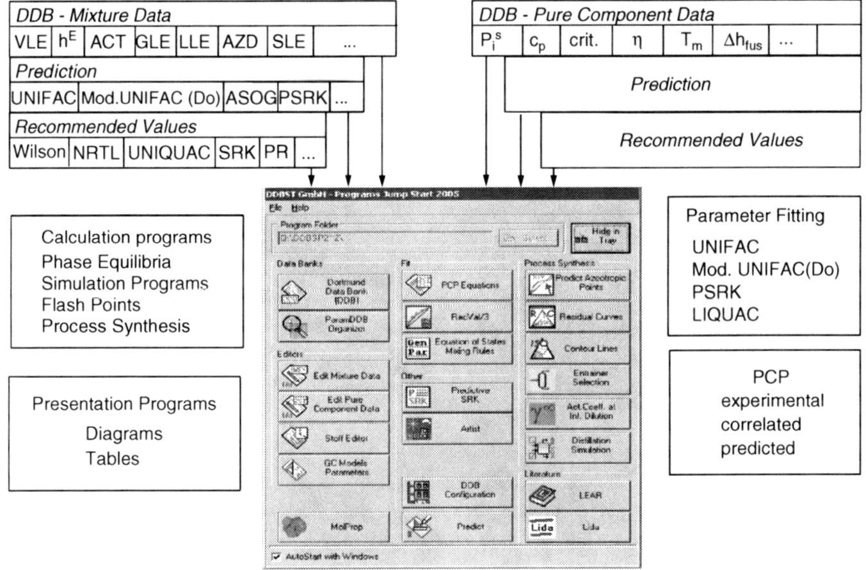

Figure 1 *The DDB Software Package.*

for the synthesis and design of separation processes. With the DDB the user has access to nearly all worldwide available experimental pure component and mixture date (phase equilibria, excess properties). At the same time the user will find the required reliable binary parameters for various g^E-models or equations of state. Furthermore, the program package allows the reliable prediction of missing data using sophisticated group contribution methods, *e.g.* modified UNIFAC, or group contribution equations of state, such as PSRK or VTPR with a large range of applicability.

For the synthesis and design of separation processes a few possible applications of industrial interest are given below. As shown in the examples selected, both procedures,

- access to the available experimental data
- application of thermodynamic models,

can be applied to solve the problem considered.

Example 1

A suitable physical solvent for the removal of the sour gases CO_2 and H_2S from natural gas (CH_4) should be selected.

Following equation (4), equations of state can be used to search for a suitable solvent from a given list of potential solvents. Besides cubic equations of state, like Soave–Redlich–Kwong or Peng–Robinson together with binary parameters $k_{12,}$ also group contribution equations of state can be applied to predict the required gas solubilities. In Figure 2 the predicted results using PSRK for methanol are shown in form of Henry constants for a limited temperature

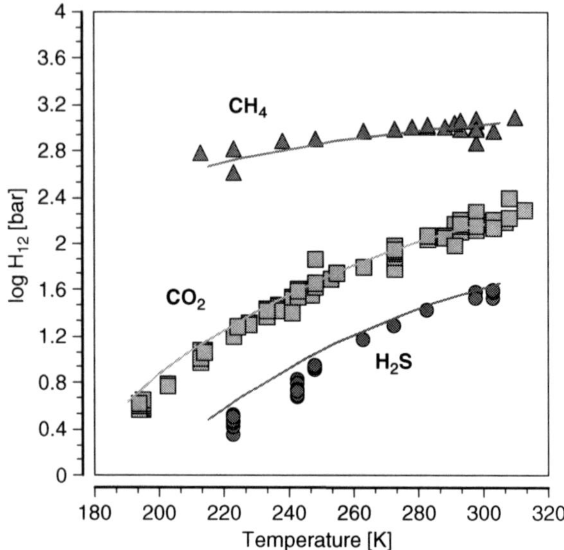

Figure 2 *Experimental and predicted Henry constants for CH_4, CO_2 and H_2S in methanol.*

range together with the experimental data taken from the DDB. As can be recognized, both procedures allow finding suitable solvents, e.g. methanol for the removal of sour gases from natural gas. Since the vapor pressure of methanol is quite high, of course one has to work at low temperature as realized in the Rectisol process.

Example 2

How can KCl be produced from Sylvinit (mainly KCl and NaCl) by crystallization?

To answer this question of course the mutual solubilities of NaCl and KCl in water have to be known. Again this information can be obtained by access to a factual data bank or with the help of an electrolyte model and the required thermodynamic properties. In Figure 3 the mutual solubilities of KCl and NaCl are shown for four different temperatures.[25] As electrolyte model the LIQUAC method was used. It can be seen that there is good agreement between the experimental and predicted solubilities at least up to temperatures of 50°C, whereby the differences in the experimental and predicted solubilities at 100°C mainly arise from the fact that the standard thermodynamic properties are well known only around 25°C. However, from the experimental or predicted solubilities it can be immediately decided how to run the crystallization process in an optimal way.

Example 3

How can pure *m*-xylene and pure *p*-xylene be obtained from *a m*/*p*-xylene mixture by melt crystallization?

Simple melt crystallization in contrast to distillation has the disadvantage that only one of the compounds can be obtained with high purity. Alternatives

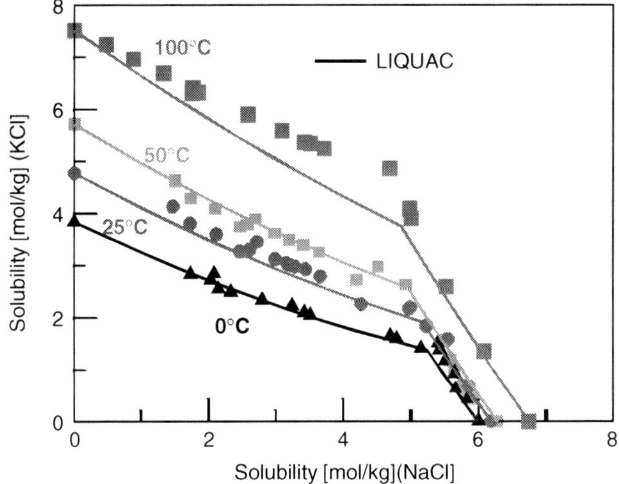

Figure 3 *Experimental and predicted mutual solubilities of KCl and NaCl in water.*

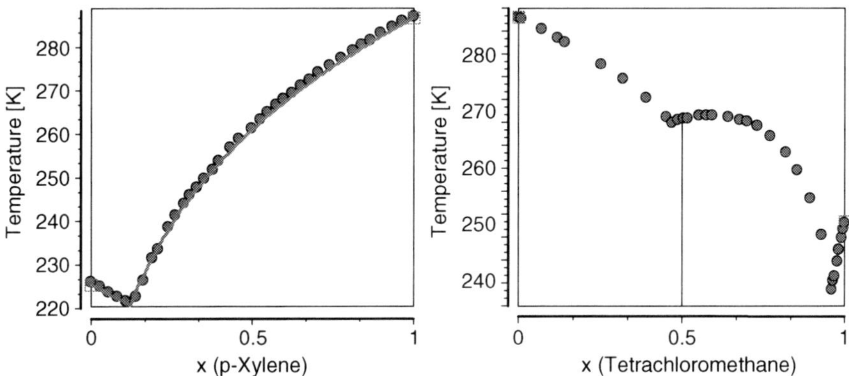

Figure 4 *Experimental SLE data of the systems: (a) p-xylene(1)-m-xylene(2); (b) tetrachloromethane (1)-p-xylene(2).*

are extractive and adductive crystallization. In the case of extractive crystallization thermodynamic models can be applied to predict the phase equilibrium behavior for the various solvents considered. Unfortunately the occurrence of complex formation, required for adductive crystallization, cannot be predicted with the help of thermodynamic models. However a simple search in the stored experimental SLE data results in finding compounds which form a complex with either *m*-xylene or *p*-xylene. Figure 4 shows the experimental and predicted SLE behavior for the system *p*-xylene-*m*-xylene and the experimental results for carbon tetrachloride with *p*-xylene. As can be seen CCl_4 forms the required complex to realize the separation of *m*/*p*-xylene mixtures by adductive crystallization.

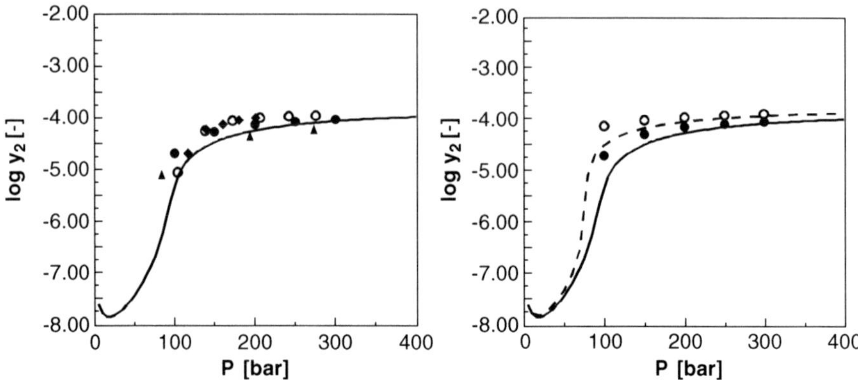

Figure 5 *Solubility of anthracene in supercritical CO_2 at 318.15 K: (a) without a co-solvent (VTPR: —, experimental ● ▲ ○ ◆); (b) with 4 mol% ethanol (VTPR: --, experimental ○); without a co-solvent (VTPR: —, experimental ●).*

Example 4

For an extraction process with supercritical CO_2 we would like to find out: what is the solubility of anthracene in CO_2 at higher pressures and whether the solubility can be increased by a co-solvent, *e.g.* ethanol?

To answer this question again both procedures can be applied. In Figures 5a and b the experimental and predicted results using Equation (10) and the group contribution equation of state VTPR[26] are shown. Again it can be seen that the predicted anthracene solubilities are in qualitative agreement with the slightly scattering experimental data. Even the effect of the co-solvent is correctly predicted.

Example 5

For environmental protection the solubility of various hydrocarbons (hexane, hexene-1, cyclohexane, *etc.*) in water as function of temperature is required.

Again the experimental data can be taken directly from the DDB or predicted with the help of group contribution methods, like UNIFAC or modified UNIFAC. For the UNIFAC methods it is well known, that the predicted hydrocarbon solubilities are not in good agreement with the experimental findings.[27] Therefore access to the DDB would be the first choice. For the temperature range 0 – 100°C the solubilities of the C_5–C_8 alkanes are shown in Figure 6. At the same time the predicted results using modified UNIFAC are given. Surprisingly good agreement between the experimental and predicted solubilities is observed. The reason is that for the prediction of the hydrocarbon solubilities the relation suggested by Banerjee[28] is used, in which the solubilities are calculated from activity coefficients at infinite dilution predicted by modified UNIFAC using the following relation[27] which contains a few additional parameters, *e.g.*:

$$\log c_i^{\mathrm{w,s}} = 1.104 \log \left[\frac{55.56}{\gamma^{\infty}_{\mathrm{org}\to\mathrm{aq},25^{\circ}\mathrm{C}}} \right] + 0.0042\ T - 2.817 \tag{11}$$

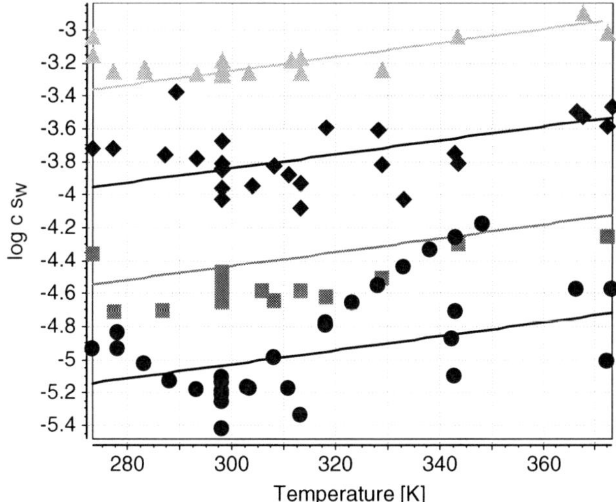

Figure 6 *Experimental and predicted solubilities of various n-alkanes in water in the temperature range 273–373 K: (▲) pentane, (◆) hexane, (■) heptane, (●) octane.*

3.5 Conclusion

As shown already for the design of special distillation processes, such as azeotropic and extractive distillation,[29] reliable experimental phase equilibrium data together with powerful predictive thermodynamic models (group contribution methods, group contribution equations of state, electrolyte models) together with a sophisticated software package are ideal tools for the synthesis of thermal separation processes. In this paper the procedure was demonstrated using various problems for the synthesis and design of absorption, crystallization and extraction processes, where in particular a reliable knowledge of the solubility of gases, liquids or solids in liquids or supercritical fluids is required.

References

1. J. Gmehling and B. Kolbe, *Thermodynamik*, Wiley-VCH, Weinheim, 1992.
2. G. Soave, *Chem. Eng. Sci.*, 1972, **27**, 1197.
3. D.Y. Peng and D.B. Robinson, *Ind. Eng. Chem. Fundam.*, 1976, **15**, 59.
4. G.M. Wilson, *J. Am. Chem. Soc.*, 1964, **86**, 127.
5. H. Renon and J.M. Prausnitz, *AIChE J.*, 1968, **14**, 135.
6. D.S. Abrams and J.M. Prausnitz, *AIChE J.*, 1975, **21**, 116.
7. Aa. Fredenslund, R.L. Jones and J.M. Prausnitz, *AIChE J.*, 1975, **21**, 1086.
8. Aa. Fredenslund, J. Gmehling and P. Rasmussen, *Vapor-Liquid Equilibria Using UNIFAC*, Elsevier, Amsterdam, 1977.
9. H.K. Hansen, M. Schiller, Aa. Fredenslund, J. Gmehling and P. Rasmussen, *Ind. Eng. Chem. Res.*, 1991, **30**, 2352.
10. U. Weidlich and J. Gmehling, *Ind. Eng. Chem. Res.*, 1987, **26**, 1372.

11. J. Gmehling, Jiding Li and M. Schiller, *Ind. Eng. Chem. Res.*, 1993, **32**, 178.
12. T. Holderbaum and J. Gmehling, *Fluid Phase Equilib.*, 1991, **70**, 251.
13. K. Fischer and J. Gmehling, *Fluid Phase Equilib.*, 1995, **112**, 1.
14. J. Ahlers and J. Gmehling, *Fluid Phase Equilib.*, 2001, **191**, 177.
15. J. Ahlers and J. Gmehling, *Ind. Eng. Chem. Res.*, 2002, **41**, 3489.
16. Jiding Li, H.-M. Polka and J. Gmehling, *Fluid Phase Equilib.*, 1994, **94**, 89.
17. J. Kiepe, O. Noll and J. Gmehling, *Ind. Eng. Chem. Res.*, 2006, **45**, 2361.
18. W. Yan, M. Topphoff, C. Rose and J. Gmehling, *Fluid Phase Equilib.*, 1999, **162**, 97.
19. Dortmund Data Bank 2005, www.ddbst.de.
20. J. Gmehling, U. Onken, W. Arlt, J. Rarey, B. Kolbe, P. Goenzhausen and U. Weidlich, *Vapor-Liquid Equilibrium Data Collection*, DECHEMA Chemistry Data Series, 31 parts, Frankfurt, since 1977.
21. J.M. Sørensen, W. Arlt, E.A. Macedo and P. Rasmussen, *Liquid-Liquid Equilibrium Data Collection*, DECHEMA Chemistry Data Series, 4 parts, DECHEMA, Frankfurt since 1979.
22. D. Tiegs, J. Gmehling, A. Medina, M. Soares, J. Bastos, P. Alessi and I. Kikic, *Activity Coefficients at Infinite Dilution*, 4 parts, DECHEMA Chemistry Data Series, Frankfurt, since 1986.
23. J. Gmehling, C. Christensen, T. Holderbaum, P. Rasmussen and U. Weidlich, *Heats of Mixing Data Collection*, DECHEMA Chemistry Data Series, 4 parts, DECHEMA, Frankfurt, since 1988.
24. J. Gmehling, J. Menke, J. Krafczyk and K. Fischer, *Azeotropic Data*, 3 parts, Wiley-VCH, Weinheim, 2004.
25. Jiding Li, Y. Lin and J. Gmehling, *Ind. Eng. Chem. Res.*, 2005, **44**, 1602.
26. J. Ahlers, T. Yamaguchi and J. Gmehling, *Ind. Eng. Chem. Res.*, 2004, **43**, 6569.
27. A. Jakob, H. Grensemann, J. Lohmann and J. Gmehling, *Ind. Eng. Chem. Res.*, submitted.
28. S. Banerjee, *Environ. Sci. Technol.*, 1985, **19**, 369.
29. J. Gmehling and C. Möllmann, *Ind. Eng. Chem. Res.*, 1998, **37**, 3112.

CHAPTER 4

Solubility of Gases in Ionic Liquids, Aqueous Solutions, and Mixed Solvents

GERD MAURER AND ÁLVARO PÉREZ-SALADO KAMPS

Applied Thermodynamics, University of Kaiserslautern, P. O. Box 30 49, D-67653 Kaiserslautern, Germany

4.1 Introduction

The solubility of gases in liquids is of interest in many applications. There have been many publications on this subject over a period of more then a century. Nevertheless, gas solubility is still an important area of research from both an experimental point of view and from a theoretical point of view. This is mainly due to the comparatively large differences in the properties of the species involved (small gas molecules on the one hand, and water, ions, large molecules, *e.g.*, ionic liquids, on the other). This chapter discusses some typical examples from recent work on the solubility of gases in various non-reacting liquids (pure ionic liquids, aqueous solutions of electrolytes, aqueous solutions of organic solvents, and aqueous solutions of an organic solvent and a strong electrolyte) as well as in chemical reacting liquids (*e.g.*, the solubility of sour gases in aqueous ammoniacal/amine solutions).[1–41] The thermodynamic fundamentals and some experimental techniques are discussed, and typical examples are given. Calculations were performed either combining an expression for the excess Gibbs energy (G^E) of the liquid solution with an equation of state (EoS), which is required for calculating the fugacities in the vapor phase, or by employing the Gibbs ensemble Monte Carlo molecular simulation method (GE – MC) which requires intermolecular pair potentials. The G^E/EoS-method at least needs information on the solubility of the single gases in the solvent and – in most cases – also on the interactions between the gas and other solute components. For example, the solubility of ammonia in aqueous solutions of sodium sulfate requires the Henry's constant for the solubility of ammonia in pure water, interaction parameters between ammonia molecules in water as well as between sodium and sulfate ions in water (which all can be derived from binary data alone), but also parameters for interactions between sodium sulfate on one side and ammonia on the other side, in water. These interaction parameters can only be determined from ternary data (in the example on the solubility of ammonia in aqueous solutions of sodium sulfate). The molecular simulation methods require at least the pair potentials for interactions between all species. Most of these pair potentials were adopted from investigations on the pure substances or were estimated by quantum chemistry approximations. They were then used without any additional parameters to predict the solubility of a gas in a liquid.

4.2 Fundamentals

In modeling the solubility of gases in liquid solutions one distinguishes between solute components (*i.e.*, gases and electrolyte components such as non-volatile salts) and solvent components. Solvent components can be volatile (*e.g.*, water and organic liquids) and non-volatile (*e.g.*, non-volatile liquids – like most ionic fluids). Often an organic liquid compound is also called a solute, particularly when its concentration is small. The vapor-liquid equilibrium is commonly described by applying Henry's law (or its extension) for a volatile solute component and Raoult's law (or its extension) for a volatile solvent component. The concentration of the dissolved gas i (as well as the concentration of any solute s) is here expressed by its molality m_s, *i.e.*, the number of moles of solute s per kilogram of solvent. The extended Henry's law for a gaseous solute i is

$$k_{H,i}^{(m)}(T,p,\tilde{x}_j)a_i^{(m)}(T,p,\tilde{x}_j,m_s) = f_i''(T,p,y_k) \tag{1}$$

$k_{\mathrm{H},i}^{(m)}(T,p,\tilde{x}_j)$ is Henry's constant of gas i (based on the molality scale) at temperature T and pressure p in the particular solvent (mixture). The composition of the solvent mixture is here expressed by (solute-free) mole fractions $\tilde{x}_j$. Henry's constant can be interpreted as the fugacity of gas i in its reference state for the liquid phase. It can also be interpreted as the vapor pressure of the "pseudo-pure component" gas i. Here, the reference state for the chemical potential of gas i in the liquid phase is a one molal solution of that gas in the solvent (or in the solvent mixture) which experiences interactions as in infinite dilution in that solvent. $a_i^{(m)}(T,p,\tilde{x}_j,m_s)$ is the activity of gas i in the liquid phase. It depends on temperature, pressure, and the composition of the liquid phase. However, the influence of pressure on that activity can usually be neglected. The activity $a_i^{(m)}(T,\tilde{x}_j,m_s)$ of a (volatile as well as of a non-volatile) solute species is expressed using the concept of activity coefficients:

$$a_i^{(m)}(T,\tilde{x}_j,m_s) = \frac{m_i}{m^0}\gamma_i^*(T,\tilde{x}_j,m_s) \tag{2}$$

where $m^0 = 1$ mol kg^{-1}. The activity coefficient $\gamma_i^*(T,\tilde{x}_j,m_s)$ of a solute species is calculated from an appropriate expression for the excess Gibbs energy of the liquid.

In Equation (1), $f_i''(T, p, y_k)$ is the fugacity of gas i in the vapor phase. The fugacity depends on temperature, pressure, and the composition of the vapor phase (expressed by mole fractions y_k). It is written as the product of pressure, vapor phase mole fraction y_i, and vapor-phase fugacity coefficient $\phi_i(T, p, y_k)$:

$$f_i''(T, p, y_k) = p y_i \phi_i(T, p, y_k) \tag{3}$$

The fugacity coefficients have to be calculated from an equation of state for the vapor phase.

The influence of pressure on Henry's constant is expressed by the Krichevsky–Kasarnovski correction:

$$k_{H,i}^{(m)}(T, p, \tilde{x}_j) = k_{H,i}^{(m)}(T, p^s(\tilde{x}_j)) \exp\left(\int\limits_{p^s(T,\tilde{x}_j)}^{p} \frac{v_i^\infty(T, \tilde{x}_j)}{RT} dp \right) \tag{4}$$

where $k_{H,i}^{(m)}(T, p^s(\tilde{x}_j))$ is Henry's constant of gas i in the solvent at the vapor pressure $p^s(T, \tilde{x}_j)$ of the solvent, and $v_i^\infty(T, \tilde{x}_j)$ is the partial molar volume of gas i at infinite dilution in the solvent. (The influence of pressure on v_i^∞ is usually neglected.)

For a solvent component i, the vapor-liquid equilibrium is expressed by the extended Raoult's law:

$$p_i^s(T)\phi_i^s(T) \exp\left(\int\limits_{p_i^s(T)}^{p} \frac{v_i(T)}{RT} dp \right) a_i(T, \tilde{x}_j, m_s) = f_i''(T, p, y_k) \tag{5}$$

The term $p_i^s(T)$ is the vapor pressure of pure solvent component i, $\phi_i^s(T)$ is the fugacity coefficient of that component at saturation, $v_i(T)$ is the molar liquid volume of pure component i, and a_i is the activity of i. (The influence of pressure on v_i as well as on a_i is usually neglected.) The reference state for the chemical potential of solvent component i in the liquid phase is the pure liquid solvent at temperature T and pressure p. The activity of a single solvent component is usually calculated from the activities of the solutes by applying the Gibbs–Duhem equation. If the solvent consists of more than one component, the calculation of the activities of those components is somewhat more laborious.[42]

There are two cases which deserve special attention. When there are chemical equilibrium reactions – *e.g.*, when the gas is dissolved "chemically" – one has to consider the chemical reaction equilibrium together with the conservation of mass criterion. When the gas is dissolved in a liquid in the presence of a solid phase (or several solid phases), the condition of solid–liquid phase equilibrium has to be considered additionally (together with the conservation of mass criterion).

4.3 Experimental Arrangements

In our research on gas solubility we have recently applied three experimental techniques.

4.3.1 Apparatus for Measuring the Solubility of a Single Gas in a Solvent at Elevated Pressures

For about 15 years, a synthetic technique has been extensively used for the reliable measurement of the solubility of a single gas in a pure solvent as well as in a solvent mixture.[1–31] Figure 1 illustrates that equipment. Its central part is a thermostated cylindrical high-pressure view cell (volume ≈ 30 cm^3) with sapphire windows on both ends. In an experiment, the pressure is determined which is required to dissolve at a constant and pre-set temperature a precisely known amount of the gas in a precisely known amount of the solvent.

In a typical experiment with a poor solvent for the particular gas (*e.g.*, carbon dioxide in water) the gas is charged first (from a gas cylinder). The mass of the gas filled into the cell is determined volumetrically, *i.e.*, from the exactly known volume of the cell and readings for temperature and pressure by means of an equation of state. Then the solvent is added, stepwise, to the cell by a high-pressure spindle press until the gas is completely dissolved in the liquid phase. The amount of solvent charged to the cell is always only slightly above the minimum amount needed to dissolve the gas completely. After equilibration, the pressure is decreased in small steps by withdrawing very small amounts of the liquid mixture from the cell until the first small stable bubbles appear. The

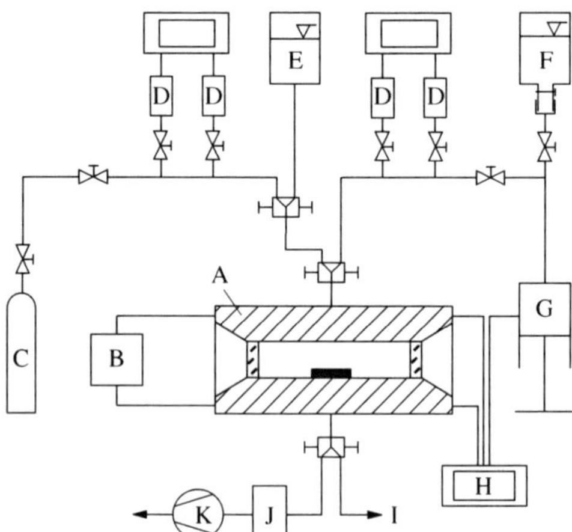

Figure 1 *The apparatus for measuring the solubility of a single gas in a solvent at elevated pressures: (A) cylindrical high-pressure equilibrium view cell with two sapphire windows and magnetic stirrer, (B) thermostat, (C) container for the gas, (D) pressure transducers, (E) tank for rinsing water, (F) tank for solvent mixture, (G) high-pressure spindle press, (H) AC-bridge with three platinum resistance thermometers, (I) solution outlet, (J) cooling trap, and (K) vacuum pump.*

pressure then attained is the equilibrium pressure to dissolve the charged amount of the gas in the remaining amount of solvent at the particular fixed temperature. As the liquid mixture is almost incompressible, the amount of that mixture and, in particular, the amount of dissolved gas, which is withdrawn from the cell, is negligibly small. The mass of the solvent that filled the cell is calculated from the volume displacement in the spindle press and the solvent density.

That equipment was used for measuring the solubility of some single gases (*e.g.*, NH_3 and the sour gases CO_2, H_2S, SO_2, and HCN) in water and aqueous solutions of strong electrolytes (*e.g.*, Na_2SO_4, $(NH_4)_2SO_4$, NaCl, and NH_4Cl) and for measuring the solubility of CO_2 in aqueous solutions of non-reacting organic solvents (*e.g.*, methanol, acetone). It was further used to measure the solubility of CO_2 (and H_2S) in aqueous solutions of an amine (*N*-methyl-diethanolamine (MDEA) and piperazine) and aqueous binary amine solutions as well as for measuring the solubility of non-reacting gases (CO_2, H_2, O_2, CO) in pure ionic liquids.

4.3.2 Apparatus for Measuring the Simultaneous Solubility of Ammonia and a Sour Gas in a Solvent at Elevated Pressures

Figure 2 illustrates the experimental arrangement used for about 20 years to examine the vapor–liquid equilibrium of chemical reacting systems.[32–41] A high-pressure thermostated cell (volume ≈2 L) is charged with a known

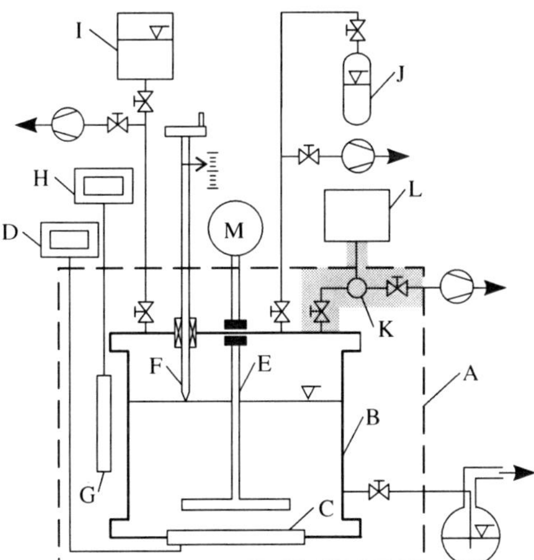

Figure 2 *The apparatus for measuring simultanously the solubility of chemical reacting gases in a solvent at elevated pressures: (A) thermostat, (B) equilibrium cell, (C) pressure gauge, (D) pressure transducer, (E) magnetically coupled hollow stirrer, (F) buoy, (G) platinum resistance thermometer, (H) AC-bridge, (I) tank with liquid solvent, (J) container for the gas (first ammonia, then the sour gas), (K) sampling valve, (L) gas chromatograph, (M) electromotor, and (gray zone) heating.*

amount (about 1 kg) of pure liquid water or a gravimetrically prepared aqueous salt solution. Then, ammonia is added to achieve the desired molality of that gas. The amount of ammonia is exactly known as it is taken from a small tank which is weighed before and after the filling procedure. Afterwards, in a similar way the sour gas (*e.g.*, carbon dioxide) is added in several steps. After each step the phases are equilibrated, temperature, pressure, and vapor-phase volume are measured, and small vapor phase samples are taken and analyzed by gas chromatography. The amounts of the volatile components in the gaseous phase are then calculated from these direct experimental data. Finally, from a mass balance, the stoichiometric composition of the liquid phase is determined.

That technique was used to measure the simultaneous solubility of ammonia and carbon dioxide (or hydrogen sulfide) in water, in aqueous solutions of a strong electrolyte (*e.g.*, Na_2SO_4, $(NH_4)_2SO_4$, NaCl, and NH_4Cl) as well as in the aqueous/organic solvent (water+methanol) with as well as without a strong electrolyte (Na_2SO_4 and NaCl).

4.3.3 Apparatus for Measuring the Solubility of a Sour Gas in Aqueous Solutions of Amines at Low Pressures

The methods described above are not suited for investigations at very low pressures as the experimental determination of the total pressure is subject to high experimental uncertainties caused by the unavoidable presence of small amounts of other gases. However, the reliable design of many gas absorption processes depends on the regeneration of the solvent, *i.e.*, the amount of remaining gas in that solvent. A typical example is the removal of CO_2 and/or H_2S from natural gases which is achieved by "chemical" absorption in, *e.g.*, an aqueous solution of *N*-methyldiethanolamine (MDEA). For the reliable determination of the solubility of sour gases in aqueous solutions of amines we use headspace gas chromatography. Figure 3 shows a diagram of the experimental arrangement. Its main components are the cell holder, the valve holder (containing a multi-position valve and the sampling system), two large buffer tanks filled with nitrogen, and the gas chromatograph.[43]

Eight sample cells (stainless-steel vials, volume $\approx$ 11–30 cm^3) are partially filled with the liquid mixture (*e.g.*, an aqueous solution of MDEA and carbon dioxide) and mounted in the cell holder. Only one of those sample cells is shown in Figure 3. The temperature is measured in the thermostated bath of the cell holder with a platinum resistance thermometer. At equilibrium, a certain partial pressure of the sour gas is attained in the gaseous phase (the headspace) of those sample cells, which is to be measured.

First, the cell is pressurized (from buffer tank A) with nitrogen to a constant pressure. Secondly, the gas phase in the cell expands to the buffer tank B, which is also pressurized to a constant pressure, and the sample loop is filled. The sample valve is then switched, and the sample is transferred to the gas chromatograph. After the sample is taken, the sampling system is purged with nitrogen. In addition, the multi-position valve makes it possible to connect each of the eight sample cells to the sample loop.

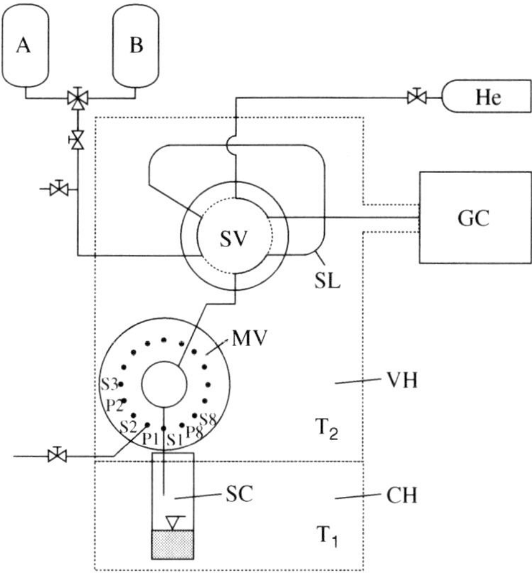

Figure 3 *The headspace chromatograph arrangement for measuring the solubility of a sour gas in aqueous solutions of amines at low pressures: (CH) liquid-thermostated cell holder (temperature T_1), (VH) liquid-thermostated valve holder (temperature $T_2 > T_1$), (A) nitrogen tank (higher pressure), (B) nitrogen tank (lower pressure), (GC) gas chromatograph, (He) helium (carrier gas), (SC) sample cell, (MV) multiposition valve, (S1–S8) sample positions, (P1–P8) purge positions, (SV) sample valve, and (SL) sample loop.*

The temperature of the valve holder is kept at 15–20 K higher than in the cell holder. Also the feed line to the gas chromatograph is thermostated to a higher temperature in order to avoid condensation in the sampling system.

The primary data collected in the headspace chromatographic measurements is the peak area of the sour gas (here carbon dioxide). From this peak area the partial pressure of carbon dioxide can be determined. The peak area is proportional to the mass of that gas in the sample loop. As the volume and temperature of the sample loop are constant, and according to the ideal gas law, the peak area of the gas is proportional to the partial pressure of the gas in the sample loop and, hence, also in the sample cell. The relation between peak area and partial pressure in the cell is determined by calibration measurements.

4.4 Experimental Results and Comparison with Predictions/Correlations

4.4.1 Gas Solubility in Ionic Liquids

The solubility of some pure gases in the ionic liquid [bmim][PF_6] is discussed here as a typical example for the solubility of a single gas in a pure liquid. Ionic liquids have been proposed as solvents/catalysts for chemical reactions, *e.g.*, in transition metal catalysis (such as hydroformylation, hydrogenation, and

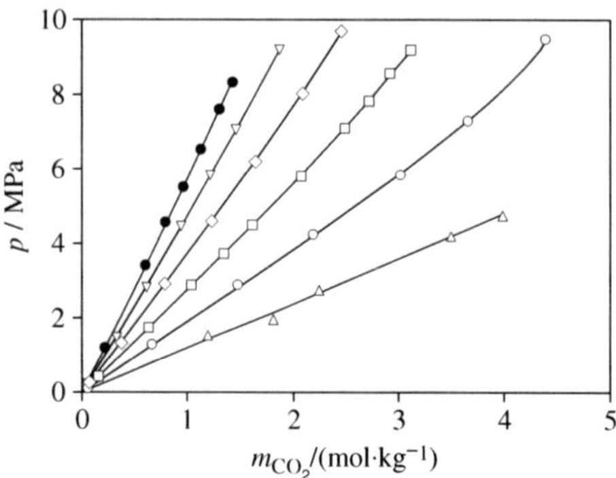

Figure 4 *Total pressure above solutions of (CO_2+[bmim][PF_6]): comparison between experimental data [(△) T≈293 K, (○) 313 K, (□) 333 K, (◇) 353 K, (▽) 373 K, (●) 393 K][23] and correlation results (curves).[23]*

oxidation). The basic design of such processes requires reliable data on the solubility of the reacting gases in pure ionic liquids as well as mixtures with ionic liquids. The solubility of CO_2, CO, H_2, and O_2 in [bmim][PF_6] was measured recently at temperatures from ~293 to ~393 K and pressures up to about 10 MPa.[23,25–27] The ionic liquid sample was carefully degassed, dried under vacuum and stored avoiding any further contact with air and humidity. The water content was less than 0.1 mass%, as measured by Karl Fischer analysis before the measurements and confirmed afterwards.

Figure 4 shows experimental results for the solubility of carbon dioxide in [bmim][PF_6] at temperatures from 293 to 393 K as the total pressure above those solutions *versus* the molality of the gas.[23] Figure 4 shows the typical behavior of physical gas solubility. In the pressure region investigated, the solubility pressure linearly increases with increasing amount of the gas in the liquid. As it is common for many (gas+solvent) systems, the solubility of carbon dioxide in [bmim][PF_6] decreases with increasing temperature.

However, for the sparsely soluble gases (O_2 and CO) there is almost no influence of temperature on the solubility in the ionic liquid [bmim][PF_6], and for the less soluble gas (H_2) the solubility increases with increasing temperature.[25–27]

Because the vapor pressure of an ionic liquid is usually negligibly small, the fugacity of the gas in the vapor phase can be replaced by the fugacity of the pure gas. At low and moderate pressures the influence of pressure on Henry's constant can be neglected. Furthermore, the activity of the gas in the liquid phase can be approximated by the molality when the solubility of the gas is small. These approximations result in

$$k_{H,i}^{(m)}(T)\frac{m_i}{m^0} = f_i''(T,p) \qquad (6)$$

Henry's constant $k_{H,i}^{(m)}$ is the only binary property which is required to correlate the solubility of a sparsely soluble gas in an ionic liquid. There are

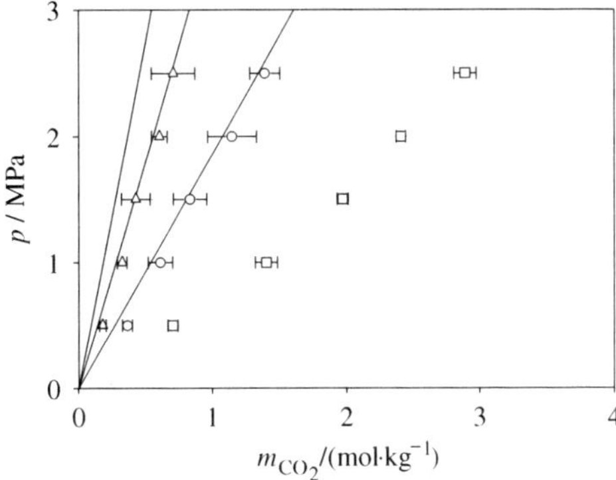

Figure 5 *Solubility of CO_2 in the ionic liquid [bmim][PF_6]: Comparison between simulation results [(□) T≈313 K, (○) 353 K, (△) 393 K][47] and results from a correlation based on experimental data (curves).*[23]

several methods to describe the influence of temperature on Henry's constant. But, more interesting at this moment is to compare the experimental results with predictions from molecular simulations. Figure 5 shows a comparison between experimental results for the solubility of CO_2 in [bmim][PF_6][23] (at temperature 313, 353, and 393 K and at pressures up to 2.5 MPa) and prediction results applying the Gibbs ensemble Monte Carlo technique[44,45] in the isothermal-isobaric ensemble (*NpT*-GEMC). Such simulations require the intermolecular pair potentials for interactions between the gas molecules and molecules of the ionic liquid. All pair potentials were taken from the literature. Each gas molecule was approximated by two or three Lennard–Jones 6–12 centers with (or without) embedded electrical point charges. The ionic liquid [bmim][PF_6] was treated as an electrolyte consisting of [bmim]$^+$- and [PF_6]$^-$-ions. The pair potentials of the ions were approximated by a simplification of the quantum-chemistry-based united atom force field (UA1) model proposed by Shah *et al.*[46] Each carbon atom of [bmim]$^+$ was treated as a Lennard–Jones center carrying also an electrical charge. The [PF_6]$^-$ -anion was treated as a single Lennard–Jones site carrying one negative electric charge. The interactions between unlike molecules were approximated using mixing rules for the Lennard–Jones size and energy parameters without any adjustable binary parameter.[27,47] The statistical uncertainty of the simulation results is typically between 5 and 15%.

For (CO+[bmim][PF_6]), (H_2+[bmim][PF_6]), and (O_2+[bmim][PF_6]) *NpT*-GEMC simulations were carried out as well.[27,47]

Generally the simulation results over estimate the solubility of the gases (CO_2, CO, H_2 and O_2) in [bmim] [PF_6] by a factor of about two.

4.4.2 Gas solubility in Aqueous Solutions of Strong Electrolytes

The influence of strong electrolytes on the solubility of some single gases (*e.g.*, NH_3, CO_2, H_2S, and SO_2) in water was investigated recently.[3–8,10–17,39] Here, only a few typical examples are discussed.

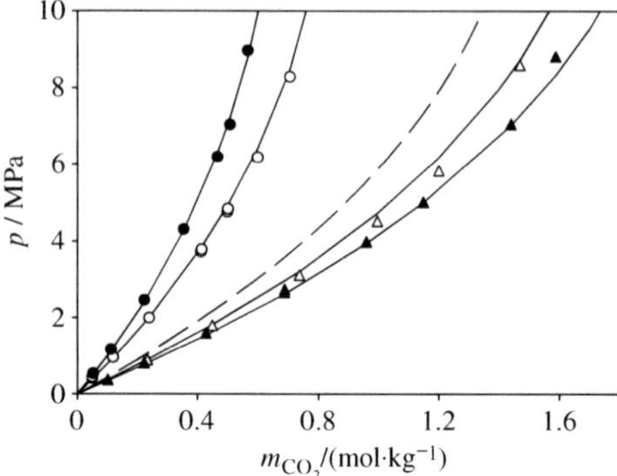

Figure 6 *Solubility of CO_2 in aqueous solutions of the single salts $NaNO_3$ and NH_4NO_3 at $T \approx 313$ K: Comparison between experimental data (symbols)[10] and correlation results (curves);[10] (– –) no salt, (○, ●) $\bar{m}_{NaNO_3} \approx 6$ and 10 mol kg^{-1}, respectively, (△, ▲) $\bar{m}_{NH_4NO_3} \approx 6$ and 10 mol kg^{-1}, respectively.*

The solubility of a gas in water (at given temperature and pressure) is strongly affected by the presence of a strong electrolyte. Adding a strong electrolyte usually results in a decrease of the gas solubility ("salting-out effect"). For example, Figure 6 shows experimental and correlation results for the solubility of carbon dioxide (a sparsely soluble gas) in water as well as in aqueous solutions of the single salts sodium nitrate and ammonium nitrate at $T \approx 313$ K.[10] Carbon dioxide is "salted out" by sodium nitrate, and it is "salted in" by ammonium nitrate. For example, for a solution containing about 0.5 mol kg^{-1} CO_2 and 10 mol kg^{-1} $NaNO_3$ (NH_4NO_3), the total pressure is about 7 MPa (1.9 MPa), whereas it is about 2.4 MPa above the salt-free solution.

The solubility of the strong electrolyte in water might also be affected by the presence of a gas, in particular, if that gas is reasonably soluble in water. For example, at $T \approx 353$ K, ~ 3 mol of Na_2SO_4 dissolve in 1 kg of pure liquid water, whereas only ~ 2 (~ 1) mol of that salt dissolve in a solution consisting of 1 kg of pure water and ~ 3.5 (~ 7.7) mol of ammonia.[4]

All of those effects have been successfully described by applying the well-known vapor–liquid and solid–liquid equilibrium conditions, and by describing the properties of the liquid and the gaseous phases by means of Pitzer's molality-scale-based Gibbs excess energy model[48,49] and the virial equation of state, respectively. That Pitzer's model is based on the unsymmetric convention and is therefore particularly suited for describing the influence of strong electrolytes on the solubility of gases in liquid water, when the amounts of the solute components (gas and strong electrolytes) is remarkably smaller than the amounts of the solvent (water). The interaction parameters in that model need to be determined from experimental data on the phase equilibrium not only for the ternary system (gas+salt+water), but also for the binary subsystems (gas+water) and (salt+water). A big advantage of Pitzer's molality-scale-based G^E-model is, that there is a lot of reliable information on these interaction parameters available in the literature.[49]

NpT-GEMC simulations were also applied to predict the solubility of some gases (*e.g.*, CO_2, H_2S, NH_3) in water as well as the influence of a strong electrolyte (namely NaCl) on that gas solubility.[50–53] The effective pair potentials for water, the gases, and the ions combine the Lennard–Jones-12–6 potential for the non-polar interactions with fixed-point charge Coulomb interactions to account for the polarity of a molecule as well as for the ionic electrostatic interactions. For water and the gases the pair potentials were selected from the literature, on condition that they are able to describe the thermodynamic properties (mainly the saturation properties) of the pure substances. For the ions they were also adopted from the literature, where they were generally fitted to various properties. In order to test the predictive capability of the simulations, the interactions between different species (gas, water, sodium, and chloride) were estimated from the pair potentials of the pure species using common mixing rules without any adjustable binary parameter. The simulation results for the solubility of CO_2 and H_2S in water favorably agree with experimental data. However, some disagreement is observed between simulation and experimental results for the solubility of NH_3 in water. Furthermore, the experimentally observed "salting-out" effect is only qualitatively predicted by the simulation.

4.4.3 Solubility of Ammonia and Sour Gases in Water and Aqueous Solutions of Strong Electrolytes

The simultaneous solubility of the basic gas NH_3 and the single sour gases CO_2, H_2S, and SO_2 in water as well as in aqueous solutions of strong electrolytes (*e.g.*, NaCl and Na_2SO_4) was experimentally investigated in recent years.[2,32–38,40] Several chemical reactions occur in the aqueous phase and need to be taken into account when a thermodynamic model is to describe such "chemical" gas solubilities. Those kind of systems are referred to as "complex" systems, because many effects occur simultaneously (*e.g.*, physical solubility of several gases, chemical reactions involving ionic species and physical interactions between molecular and/or ionic species) resulting in strongly non-linear equation systems. The thermodynamic framework used to describe the gas solubility in aqueous solutions of strong electrolytes was successfully extended to account for chemical reactions[2,34–38,40] (*cf.*, Edwards *et al.*[54]).

As an example, the total pressure and the partial pressures of ammonia and carbon dioxide above an aqueous solution of ammonia ($\bar{m}_{NH_3} \approx 1.5$ and 4 mol kg^{-1}), sodium sulfate ($\bar{m}_{Na_2SO_4} = 1\,mol\,kg^{-1}$), and carbon dioxide at $T \approx 393$ K are plotted in Figure 7 against the stoichiometric molality of carbon dioxide ($\bar{m}_{CO_2}$) in the liquid phase. The symbols represent experimental results.[35] The full curves represent prediction results from a model, which is based only on experimental information on the phase equilibrium of the binary and ternary sub-systems (salt+water, gas+water, gas+salt+water, and gas 1+gas 2+water).[35,55,56] The dashed curves represent correlation results for the salt-free system $NH_3+CO_2+H_2O$.[34,55,56]

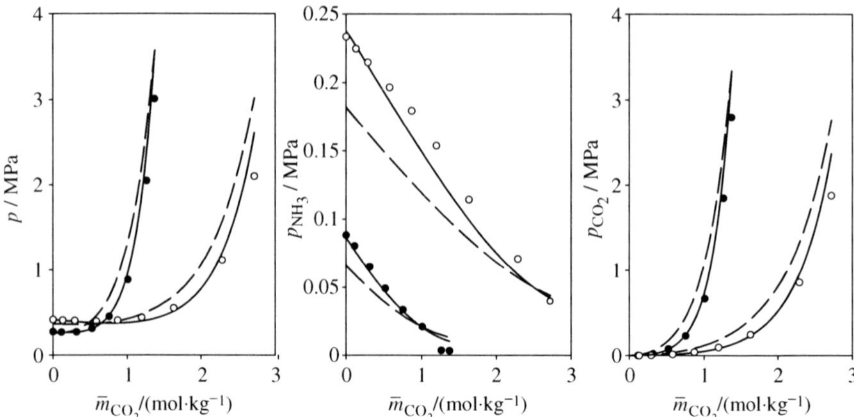

Figure 7 *Total pressure and partial pressures of ammonia and carbon dioxide above liquid mixtures of $CO_2+NH_3+Na_2SO_4+H_2O$ at ~393 K and $\bar{m}_{Na_2SO_4} \approx 1$ mol kg^{-1}: Comparison between experimental data [(●) $\bar{m}_{NH_3} \approx 1.5$ mol kg^{-1}, (○) 4 mol · kg^{-1}] and prediction results (full curves).[35,55,56] The dashed curves represent correlation results for the system $CO_2+NH_3+H_2O$.[34,55,56]*

Adding carbon dioxide to an ammoniacal, sodium sulfate containing solution at constant temperature at first results in a slight decrease of the total pressure. After passing a minimum, a steep increase in the total pressure is observed. The partial pressure of ammonia decreases with increasing amount of the sour gas in the liquid phase as more and more ammonia is converted into ionic, non-volatile species (ammonium and carbamate ions). The partial pressure of carbon dioxide at first is very small, *i.e.*, carbon dioxide is almost completely dissolved chemically (as bicarbonate, carbonate, and carbamate ions). But it increases rapidly when ammonia has been spent in the liquid phase by chemical reactions.

The behavior of the quaternary system is very similar to that observed for the salt-free system. But sodium sulfate causes significant effects. As can be seen from Figure 7, ammonia is "salted-out" by sodium sulfate, *i.e.*, the partial pressure of ammonia above the salt-containing solution is larger than that above the salt-free solution. The effect decreases with increasing amount of the sour gas in the liquid phase. However, carbon dioxide is "salted-in", *i.e.*, the partial pressure of carbon dioxide above the salt-containing solution is below that above the salt-free solution. Without ammonia, carbon dioxide is "salted-out" by sodium sulfate.[3] The change from "salting-out" to "salting-in" in the presence of ammonia is due to the strong influence of sodium sulfate on the chemical reaction equilibrium in the liquid phase. The partial pressure of water is reduced by sodium sulfate, but it is only slightly changed when both gases are dissolved. For the total pressure, "salting-out" is observed at low carbon dioxide concentrations – mainly due to the enhancement of the partial pressure of ammonia – whereas at higher molalities the increasing "salting-in" of carbon dioxide and decreasing "salting-out" of ammonia results in a lower pressure than that above the salt-free solution. Surprisingly, all of these effects are quantitatively predicted by the model.

4.4.4 Solubility of Sour Gases in Aqueous Solutions of Amines

The removal of sour gases, *e.g.*, CO_2 or H_2S, from flue gas or natural gas is mostly achieved by "chemical" absorption in aqueous solutions of single amines (*e.g.*, *N*-methyldiethanolamine, MDEA) or amine mixtures (*e.g.*, MDEA and piperazine). The competitive chemical absorption of carbon dioxide and hydrogen sulfide is kinetically controlled. However, deviation from equilibrium provides the driving force in a kinetically controlled process. Therefore, the reliable design of the absorption and desorption columns at first requires the knowledge of vapor–liquid equilibrium (at both elevated and low pressures).[9,18–22,57]

As a typical example we discuss here the solubility of CO_2 in aqueous solutions of MDEA. In Figure 8, the total pressure and the partial pressure of carbon dioxide above an aqueous solution of MDEA ($\bar{m}_{CO_2}$) at $T \approx 313$, 353, and 393 K are plotted against the stoichiometric molar ratio of carbon dioxide to MDEA in the liquid phase.

Owing to the basic character of the amine, the vapor–liquid equilibrium behavior of the system CO_2+MDEA+H_2O is similar to that observed for the system $CO_2+NH_3+H_2O$. However, MDEA has an almost negligibly small vapor pressure. Therefore, when CO_2 is added to an aqueous solution of MDEA, the total pressure does not go through a minimum, and at first, it only very slightly increases with increasing amount of the sour gas in the liquid. In particular, the partial pressure of carbon dioxide at first (*i.e.*, for low gas loadings $\bar{m}_{CO_2}/\bar{m}_{MDEA}$) is very small, *i.e.*, carbon dioxide is almost completely dissolved chemically (as

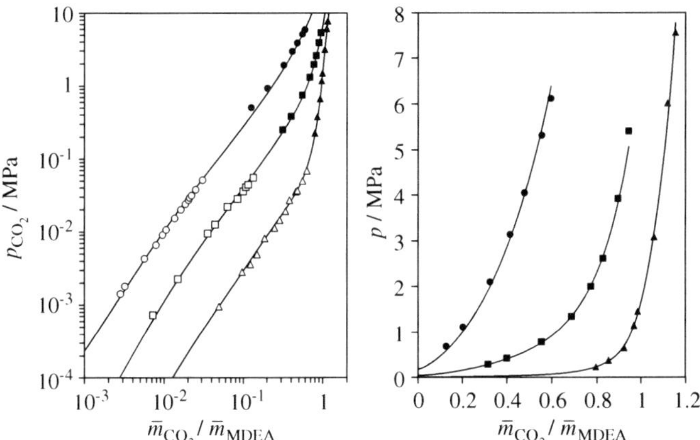

Figure 8 *Partial pressure of carbon dioxide and total pressure above liquid mixtures of* CO_2*+MDEA+*H_2O *at* $\bar{m}_{MDEA} \approx 8$ *mol kg*$^{-1}$*: comparison between experimental data [(▲, △) T ≈ 313 K, (■, □) 353 K, (●, ○) 393 K] and correlation results (curves).*[18,57]

bicarbonate and carbonate ions). But it increases rapidly (for higher gas loadings) when, in the liquid phase, MDEA has been spent by chemical reactions, *i.e.*, when MDEA has been protonated. In the very low gas loading region, the vapor–liquid equilibrium is mainly influenced by the chemical reactions. However, in the high gas loading region, it is also highly influenced by the physical interactions (in particular, between CO_2 and the ionic species). Therefore, the parameters describing those interactions were first estimated from gas solubility data in the high gas loading region alone, *i.e.*, in the elevated pressure region (0.2 MPa < p < 10 MPa).[9,18] The filled symbols in Figure 8 (right) represent these experimental results for the total pressure. Secondly, some interaction parameters were tuned by also taking into account gas solubility data in the low gas loading region, *i.e.*, in the low-pressure region (p_{CO_2} < 70 kPa).[57] The empty symbols in Figure 8 represent these experimental results for the partial pressure of carbon dioxide. The model accurately (almost within experimental uncertainty) describes the experimental results for the solubility of CO_2 in aqueous solutions of MDEA for both low and high gas loadings (*cf.*, Figure 8).[57]

4.4.5 Gas Solubility in Mixed Solvents (Water+Organic Compound)

The solubility of some single gases (*e.g.*, carbon dioxide) in binary solutions of water+an organic compound (*e.g.*, methanol, acetone, dimethylformamide) has been recently experimentally investigated.[24,31] In particular, the whole solvent composition range was covered, from pure water to the pure organic component.

As an example, the total pressure above a solution of (carbon dioxide+water+methanol) at $T \approx 354$ K is plotted in Figure 9 against the molality of carbon dioxide in the liquid solvent mixture.[24]

The dashed curve gives the results of a former correlation of the experimental data on the solubility of carbon dioxide in pure water,[3] which – as was already

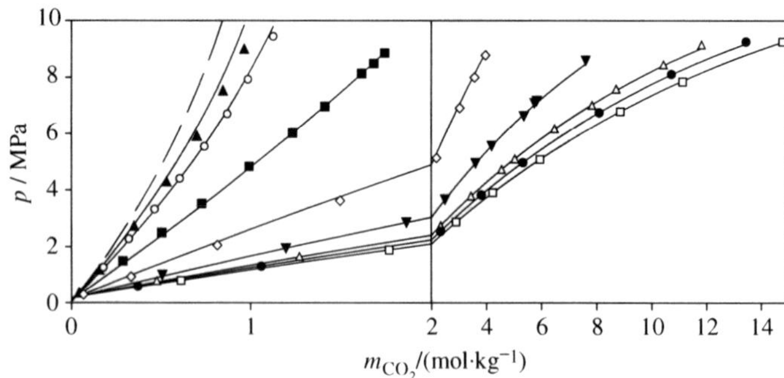

Figure 9 *Total pressure above solutions of (CO_2 + CH_3OH + H_2O) at $T \approx 354$ K and different mole fractions of methanol in the gas-free solvent mixture: [(▲)$\tilde{x}_M \approx 0.05$, (○) 0.1, (■) 0.25, (◇) 0.5, (▼) 0.75, (△) 0.9, (●) 0.95, (□) 1] experimental results;[24] (—) correlation results;[42] (– – –) solubility of CO_2 in pure water, correlation results.[3]*

mentioned – is based on Pitzer's molality-scale-based equation for the Gibbs excess energy. In that equation, only water is treated as a solvent component, whereas all other components are treated as solutes. Pitzer's equation belongs to the group of osmotic virial equations and therefore, in principle, it does not allow for a change in the solvent. But an extension of that equation to solvent mixtures has been recently presented, which overcomes that restriction.[42] In this equation both water and the organic compound are treated as solvent components, and it is possible to calculate the activities of all present (*solute* and *solvent*) species. The model requires information on the vapor–liquid equilibrium of the binary system (water+methanol) and on the solubility of the gas in the mixtures of (water+methanol). Applying that particular thermodynamic framework[42] allows one to explicitly take into account the influence of temperature, *pressure*, and *solvent mixture composition* on the Henry's constant of the gas in the solvent mixture. Therefore, that model allows for an accurate correlation of the solubility of gases in solvent mixtures up to rather high gas solubilities (and solubility pressures).

4.4.6 Gas Solubility in Mixed Solvents (Water+Organic Compound+Strong Electrolyte)

The influence of some strong electrolytes (*e.g.*, NaCl, Na_2SO_4) on the solubility of some single gases (*e.g.*, CO_2) in binary solutions of water+an organic compound (*e.g.*, methanol, acetone) has also recently been experimentally investigated.[29–31]

As an example, the total pressure above a solution of (CO_2+NaCl+H_2O+CH_3OH) at $\tilde{x}_M \approx 0.25$ and $\bar{m}_{NaCl} \approx 2\,mol\,kg^{-1}$ is plotted in Figure 10 against the molality of carbon dioxide in the liquid phase. The symbols represent experimental results at $T \approx 314$ and 395 K.[29]

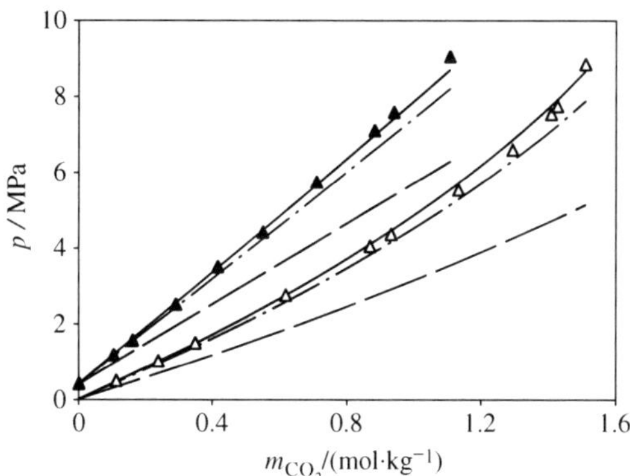

Figure 10 *Total pressure above (CO_2+NaCl+CH_3OH+H_2O) at $\tilde{x}_M \approx 0.25$, $\bar{m}_{NaCl} \approx 2$ mol kg^{-1}: [(△) $T \approx 314$ K, (▲) 395 K] experimental results;[29] (–·–) prediction;[29] (—) correlation;[29] (– –) salt-free system, correlation results,[42] based on experimental data from ref. 24.*

The dashed curves give the total pressure above the salt-free system (CO_2+H_2O+CH_3OH).[24,42] Like in the methanol-free system (CO_2+NaCl+H_2O),[7] carbon dioxide is "salted-out" by sodium chloride, *i.e.*, the gas solubility is reduced by the presence of the salt. The aforementioned extension of Pitzer's G^E-model to solvent mixtures has been successfully applied to predict/correlate the influence of NaCl on the solubility of CO_2 in mixtures of (H_2O+CH_3OH). The results of that prediction/correlation are shown in Figure 10. Besides the experimental information already required to describe the solubility of CO_2 in the salt-free mixtures of (H_2O+CH_3OH), that model requires data on the solubility of the salt in (H_2O+CH_3OH), *i.e.*, in particular on the temperature and solvent mixture composition dependent solubility product of that salt in those mixtures, and on the solubility of the gas in aqueous solutions of the salt, which were all taken from the literature.

References

1. B. Rumpf and G. Maurer, *Fluid Phase Equilib.*, 1992, **81**, 241.
2. B. Rumpf, F. Weyrich and G. Maurer, *Fluid Phase Equilib.*, 1993, **83**, 253.
3. B. Rumpf and G. Maurer, *Ber. Bunsenges. Phys. Chem.*, 1993, **97**, 85.
4. B. Rumpf and G. Maurer, *Ind. Eng. Chem. Res.*, 1993, **32**, 1780.
5. B. Rumpf and G. Maurer, *Fluid Phase Equilib.*, 1993, **91**, 113.
6. B. Rumpf and G. Maurer, *J. Solution Chem.*, 1994, **23**, 37.
7. B. Rumpf, H. Nicolaisen, C. Öcal and G. Maurer, *J. Solution Chem.*, 1994, **23**, 431.
8. B. Rumpf, H. Nicolaisen and G. Maurer, *Ber. Bunsenges. Phys. Chem.*, 1994, **98**, 1077.
9. J. Kuranov, B. Rumpf, N.A. Smirnova and G. Maurer, *Ind. Chem. Eng. Res.*, 1996, **35**, 1959.
10. B. Rumpf, J. Xia and G. Maurer, *J. Chem. Thermodyn.*, 1997, **29**, 1101.
11. B. Rumpf, J. Xia and G. Maurer, *Ind. Eng. Chem. Res.*, 1998, **37**, 2012.
12. J. Xia, B. Rumpf and G. Maurer, *Fluid Phase Equilib.*, 1999, **155**, 107; *Corrigendum*, 2000, **168**, 283.
13. J. Xia, B. Rumpf and G. Maurer, *Ind. Eng. Chem. Res.*, 1999, **38**, 1149.
14. J. Xia, B. Rumpf and G. Maurer, *Fluid Phase Equilib.*, 1999, **165**, 99.
15. J. Xia, Á. Pérez-Salado Kamps, B. Rumpf and G. Maurer, *Ind. Eng. Chem. Res.*, 2000, **39**, 1064.
16. J. Xia, Á. Pérez-Salado Kamps, B. Rumpf and G. Maurer, *Fluid Phase Equilib.*, 2000, **167**, 263.
17. J. Xia, Á. Pérez-Salado Kamps, B. Rumpf and G. Maurer, *J. Chem. Eng. Data*, 2000, **45**, 194.
18. Á. Pérez-Salado Kamps, A. Balaban, M. Jödecke, G. Kuranov, N.A. Smirnova and G. Maurer, *Ind. Eng. Chem. Res.*, 2001, **40**, 696.
19. Á. Pérez-Salado Kamps, B. Rumpf, G. Maurer, Y. Anoufrikov, G. Kuranov and N.A. Smirnova, *AIChE J.*, 2002, **48**, 168.
20. Y. Anoufrikov, Á. Pérez-Salado Kamps, B. Rumpf, N.A. Smirnova and G. Maurer, *Ind. Eng. Chem. Res.*, 2002, **41**, 2571.

21. J. Xia, Á. Pérez-Salado Kamps and G. Maurer, *Fluid Phase Equilib.*, 2003, **207**, 23.
22. Á. Pérez-Salado Kamps, J. Xia and G. Maurer, *AIChE J.*, 2003, **49**, 2662.
23. Á. Pérez-Salado Kamps, D. Tuma, J. Xia and G. Maurer, *J. Chem. Eng. Data*, 2003, **48**, 746.
24. J. Xia, M. Jödecke, Á. Pérez-Salado Kamps and G. Maurer, *J. Chem. Eng. Data*, 2005, **49**, 1756.
25. J. Kumełan, Á. Pérez-Salado Kamps, D. Tuma and G. Maurer, *Fluid Phase Equilib.*, 2005, **228–229**, 207.
26. J. Kumełan, Á. Pérez-Salado Kamps, D. Tuma and G. Maurer, *J. Chem. Eng. Data*, 2006, **51**, 11.
27. J. Kumełan, Á. Pérez-Salado Kamps, I. Urukova, D. Tuma and G. Maurer, *J. Chem. Thermodyn.*, 2005, **37**, 595; *Corrigindum ibid.* in press.
28. J. Kumełan, Á. Pérez-Salado Kamps, D. Tuma and G. Maurer, Solubility of CO_2 in the ionic liquid [hmim][Tf_2N], *J. Chem. Thermodyn.*, 2006, **38** (in press).
29. Á. Pérez-Salado Kamps, M. Jödecke, J. Xia, M. Vogt and G. Maurer, *J. Chem. Eng. Res.*, 2006, **45**, 1505.
30. Á. Pérez-Salado Kamps, M. Jödecke, M. Vogt, J. Xia and G. Maurer, *J. Chem. Eng. Res.*, 2006, **45**, 3673.
31. M. Jödecke, Experimentelle und theoretische Untersuchungen zur Löslichkeit von Kohlendioxid in wässrigen, salzhaltigen Lösungen mit organischen Komponenten, 2004. Ph.D. Dissertation, University of Kaiserslautern, Germany.
32. G. Müller, E. Bender and G. Maurer, *Ber. Bunsenges. Phys. Chem.*, 1988, **92**, 148.
33. U. Göppert and G. Maurer, *Fluid Phase Equilib.*, 1988, **41**, 153.
34. F. Kurz, B. Rumpf and G. Maurer, *Fluid Phase Equilib.*, 1995, **104**, 261.
35. V. Bieling, F. Kurz, B. Rumpf and G. Maurer, *Ind. Eng. Chem. Res.*, 1995, **34**, 1449.
36. F. Kurz, B. Rumpf and G. Maurer, *J. Chem. Thermodyn.*, 1996, **28**, 497.
37. F. Kurz, B. Rumpf, R. Sing and G. Maurer, *Ind. Eng. Chem. Res.*, 1996, **35**, 3795.
38. B. Rumpf, Á. Pérez-Salado Kamps, R. Sing and G. Maurer, *Fluid Phase Equilib.*, 1999, **158–160**, 923.
39. R. Sing, B. Rumpf and G. Maurer, *Ind. Eng. Chem. Res.*, 1999, **38**, 2098.
40. Á. Pérez-Salado Kamps, R. Sing, B. Rumpf and G. Maurer, *J. Chem. Eng. Data*, 2000, **45**, 796.
41. D. Schäfer, Untersuchungen zur Löslichkeit von Ammoniak und Kohlendioxid in organisch-wässrigen Lösungsmittelgemischen, 2003. Ph.D. Dissertation, University of Kaiserslautern, Germany.
42. Á. Pérez-Salado Kamps, *Ind. Eng. Chem. Res.*, 2005, **44**, 201.
43. N. Asprion, H. Hasse and G. Maurer, *J. Chem. Eng. Data*, 1998, **43**, 74.
44. A. Panagiotopoulos, *Z. Mol. Phys.*, 1987, **61**, 813.
45. A. Panagiotopoulos, N. Quirke, M. Stapleton and D.J. Tildesley, *Mol. Phys.*, 1988, **63**, 527.

46. J.K. Shah, J.F. Brennecke and E.J. Maginn, *Green Chem.*, 2002, **4**, 112.
47. I. Urukova, J. Vorholz and G. Maurer, *J Phys. Chem. B*, 2005, **109**, 12154; *Corrigindum ibid.* in press.
48. K.S. Pitzer, *J. Phys. Chem.*, 1973, **77**, 268.
49. K.S. Pitzer, in *Activity Coefficients in Electrolyte Solutions*, K.S. Pitzer (ed), CRC, Boca Raton, FL, 1991, pp. 75–153.
50. J. Vorholz, V.I. Harismiadis, B. Rumpf, A.Z. Panagiotopoulos and G. Maurer, *Fluid Phase Equilib.*, 2000, **170**, 203.
51. J. Vorholz, B. Rumpf and G. Maurer, *Phys. Chem. Chem. Phys.*, 2003, **4**, 4449.
52. J. Vorholz, Computersimulation der Löslichkeit einiger Gase in Wasser und in wässrigen Lösungen von Natriumchlorid, 2001, Ph.D. Dissertation, University of Kaiserslautern, Germany (Fortschritt-Berichte VDI Reihe 3 Nr. 695. Düsseldorf: VDI-Verlag 2001, ISSN 0178-9503, ISBN 3-18-369503-0).
53. J. Vorholz, V.I. Harismiadis, A.Z. Panagiotopoulos, B. Rumpf and G. Maurer, *Fluid Phase Equilib.*, 2004, **226**, 237.
54. T.J. Edwards, G. Maurer, J. Newman and J.M. Prausnitz, *AIChE J.*, 1978, **24**, 966.
55. U. Lichtfers, Spektroskopische Untersuchungen zur Ermittlung von Speziesverteilungen im System Ammoniak-Kohlendioxid-Wasser, 2000, Ph.D. Dissertation, University of Kaiserslautern, Germany.
56. U. Lichtfers and B. Rumpf, An infrared spectroscopic investigation on the species distribution in the system NH_3+CO_2+H_2O, in G. Maurer (ed), *Thermodynamic Properties of Complex Fluid Mixtures*, Research Report, Deutsche Forschungsgemeinschaft DFG, Wiley-VCH (ISBN 3-527-27770-6), 2004, pp. 92–119.
57. V. Ermatchkov, Phasengleichgewichte in den komplexen, chemisch reagierenden Systemen NH_3+SO_2+H_2O+Salze und CO_2+H_2O+MDEA/Piperazin, Ph.D. Dissertation, University of Kaiserslautern, Germany, 2006.

CHAPTER 5

Solubility Phenomena in "Green" Quaternary Mixtures (Ionic liquid + Water + Alcohol + CO_2)

MANUEL NUNES DA PONTE[1] AND LUÍS P.N. REBELO[2]

[1] *REQUIMTE, Departamento de Quimica, FCT, Universidade Nova de Lisboa, Quinta da Torre, 2829-516, Caparica, Portugal*

[2] *Instituto de Tecnologia Quimica e Biologica, ITQB 2, Universidade Nova de Lisboa, Av. da Republica, Apartado 127, 2780-901, Oeiras, Portugal*

5.1 Introduction

Coupling traditional and alternative green solvents in order to design benign solvents for sustainable technology has been a recent focus of attention in Green Chemistry. Among alternative solvents, supercritical carbon dioxide and room temperature ionic liquids have been the most extensively studied.

Issues of solubility and immiscibility are central to any problem of solvent design. It was indeed immiscibility and asymmetry of solubilities that led Brennecke and collaborators[1] to propose high-pressure carbon dioxide + an ionic liquid as an interesting combination to carry out chemical processes. Those authors reported that mixtures of supercritical CO_2 with the ionic liquid [C_4mim][PF_6] remained biphasic up to high pressures, and that while carbon dioxide dissolved significantly into the [C_4mim][PF_6]-rich liquid phase, no ionic liquid dissolved in the gas phase.

The enormous variety of ionic liquids, due to the vast number of cations and anions that may be combined, makes it likely that many different kinds of phase equilibrium in mixtures with carbon dioxide are possible. For instance, high solubility of some ionic liquids in CO_2 has recently been discovered.[2]

For those ionic liquids that remain insoluble in carbon dioxide, like the imidazolium-based ones, supercritical CO_2 may be used to extract dissolved compounds.[3]

5.2 Liquid–Liquid Equilibria: Co-Solvent Effects in Ternary Mixtures

Many ionic liquids exhibit some degree of hydrophobicity. Liquid–liquid immiscibility with various alcohols is also common.

Swatloski *et al.*[4,5] and Najdanovic–Visak *et al.*[6] discovered that the addition of water to binary mixtures of imidazolium-based ionic liquids with several alcohols, which show extensive immiscibility areas, increased mutual solubility, until a single phase is formed. Najdanovic–Visak *et al.*[7] have also shown that this surprisingly large co-solvent effect extends over wide ranges of temperatures and compositions. This work has also included the study of pressure and isotope-substitution effects on phase transitions. The ratio of water to alcohol content is the tool for fine-tuning desired situations of total miscibility, partial miscibility, or almost complete phase separation.

This effect is felt even for alcohols with a sufficiently long alkyl chain to induce immiscibility with water, such as 2-methylpropanol[7] or 1-butanol.[8] In these cases, the ternary phase diagrams of mixtures of IL + water + an alcohol show wide areas of total miscibility, although all three corresponding binary mixtures (IL + water, IL + alcohol, water + alcohol) exhibit liquid–liquid immiscibility.

For instance, the mixture [C_4mim][NTf_2] + water and/or 1-butanol demonstrates behaviour similar to the other studied mixtures which involve ILs, alcohols and/or water. Besides displaying upper critical solution temperatures (UCST) and asymmetry of liquid–liquid phase diagrams when plotted on a mole fraction basis, there is an excellent water/1-butanol co-solvent effect. As in the case of [C_4mim][PF_6]+ethanol[6] mixtures, there is a sharp drop in the demixing temperature as water is added to the [C_4mim][NTf_2] + 1–butanol mixture. This is true up to a minimum temperature, where further addition of water induces an increase in the demixing temperature. For some concentrations of ternary mixtures, a transition from a liquid phase to solid phase has been observed.

These water + alcohol + ionic liquid mixtures, due to the variety of their phase transitions, are especially appropriate solvents to carry out reactions where the rate may be controlled by switching, for instance, from two phases to one phase, by small composition changes. The separation of reaction products from the homogeneous reaction mixture will, however, become more complicated than in biphasic reactions.

5.3 Liquid–Liquid-Vapour Equilibria: The CO_2 Anti-Solvent Effect

More than 50 years ago, Francis[9] published an extensive account of phase behaviour of binary and ternary mixtures containing liquid carbon dioxide. He lists 21 systems where completely miscible liquids are separated into two liquid phases by the introduction of CO_2, and many others where partial immiscibility of two liquids is enhanced.

This kind of behaviour was first reported for ionic liquid-containing mixtures by Scurto *et al.*[10] They have presented phase behaviour results of [C_4mim][PF_6] + methanol + CO_2. The ionic liquid is completely miscible with methanol, but their data show the formation of an additional liquid phase at relatively low pressures, leading to three-phase, liquid–liquid–gas equilibrium. For the mixtures with lower content in ionic liquid, those authors have shown that an increase in pressure leads to the disappearance of the intermediate liquid phase, by merging with the upper gas phase, through a critical point. The pressures measured for these critical points were identical to those for the ionic liquid + methanol binaries. In fact, the resulting fluid phase did not contain any ionic liquid, which was retained in the lower liquid phase.

The Scurto *et al.*[11] have also found that the introduction of gaseous or liquid carbon dioxide into a mixture of water and an ionic liquid can cause the separation of both hydrophobic and hydrophilic ionic liquids from aqueous solution, with the formation of an intermediate third phase. Zhang *et al.*[12] reached similar conclusions in their study of the tri-phase behaviour of 1-*n*-butyl-3-methyl-imidazolium tetrafluoroborate mixed with water and carbon dioxide at elevated pressures.

The behaviour discovered by Scurto *et al.* is therefore an expression in ionic liquid systems of the long-known capability of carbon dioxide as an inducer of immiscibility. The importance of this effect is that carbon dioxide may be used, at relatively low pressures, to control the number of phases in a reactive system where the reaction takes place in the ionic liquid.

A thermodynamic analysis of this effect can be based on a succession of phase diagrams for ternary diagrams, as shown by Najdanovic–Visak *et al.*[13] At low pressures, simple liquid–vapour equilibrium is observed, and the vapour side is essentially pure CO_2. An increase in CO_2 pressure induces dissolution of carbon dioxide in the liquid phase and eventually the separation of a second, carbon dioxide-rich liquid, generating three-phase LLV equilibrium. If carbon dioxide continues to be added to the system, the upper liquid L_2 will be further and further depleted in ionic liquid. The phase behaviour of the two upper phases, liquid L_2 and vapour, should approach the behaviour of a mixture of all components except the ionic liquid, as was remarked in the case of the methanol-based mixtures studied by Scurto *et al.*[10], where the L_2 + V part of mixture behaved as a binary CO_2 + methanol.

5.4 Quaternary Systems IL + Water + Alcohol + CO_2

Najdanovic–Visak *et al.*[14] have shown that it is possible to separate [C_4mim][PF_6] from water + ethanol mixtures using CO_2. As carbon dioxide is added to [C_4mim][PF_6] + ethanol + water, a third phase starts to form between the liquid and gas phases, in a similar fashion to what Scurto *et al.* reported on [C_4mim][PF_6] + methanol. At higher pressures, critical points involving the intermediate liquid phase and the vapour were observed, for different water–ethanol molar ratios. This study opened the way to use the above mentioned co-solvent effect of water + alcohol in mixtures with ionic liquid mixtures to carry out reactions in a homogeneous phase, as high pressure carbon dioxide can later be used to re-separate the phases.

In the case presented by Najdanovic–Visak *et al.*[14], the phase behaviour of the quaternary IL + CO_2 + water + ethanol may be analysed in the same way as above, but now as a pseudo-ternary. Especially in the vicinity of the critical points (L_2=V) observed by these authors, the L_2 + V part of the overall mixture might be interpreted in terms of the ternary phase diagram for CO_2 + water + ethanol.

As in this case there is one more degree of freedom than in the methanol + IL mixtures, there are more experiments where no critical point is observed, because the compositions are not the right ones, and the high pressure cell is filled with liquid L_2 (with disappearance of V) at pressures lower than the critical. When critical points were indeed observed, those authors concluded, by comparison with the critical points for the pure ternary mixture (without IL) that water is taken out preferentially from ethanol-rich IL mixtures, while for water-rich initial mixtures, ethanol is preferentially withdrawn from the ionic liquid. This effect might be related to the more recent finding by Najdanovic–Visak *et al.*[7] that pressure increases mutual solubility of [C_4mim][PF_6] and ethanol, but decreases it in the case of the same ionic liquid and water.

In a recent study spanning a much wider range of pressures, Najdanovic–Visak *et al.*[8] found that the system [C_4mim][NTf_2] + water + i-butanol follows the same pattern when mixed with high pressure carbon dioxide. An interesting observation in this study was that the demixing pressure increases along with the increase of water to butanol fraction, that is, a higher water content turns the system more immiscible. In fact, it was concluded that the pressure of the phase transitions is independent of the composition in ionic liquid, and it is essentially driven by the water to alcohol ratio. This strongly suggests that the phases above the lower IL-rich one behave as if no ionic liquid was present.

However, when the phase transition ($L_1 + L_2 + V \rightarrow L_1 + L_2 = V$) pressures are represented as function of water/alcohol overall molar ratios, as in the Figure, the quaternary systems IL + water + alcohol + CO_2 show a very different slope from the corresponding ternaries water + alcohol + CO_2. These diagrams can only be reconciled with the above suggestion if the ionic liquid-rich phase L_1 acts as a selective reservoir for the component present in higher concentration (water or alcohol, depending on the overall composition). In this case, the overall water to alcohol ratios, corresponding to the material present in all phases in the cell, could be very different from the actual ratio in the upper phases L_2 and V. Further phase equilibrium studies, including direct analysis of compositions of phases, are needed in order to clarify this point (Figure 1).

5.5 A Cascade of Phase Changes as Switching Devices for Integrated Reaction + Separation

The cascade of phase changes in aqueous ionic liquid solutions induced by an alcohol and carbon dioxide can be used to allow reaction cycles to proceed. A

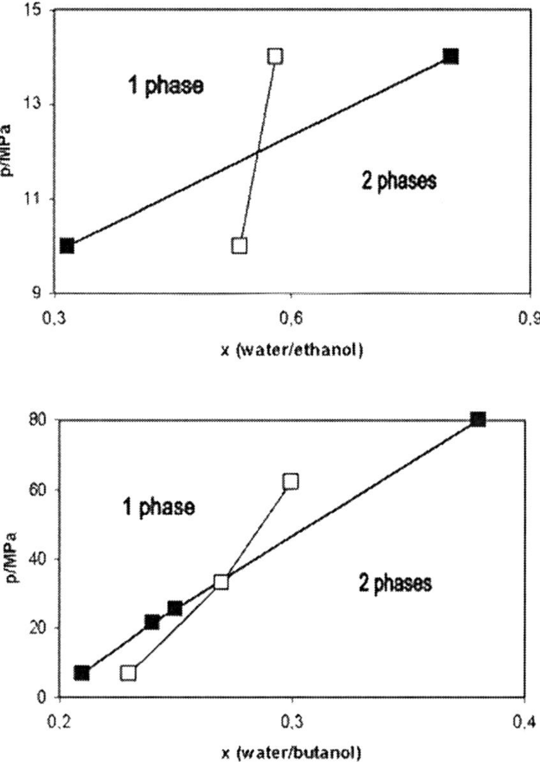

Figure 1 *Comparison between phase transition ($L_1 + L_2 + V \rightarrow L_1 + L_2 = V$) pressures as function of water/alcohol over all molar ratios, of quaternary systems IL + water + alcohol + CO_2 (closed symbols) and the corresponding ternaries water + alcohol + CO_2 (open symbols). Upper Panel: IL = [C_4mim][PF_6], alcohol = ethanol; lower panel, IL = [C_4mim][NTf_2], alcohol=1-butanol*

reaction usually carried out in biphasic water + IL conditions can benefit from homogeneous, monophasic conditions, for increased rates, by addition of ethanol, without losing the advantages of biphasic systems for catalyst recycling and product separation, because carbon dioxide can then be used to extract the reaction products and regenerate the ionic liquid phase.

Najdanovic–Visak *et al.*[14] have chosen the (usually slow) epoxidation of isophorone by hydrogen peroxide, catalysed by sodium hydroxide, as a model reaction to carry out a proof of principle experiment.

Bortolini *et al.*[15] carried out this epoxidation, and the epoxidation of several other electrophilic alkenes, dissolved in [C_4mim][PF_6], by contact with an aqueous solutions of hydrogen peroxide. These reactions were performed in biphasic conditions, due to the above-mentioned immiscibility of water and the ionic liquid. Ethyl acetate was used to extract the products from the reaction mixture. This solvent is partially miscible with [C_4mim][PF_6], which may lead to product contamination. Later, Bortolini *et al.*[16] contacted their two-phase reaction mixture (for a different substrate, 2–cyclohexen–1–one), with supercritical CO_2, at 313 K, but at the pressure of 20 MPa. Their results are essentially solubilities of the epoxide in the carbon dioxide-rich phase. They concluded that extraction of the reaction product in those conditions is viable. Carbon dioxide in this case is merely used as an extraction agent.

Najdanovic–Visak *et al.*[14] used a different approach. They carried out the reaction in a homogeneous phase, by addition of ethanol to the water + ionic liquid immiscible system. These homogeneous conditions allowed a much faster reaction process, with a 74% yield after 1 h, which compares with less than 30% in the same interval, in the biphasic conditions of Bortolini *et al.*[15] An added flow of carbon dioxide at 12 MPa and 313 K totally removed the product from the reaction mixture. It was concluded that the successive phase switches induced by ethanol and carbon dioxide provided advantageous reaction conditions.

Acknowledgments

Support from Fundação para a Ciência e Tecnologia (Lisbon, Portugal), through grants POCTI/EQU/35437/00 and POCTI/QUI/38269/2001, is gratefully appreciated.

References

1. L.A. Blanchard, D. Hancu, E.J. Beckman and J.F. Brennecke, *Nature*, 1999, **399**, 28.
2. J.W. Hutchings, K.L. Fuller, M.P. Heitz and M.M. Hoffmann, *Green Chem.*, 2005, **7**, 475.
3. L.A. Blanchard and J.F. Brennecke, *Ind. Eng. Chem. Res.*, 2001, **40**, 287.
4. R.P. Swatloski, A.E. Visser, W.M. Reichert, G.A. Broker, L.M. Farina, J.D. Holbrey and R.D. Rogers, *Chem. Commun.*, 2001, **20**, 2070.
5. R.P. Swatloski, A.E. Visser, W.M. Reichert, G.A. Broker, L.M. Farina, J.D. Holbrey and R.D. Rogers, *Green Chem.*, 2002, **4**, 81.
6. V. Najdanovic-Visak, J.M.S.S. Esperança, L.P.N. Rebelo, M. Nunes da Ponte, H.J.R. Guedes, K.R. Seddon and J. Szydlowski, *Phys. Chem. Chem. Phys.*, 2002, **4**, 1701.
7. V. Najdanovic-Visak, J.M.S.S. Esperança, L.P.N. Rebelo, M. Nunes da Ponte, H.J.R. Guedes, K.R. Seddon, H.C. Sousa and J. Szydlowski, *J. Phys. Chem. B.*, 2003, **107**, 12797.
8. V. Najdanovic-Visak, L.P.N. Rebelo and M. Nunes da Ponte, *Green Chem.*, 2005, **7**, 443.
9. A.W. Francis, *J. Phys. Chem.*, 1954, **58**, 1099.
10. A.M. Scurto, S.N.V.K. Aki and J.F. Brennecke, *J. Am. Chem. Soc.*, 2002, **124**, 10276.
11. A.M. Scurto, S.N.V.K. Aki and J.F. Brennecke, *Chem. Commun.*, 2003, 572.

12. Z.F. Zhang, W. Wu, H. Gao, B. Han, B. Wang and Y. Huang, *Phys. Chem. Chem. Phys.*, 2004, **6**, 5051.
13. V. Najdanovic-Visak, A. Serbanovic, J.M.S.S. Esperança, H.J.R. Guedes, L.P.N. Rebelo and M. Nunes da Ponte, *in Ionic Liquids III A*, Chapter 23, R. Rogers and K.R. Seddon (eds), American Chemical Society Symp. Ser. Vol 901, 2005, 301.
14. V. Najdanovic-Visak, A. Serbanovic, J.M.S.S. Esperança, H.J.R. Guedes, L.P.N. Rebelo and M. Nunes da Ponte, *Chem. Phys. Chem.*, 2003, **4**, 520.
15. O. Bortolini, V. Conte, C. Chiappe, G. Fantin, M. Fogagnolo and S. Maietti, *Green Chem.*, 2002, **4**, 94.
16. O. Bortolini, S. Campestrini, V. Conte, G. Fantin, M. Fogagnolo and S. Maietti, *Eur. J. Org. Chem.*, 2003, **24**, 4804.

CHAPTER 6

The Solubility of Gases in Water and Seawater

RUBIN BATTINO[1] AND H. LAWRENCE CLEVER[2]

[1] *Department of Chemistry, Wright State University, Dayton OH 45435, USA*

[2] *Department of Chemistry, Emory University, Atlanta GA 30322, USA*

6.1 Introduction

The solubility of gases in liquids has been a subject of great practical and theoretical importance since the start of modern science. The modern era of gas solubilities started with the contributions of William Henry in the early 1800s, and continued with Robert Bunsen in the mid-1800s, and of I. M. Sechenov and of Wilhelm Ostwald in the late 1800s. The contributions of Joel H. Hildebrand in the first half of the 20th century initiated our modern knowledge of gases in liquids. The past 50 years has seen more work and workers than can be individually acknowledged here.

From the theoretical standpoint gas solubilities provide a wonderful probe for the properties of solutions and the liquid state. There is a large range of gas solubility. For example, the mole fraction solubility at 1 bar of gases in water ranges from less than 10^{-5} (*i.e.*, a few molecules of SF_6 for each 1 million water molecules) to about 0.2 (*i.e.*, one molecule of ammonia to four molecules of water). There are both polar and non-polar gases available with a wide range of dipole moments for testing intermolecular interactions. The polarizability (size alone) can be studied using simple spherical molecules in the noble gas series and CH_4, CF_4, and SF_6. Organic gas molecules vary in size and shape, which can make for interesting studies. Some polar gases are chemically reactive with water. The solubility of these gases is given as a bulk solubility (sum of all species derived from the solute gas). With dilute solutions of mole fraction 10^{-4} or less, theoretical treatments become easier as the solution behavior approaches that of an "ideal" solution.

To appreciate the practical importance of gas solubility one needs to do no more than take a breath – the processes by which oxygen gets to the cells which use it and carbon dioxide is expelled involve several dissolutions and evolutions of the gases. Biological oxygen demand (BOD) measurements are crucial in dealing with natural and waste waters. The amount of oxygen in the blood is closely monitored in many surgical interventions. The ocean is a sink (or sometimes a source) of atmospheric gases, including carbon dioxide, methane, ethane, and the chlorofluorocarbons. Many industrial processes require knowledge of gas solubilities. For this usefulness in theory and practicality, it is important to know how to measure gas solubilities, and also about the significant factors that determine the measured solubility. The experimental determination of the solubility of a gas in a liquid requires several steps: (i) purification and characterization of the solute gas and liquid solvent; (ii) thorough degassing of the solvent; (iii) equilibration of the gas and liquid phases under the conditions of known constant temperature and pressure; (iv) measurements that allow determination of the composition of both gas and liquid phases; and (v) treatment of the experimental data to obtain a reliable measure of the equilibrium state (solubility). Any paper reporting data on the solubility of a gas in a liquid should include an adequate description of these steps, and comparison measurements on a standard system (such as oxygen in water) to allow the user to judge the reliability of the data. There are several reviews,[1–9] which can be consulted for additional detail on experimental methods, solubility units, and thermodynamic considerations.

This particular chapter is concerned with gas solubilities in water and seawater. Wilhelm *et al.*[10] reviewed the solubility of gases in water some 30 years ago, and Battino[11] specifically reviewed the high-precision solubility of gases in water. Owing to the interest of the solubility of pharmaceuticals in water, there are many data for these systems, and many correlational studies involving hundreds of compounds, but with very few gases – these systems are ignored in this chapter. After a brief discussion of solubility units and some other related matters, we present some tables of gas solubilities in water and seawater, followed by annotated bibliographies on solubilities in water and seawater.

6.2 Quantities Used as a Measure of Gas Solubility

A number of quantities, three are defined below, are in use to express the solubility of a gas in a liquid. We strongly urge that workers reporting new solubility measurements give their results in one or more of these three quantities since they appear to be of the most general application. They are: mole fraction at a standard pressure (such as 1 bar), the Ostwald coefficient, and the Henry's law constant – both of the latter with a statement of the applicable pressure range. A number designation of 1 for the solvent and 2, 3, ... for the solutes is used throughout this chapter.

The SI standard pressure is 10^5 Pa or 1 bar. Most low pressure gas solubility values in the literature are referenced to a standard pressure of 1 atm (1.01325 bar). We will use the standard pressure of 1 atm in our tables. The values can be converted to the 1 bar-standard pressure by multiplying by 1.01325 bar

$(1\ \text{atm})^{-1}$. We present here just three of the concentration units used – for additional information see, *e.g.*, Clever and Battino.[4,9]

(*a*) *Mole fraction (amount fraction)*. The mole fraction of component 2 in a solution of c components is

$$x_2 = n_2/\sum n_i \tag{1}$$

where n_i is the amount of substance of components $i = 1, 2, \ldots, c$. For two components this is $x_2 = n_2/(n_1 + n_2)$. The temperature and the gas partial pressure must be specified.

(*b*) *Ostwald coefficient*. The Ostwald coefficient of gas solubility is discussed by Battino.[12] He points out the interrelations and limitations of four definitions of the Ostwald coefficient. Using Battino's notation, the definitions of the Ostwald coefficient are

$$L_{\mathrm{V}}^{0} = (V_{\mathrm{g}}/V_{\mathrm{L}}^{0})_{\text{equil}} \tag{2}$$

where V_{g} is the volume of gas absorbed by a volume V_{L}^{0} of pure liquid at a specified temperature and a total pressure p,

$$L_{V} = (V_{\mathrm{g}}/V_{\mathrm{L}})_{\text{equil}} \tag{3}$$

in which V_{g} is the volume of gas absorbed by a volume V_{L} of saturated solution at a specified temperature and total pressure p. For dilute solutions and for a precision of the order of 1%, L_{V}^{0} and L_{v} differ negligibly,

$$L_{\mathrm{c}} = (C_{\mathrm{g}}^{\mathrm{L}}/C_{\mathrm{g}}^{\mathrm{V}})_{\text{equil}} \tag{4}$$

where $C_{\mathrm{g}}^{\mathrm{L}}$ is the amount concentration of gas in the liquid phase in mol L^{-1}, $C_{\mathrm{g}}^{\mathrm{V}}$ is the amount concentration of gas vapor phase in mol L^{-1}, and these concentrations are determined at the specified temperature and total pressure of the measurement,

$$L_{c}^{\infty} = \lim_{C_{\mathrm{g}}^{\mathrm{L}} \to 0} (C_{\mathrm{g}}^{\mathrm{L}}/C_{\mathrm{g}}^{\mathrm{L}})_{\text{equil}} \tag{5}$$

where the terms have the same meaning as the preceding definition, except this is the Ostwald coefficient in units of amount concentration at the limit of infinite dilution. (The *amount concentration c* is defined as $c_2 = n_2/V$, where V is the solution volume.) The differences among these definitions are negligible as long as one can assume ideal gas behavior without expansion of the solvent upon dissolving the gas. Authors should make clear which definition they use and how the experimentally measured data were converted to the Ostwald coefficient.

(*c*) *Henry's Law Constant*. The basic thermodynamic quantity that describes the solubility of gas 2 in a solvent 1 of saturation vapor pressure p_1^{V} at a specified temperature is the Henry's law constant

$$H_{2,1}\,(p_1^{V}) = \lim(f_2/x_2) = \phi_2 p_2/y_2 x_2 \tag{6}$$

The limit is taken as the pressure tends to zero. This describes gas 2 with fugacity f_2 and fugacity coefficient ϕ_2 in equilibrium with solvent 1 in the saturated solution with amount (mole) fraction x_2 and activity coefficient y_2.

To evaluate $H_{2,1}$ (p_1^{V}) from the data of an experiment requires the Poynting correction and evaluation of the fugacity coefficient of 2 in the gas phase and activity coefficient of 2 in the saturated solution. A rigorous evaluation of the Henry's law constant has been carried out for only a few systems. Wilhelm[6,7] has discussed in detail the conversion to a thermodynamic constant. Researchers need to keep in mind that requirements under which the Henry's law constant is a true thermodynamic constant.

6.3 Oxygen Solubility in Water

The solubility of oxygen in water at 298.15 K may be our most reliably known solubility value. *The oxygen + water system is recommended as a test system for any apparatus*. For reference, Table 1 gives the solubility of oxygen in water[13] in three units at a temperature of 298.15 K.

6.4 Two Related Experiments that Complement Gas Solubility Data

6.4.1 Partial Molar Volumes

The partial molar volume is defined as

$$V_2 = (\partial V/\partial n_2)_{\mathrm{T,P,n}} \tag{7}$$

and is a particularly useful parameter. Partial molar volumes may be obtained by volume or density measurements, and also by the pressure dependence of the chemical potential:

$$(\partial \mu_2/\partial P)_{\mathrm{T}} = V_2 \tag{8}$$

The major reviews on this subject were done by Handa *et al.*[14,15].

Table 1 *The solubility of oxygen in water at 298.15 K, and an oxygen partial pressure of either 0.100000 MPa (1 bar) or 0.101325 MPa (1 atm)*

Quantity (Unit)	
x_2 (atm)	2.2996×10^{-5}
x_2 (bar)	2.2695×10^{-5}
L_{v} (cm^{-3})	0.031153
$H_{2,1}$ (MPa)	4.4038×10^{3}

6.4.2 Enthalpy Changes on Solution

Using the Gibbs–Helmholtz equation

$$(\partial(\Delta G^0/T)/\partial T)_P = -\Delta H^0/T^2 \tag{9}$$

it is possible to determine enthalpy changes upon solution when solubilities are measured as a function of temperature. Some of the relevant papers involving the direct calorimetric determination of ΔH are[16–20] and the reader may consult them for information in this field.

6.5 Treatment of Data

6.5.1 Corrections for Non-Ideality

The rigorous thermodynamics approach to the solubility of gases in liquids has been dealt with in detail by Wilhelm.[6–7,21] His writings should be consulted for details on data reduction in a consistent and rigorous manner. Generally, for precisions of 1% or poorer it is not necessary to correct for non-ideality. However, for precision measurements these corrections are essential.

6.5.2 Temperature Dependence of Solubility – Fitting Equations

The basic equation relating changes in thermodynamic functions to solubility, which is really an equilibrium constant, is

$$\Delta G^0 = -RT \ln K = -RT \ln x_2 \tag{10}$$

where x_2 is the equilibrium mole fraction solubility at 1 atm or 1 bar partial pressure of gas depending on the standard state. Two fitting equations for ln x_2 as a function of temperature have proven to be both useful and popular.

The Clarke and Glew[22] version is a variation of the van't Hoff form and is

$$\ln x_2 = A + B/(T/\mathrm{K}) + C \ln(T/\mathrm{K}) + D(T/\mathrm{K}) \tag{11}$$

Weiss[23] found that the coefficients would be of approximately equal magnitude if the Kelvin temperature was divided by 100. Thus, the Clarke–Glew–Weiss or CGW fitting equation is

$$\ln x_2 = A + B/(T/100\ \mathrm{K}) + C \ln(T/100\ \mathrm{K}) + D(T/100\ \mathrm{K}) \tag{12}$$

If we let $\tau = T/100$ K, then we get

$$\ln x_2 = A + B/\tau + C \ln \tau + D\,\tau \tag{13}$$

Benson and Krause[24] first proposed fitting the logarithm of the Henry's law constant as a power series in $1/T$ as in

$$\ln H_{2,1} = a + b/(T/\mathrm{K}) + c/(T/\mathrm{K})^2 + d/(T/\mathrm{K})^3 \tag{14}$$

where $H_{2,1}$ is the Henry's law constant. For those interested, the changes in the thermodynamic functions for these two equations may be found in the introductory material in Battino.[25]

6.5.3 Pressure Fitting Equations

Henry's law describes the relationship between solubility and pressure. This has been described earlier under solubility units,

$$\lim_{x_2 \to 0} (f_2/x_2) = H_2 \tag{15}$$

where f_2 is the fugacity and H_2 is the Henry's law constant. Note that the Henry's law constant is valid only in the limit of infinite dilution or zero partial pressure of gas. In the pressure range, where Henry's law accurately describes the behavior of the solution, $f_2 = x_2H_2$. Under ideal conditions where we can set $f_2 = P_2$, and $P_2 = x_2H_2$. This equation represents the solubility of gases in liquids quite well over a surprisingly large range of pressures.

The Krichevsky–Kasarnovsky[26] equation is better than Equation (15) and is

$$\ln f_2/x_2 = \ln K_2 + [V_2^0\,(p - p_{\text{sat}})]/RT \tag{16}$$

where V_2^0 is the partial molar volume at infinite dilution, p_{sat} is the saturation vapor pressure of the solvent at temperature T, and p is the total pressure. The Krichevsky and Il'inskaya[27] equation is better than Equation (16) because the constant A takes into account solvent–solute interactions, and is

$$\ln f_2/x_2 = \ln K_2 + (A/RT)(x_1^2 - 1) + [V_2^0\,(p - p_{\text{sat}})]/RT \tag{17}$$

where A is an empirically evaluated constant.

6.5.4 Salt Effects

The solubility of a gas in an aqueous electrolyte solution is often reported as the Sechenov salt-effect parameter, a relationship pointed out by Sechenov[28] over 100 years ago. The general empirical Sechenov equation is

$$\log(z_2^*/z_2) = k_{\text{syz}}\, y \tag{18}$$

where the solubility of gas 2 is expressed by the quantities z_2^* in pure water and z_2 in the electrolyte solution, and the salt composition is expressed in the quantity y. See Clever[29d] (pp. xxix–xliii), and Clever[30] for details on equations for salting-out effects.

6.6 The Solubility of Gases in Water

There now exists a large body of data believed to be of first-class reliability for the solubility of gases in water near atmospheric pressure and over the 273–323

Table 2 *Recommended high-precision values of the solubility of 10 gases in water at 0.101325 MPa (1 atm) partial gas pressure and 298.15 K. Thermodynamic changes for the solution process are also given*

Gas	$10^5\,x_2$	$10^2\,L$	ΔH_2^0 (kJ mol^{-1})	ΔS_2^0 (J K^{-1} mol^{-1})	ΔC_P (J K^{-1} mol^{-1})
He	0.70789	0.9590	−0.54	−100.4	122
Ne	0.82241	1.1141	−3.64	−109.6	143
Ar	2.5306	3.4244	−11.92	−128.0	195
Kr	4.5463	6.1433	−15.34	−134.6	218
Xe	7.9500	10.715	−19.06	−142.4	250
N_2	1.1774	1.5940	−10.45	−129.4	214
CO	1.7744	2.4023	−10.78	−127.1	215
CH_4	2.5523	3.4559	−13.19	−132.2	237
C_2H_6	3.4006	4.6061	−19.43	−150.7	–
C_2H_4	8.899	12.011	−16.40	−133	239

Notes: Helium, neon, argon, krypton and xenon data are from ref. 32; Nitrogen data from ref. 33; carbon monoxide data from ref 34; methane data from ref. 35; ethane data from ref. 36; and ethene data from ref. 37.

Table 3 *Recommended values of the solubility of oxygen in water*[13] *at 0.101325 MPa (1 atm) partial pressure of gas at several temperatures. Also given are thermodynamic changes for the solution process. The Henry's law constants at several temperatures are in a sub-table*

T(K)	$10^5\,x_2$	$10^2\,L$	ΔH_2^0 (kJ mol^{-1})	ΔS_2^0 (J K^{-1} mol^{-1})	ΔC_P (J K^{-1} mol^{-1})
273.15	3.9594	4.9261	−17.43	−143	239
278.15	3.4651	4.3900	−16.26	−140	230
283.15	3.0736	3.9628	−15.13	−137	222
288.15	2.7603	3.6194	−14.04	−134	215
293.15	2.5071	3.3414	−12.99	−132	207
298.15	2.3009	3.1152	−11.97	−129	200
303.15	2.1317	2.9304	−10.98	−126	194
308.15	1.9923	2.7793	−10.03	−124	188
313.15	1.8769	2.6560	−9.11	−121	182
318.15	1.7813	2.5557	−8.21	−118	176
323.15	1.7021	2.4748	−7.35	−116	170
328.15	1.6365	2.4104	−6.51	−113	165

T (K)	$10^{-9}\,H_{2,1}(T,P_{S,1})$ (Pa)	T (K)	$10^{-9}\,H_{2,1}(T,P_{S,1})$ (Pa)
273.15	2.5591	303.15	4.7533
278.15	2.9242	308.15	5.0859
283.15	3.2966	313.15	5.3985
288.15	3.6708	318.15	5.6882
293.15	4.0415	323.25	5.9531
298.15	4.4038	328.15	6.1917

Table 4 *Annotated Bibliography of the Solubility of Gases in Water*

Gas	*Comment*	*Refs*
O_2	Evaluation. 1 atm; 273–348 K	25a, pp. 1–5
O_3	Evaluation. Papers from 1872 to 1980	25b, pp. 474–480
H_2	Evaluation. 1 atm; 273–353 K	38a, pp. 1–3
H_2	Evaluation. >2 bar	38b, pp. 303–304
He	Evaluation. 1 atm; 273–348 K	39a, pp. 1–4
He	Evaluation. Up to 1013 bar; 273–589 K	39b, p. 257
Ne	Evaluation. 1 atm; 273–348 K	39a, pp. 124–126
Ar	Evaluation. 1 atm; 273–348 K	40a, pp. 1–7
Ar	Evaluation. Up to 126 bar	40b, p. 256
Kr	Evaluation. 1 atm; 273–353 K	41a, pp. 1–3
Xe	Evaluation. 1atm; 273–348 K	41a, pp. 134–136
Rn	Evaluation. 1 atm; 273–373 K	41a, pp. 227–229
Cl_2	Evaluation. 1 atm; 283–333 K. "Bulk" solubility	42, pp. 333–334
Cl_2O	Data from 3 papers	42, pp. 449–453
ClO_2	Evaluation. 1 atm; 283–333 K	42, p. 454
SO_2	Evaluation. 1 atm; 278–328 K. "Bulk" solubility	42, pp. 3–5
H_2S	Evaluation. 1 atm; 273–603 K. "Bulk" solubility	43, pp. 1–3
D_2S	Evaluation. 1 atm; 278–323 K	43, p. 329
H_2Se	Evaluation. 1 atm; 288–308 K	43, pp. 330–331
SF_6	Evaluation. 1 atm; 273–473 K	44, pp. 227–228
N_2	Evaluation. 1 atm; 273–348 K	29a, pp. 1–4
N_2	Evaluation. >2 bar; 298–398 K	29b, p. 333
N_2O	Evaluation. 1 atm; 273–313 K	45a, pp. 1–2
NO	Evaluation, 1 atm; 273–358 K	45a, pp. 260–261
NF_3	Evaluation. 1 atm; 278–323 K	44, p. 224
N_2F_4	Evaluation, 1 atm. 288–318 K	44, p. 224
NH_3	Evaluation, up to 2 MPa; 273–374 K "Bulk" solubility	46,47
CO	Evaluation, 1 atm; 273–328 K	48a, pp. 1–2
CO_2	Evaluation. 1 atm; model for data from 1855 to 1982. "Bulk" solubility	49a, pp. 1–4
CO_2	Evaluation. Upto 647 K; "Bulk" solubility	49b, pp. 5–12
CH_4	Evaluation. 1 atm; 273–523 K	35a, pp. 1–6
CH_4	Evaluation. 0.5–200 MPa. 298–627 K	35, pp. 24–28
CHF_3	Evaluation. 1 atm; 278–328 K	44, p. 327
CF_4	Evaluation. 1 atm; 273–323 K	44, p. 289
CH_3F	Evaluation, 1 atm; 273–353 K	44, pp. 337–338
C_2H_6	Evaluation. 1 atm; 273–323 K	36a, pp. 1–2
C_2H_6	Evaluation. >2 bar; 310–673 K	36b, p. 16
$C_2H_4F_2$	Evaluation. 1 atm; 273–338 K	44, pp. 375–376
C_2H_4	Evaluation. 1 atm; 278–323 K	50a, pp. 1–2
C_2H_4	Evaluation. >2 bar; 298–361 K	50b, pp. 14–16
C_2F_4	Evaluation. 1 atm; 273–343 K	44, p. 399
C_3H_8	Evaluation. 1 atm; 273–348 K	51, pp. 1–2
C_4H_{10}	Evaluation. 1 atm; 273–353 K	51, pp. 16–17
i-C_4H_{10}	Evaluation. 1 atm; 278–318 K	51, p. 34
c-C_4F_8	Evaluation. 1 atm; 278–318 K	44, pp. 292–293

Table 5 *Annotated bibliography of the solubility of gases in seawater*

Gas	*Comment*	*Refs.*
O_2	Evaluation. 1 atm; 274–309 K. Effect of pressure discussed	25c, pp. 41–43
H_2	Evaluation. 1 atm; 273–303 K	38c, pp. 17–18
He	Evaluation. 1 atm; 272–313 K	39c, p. 16
Ne	Evaluation. 1 atm; 274–303 K	39c, p. 138
Ar	Evaluation. 1 atm; 274–308 K	40c, pp. 27–28
Kr	Evaluation. 1 atm; 274–313 K	41c, p. 9
SO_2	Evaluation. 1 atm; 279–303 K; "Bulk" solubility	42, p. 34
H_2S	1 atm; 273–303 K; "Bulk" solubility	52, p. 259–268
SF_6	1 atm; 272–313 K	53, pp. 175–187
N_2	Evaluation. 1 atm; papers from 1880–1964	29b, pp. 31–33
N_2O	Evaluation. 1 atm; 272–313 K	45, pp. 23–24
CO	1 atm; 274–304 K	48b, pp. 15–16
CO_2	Evaluation. 1 atm; 293–333 K; up to 15 bar. "Bulk" solubility	49c, pp. 76–78
CH_4	Evaluation. 1 atm; 274–303 K	35b, pp. 50–51
CF_4	1 atm; 288–303 K	54, pp. 167–169
$CClF_3$	1 atm; 272–313 K	55, pp. 1151–1161
CCl_2F_2	1 atm; 273–313 K	56, pp. 1485–1497
CCl_3F	1 atm; 273–313 K	56, pp. 1485–1497
C_4H_{10}	1 atm; 276–292 K. Synthetic seawater	51, p. 120

K temperature interval. These data have been measured using the high-precision Benson–Krause apparatus,[13,24,31] and the data treated by the guidelines of Wilhelm.[6–7,21] Henry's law constants and Ostwald coefficients are given for 10 gas+water systems in Table 2 at 298.15 K. In addition to solubility data, Table 2 contains enthalpy, entropy, and heat capacity changes for the solution process at 298.15 K. Incidentally, many of these values have been confirmed by direct calorimeteric measurements. (Note that the complete data set is given in ref. 9, pp. 139–144.)

Table 3 gives oxygen solubilities in water from 273.15 to 318.15 K at 5° intervals, along with the changes in thermodynamic properties on solution. In addition, in Table 3 we present the Henry's law constant as a separate sub-table.[13] Tables 4 and 5 are annotated bibliographies of the solubilities of gases in water and seawater, respectively. Also, see Wilhelm *et al.*,[10] for a 1977 review article on this subject.

6.7 Annotated Bibliography of the Solubility of Gases in Water

Table 4 is an annotated bibliography of the solubility of gases in water. "Evaluation" means that that there is a critical evaluation of that gas/water system in the *Solubility Data Series* volume cited. These critical evaluations provide an analysis of all of the relevant papers for that system, along with recommended values up to the time of publication of the respective volume.

Also, for some important systems for which no evaluation has been prepared references to recent important papers are given. The reference number refers to the list of references.

6.8 Annotated Bibliography of the Solubility of Gases in Seawater

Table 5 contains an annotated bibliography of the solubility of gases in seawater up to the time of publication of this volume.

6.9 Summary

The solubility of gases in water and seawater is of much importance. In this chapter we discussed the solubility of gases in liquids in general, and the solubility of gases in water and seawater in particular. Tables of relevant data were presented as well as annotated bibliographies. Solubility data for 10 important gases were also presented.

References

1. A.E. Markham and K.A. Kobe, *Chem. Rev.*, 1941, **28**, 519.
2. R. Battino and H.L. Clever, *Chem. Rev.*, 1966, **66**, 395.
3. A.S. Kertes, O. Levy and G.Y. Markovits, Solubility, in *Experimental Thermochemistry*, B. Vodar and B. LeNaindre (eds), vol II, Butterworth, London, 1974, Chapter 15.
4. H.L. Clever and R. Battino, The solubility of gases in liquids, in *Solutions and Solubilities*, M.J.R. Dack (ed), *Techniques of Chemistry*, A. Weissberger, (series. ed), vol. VIII, Part II, Wiley, New York, 1975, Chapter 7.
5. C.L. Young, R. Battino and H.L. Clever, The solubility of gases in liquids: introductory information, in *Helium and Neon, IUPAC Solubility Series*, H. L. Clever (ed), vol 1, Pergamon Press, Oxford, 1979, pp. xv–xxi.
6. E. Wilhelm, *The solubility of gases in liquids thermodynamic considerations*, in *Nitrogen and Air, IUPAC Solubility Series*, R. Battino (ed), vol 10, Pergamon Press, Oxford, 1982, pp. xx–xxviii.
7. E. Wilhelm, *CRC Crit. Rev. Anal. Chem.*, 1985/1986, **16**, 129.
8. P.G.T. Fogg and W. Gerrard, in *Methods of Measuring the Solubilities of Gases in Liquids*, Wiley, New York, 1991, Chapter 2.
9. H.L. Clever and R. Battino, The solubility of gases in liquids, in The Experimental Determination of Solubilities, G.T. Hefter and R.P.T. Tomkins (eds), Wiley, Chicester, 2003, *Wiley Series in Solution Chemistry*, Vol 6, P. Fogg (ed), Chapter 2.1, pp. 101–150.
10. E. Wilhelm, R. Battino and R.J. Wilcock, *Chem. Rev.*, 1977, **77**, 219.
11. R. Battino, *Rev. Anal. Chem.*, 1989, **9**, 131.

12. R. Battino, *Fluid Phase Equilibr.*, 1984, **15**, 231.
13. T.R. Rettich, R. Battino and E. Wilhelm, *J. Chem. Thermodyn.*, 2000 **32**, 1145.
14. Y.P. Handa and G.C. Benson, *Fluid Phase Equilibr.*, 1982, **8**, 161.
15. Y.P. Handa, P.J. D'Arcy and G.C. Benson, *Fluid Phase Equilibr.*, 1982 **8**, 181.
16. S.J. Gill and I. Wadsö, *J. Chem. Thermodyn.*, 1982, **14**, 905.
17. S.F. Dec and S.J. Gill, *Rev. Sci. Instrum.*, 1984, **55**, 765.
18. G. Olofsson, A.A. Ushodj, E. Qvarnstrom and I. Wadsö, *J. Chem. Thermodyn.*, 1984, **16**, 1041.
19. S.F. Dec and S.J. Gill, *J. Solution Chem.*, 1984, **13**, 27.
20. S.F. Dec and S.J. Gill, *J. Solution Chem.*, 1984, **14**, 417.
21. E. Wilhelm, *J. Thermal. Anal.*, 1997, **47**, 545.
22. E.C.W. Clarke and D.N. Glew, *Trans. Faraday Soc.*, 1966, **62**, 539.
23. R.F. Weiss, *Deep-Sea Res. Oceanogr. Abstr.*, 1970, **17**, 721.
24. B.B. Benson and D. Krause, *J. Phys. Chem.*, 1976, **64**, 689.
25. (a) R. Battino, (b) L.A. Roth and (c) C.-T. A. Chen in *Oxygen and Ozone, IUPAC Solubility Series*, R. Battino (ed), vol 7, Pergamon Press, Oxford, 1981.
26. I.R. Krichevsky and J.S. Kasarnovsky, *J. Am. Chem. Soc.*, 1935, **57**, 2168.
27. I. Krichevsky and A. Il'inskaya, *Acta Physiochim., URSS*, 1945, **20**, 327.
28. I.M. Sechenov, *Z. Phys. Chem., Stoechiom. Verwandtschaftsl.*, 1889, **4**, 117.
29. (a) R. Battino, (b) C.L. Young and (c) C.-T.A. Chen and (d) H.L. Clever, in *Nitrogen and Air, IUPAC Solubility Series*, R. Battino (ed), vol 10, Pergamon Press, Oxford, 1982.
30. H.L. Clever, *J. Chem. Eng. Data*, 1983, **28**, 340.
31. B.B. Benson and D. Krause, *J. Solution Chem.*, 1989, **18**, 835.
32. D. Krause and B.B. Benson, *J. Solution Chem.*, 1989, **18**, 823.
33. T.R. Rettich, R. Battino and E. Wilhelm, *J. Solution Chem.*, 1984, **13**, 335.
34. T.R. Rettich, R. Battino and E. Wilhelm, *Ber. Bunsenges. Phys. Chem.*, 1982, **86**, 1128.
35. (a) R. Battino and (b) D.A. Wiesenburg, in *Methane, IUPAC Solubility Series*, H.L. Clever and C.L. Young (eds), Pergamon Press, vol 27/28, Oxford, 1989, pp. 1–5.
36. (a) R. Battino and (b) C.L. Young, in *Ethane, IUPAC Solubility Series*, vol. 9, W. Hayduk (ed), Pergamon Press, Oxford, 1985, 1–5.
37. T.R. Rettich, R. Battino and E. Wilhelm, communication from authors.
38. (a) R. Battino, (b) C.L. Young and (c) D.A. Wiesenburg, in *Hydrogen and Deuterium, IUPAC Solubility Series*, vol 516, C.L. Young (ed), Pergamon Press, Oxford, 1981.
39. (a) R. Battino, (b) C.L. Young and (c) H.L. Clever, in *Helium and Neon, IUPAC Solubility Series,* vol 1, H.L. Clever (ed), Pergamon Press, Oxford, 1979.
40. (a) R. Battino, (b) C.L. Young and (c) C.-T.A. Chen, in *Argon, IUPAC Solubility Series*, vol 4, H.L. Clever (ed), Pergamon Press, Oxford, 1980.

41. (a) R. Battino and (b) H.L. Clever, in *Krypton, Xenon and Radon, IUPAC Solubility Series*, vol 2, H.L. Clever, (ed), Pergamon Press, Oxford, 1979.
42. R. Battino in *Sulfur Dioxide, Chlorine, Fluorine and Chlorine Oxides, IUPAC Solubility Series*, vol 12, C.L. Young (ed), Pergamon Press, Oxford, 1983.
43. P.G.T. Fogg, in *Hydrogen Sulfide, Deuterium Sulfide and Hydrogen Selenide, IUPAC Solubility Series*, vol 32, P.G.T. Fogg and C.L. Young (eds), Pergamon Press, Oxford, 1988.
44. R. Battino and H.L. Clever, in *Gaseous and Volatile Fluorides*, J. Phys. Chem. Ref. Data, 2005, **33**, 224 (*IUPAC Solubility Series*, vol 80), H.L. Clever (ed).
45. (a) R. Battino and (b) C.L. Young, in *Oxides of Nitrogen, IUPAC Solubility Series*, vol 8, C.L. Young (ed), Pergamon Press, Oxford, 1981.
46. R. Tillman-Roth and R.J. Combs, *J. Phys. Chem. Ref. Data*, 1998, **27**, 45.
47. P.E. Field and R.J. Combs, *J. Solution Chem.*, 2002, **31**, 719.
48. (a) R. Battino and (b) D.A. Wiesenburg, in *Carbon Monoxide, IUPAC Solubility Series*, vol 43, R.W. Cargill (ed), Pergamon Press, Oxford, 1990.
49. (a) J.J. Carroll and A.E. Mather, (b) R. Crovetto and (c) D.A. Wiesenburg, in *Carbon Dioxide in Water and Aqueous Electrolyte Solutions, IUPAC Solubility Series*, vol 62, P. Scharlin (ed), Oxford University Press, Oxford,1996.
50. (a) R. Battino and (b) W. Hayduk, in *Ethene, IUPAC Solubility Series*, vol 57, W. Hayduk (ed), Oxford University Press, Oxford, 1994.
51. R. Battino, in *Propane Butane and 2-Methylpropane IUPAC Solubility Series*, W. Hayduk, (ed), vol. 24, Pergamon Press, Oxford, 1986.
52. A.A. Douabul and J.P. Riley, *Deep-Sea Res.*, 1979, **26**(3A), 259.
53. J.L. Bullister, D.P. Wisegarver and F.A. Menzia, Deep-Sea Res. 2002, **I49**, 175.
54. P. Scharlin and R. Battino, *J. Chem. Eng. Data*, 1995, **40**, 167.
55. X. Bu and M.J. Warner, *Deep Sea Res. I: Oceanogr Res. Papers*, 1995, **42**(7), 1151.
56. M.J. Warner and R.F. Weiss, *Deep-Sea Res. A: Oceanogr. Res. Papers*, 1985, **32**(12A), 1485.

CHAPTER 7

Isotope Effects on Solubility

W. ALEXANDER VAN HOOK[1] AND LUÍS P.N. REBELO[2]

[1] *Chemistry Department, University of Tennessee, Knoxville TN 37996-1600, USA*

[2] *Instituto de Tecnologia Quimica e Biologica, ITQB 2, Universidade Nova de Lisboa, Av. da Republica, Apartado 127, 2780-901, Oeiras, Portugal*

7.1 Introduction

The effects of isotopic substitution on the chemical and physical properties of molecules are usually modest. That is because almost all the chemistry, and a good fraction of the physics depends principally on electronic structure, that is to say on the system wave function, *i.e.* on charge–charge interactions. To an excellent approximation the electronic structure is independent of the distribution of nuclear mass, which is the Born–Oppenheimer approximation, and the function used to describe the dependence of intramolecular (and by extension intermolecular) energy on nuclear position is isotope independent. That function maps as the potential energy surface (PES) and a thorough understanding of PES is prerequisite to a complete theoretical description of the system. In contrast to PES, the total energy (which includes both kinetic and potential energies), and the associated free energy, are isotope dependent. That is because the kinetic energy terms used to describe motion on PES are mass-dependent. Kinetic energy is related to molecular structure, to mass and mass distribution, and to temperature, in well-understood ways. Thus, one reason for interest in isotope effects on physical properties, including solute and solvent isotope effects on solubility, is that these sometimes small and often subtle differences afford a useful probe to investigate PES. At a more applied level isotope shifts in the solubilities of gases, salts, and molecular solids, and isotope shifts in solubilities of liquids and their critical solution temperatures (both upper and lower) can be exploited for the design of isotope separation schemes.

7.2 Theoretical Background

Solubility isotope effects, like all thermodynamic isotope effects, are pure quantum effects, and can be described using the modern quantum mechanical theory of condensed phase isotope effects developed by Bigeleisen[1] and elaborated by Stern *et al.*[2] The theory begins by considering the difference in chemical potential for an isotopomer pair, μ' and μ, on transfer from some reference state (normally the low-pressure ideal gas, or the pure liquid or pure solid, in either case superscripted "o") to solution. By convention the prime symbolizes the lighter isotopomer, the symbol Δ is reserved for isotopic differences, *i.e.* $\Delta\mu^{\mathrm{o}}=\mu^{\mathrm{o}\prime}-\mu^{\mathrm{o}}$, and δ describes the change in state, *i.e.* (reference = solution). The quantity of interest is thus $\delta\Delta\mu=(\mu_{\mathrm{S}}'-\mu_{\mathrm{S}})-(\mu^{\mathrm{o}\prime}-\mu^{\mathrm{o}})$, where subscript "S" denotes solution. In solution the chemical potential of, say, the lighter isotopomer of the component of interest, may be written to good enough approximation as $\mu_{\mathrm{S}}'=\mu_{\mathrm{S}}'^{*}+\mathrm{RT}\ln\gamma'x'$, where $\mu_{\mathrm{S}}'^{*}$ is the standard state partial molar free energy of the compound of interest in the solution, γ' is its activity coefficient, and x' the mole fraction. It is sometimes convenient to use different concentration scales, *e.g.* molality, partial pressure, *etc.*, in which case appropriate changes in notation are applied. At equilibrium the partial molar free energies of the solute in the reference and solution states are equal, so $\delta\Delta\mu=0=(\mu_{\mathrm{S}}'^{*}-\mu_{\mathrm{S}}^{*})-(\mu^{\mathrm{o}\prime}-\mu^{\mathrm{o}})+RT\ln(\gamma'x'/\gamma x)$. For ideal solutions $\gamma'=\gamma=1$ and the isotope effect on solubility is a direct measure of the isotope effect on the free energy of transfer between the reference and solution standard states:

$$\delta\Delta\mu(^{\infty,\mathrm{o}})/\mathrm{RT}=[(\mu_{\mathrm{S}}'^{\infty}-\mu_{\mathrm{S}}^{\infty})-(\mu^{\mathrm{o}\prime}-\mu^{\mathrm{o}})]/RT=\ln(x'/x) \quad (1)$$

Equation (1) refers to the Henry's law standard state in the solution, which is infinite dilution (hence the change in notation from superscript "*" to superscript "∞"). At higher concentrations corrections for nonideality must be included, *i.e.* the-$\ln(\gamma'/\gamma)$ term.

The theory of isotope effects in condensed phases[1–5] points out the equivalence of $\delta\Delta\mu/RT$ and $\ln[(s/s')f^{\mathrm{o}}/(s/s')f^{*}]$. The $(s/s')f$'s are the (quantum-mechanical over classical) partition function ratios which describe the solution and reference phases. The introduction of reduced partition functions emphasizes the fact that these isotope effects, like all thermodynamic isotope effects, are pure quantum phenomena. The calculation of the isotopic free energy differences, then, requires evaluation of the reduced partition function ratios in the reference and solution standard states. That important step generally involves some type of approximate model calculation. A common approach follows Stern *et al.*[2] who applied a lattice cell model in the harmonic approximation to treat the $3n$ normal modes per molecule in the solution and reference phases. At a given temperature, say T_{C},

$$\frac{\delta\Delta\mu}{RT_{\mathrm{c}}}=\ln\left(\frac{f^{\mathrm{o}}}{f^{\infty}}\right)$$

$$=\ln\left\{\prod_{i=1}^{3n}\frac{(u_i/u_i')^{0}}{(u_i/u_i')^{\infty}}\frac{\exp\left((u_i'-u_i)^{\mathrm{o}}/2\right)}{\exp\left((u_i'-u_i)^{\infty}/2\right)}\frac{\left(\frac{1-\exp(-u_i')}{1-\exp(-u_i)}\right)^{\mathrm{o}}}{\left(\frac{1-\exp(-u_i')}{1-\exp(-u_i)}\right)^{\infty}}\right\} \quad (2)$$

Equation (2) assumes the Henry's law infinite dilution reference for the solution. Also $u_i = hc\nu_i/kT_c$, the ν_I's are the frequencies (in wave numbers) of the internal ($3n-6$, $3n-5$ for linear molecules, or 0 for monatomics) and external (6, 5, or 3, respectively) vibrational modes. In the special case where the vibrational frequencies can be separated into a low-frequency group (generally the 6 (5 or 3) external modes assigned to translational and librational motion), and a high-frequency one (the $3n - 6(5)$ internal vibrations), Equation (2) reduces to a convenient approximate relation for the isotopic double difference of chemical potentials (the so-called *AB*-approximation):

$$\begin{aligned}\frac{\delta\Delta\mu}{RT_c} &= \frac{1}{24}\left(\frac{hc}{kT_c}\right)^2 \sum_{ext}^{6}\left[\left(\nu_i'^2 - \nu_i^2\right)^{o} - \left(\nu_i'^2 - \nu_i^2\right)^{\infty}\right] \\ &+ \frac{1}{2}\left(\frac{hc}{kT_c}\right) \sum_{\text{int}}^{3n-6}\left[\left(\nu_i' - \nu_i\right)^{o} - \left(\nu_i' - \nu_i\right)^{\infty}\right] \\ &= A/T^2 + B/T\end{aligned} \tag{3}$$

7.3 Liquid–Liquid Equilibria

Isotope effects on the mutual solubility of liquids have been by far the most studied of solubility isotope effects. Consider a binary liquid mixture of components 1 and 2, which phase-separates into two liquid solutions in equilibrium one with the other, the first (A) richer in component 1, the second (B) in component 2. Focusing on component 2, the one isotopically substituted, its standard state in A is expressed using the Henry's law standard state (infinite dilution), while in solution B, the standard state is more reasonably taken as the pure liquid (Raoult's law). In gross approximation, then, the transfer free energy of interest is that from the Raoult to Henry standard state. Remember, however, that in their standard states, solutions are necessarily ideal, while in contrast the very observation of liquid–liquid demixing is *prima facae* evidence of nonideality. Nonideality is most conveniently described using excess partial molar free energies, $\delta\Delta\mu_2{}^{ex} = RT \ln \gamma_2'/\gamma_2$, $g^{ex} = x_1\mu_1{}^{ex} + x_2\mu_2{}^{ex}$. Thus, proper understanding of isotope effects on liquid–liquid equilibria (or liquid–liquid demixing) requires a marriage of the theory of condensed phase isotope effects with model theories which treat the excess free energy of nonideal solutions.[6–8] Thus, for example, by using the Guggenheim[6] or modified Flory–Huggins[7,8] theories to describe critical immiscibility and liquid–liquid demixing, one obtains an equality which relates the isotope shift in the UCST and/or LCST, ΔT_c, with the g^E-model parameters.[8] (UCST and LCST, the upper and lower critical solution temperature, respectively, are those temperatures above or below which, respectively, the solution is homogeneous at all concentrations):

$$-\frac{\Delta T_c}{T_c} = \frac{1}{(\chi_c - d_o + d_2(1 + \ln T_c))}\frac{\delta\Delta\mu_2}{RT_c} \tag{4}$$

In Equation (4), χ_c is the isotope-independent critical value of the temperature-dependent interaction parameter that triggers phase separation, $\chi = d_o + d_1/T + d_2 \ln T$ (χ_c varies between 2 for strictly symmetrical mixtures, and 0.5 for solutions of infinitely long polymers dissolved in small molecule solvents). In summary, Equations (2)–(4) establish the relationship between the isotope shift on UCST and/or LCST with the isotope dependent normal mode frequency shifts which occur on transfer from the reference state to solution.

7.3.1 Small Molecule Solutions Including Aqueous Systems

Fenby *et al.*[9] and Rabinovich[10] have given useful reviews of experimental data on deuterium isotope effects on liquid–liquid (L–L) equilibria (deuterium isotope effects) prior to 1980, but without detailed theoretical discussion. The data show that deuteration on water (or alcohols or amines) enlarges the two-phase region(s) (either lowering LCST, increasing UCST, or both). The effects are substantial; it is common to observe temperature shifts as large as 10 or 20 K, or even more. If one assumes that it is hydrogen-bonding effects which are predominantly responsible for UCST partial miscibility isotope effects, then, one would expect a significant decrease in the magnitude of ΔT_c as T_c itself increases. That idea is supported by the regular trend observed for more than 20 binary aqueous systems on H_2O/D_2O substitution; the ~25 K isotopic shift observed at 270 K becomes negligible at 470 K. The importance of hydrogen bonding can also be appreciated by noting that in studies[6,11] on methanol/cyclohexane mixtures there is an increase in the UCST of about 1.2 K per H/D substituted hydrogen-bond. In contrast, H/D substitution at other kinds of hydrogen leads to a decrease in T_c of 0.3 K/atom replaced. For systems where isotope exchange is possible the interpretation becomes more complicated.[11,12] The (methanol + cyclohexane) system was chosen by Singh and Van Hook[6] and independently by Schon *et al.*[11] for detailed studies involving deuterium substitution on each component (Table 1). The laboratory to laboratory differences in the critical demixing temperatures are almost certainly due to trace contamination with water.[13] The SVH results were successfully interpreted using the model based on Equations (2)–(4), above, using the Guggenheim theory of symmetrical mixtures. The data are consistent with shifts in the methanol librational and OH stretching frequencies of approximately (-385 cm^{-1}) and ($+100$ cm^{-1}), respectively, on transfer of methanol from the pure liquid state to infinite dilution in cyclohexane, and to a much more modest shift of ($+7$ cm^{-1}) for the CD stretching frequencies of cyclohexane on its transfer from pure liquid to infinite dilution in methanol. Ethylene glycol (OH/OD) + nitromethane mixtures were similarly studied and interpreted.[14] Recently the UCST demixing envelopes describing the H/D isotope effects for (nitromethane + pentanol) and (nitromethane + isobutanol) solutions were carefully determined,[15] (Figure 1). For these solutions the T_c shifts are very small (1.4 K and less) but in all cases it was found that deuteration leads to the enlargement of the limited miscibility region.

Table 1 *(a) Isotope effects on UCST's of methanol+cyclohexane mixtures*

	$[T_C(a)/K$ or $(T_C(a)\text{-}T_C)/K]^6$	$[T_C(a)/K$ or $(T_C(a)\text{-}T_C)/K]^{11}$
(a) C_6H_{12}/CH_3OH	318.81	318.29
C_6D_{12}/CH_3OH	3.9	4.9
C_6H_{12}/CH_3OD	−2.5	−1.9
C_6D_{12}/CH_3OD	---	3.4
C_6H_{12}/CD_3OD	0.2	6.6

(b) Solute and solvent solubility isotope effects for (benzene–water) solutions at 306.2 K,[16] and solute isotope effects for methane,[35] nitrogen,[28] oxygen,[28] and CO_2,[36] water solutions. (Isotopbote effects on free energies of transfer, ideal gas to solution in the limit of infinite dilution.)

Solute	*Solvent*	$\delta\Delta\mu(*,o)/RT$
Water-rich side		
C_6H_6	H_2O	
C_6H_6	D_2O	−0.088
C_6D_6	H_2O	
C_6D_6	D_2O	−0.109
C_6H_6	H_2O	
C_6D_6	H_2O	0.061
C_6H_6	D_2O	
C_6D_6	D_2O	0.040
H_2O	H_2O	
D_2O	D_2O	0.131[a]
Benzene-rich side		
H_2O	C_6H_6	
D_2O	C_6H_6	−0.014
C_6H_6	C_6H_6	
C_6D_6	C_6D_6	−0.027[a]
CH_4	H_2O	
CD_4	H_2O	−0.014
$^{12}CH_4$	H_2O	
$^{13}CH_4$	H_2O	+0.0006
$^{29}N_2$	H_2O (273 K)	
$^{28}N_2$	H_2O	+0.00085
$^{34}O_2$	H_2O (273 K)	
$^{32}O_2$	H_2O	+0.00080
$^{12}C^{16}O_2$	H_2O (273 K)	
$^{13}C^{16}O_2$	H_2O	−0.00012
$^{12}C^{16}O_2$	H_2O (273 K)	
$^{12}C^{18}O^{16}O$	H_2O	0.0008

[a] Vapor pressure isotope effect, $\ln(P'/P)$.

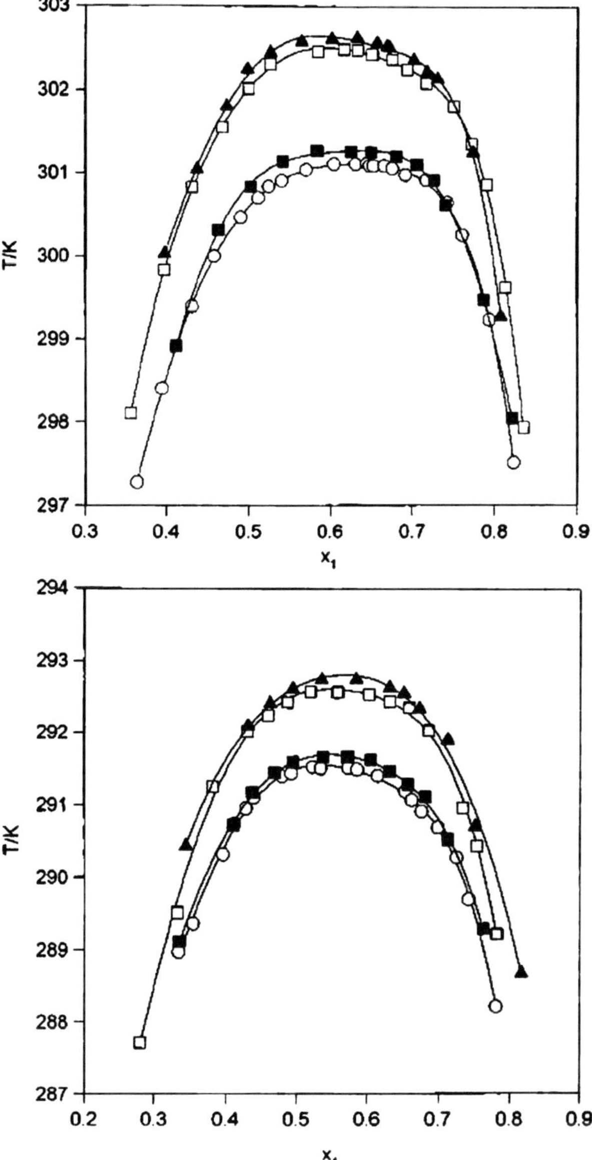

Figure 1 *Phase diagrams for variously deuterated (nitromethane(1) + pentanol), top, and (nitromethane(1) + isobutanol) solutions, bottom. Open circles = (NME(H_3) + alcohol(OH)); solid squares = (NME(H_3) + alcohol(OD)); open squares = (NME(D_3) + alcohol(OH)); solid triangles = (NME(D_3) + alcohol(OD)) (From Milewska and Szydlowski, J. Chem. Eng. Data, 1999,* **44***, 505, with permission.)*

At temperatures well below UCST, solubilities of hydrocarbons in water or water in hydrocarbons drop to very low values. The solutions are very nearly ideal in the Henry's law sense, and the isotope effects on solubility can be directly interpreted as the isotope effect on the standard state partial molar free energy of transfer. Good examples include the aqueous solutions of benzene, cyclohexane, toluene, and tetrachloromethane where solute and solvent isotope effects in the neighborhood of room temperature have been thoroughly studied.[16,17] Sample results of interest (for benzene/water) are included in Table 1. The data are sufficient to enable the use of a model calculation to deduce both the shift in the position of the benzene PES on phase change and the change in shape of its isotope sensitive vibrations. See Figure 2 for a schematic diagram.

A dramatic isotope effect on the solubility diagram has been reported for solutions of 3-methylpyridine + water (H/D).[18,19] At atmospheric pressure a 70 K ($T_{UCST} - T_{LCST}$) closed-immiscibility loop is observed for (3-methylpyridine + D_2O) and the immiscibility gap is modestly pressure dependent (Figure 3). Viewed on the (p,T) projection, the phase diagram shows a characteristic

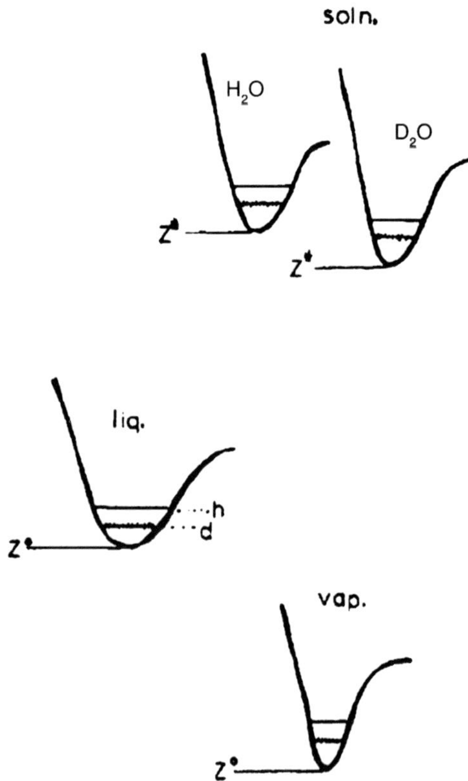

Figure 2 *Schematic diagram illustrating the shift in the position of the benzene PES on its phase change to liquid or aqueous solution, and the change in shape of its isotope sensitive vibrations on the transfer (Modified from Dutta–Choudhury, Miljevic, and Van Hook, J. Phys.Chem., 1982,* **86**, *1711, with permission.)*

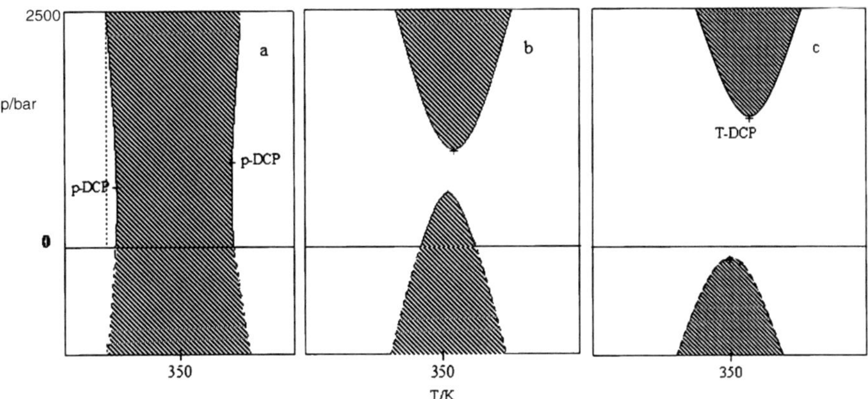

Figure 3 *Schematic changes in the phase diagram of (water(H/D) + 3 methylpyridine) on isotopic substitution. (a) Solvent = D_2O. (b) Solvent = 20 wt.% D_2O. (c) Solvent = H_2O. T-DCP and p-DCP designate temperature-double critical points and pressure-double critical points, respectively. The dotted curves designate metastable critical lines in the negative pressure region. The 2-phase regions are ~70 K wide at 2500 bar (From Visak et al. J. Phys. Chem. B., 2003,* **107***, 9837, with permission.)*

"hour-glass" shape (Figure 3a). On addition of H_2O, however, the gap shrinks, becomes more and more pinched at the waist, and at 20 wt% D_2O (Figure 3b) it dissapears completely as the phase diagram changes from this "hour-glass" to a shape that resembles the UCST/LCST configuration in the (T,x) plane – two immiscible domes, an upper and a lower. With continued addition of H_2O, holding (3-methylpyridine + water) at the critical concentration, the upper and lower immiscible branches move further and further apart, until finally at high enough H_2O/D_2O (17 wt% D_2O) the low-pressure branch is no longer present at atmospheric pressure, dropping below the p=0 isobar. For (3-methylpyridine + H_2O) the miscibility gap between the upper and lower branches amounts to ~160 MPa (Figure 3c). The phenomenon corresponds to an impressive pressure shift of many hundreds of atmospheres merely upon (H/D) solvent isotopic substitution. The numerical value of the effect was established by combining the pioneer work of Schneider[18] in the high-pressure region of the diagram with the recent demonstration of Visak *et al.*[19] that (3-methylpyridine + H_2O) starts to phase-separate at a negative pressure (isotropic tension) of −20 MPa.

Very large isotope effects like those described above seem to be limited to the hypercritical regions of phase diagrams, *i.e.* not too far from thermodynamic divergences of the type $(\mathrm{d}p/\mathrm{d}T)_c=\infty$ or $(\mathrm{d}T/\mathrm{d}p)_c=\infty$ (*i.e.* pressure-double critical points (p-DCP) or temperature-double critical points (T-DCP), respectively). The combination of an extended Flory–Huggins-type model[7] with the quantum-mechanical theory of isotope effects, as previously described by *e.g.* Equations (2) and (3), shows that one merely needs to invoke reasonable frequency shifts to successfully predict the extraordinary variations observed in the L–L phase diagram of (3-methylpyridine + water) solutions.[20] More specifically, a harmonic cell model calculation which begins by appropriately modifying and combining Equations (2) and (3) and then uses red shifts of $-(250 \pm 30)\ cm^{-1}$ in each of the three librational modes of water, together with an overall blue shift of $+(400 \pm 25)\ cm^{-1}$ in the internal frequencies on transfer of water from the pure liquid to infinite dilution in 3-MP, yields (*e.g.* at 200 MPa) $\Delta T_L = +20$ K and $\Delta T_U = -39$ K to be compared with experimental values of +25 K and −33 K, respectively. Considering the complexity of the phase diagram and the simplicity of the model, the agreement is rewarding.

A similar conclusion was drawn in a recent study[21] of H_2O/D_2O isotope effects on dimixing of the aqueous organo–ionic solution, water (H/D) + 1-butyl-3-methylimidazolium tetrafluoroborate, {water(H/D) + [bmim][BF_4]}. For this system only UCST shows, and the isotope effect on the critical temperature is modest, $\Delta T_U = -3.7$ K at atmospheric pressure. Nonetheless, the calculated frequency shifts are similar to those found in water ((H/D) + 3-MP): the three averaged librational modes of water undergoing a red shift of $-265\ cm^{-1}$, while the overall internal frequency shift is estimated to be +560 cm^{-1} to the blue, all on the transfer of water from pure liquid to infinite dilution in [bmim][BF_4]. Notice these shifts are somewhat larger for the (water+ionic liquid) than they are for (water+3-MP), suggesting that the role of hydrogen bonding is enhanced in the first case.

7.3.2 Polymer Systems and Polymer Solutions

In the previous section, we witnessed examples of enhanced isotope effects on demixing phase diagrams in the neighborhood of thermodynamic divergences. We learned that the molecular origin of solubility IEs lies in the shifts in normal mode vibrational frequencies on isotopic substitution and thermodynamic state transfer. Typically, IEs are inversely proportional to the temperature raised to some power Equation (3), and directly proportional to the total isotopic and phase frequency shift, $\Delta\delta\nu_i$. It follows that liquid–liquid demixing isotope effects will be appreciably enhanced at very low temperature, or, should one be restricted to higher temperatures by freezing point considerations, to molecules where many (isotopically substituted) oscillators are found. The first case is realized for $^3He/^4He$ and H_2/D_2 mixtures, the second by polymer/polymer and polymer/solvent mixtures.

It has been long established that $^3He/^4He$ liquid mixtures phase separate at temperature below 0.9 K.[10] (see Figure 4), and a theoretical explanation was advanced by Prigogine.[22] Similarly, solid H_2/D_2 mixtures show phase separation, but mixtures of the isotopic liquids do not phase separate, although they do show appreciable nonideality. No other "small molecule" mixtures of isotopomers exhibit liquid–liquid demixing, but in polymer mixtures, the excess properties are sufficiently large (for long chains) that binary mixtures of perprotonated and perdeuterated chains can, and do, phase separate (UCST type diagram). Such phase separation was observed by Bates and Wignall[23] and Bates and Wilthuis[24] in polystyrene(H/D), polybutadiene(H/D), and

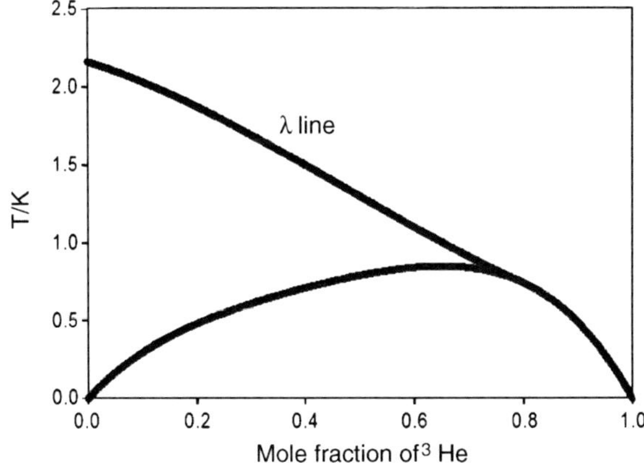

Figure 4 *Curves showing phase separation and the λ-line for liquid mixtures of ^{3}He and ^{4}He. The liquid is superfluid at temperatures below the λ-line.*

polyethylene–polypropylene(H/D) mixtures. By an analysis offered by Singh and Van Hook[25], extending Prigogine's ideas, the Helmholtz excess free energy of the solution, A^{ex}, can be written as

$$A^{ex} = \phi_1\phi_2(N\Gamma_i r_i/2)(\Delta V/V)(u_H\text{-}u_D) \tag{5}$$

In Equation (5), ϕ_1 and ϕ_2 are volume fractions of isotopomers 1 and 2, N is the number of monomer units per molecule, r_i the number of H/D substituted bonds per monomer, Γ_i the Gruneisen coefficient, $\Gamma_i = -\partial\ln(u_i)/\partial\ln(V)$, for the effective frequency (which, for these polymers is the CH(CD) stretch), $\Delta V/V$ is the molar volume isotope effect, and the u_i's are reduced frequencies, $u_i = h\nu_i/kT$. The thermodynamic conditions for phase separation dictate that the system first phase separates at $\phi_1 = \phi_2 = 0.5$ when

$$(N\Gamma_i r_i/2)(\Delta V/V)u_H(1 - u_H/u_D) \geq 2 \tag{6}$$

The μ values are reduced masses for the CH(CD) oscillators. Note that the effects are cumulative *via* the increasing number of substituted bonds, Nr_i. For instance, in the case of polybutadiene, $r_i = 6$, the critical polymerization number for H/D demixing is found to be 1.2×10^3 monomer units.

H/D isotope effects on demixing of (polymer + small molecule) solutions have been extensively investigated[26] since the early 1990s. Both solute (polymer) and solvent isotope effects have been measured but most studies have investigated the effects of H/D solvent (small molecule) substitution. The polymer studied most often has been polystyrene, and solvents have included acetone, propionitrile, cyclohexane, methylcyclopentane, *etc*. Typically, the phase diagrams exhibit both UCST and LCST branches at low and high temperatures, respectively. Therefore, it is possible to reach the hypercritical region by proper manipulation of the polymer chain length, and/or pressure,

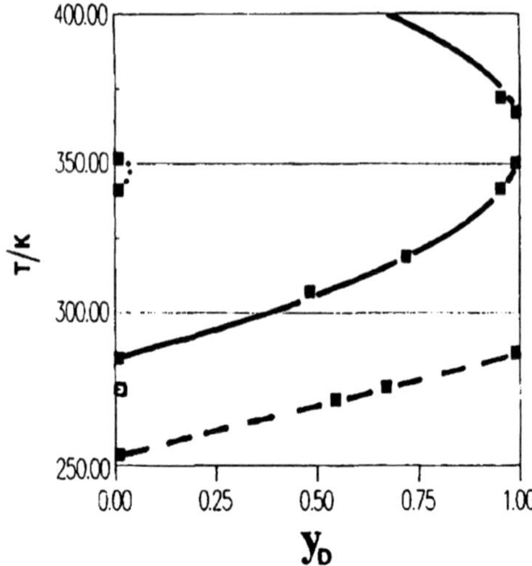

Figure 5 *Solvent isotope effects on upper critical solution (mostly) and lower critical transition temperatures for some (polystyrene (d) + acetone (h/d)) solutions, y_D = $(CD_3)_2CO/((CD_3)_2CO + (CH_3)_2CO)$. Dotted line through solid squares = PS(22 kg mol^{-1}); solid line through solid squares = PS(13.5 kg mol^{-1}); dashed line through solid squares = PS(7.5 kg mol^{-1}); open square = PS(11.6 kg mol^{-1}) (From Luszczyk et al., Macromolecules, 1995,* **28***, 745, with permission.).*

and/or solvent H/D fraction. Again, large isotope effects are expected. Figure 5 is a striking example of the occurrence of such phenomena.[26c]

7.4 Solubility of Gases in Liquids

Solute and solvent isotope effects for gases dissolved in liquids have been reviewed by Rabinovich,[10] more recently by Scharlin and Battino[27] and Jancso.[28] While there is a good deal of data which compares the solubilites of H_2 and D_2 in various solvents (see Table 2 for examples), by far the most extensive information compares the solubility of various gases in H_2O and D_2O (see Table 3 for examples).

Table 2 shows the solubility of D_2 to be considerably higher than H_2 in all solvents investigated. The isotope effect increases sharply as temperature falls (although the temperature dependences are not included in Table 2), but the economics of extracting D_2 or HD from their mixtures with H_2 are unfavorable compared with distillation of the liquids,[29] even when using liquid nitrogen as the extractant at 67 K where the solubilities differ by 30%. Solute solubility isotope effects for nitrogen, oxygen, methane, and carbon dioxide are included in Table 3. The effects are quite small for substitution at non-hydrogenic positions. The inverse isotope effects (*i.e.* lighter isotope more soluble than the

Table 2 *Ratio of Solubilities of H_2 and D_2 in various solvents*[10]

Solvent	*T(K)*	$x(D_2)/x(H_2)^a$
H_2O	292	1.027
NH_3	240	1.029
SO_2	293	1.033
Ar	87	1.165
N_2	67	1.297
CH_4	112	1.084
CS_2	298	1.020
Octane	308	1.020
Benzene	308	1.027

[a] x=mole fraction of gas in solution in the Henry's law limit.

Table 3 *Solvent isotope effects on gas solubility. Thermodynamics of transfer at 298.15 K,*[27] $\Delta Y^* = Y^*(H_2O)\text{-}Y^*(D_2O)$

Solute	$\Delta G^*/(J\ mol^{-1})$	$\Delta H^*/(kJ\ mol^{-1})$	$T\Delta S^*/(kJ\ mol^{-1})$
^4He	210	2.62	2.41
Ne	163	2.06	1.90
Ar	180	0.24	0.06
Kr	145	−0.18	−0.32
H_2	546		
D_2	597	0.23	−0.37
N_2	159	0.88	0.72
O_2	231	0.24	0.01
CH_4	144	1.84	1.70
C_2H_6	123	1.86	1.74
C_3H_8	147	1.72	1.57
CCl_2F_2	124	1.81	1.69
$CClF_3$	505	−4.08	−4.58
c-C_4F_8	971	9.74	8.77
CF_4	111	5.81	5.69
SF_6	221	2.05	1.83

heavy) observed for substitution at the carbon atoms of methane and carbon dioxide are of special interest.

The aqueous solvent isotope effects reported in Table 3 are the result of careful gas solubility measurements by Scharlin and Battino[27] covering the range ($288 < T$(K)< 318). Thermodynamic analysis yielded standard state free energies, and Van't Hoff enthalpies, entropies and heat capacities of transfer. The authors' analysis showed that all gases studied are structure makers near room temperature. The solvent isotope effects were employed using the Ben Naim theory of hydrophobic solvation[30] to estimate the change in the average number of hydrogen bonds induced by the solvation process, and hence the water structural changes, (solution-pure solvent).

The extension of these ideas to the theoretical treatment of isotope separation by gas chromatography is straightforward.[4,31] The isotope effects observed in chromatography are governed by the isotopic ratio of Henry's law constants (for gas-liquid separations), or adsorption constants (for gas-solid separations), and gas chromatography has been widely used for isotope separation and for the rapid and convenient analysis of mixtures of isotopomers.[32]

7.5 Solubility of Ionic Solids in H_2O/D_2O

Solvent isotope effect data on electrolyte solubilities in H_2O and D_2O have been collected by Rabinovich[10] and (in part) critically compiled by Jancso and Van Hook.[33] It is customary to express concentrations in aqueous electrolyte solutions using the aquamolality scale (L=moles salt per 55.508 mol solvent (1000 g for H_2O)). Some typical solubilities (298.15 K) are listed in Table 4. Of the 47 salts discussed by Rabinovich all but four are less soluble in D_2O than in H_2O. For those salts whose solubility increases with temperature, which is the ordinary behavior, the isotope effects decrease with temperature. Writing the standard state partial molar free energy of pure solid salt as $\mu^{o}{}_{(SALT)}$ and its standard state in solution as $\mu^{o}{}_{(H\text{ or }D)}$ we have on comparing the saturated solutions in H_2O and D_2O,

$$\mu^{o}{}_{(SALT)} = \mu^{o}{}_{(H)} + RT\ln(\gamma_H L_H) = \mu^{o}{}_{(D)} + RT\ln(\gamma_D L_D) \quad (7a)$$

and keeping in mind that the activity coefficients are a function of concentration, we obtain

$$-(\mu^{o}{}_{(H)} - \mu^{o}{}_{(D)}) = RT[\ln(L_H/L_D) + \ln(\gamma_H/\gamma_D)_{LD} + \int(\mathrm{d}\ln\gamma_H/\mathrm{d}L_H)\,\mathrm{d}L_H \quad (7b)$$

The integral extends from L_H to L_D. The equations demonstrate that the solvent isotope effect on salt solubility offers a convenient way to determine the solvent

Table 4 *Solvent isotope effects on solubility of selected electrolytes in H_2O and D_2O,[10] ΔL, % = 100 $(L_H - L_D)/L_H$. L_H and L_D are moles anhydrous salt per 55.508 moles H_2O and D_2O, respectively*

Salt	*T(K)*	L_H	ΔL, (%)
NaCl	293	6.13	6.7
NaBr (2 aq)	293	8.89	1.5
NaBr	323	11.38	3.3
NaI (2 aq)	293	11.26	1.9
NaI	373	20.1	1.7
KCl	293	4.61	9.8
KBr	298	5.75	1.1
KI	298	8.90	9.2
$BaCl_2$ (2 aq)	298	1.78	13
$BaCl_2$	398	3.01	9.3
$HgCl_2$	298	0.27	25
$PbCl_2$	298	0.039	36
Na_2SO_4 (10 aq)	298	1.96	1.6
Na_2SO_4	323	3.26	1.2

isotope effect of the standard state partial molar free energies of the salt in solution provided the concentration dependence of its activity coefficient in one of the solvents, most likely H_2O as written, is available at high concentration.

The largest solubility isotope effects are found for sparingly soluble salts. Thus, for example, lead chloride and potassium bichromate are 36 and 33.5% less soluble in H_2O than D_2O at 298.15 and 278.15 K, respectively. For the more soluble salts, NaCl and KCl, the values are 6.4% and 9.0%, both at 298.15 K. Interestingly LiF and LiCl(aq) have inverse solubility isotope effects of 13% and 2% respectively at 298.15 K. Recall that lithium salts are commonly designated as "structure makers". Almost all other elctrolytes are "structure breakers". Van Hook[34] has calculated the pressure coefficient of solubility, $\theta = (1/L)(dL/dP)_T$, and its solvent isotope effect, $\Delta\theta/\theta$, for alkali metal halides at 298.15 K. Due consideration of the effects of nonideality is required. For NaCl at 298 K, $\theta = 4.6 \times 10^{-5}\text{atm}^{-1}$ and $\Delta\theta/\theta = -0.048$. Thus increased pressure, like increased temperature, tends to decrease the aqueous solvent solubility isotope effect.

Acknowledgments

Support from the Ziegler Research Fund, University of Tennessee (AVH) and the *Fundação para a Ciência e Tecnologia* grant (LPR) # POCTI/EQU/35437/00 are gratefully appreciated.

References

1. J. Bigeleisen, *J. Chem. Phys.*, 1961, **34**, 1485.
2. M.J. Stern, W.A. Van Hook and M. Wolfsberg, *J. Chem. Phys.*, 1963 **39**, 3179.
3. G. Jancso and W.A. Van Hook, *Chem. Rev.*, 1974, **74**, 689.
4. W.A. Van Hook, *Condensed Matter Isotope Effects*, in *Isotope Effects in Chemistry and Biology*, A. Kohen and H.H. Limbach, Taylor and Francis, CRC, Boca Raton, FL, 2005, 119.
5. G. Jancsó, L.P.N. Rebelo and W.A. Van Hook, *Chem. Rev.*, 1993, **93**, 2645.
6. R.R. Singh and W.A. Van Hook, *J. Chem. Phys.*, 1987, **87**, 6097.
7. L.P.N. Rebelo, *Phys. Chem. Chem. Phys.*, 1999, **1**, 4277.
8. (*a*) V. Najdanovic-Visak, J.M.S.S. Esperança, L.P.N. Rebelo, M. Nunes da Ponte, H.J.R. Guedes, K.R. Seddon, H.C. de Sousa and J. Szydlowski, *J. Phys. Chem. B.*, 2003, **107**, 12797; (*b*) A. Siporska, J. Szydlowski and L.P.N. Rebelo, *Phys. Chem. Chem. Phys.*, 2003, **5**, 2996; (*c*) L.P.N. Rebelo, V. Najdanovic-Visak, Z.P. Visak, M. Nunes da Ponte, J. Szydlowski, C.A. Cerdeiriña, J. Troncoso, J.L. Romaní, J.M.S.S. Esperança, H.J.R. Guedes and H.C. de Sousa, *Green Chem.*, 2004, **6**, 369.
9. D. Fenby, Z.S. Kooner and J.R. Khurma, *Fluid Phase Equilibrium*, 1981 **7**, 327.

10. I.B. Rabinovich, *Influence of Isotopy on the Physicochemical Properties of Liquids*, Consultants Bureau, New York, 1970.
11. W. Schön, R. Wiechers and D.J. Wörmann, *J. Chem. Phys.*, 1986, **85**, 2922.
12. D.V. Fenby, J.R. Khurma and Z.S. Kooner, *Aust. J. Chem.*, 1983, **36**, 215.
13. R.R. Singh and W.A. Van Hook, *J. Chem. Thermodynamics*, 1986 **18**, 1021.
14. M.V. Salvi and W.A. Van Hook, *J. Phys.Chem.*, 1990, **94**, 7812.
15. A. Milewska and J. Szydlowski, *J. Chem. Eng. Data*, 1999, **44**, 505.
16. M.K. Dutta-Choudhury, N. Miljevic and W.A. Van Hook, *J. Phys.Chem.*, 1982, **86**, 1711.
17. P. Backx and S. Goldman, *J. Phys.Chem.*, 1981, **85**, 2975.
18. G. Schneider, *Zeitschrift Phys. Chem. Neue Folge*, 1963, **37**, 333; *ibid*: **39**, 187.
19. Z.P. Visak, L.P.N. Rebelo and J. Szydlowski, *J. Phys. Chem. B.*, 2003 **107**, 9837.
20. Z.P. Visak, J. Szydlowski and L.P.N. Rebelo, *J. Phys. Chem. B.*, 2006, **110**, 00-000, in press and published in advance on the web.
21. L.P.N. Rebelo, V. Najdanovic-Visak, Z.P. Visak, M. Nunes da Ponte, J. Szydlowski, C.A. Cerdeiriña, J. Troncoso, L. Romaní, J.M.S.S. Esperança, H.J.R. Guedes and H.C. de Sousa, *Green Chem.*, 2004, **6**, 369.
22. I. Prigogine, *The Molecular Theory of Solutions*, North-Holland, Amsterdam, 1957.
23. F.S. Bates and G.D. Wignall, *Macromolecules*, 1986, **19**, 932.
24. F.S. Bates and P. Wilthuis, *J. Chem. Phys.*, 1989, **91**, 3258.
25. R.R. Singh and W.A. Van Hook, *Macromolecules*, 1987, **20**, 1855.
26. (*a*) J. Szydlowski and W.A. Van Hook, *Macromolecules*, 1991, **24**, 4883; (*b*) J. Szydlowski, L.P.N. Rebelo and W.A. Van Hook, *Rev. Sci. Instrum.*, 1992, **63**, 1717; (*c*) M. Luszczyk, L.P.N. Rebelo and W.A. Van Hook, *Macromolecules*, 1995, **28**, 745; (*d*) M. Luszczyk and W.A. Van Hook, *Macromolecules*, 1996, **29**, 6612; (*e*) H.C. de Sousa and L.P.N. Rebelo, *J. Polym. Sci. B: Polym. Phys.*, 2000, **38**, 632; (*f*) A. Siporska, J. Szydlowski and L.P.N. Rebelo, *Phys. Chem. Chem. Phys*, 2003, **5**, 2996.
27. (a) P. Scharlin and R. Battino, *J. Solution Chem.*, 1992, **21**, 67; (b) *Fluid Phase Equilibria.*, 1994, **94**, 137.
28. G. Jancso, *Nukleonika*, 2002, **47**, S53.
29. (*a*) T. Ishida and Y. Fujii, *Enrichment of Isotopes*, in *Isotope Effects in Chemistry and Biology*, A. Kohen and H.H. Limbach, (eds), Taylor and Francis, CRC, Boca Raton, FL, 2005, 41; (*b*) W.A. Van Hook, Isotope separation, in *Handbook of Nuclear Chemistry*, A. Vartes, A. Nagy and Z. Klencsar (eds), 2003, **5**, 177.
30. (*a*) A. Ben Naim, *J. Phys. Chem.*, 1975, **79**, 1268; (*b*) A. Ben Naim, *Water and Aqueous Solutions; Introduction to a Molecular Theory*, Plenum, New York, 1974; (*c*) A. Ben Naim, *The Hydrophobic Interaction*, Plenum, New York, 1979.
31. W.A. Van Hook, *Adv. Chem. Series*, 1969, **89**, 99.
32. C.N. Filer, *J. Labelled Compound Radiopharm.*, 1999, **42**, 169.

33. G. Jancso and W.A. Van Hook, in *Solubility Data Series*, J.W. Lorimer, (ed), *Vols. 3, 30, 44, 47*, Pergamon, New York, 1979, 1987, 1990, and 1991.
34. W.A. Van Hook, *Fluid Phase Equilibria*, 1980, **4**, 287.
35. Z. Bacsik, Z. Lopes, J.N.C. Gomez, M.F.C. Jancso, G. Mink, J. Padua and A.A.H. , *J. Chem. Phys.*, 2002, **116**, 10816.
36. J.C. Vogel, P.M. Grootes and W.G. Mook, *Z. Physik*, 1970, **230**, 225.

CHAPTER 8

Solubility of Organic Solids for Industry

URSZULA DOMAŃSKA
Physical Chemistry Division, Faculty of Chemistry, Warsaw University of Technology, ul. Noakowskiego 3, 00-664 Warsaw, Poland

8.1 Introduction

The last few years have witnessed tremendous progress in the experimental and theoretical description of solid–liquid equilibria (SLE) and liquid–liquid equilibria (LLE) generally called solubility. Without a doubt, a complete picture of the solubility and the relevant parameters which govern this equilibrium would be incredibly useful in studying and improving industrially relevant crystallization and extraction processes. For example, natural gas hydrates-crystal structures in which methane molecules are trapped in a lattice-containing water causing blockages and holding up production. In the patent-pending technology project, funded by the U.K. Department of Trade and Industry and other industry partners, instead of trying to prevent hydrate formation, they encourage its crystallization with specially designed hydrates that can be transported as a stable, smoothly flowing stream. This has reduced the operating and capital cost of pipelines and has decreased the operating pressure.[1]

SLE experimental diagrams have been shown to play a dominant role in the development of crystallization processes.[2–4] Experiences in process synthesis and development, which covers petrochemical, fine chemical, pharmaceutical, polymer, and ionic liquids processes show that projects on crystallization and solids processing need not only SLE diagrams but also crystallization kinetics and mass-transfer limitations to optimize the final process design. Kinetic data required, include the "metastable" zone width and nucleation and growth rates as a function of super saturation. Both have to be experimentally measured.[2] An understanding of SLE requires knowledge of the solubility of the desired product in the selected solvent or binary solvent mixture. In every process, a knowledge of the pressure, temperature and the over-all composition of the mixture (especially when the solubility maybe affected by the presence of impurities), has to be very wide. The presence of impurities and other components may inhibit crystallization of the desired product, or induce crystallization of other components with the product.

Recent advances in Chemistry, Biochemistry, Pharmaceutical, and Polymer Sciences have created new challenges for the chemical processing industries regarding the production of high-molecular-weight chemicals, or low-molecular-weight crystal-polymers, which are normally recovered as solids *via* crystallization. For example, last year researchers developed a new class of synthetic antifouling macromolecules (chimeric peptidomimetic polymers) for possible use in a variety of medical implants, including cardiac stents and biosensors, as well as in water-processing equipment to prevent biofouling.[1] Cyclodextrins are becoming important and useful compounds. A cyclodextrin is a ring-shaped chain of glucose molecules with a central cavity that can enclose other molecules, such as odours, and then release them. New drugs (*e.g.*, molecular breeding and DNA shuffling) need new technology which can reduce the time, cost and environmental burdens. The purification systems are usually commissioned to handle enhanced-medical requirements. From a market perspective, specialty enzymes, which include those used in pharmaceutical, diagnostic and research markets will continue to benefit from advances in biotechnology that facilitate new application development. Food and beverages will continue to be the largest enzyme market. By type, the most widely used products will continue to be carbohydrases and proteases. Last year TransForm Pharmaceuticals, Inc. (Lexington, MA) in the collaboration with the University of South Florida have designed and discovered pharmaceutical *c*-crystal structures – a new category of drug molecules created by combining two, or more distinct solid compounds that are acceptable for use in pharmaceuticals. The solubility of the new drug, especially in aqueous media, is better than the competitor drugs, and this translates into improved absorption and bioavailability.

Previously vapour–liquid equilibrium (VLE) has played a key role in the design of separation (distillation) processes. The production of high-molecular-weight chemicals needs the thermodynamic basis of SLE in binary and ternary mixtures at normal and high pressure for the design and synthesis of crystallization processes. In this case, the use of phase diagrams to visualize regions in composition space where the system exists as a single phase or a mixture of multiple phases allows better understanding of the thermodynamic limitations imposed by the phase behaviour. By combining SLE data with chemical engineering operations such as cooling, heating, evaporation, co-solvent addition, one can devise separation schemes to obtain one or more solid products in a pure form. Solubility theory could be used to evaluate operating conditions for process improvement potential at elevated pressures and temperatures. Usually the product is formed in solution and then filtered from other solids such as heterogeneous catalyst, solid absorbents, or byproduct. Solubility is often a major determinant in the economic effectiveness of a particular process step.

On the other hand, new filters are designed for solid filtration, and especially in higher-temperature applications. Very often the month's topic in the Chemical Engineering Journals is "separation", or "solids handling" with many advertisements and articles on new equipments/filters, the principles of operation and the key process variables of centrifuge operation and selection,[5] or continuous crystallizers with their wide range of applications in the chemical process industries.[6] Production of solid state materials is a key technology for most chemical processes. Statistics for industrial processes suggests that over 50% of products manufactured by the world's major chemical companies involve solid particles.

Thus it is clear that SLE data-bases are vital to many industrial processes and the potential benefits arising from a detailed picture of solubilities, warrants further industrial's investigation and study. For example, a new pulsation mode was used to replace a traditional batch sinusoidal mode of solid–liquid extraction of andrographoline from plants.[7] The knowledge of SLE or LLE is only the first step of new technology.

This chapter reviews the solubility data of many organic compounds, the fundamentals of solubility correlation and prediction methods and how to apply these models to process improvement. Although, much has been learned over the past few decades, a complete and thorough understanding of solubility processes has yet to be achieved. Likewise, the current wealth of knowledge obtained regarding solid compound formation, or synergistic effect of the solubility in binary solvent mixture have not been without disagreement and controversy. Several studies discussed in the following chapter arrive at wholly different conclusions as to which kinetics and intermolecular interaction dominate for a particular system.

Firstly, a brief background will be presented concerning the solubility of different organic substances in binary and ternary systems connected with different industrial problems as well as a short description of the experimental methods. The main part of this section deals with the solubility of organic substances. The second section of this chapter will provide the reader with a short description of the definition of SLE, typical correlation equations and the most popular predictive methods. Thirdly, a review of the existing literature regarding the influence of the high pressure on phase diagram will be given and important findings discussed. The final sections examine polymer and ionic liquid solubilities. In particular, research and results by various investigators will be presented and various conclusions drawn.

8.2 Solubility in Binary Systems

In this section, the equilibria between two phases, *i.e.* solid substances with fixed compositions and liquid mixtures, or two liquid phases will be considered. The solid substances may be the pure components, says A for component "1" and B for component "2", or stechiometric compounds with the general formula A_mB_n, also call C. The phase behaviour of a system can be represented using phase diagrams in the temperature–pressure-mole fraction (or weight fraction) space, which is more conveniently referred to as composition space. Since small differences in pressure do not have a significant effect on solid–liquid phase behaviour, it is generally sufficient to consider isobaric SLE, or LLE phase diagrams.

Basic principles, phase rule, stability conditions and critical phenomena, differential equations for coexisting phases and the classification of the SLE, and LLE phase diagrams have been published repeatedly.[8–10] A particularly good description is given by Haase and Schönert[11] in the International Encyclopaedia of Physical Chemistry and Chemical Physics. The well-known equations, description of phase equilibria, or types of phase diagrams described by Haase and Schönert, will not be repeated here.

For a given solid-solvent system, the solubility always increases with increasing temperature. For a given solvent at constant temperature if two solids have similar enthalpy of fusion, then the solid with the lower melting temperature has the higher solubility. Similarly, if two solids have close melting temperature then the one with the lower enthalpy of fusion has the higher solubility.

The understanding of the shape of phase diagrams of organic compounds was only made clear with the advent of X-ray diffraction techniques in the 1940s. The form of the phase diagram is dependent on the structure of these component molecules, especially for solid solutions of organic substances. In many cases solubility rules are of great practical value, especially in the field of solid state chemistry, when an impurity molecule is to be introduced into a crystal, or a pure substance is to be obtained. A detailed investigation into phase diagrams involving mixed molecular crystals for organic compounds and polymers was done by Kitaigorodsky.[12] This knowledge is of great practical significance in areas such as biopolymers, molecular biology, enzymes or supramolecular organic compounds. In organic compounds, inter- or intramolecular hydrogen bonds can promote the formation of strongly interacting pairs of molecules (*i.e.* β-naphthol), islands of several molecules bound by these bonds, chain of molecules, ordinary or double layers (glycin), and the formation of a three-dimensional framework. In many organic crystals the molecules are bound by hydrogen bonds in pairs. The replacement of such a pair by two molecules containing no hydroxyl, or amino groups is accompanied by a loss of lattice energy and has an influence on solubility.

Another basic rule of solubility concerns molecular molecules with permanent dipole moments. As far as it is known from the experimental data, dipole moments do not affect solubility, provided that molecules in the crystal are oriented so that two (or more) molecules form an island with no net dipole moment.[12] It can be expected that solubility is not possible in only one case, namely, when the crystal as a whole possesses an electrical moment. This is the case where all the dipole moments of all the molecules have the same direction. Under those conditions, a solid solution in the solid–liquid phase diagram is not expected. Other conditions, necessary for the formation of a continuous series of solid solutions in organic systems, were discussed by Kitaigorodsky.[12]

Organic molecular compounds such as (amine+an alcohol)[13–16] systems are often found in nature. Unlike the metallic or inorganic systems, simple stoichiometric compositions (with two exceptions: clathrate compounds, hydrates and solvates (crystals with few molecules of water)) are usually obtained. Most molecular compounds are formed by molecules which do not posses geometrical similarity and involve strong intermolecular interactions, such as hydrogen bonding. Therefore, the phase diagrams of organic substances with potential to form molecular compounds are generally very simple. They are diagrams of the eutectic type, without solubility, and in the case of molecules with similar shape, diagrams of the eutectic, or peritectic type, with limited solubility. Phase diagrams with continuous solubility are extremely rare since, in addition to the geometrical similarity of the molecules, their formation requires isomorphism of crystalline structures.

The importance of solid–solid phase transition, observed for many organic substances and polymers as a characteristic inflection on the liquidus curve, have been interpreted by Ubbelohde[17] as; a transition (especially in solids that involve orientational disordering), a lambda transformation (changing between polymorphs which have independent crystal structure) and a transformation by way of hybrid crystals. The important effect of impurities on transformation temperatures and the changing shape of a hysteresis loop in solid–solid transformations of pure substance, was also disscussed.[17]

Solubility is strongly dependent on both intramolecular forces (solute–solute and solvent–solvent) and intermolecular forces (solute–solvent). The well known and ancient adage *similia similibus solvuntur (T. Lucreti Cari, De rerum natura)* implies that, in the absence of specific interactions, the intermolecular forces between chemically similar species leads to a smaller endothermic enthalpy of solution than those between dissimilar species.

8.2.1 Solid–Liquid Equilibria in Binary Systems

The experimental methods used for the determination of phase diagrams at constant pressure fall into two main categories, namely – static and dynamic. In static methods, the phases are brought into equilibrium at constant temperature and pressure and are mixed in a closed vessel for 24–48 h and then the liquid phase is separated and analysed. For the liquid-phase equilibria, this method is the most frequently used and is called the horizontal method. In the "dynamic" (vertical, or synthetic) method, a mixture of known overall composition is subjected to a gradual change in temperature. The transition from solid to liquid phase is observed visually or by optical methods. Usually the transition from one state to another is accompanied by a heat effect and by changes in other physical properties, which are measured by different techniques, such as differential scanning calorimetry, (DSC). The observed plots of measured property against temperature (cooling, or heating curves) are translated into a phase diagram.

The most common types for the organic components of binary system phase diagrams of two organic substances A and B (where A can denote a solvent) are depicted in Figure 1 (a–h). Area L represents the unsaturated solution, area $A+L$, or $B+L$, or $C+L$ correspond to the coexistence of crystals of substance A, or B, or compound C and the saturated solution. Area $A+B$, or $A+C$, or $C+B$, corresponds to mixtures of crystal of substances A, B, or compound C.

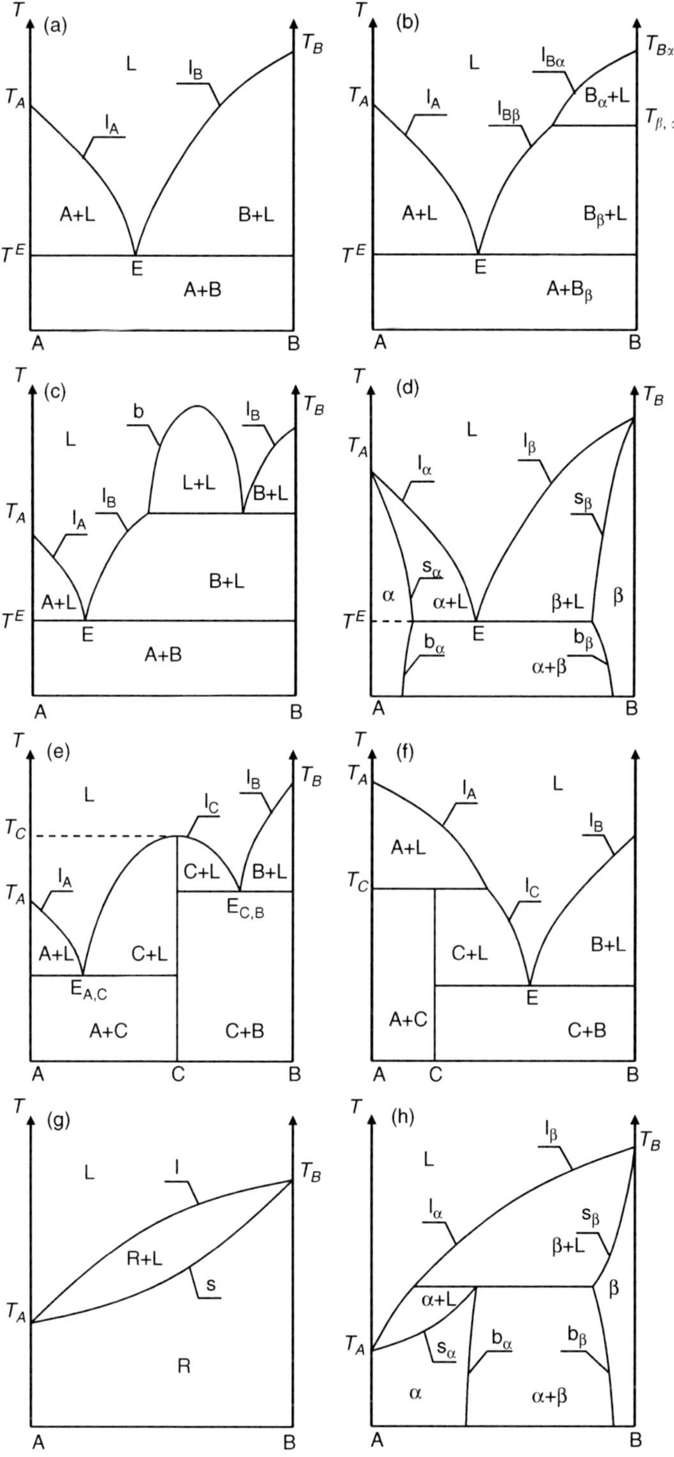

(a)
T
T_A
T_B
L
l_B
l_A
A+L
B+L
T^E
E
A+B
A
B
(b)
T
T_A
$T_{B\alpha}$
$T_{\beta,\alpha}$
L
l_A
$l_{B\alpha}$
$l_{B\beta}$
B_α+L
A+L
B_β+L
T^E
E
A+B_β
A
B
(c)
T
T_A
T_B
L
b
l_B
L+L
B+L
l_A
A+L
T^E
E
A+B
A
B
(d)
T
T_A
T_B
L
l_α
l_β
s_β
s_α
α
α+L
β+L
β
T^E
E
b_α
b_β
α+β
A
B
(e)
T
T_C
T_A
T_B
L
l_B
l_C
l_A
C+L
B+L
$E_{C,B}$
A+L
$E_{A,C}$
A+C
C+B
A
C
B
(f)
T
T_A
T_C
l_A
L
l_B
A+L
l_C
C+L
B+L
E
A+C
C+B
A
C
B
(g)
T
T_A
T_B
L
l
R+L
s
R
A
B
(h)
T
T_A
T_B
L
l_β
l_α
s_β
β+L
β
α+L
s_α
b_α
b_β
α
α+β
A
B

The curve separating regions L and $A+L$, or L and $B+L$ is the solubility curve for substances A and B, respectively. The curves intersect at point E, where solid A, solid B (or compound C) and a solution with composition E, saturated with both substances, are in equilibrium. Points T_A and T_B are the melting points of pure components A and B, respectively. The simple eutectic type means the components do not form a compound and are completely miscible in the liquid phase and not miscible in the solid phase [Figure 1(a)].

If the components do not form a compound and are partially miscible in the liquid phase, a new liquid phase appears with an upper critical solution temperature (UCST). The two solutions at constant temperature are termed conjugated. Area $L+L$ represents two liquid phases, see Figure 1(c). If the components are partially miscible in the solid phase [Figure 1(d)] the heterogeneous solid solutions α and β can be observed.

Industry-related operations, such as crystallization, heating, cooling, stream combination, splitting, solvent addition or solvent removal, can be represented on top of the phase diagram as a movement in composition space. Such a representation allows one to quickly identify thermodynamic limitations and select a set of movements to create a feasible process.

An enormous amount of research has been conducted on the solubility of the organic compounds in different organic solvents for industrial use. The section that follows will serve as a useful introduction to SLE measurements and will provide a foundation for later discussions on the correlation and prediction methods.

A few databases[9,18–20] comprising an extensive collection of data on SLE of binary and ternary mixtures do exist. The two most important are: (a) the database contains 1175 records covering 377 organic compounds;[18] 994 systems representing 945 binary systems and 49 ternary mixtures; (b) the SLE database containing 14,590 data sets, mainly of organic compounds.[19] The special collection data of solubility of carboxylic acids of more than 500 mixtures[20] is included in ref. 18.

Knowledge of the thermodynamic properties and especially solubilities of n-alkanes is important for chemical engineering design. Presently, it is of particular importance in coal-liquefaction and petrochemical processes. Most petroleum reservoir fluids contain heavy hydrocarbons which may precipitate as waxy solids when conditions of temperature and pressure change. Most of the experimental work on the formation of waxes has been restricted to the solubility of measurements of long-chain alkanes, acids and esters. Problems

Figure 1 *Temperature-composition types of solid–liquid phase diagram in binary systems: (a) simple eutectic type; (b) eutectic type system with solid–solid phase transition of compound B; (c) eutectic type system with partially miscible components in the liquid phase; (d) eutectic type system with partially miscible components in the solid phase; (e) eutectic type system with a compound stable up to the melting point; (f) eutectic type system with a compound unstable at the melting point (incongruently melting compound); (g) system where components are completely miscible in the solid and liquid phase; and (h) the peritectic system.*

associated with wax formation and deposition is a major concern in production and transportation of hydrocarbon fluids.

Data on simple eutectic systems [Figures 1(a), or (b)] have been published for many *n*-alkanes with solvents such as hydrocarbons, cyclohydrocarbons, aromatic hydrocarbons, alcohols and esters. For example, (*n*-alkane $C_{12,16,20,24,28}$+decane) binary mixtures exhibit simple eutectic mixtures[21] but longer chain alkanes show solid solutions, e.g. (C_{30}+C_{26}),[21] or (C_{17}+C_{19}),[22] or (C_{30}+C_{35}).[23] The systematic SLE studies of many binary *n*-alkane mixtures including the crystallographic measurements of different phases were investigated by Dirand and co-workers.[24] The influence of size and shape on the solubility and the role of the combinatorial entropy of hydrocarbons in different hydrocarbons was determined by Domańska and Kniaź[25] and Kniaź.[26] Solubilities of *n*-alkanes C_{18}–C_{28} in hydrocarbons, cyclohydrocarbons, benzene and many different solvents (simple eutectic mixtures), were measured by Domańska and co-workers[27–29] and Snow and co-workers.[30]

The phase diagram representing the solubility of octadecane (C_{18}) in ethanol is given in Figure 1(c). The immiscibility gap was observed for weight fractions of octadecane higher than 17 wt%.[31]Complete miscibility and simple eutectic phase diagrams were observed for *n*-alkanes (C_8, C_{10}, C_{12}, C_{16}) with alcohols (butan-1-ol, octan-1-ol and hexadecan-1-ol).[32–35] Systematic studies on solubilities of hydrocarbons: teteracosane[36] and hexacosane[37] in pentan-1-ol, hexan-1-ol, heptan-1-ol, octan-1-ol, nonan-1-ol, decan-1-ol, undecan-1-ol, dodecan-1-ol exhibit simple eutectic properties. Many other binary mixtures of *n*-alkanes with alcohol (cyclododecanol), or acids (hexadecanoic acid, dodecanoic acid), or esters (methyl hexadecanoate, methyl, or ethyl octadecanoate, methyl nonadecanoate), measured by Gioia Lobbia and co-workers[38–40] exhibit simple eutectic mixtures.

The so-called hydrophobic effect – *i.e.* the tendency of non-polar solute molecules to avoid aqueous solutions – is considered as one of the most important factors in the solubility of hydrocarbons in water. Sixty years ago it was pointed out by Frank and Evans[41] that the solubility of hydrocarbons in water is low; much lower than the solubility in organic solvents in spite of the fact that the dissolution in water, at temperatures below 298 K, is favoured by a negative enthalpy change. More than 30 years later it was confirmed that the dissolution of hydrocarbons in water, at low temperatures, is an exothermic process.[42,43] Hydrocarbon molecules are hydrated by energy-low and entropy-low, bulky water structures and hydration process occurs as highly cooperative solute–solvent equilibria. These are very small solubilities, for example the solubility of anthracene in water is 0.4×10^{-8} mole fraction at temperature $T = 297.75$ K, and for fluorene it is 0.24×10^{-6} at $T = 291$ K.[44] *n*-Alkanes mutual solubility with water (LLE) has been reported by Mączyński and co-workers.[45,46]

Although, the solubility of aromatic hydrocarbons such as biphenyl, naphthalene, fluorene, phenanthrene, acenaphthene, pyrene in organic solvents is reasonably high, the activity coefficients are greater than one ($\gamma > 1$), indicating that the solubility is lower than the ideal solubility.[47–49] Solubility in such systems are determined by the low-energy interaction of the type: n–π, or π–π. Solid–liquid-phase diagram of (fluorine+benzene), or (dibenzofuran+benzene) and many others exhibit simple eutectic behaviour.[50] When there is a polar hetero-atom in the organic molecule, the interaction with a solvent is more complicate and the SLE-phase diagram is of type presented in Figure 1(g) [solid solution, *i.e.* (fluorine+dibenzofuran)],[51] or of type presented in Figure 1(d) [eutectic system with partial miscibility in the solid phase (fluorine+dibenzothiophene)],[51] or of type presented in Figure 1(h) [peritectic mixtures, *i.e.* (dibenzothiophene-dibenzofuran)].[51]

Systematic studies of SLE of organic compounds in polar, or non-polar solvents highlights an important point that non-polar solutes, solutes with stable intramolecular hydrogen bonds (*i.e.* 2-acetyl-1-naphthol, or 2-benzoyl-1-naphthol, or 5-methyl-2-nitrophenol) or compounds forming stable complexes in solution, such as cyclic dimers of alkanoic and benzoic acids are more soluble in non-polar solvents. Furthermore, their solubilities in hexane and in alcohol are very similar. In many of these cases, the non-specific forces are responsible for the solubility, because the intramolecular hydrogen bond is very stable and is not broken by solute–solvent hydrogen bonding. Another group, includes compounds capable of forming intermolecular hydrogen bonding with solvents (*i.e.* 1-acetyl-2-naphthol, or 1-benzoyl-2-naphthol, or 4-nitro-5-methylphenol, or 2,5-xylenol). These compounds show a higher solubility, particularly in polar solvents. The results highlight the role played by intra- and inter-molecular hydrogen bonding in the process of dissolution.[52,53]

Solid–liquid phase equilibria form the basis for crystallization processes which in turn are used in chemical and petrochemical industry for the separation of mixtures. Crystallization, in particular, is used for the separation of thermo-labile components, or of components such as positional isomers with very similar vapour pressures for which the separation factor is approximately equal to unity and cannot be influenced by selective solvents. From the industrial point of view the separations of isomers of xylenes, or diethoxymethane with 2,2-dimethoxybutane (or 1,1-diethoxyethane) (measured by Gmehling,[54,55]) as well as *o*-, *m*-, *p*-cresol, or 2,3-xylenol, 2,4-xylenol, 2,5-xylenol, 2,6-xylenol, 3,4-xylenol and 3,5-xylenol,[56] is very important.

The problems associated with waste water and environmental issues highlight the importance of knowing the solubilities of organic compounds in water. Solubilities of hydrocarbons in water and seawater were published recently by the IUPAC-NIST, Solubility Data Series – see ref. 57 and information included.

A large amount of solubility data at 298.15 K of organic substances in organic solvents was published by Acree, Jr. and co-workers. For example, the solubilities of pyrene, anthracene, carbazole in cyclohexane, heptane, isooctane, tetrahydropyran, chloroalkane and dibutyl ether[58–60] was presented as helpful data for chemical engineers designing chemical separation and purification processes. Such data are also important for chemists when selecting solvents for chemical reactions.

Knowledge of the solubility characteristics of long-chain, aliphatic alcohols in organic solvents is of great importance to fats, cosmetic and oil technology and research. Simple eutectic systems with two solid–solid first-order phase transitions, similar to that presented in Figure 1(b), was exhibited by the long-chain 1-alkanols (C_{14}, C_{16}, C_{18}, C_{20}) with *n*-alkanes (C_7–C_{16}), cyclohexane, benzene, toluene and propan-2-ol.[61–67] In the homologous series of *n*-alkanes, or *n*-alkanols the SLE curves depend mainly on the enthalpy of melting and on the melting point of the particular solute, because both of these properties are large. Usually, the liquidus curves for the longer chain compounds are found at higher temperatures on SLE diagram.

In recent years, with the growing demanded for refrigeration and air-conditioning, cool storage systems have been installed in increasing numbers. A cool storage system removes heat from a thermal storage medium during periods of low cooling demand. The stored cooling is later used to meet an air-conditioning or process-cooling load. The most common cool storage media are water, ice and other substances exhibiting convenient solid–liquid phase transitions. Data obtained from many investigations show that waxes, or components of waxes such as long-chain *n*-alkanes, or *n*-alkanols, or fatty acids, or fatty acid's esters may be used for cooling storage. Several investigations have been undertaken on a number of binary mixtures of these compounds to elucidate the phase transitions behind the solid–liquid phase diagram.

The liquidus curves for a number of fatty acids (stearic acid, palmitic acid), their mixtures and mixtures with *n*-alkanes were published more than 50 years ago.[68] Solubilities of fatty acids in many other organic solvents were presented as well.[69–72] Solubilities of three ethylene glycol monoesters of stearic, eicosanoic and behenic acids have been measured in 25 pure solvents and in 28 binary solvent systems[73] as a supplement to the known data of alkyl esters of palmitic and stearic acids.[74] Solubility measurements of fatty acids and their esters did not shown the solid–solid phase transition in these compounds, which was, however, observed much earlier for the pure compounds by the calorimetric study.[75–77]

The substituted aromatic groups and flexible chains of drug molecules, exhibiting dipolar forces, hydrogen bonding, steric interferences and ionic interactions, make the solubility measurements in drug research and development a formidable task. Organic compounds as benzoic acid, toluic acids, *p*-hydroxybenzoic acid, methyl *p*-hydroxybenzoate [78–80] or imidazole,[81–84] which serve as models of drug molecules are soluble in water, alcohols and many organic solvents.

A great number of industrial separation processes are concerned with liquid mixtures containing aromatics (benzene, toluene, *p*-xylene, alkylbenzenes) and saturated hydrocarbons (hexane, heptane, octane, decane, dodecane). In solid–liquid, or liquid–liquid extraction processes, solvents such as sulfur dioxide, *N*-methyl-2-pyrrolidinone (NMP), *N*-formylmorpholine, dimethyl sulfoxide, and tetrahydrothiophene 1,1-dioxide (sulfolane) have been used extensively. Sulfolane is a particularly important solvent and has been the subject of many studies.[85–91] Three different types of phase diagrams for the systems of (sulfolane+1,4-dioxane, or nitrobenzene, or carbon tetrachloride) were obtained; one is of a simple eutectic type [see Figure 1(a)], the second exhibits a solid molecular compound [see Figure 1(e)] and the third one exhibits a miscibility gap from 0.08 to 0.58 mole fraction of sulfolane and a compound which melts incongruently into the immiscible liquid.[87–89]

NMP is a well-known solvent with a high selectivity for the separation of aromatic hydrocarbons from aliphatic hydrocarbons.

The SLE phase diagrams for (NMP+tetrachloromethane),[92] (NMP+*m*-cresol, or *p*-cresol, or *o*-cresol),[93–95] (NMP+2,5-dimethylphenol, or 3,4-dimethylphenol, or 2,6-dimethylphenol),[96–98] (NMP+trichloromethane),[99] (NMP+benzene, or toluene, or ethylbenzene, or propylbenzene, or mesitilene, or dichloromethane, or 1,1,1-trichloroetane, or chlorobenzene, or 1,2-dichlorobenzene, or 1,3-dichlorobenzene, or 1,2,4-trichlorobenzene)[100] have been reported in the literature. Recently, the solubility of NMP in alcohols,[101,102] in ketones,[103] in ethers[104]and in phenols[105] have been presented. NMP is an interesting compound not only because of the large carboxylic group, but also because of the specific interactions of the nitrogen atom and also of a hydrogen atom of the methyl group with a solvent molecule. For example, the interactions between NMP and benzene or phenol or dimethylphenols are believed to occur *via* complex formation between the two species and, or hydrogen bonding resulting in one (benzene, 2,6-dimethylphenol), or two congruently melting compounds in the phase diagrams.

The use of crown ethers and cryptands in synthesis and different chemical processes is becoming increasingly important. These wide applications are based on the ability of crown ethers to form well-defined complexes with a large variety of ligands in different solvents. The thermodynamics of ligand binding process for crown ethers have been studied extensively,[106–108] however very few thermodynamic results are available for these compounds in their pure state. SLE and LLE were measured for 1,3,5-trioxane and 18-crown-6, 12-crown-4, dibenzo-18-crown-6 and dibenzo-24-crown-8 in many organic solvents.[109–113] Phase diagrams for benzene, cyclohexane, tetrachloromethane, alcohols, alkynes or alkenes are of simple eutectic type [Figure 1(a)]. Stronger solute–solvent interaction was observed for mixtures of 18-crown-6 with dimethylsulfoxide and with 1,1,1-trichloroethane. The result was a phase diagram of the eutectic type with a congruently melting compound [see Figure 1(e)].[113] Contrary to these mixtures, non-miscibility in the liquid phase was observed for the 1,3,5-trioxane, 12-crown-4 and 18-crown-6 with *n*-alkanes from hexane to hexadecane.[109–112]

In summary, the studies previously discussed provide valuable information and insight regarding the interactions in facile binary systems and the polar group complexation between solute and solvent. Several of these studies have shown that the solid–liquid phase diagram dependence on interaction probability can result in all the types of phase diagrams shown in Figure 1.

8.2.2 Liquid–Liquid Equilibria in Binary Systems

The determination of LLE in systems containing substances below the boiling point under normal conditions is not very difficult. As there is no universal analytical method, each system must be considered individually. The LLE in binary systems or the tie-lines in ternary systems is mostly measured by the "direct analytical method". In this method, the heterogeneous mixture is first stirred intensely for a prolonged period of time. Then the conjugate phases are left for 24–48 h to separate. After standing, samples are taken from the individual phases and analysed at constant temperature and pressure. The more specific methods have been described by Novák et al.[10] The visual methods, described in the previous section for SLE are also very popular. A similar procedure of studying the disappearance and formation of turbidity with increasing and decreasing temperature, respectively, at constant pressure and concentration is used in the determination of LLE. The average temperature of these two measurements is usually used. In many mixtures only the temperature of formation of turbidity is observed, (called the "dew point method").

The thermodynamics and the conditions for the equilibrium of coexisting phases and the six possible shapes of the equilibrium curves for LLE have been well documented.[10] For the organic compound mixtures, a phase diagram with an UCST (see point K in Figure 2), is usually observed. For these mixtures the mutual miscibility or solubility, increases with increasing temperature. The critical solution temperature is the temperature at which the properties of both liquid phases are identical at constant pressure.

The examples from text books also show phase diagrams with lower critical solution temperatures (LCST) (*e.g.* triethyl amine+water), or phase diagrams with closed limiting miscibility curves (*e.g.* nicotine, or tetrahydrofuran, or 2,4-dimethyl pyridine+water and glycerine+benzyl ethyl amine). The phase diagram exhibiting both upper and lower, (but separated) critical temperature was found for aromatic hydrocarbons with sulfur.

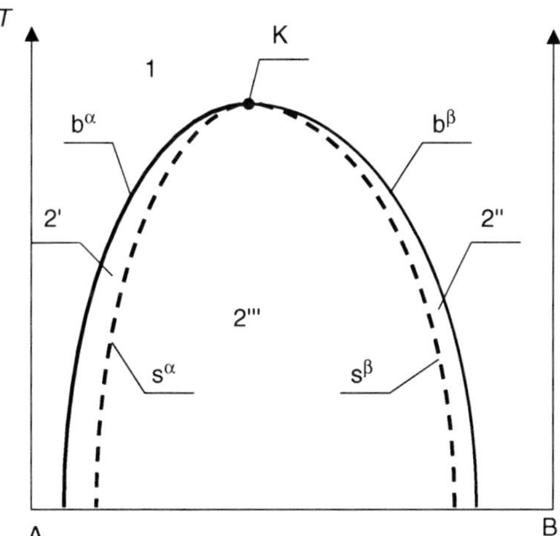

Figure 2 *The most common shape of the equilibrium curve for LLE in binary system ($p =$ const); b^α-b^β, binodal curve; s^α-s^β, spinodal curve; 1-phase area; the area $2'$ and $2''$ represent the metastable phases and $2'''$ binary liquid phases.*

The first data bank on LLE were given by Seidell[114] and Sørensen and Arlt.[115] At present 14,830 data sets for low boiling substances can be found in the Dortmund Data Bank[116] and more than 2000 at NIST.[117,118]

Owing to the rising cost of energy, new separation processes based on extraction are becoming more attractive and LLE data is becoming more important.

Systematic studies on LLE of *n*-alkanes and cycloalkanes (*i.e.* pentane, octane, butane, 2-methylpentane, 2,2,4-trimethylhexane, cyclohexane or 3-methylcyclopentane)[115] have shown immiscibility with water; with alcohols [*i.e.* (methanol, or ethanol+*n*-alkane, C_4–C_{16})];[116,119] with esters [*i.e.* (methyl ethanoate+nonane)];[116] with ketones [*i.e.* acetone+hexane)];[116] with ethers and crown ethers [*i.e.* (1,3,5-trioxane, or 12-crown-4, or 18-crown-6+*n*-alkanes, C_6–C_{16})];[109–112] with dimethyl carbonate (*i.e. n*-alkanes, C_{10}, C_{12}, C_{14}, C_{16}, C_{18}, C_{20}, C_{22}, C_{24});[120,121] with acetonitrile, ethanenitrile, aniline, furfural and with many other polar compounds. Furfural does not mix also with isooctane and 2,2,4-trimethylpentane.[115]

Most of the longer chain alcohols (*i.e.* butan-1-ol, hexan-1-ol, decan-1-ol, pentan-2-ol, 2,2-dimethylcyclohexanol)[115] do not mix with water. This is also true for aromatic hydrocarbons (*i.e.* benzene, toluene),[116] chloroform, ketones (*i.e.* 2-butanone),[116] esters (*i.e.* butyl acetate)[116] and branch chain ethers (*i.e.* ethyl 1,1-dimethylethyl ether, methyl 1,1-dimethylpropyl ether).[122] A miscibility gap is found in the mixture (methanol+carbon disulfide.[115]

Sulfolane has been successfully used in big scale separation plants (SHELL process) and many experimental data are published for typical mixtures with sulfolane. The immiscibility has been observed for the mixtures with *n*-alkanes (*i.e.* hexane, heptane, octane), cycloalkanes and alkan-1-ols (*i.e.* propan-1-ol, butan-1-ol, octan-1-ol, tetradecan-1-ol).[123,124] The UCST increases with the length of the *n*-alkane and alcohol. For alcohols, the coexistence curves shift to higher mole fractions of sulfolane in solutions with increasing chain of alkan-1-ols. Sulfolane is a globular molecule in which the negative end of its large dipole moment is exposed, and hence cannot act as proton acceptor/donor. Owing to the steric hindrance of its globular shape sulfolane does not easily interact with other compounds. In its pure form, is even a weakly structured substance just below its melting point (mesophase crystal). Owing to these properties, mixtures of sulfolane show very interesting interactions with different solvents. Mixtures with benzene, tetrachloromethane or 1,4-dioxane are almost regular as the volumes or dielectric constants on mixing scarcely exhibit any noticeable deviations from ideality. Activity coefficients of components obtained from SLE measurements are also close to unity. Nevertheless, some SLE diagrams do indicate the presence of complexes. These have been attributed to a favourable packing rather than to strong interactions.[124] On the other hand, some mixtures (with miscibility gaps) show strong positive deviations from Raoult's law.

8.3 Solubility in Ternary Systems

In all solubility measurements presented so far, it has been assumed that the interaction between solute and solvent, or two liquids is responsible for the whole process. Hence, the role and kind of interactions in the solution process is usually unclear. In ternary systems the additional interactions are responsible for additional effects such as positive, or negative synergistic effects on solubility.

8.3.1 Solubility of Solids in Binary Solvent Mixtures

The description of ternary mixtures has to be presented as an equilibrium surfaces on Gibbs phase diagram as it is shown in Figure 3(a): solute "1" is dissolved in the mixture of solvent "*A*" and "*B*" as a function of temperature. However, these results can only be obtained and understood in terms of a solute mole fraction x_1, as a function of temperature $x_1 = f(T)$ for a certain mole fraction of binary solvent x_B^0 (solute free).

Thus the solubility curves represent points 1,2,3 on Gibbs phase diagram. Usually it is very difficult to measure very low solubilities, presented in Figure 3(a), as a black area. The isotherms of the solubility: $x_1 = f(x_B^0)$ at constant temperature T may occur as one of the five types. The straight line "1" represents complete additivity and is observed very rarely and only in non-polar mixtures. The curve "2" describes most ternary mixtures and represents small deviations from additivity. The most interesting isotherms, from the chemical industries point of view, are isotherms "3" and "5" exhibiting maximum (positive synergistic effect), or minimum (negative synergistic effect) of solubility. From the mathematical point of view the isotherm "4" may exist but it has never been seen by the author. Hundreds of mixtures have been measured by Domańska and co-workers and a summary of the types found is given here. The deviations from the additivity (curve "1") can be explained by the non-ideality of the binary solvent mixture. Examples of systems exhibiting isotherms type "1", "2" , "3" and "5" are presented in Tables 1–4, respectively.

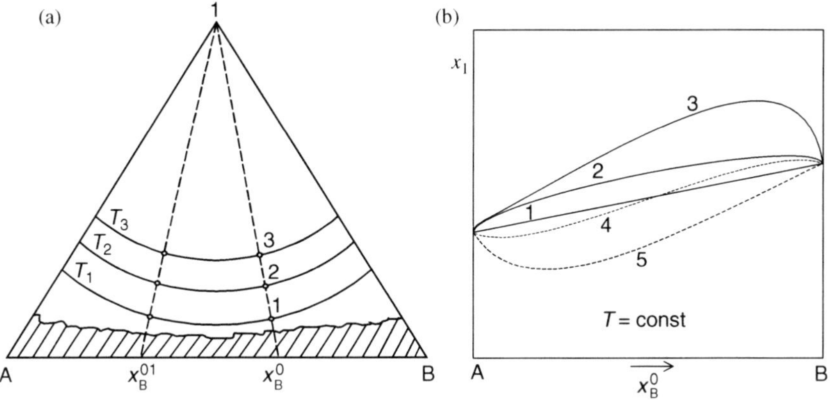

Figure 3 *(a) Gibbs diagram for solute "1" and binary solvent mixture "A" and "B"; x_B^0 is mole fraction of solvent B in the mixture A+B (solute free). (b) Isotherms of the solubility of solute "1" in function of x_B^0.*

Table 1 *Systems showing complete additivity*

Substance	*Binary solvent*	*Ref.*
2-Acetyl-1-naphthol	Cyclohexane+heptane	125
5-Methyl-2-nitrophenol	Decane+benzene Hexane+hexadecane Cyclohexane+tetrachloromethane	127
Octadecane, nonadecane	Cyclohexane+heptane	126
o-Toluic acid	Cyclohexane+heptane	70
Stearic acid, arachidic acid, behenic acid	Cyclohexane+heptane	70
Hexadecan-1-ol, octadecan-1-ol, eicosan-1-ol	Cyclohexane+heptane	128

Table 2 *Systems showing small deviations from additivity*

Substance	*Binary solvent*	*Ref.*
5-Methyl-2-nitrophenol	Methanol, or butan-1-ol, hexan-1-ol+ethyl acetate; hexan-1-ol+butyl acetate; benzene+carbon tetrachloride; cyclohexane+decane, or benzene; benzene+ethyl acetate	52
Octadecane, nonadecane	Cyclohexane+ethanol	126
Eicosane	Trichloroethylene or tetrachloroethylene+propan-2-ol; cyclohexane+propan-2-ol	70
o-Toluic acid	Cyclohexane+ethanol, or propan-2-ol	70
Arachidic acid	Cyclohexane+methylene iodide	71
Ethylene glycol-monoarachidic acid ester	Isobutyl methyl ketone+propan-2-ol; ethanol, or propan-2-ol, or butan-2-ol+butyl acetate	73
1-Acetyl-2-naphthol, or 1-Benzoil-2-naphthol	Hexane+ethanol or butan-1-ol	125 129
2,5-Dimethylphenol	Methanol, or butan-1-ol, or hexan-1-ol+ethyl acetate; hexane+butan-1-ol	130

The substances described as a group I in the previous section, *i.e.* molecules with a strong intramolecular hydrogen bonding, non-polar substances, or substances exhibiting strong association in the solution, have been mixed with non-polar solvents and shown to exhibit isotherm "1". For example, 2-acetyl-1-naphtol shows very similar solubilities (especially at low temperatures) in *n*-alkanes (C_{18}, C_{19}, C_{20}), in cyclohexane, in heptane and in a binary solvent mixture (cyclohexane+heptane).[125,126]

Isotherm of type "2" has been observed in most of the tested systems. 5-Methyl-2-nitrophenol has shown small deviations from additivity in the mixtures of (alcohol+ester), or (hydrocarbon+benzene) and (ester+benzene). Also 5-methyl-4-nitrophenol has shown the same effect in alcohol, or (ketone+water), or (alcohol+ketone), or (ketone, or ether+an ester) binary solvent mixtures. Some of measured systems are presented in Table 2.

For all systems discussed above, the solubility in the binary solvent lies between the solubilities in the pure solvents on the graph $T = f(x_1)$. The

Table 3 *Systems showing positive synergistic effects*

Substance	*Binary solvent*	*Ref.*
2-Acetyl-1-naphthol	Cyclohexane, or hexane+ethanol, or butan-1-ol	125
2-Benzoil-1-naphthol	Hexane+butan-1-ol	129
5-Methyl-2-nitrophenol	Hexane+ethanol, or butan-1-ol, or hexan-1-ol, or octan-1-ol; decane+butan-1-ol, or hexan-1-ol, or octan-1-ol; hexadecane+octan-1-ol; cyclohexane+methylene iodide	131
Naphthalene	Hexane+ethanol, or butan-1-ol, or hexan-1-ol, or octan-1-ol; cyclohexane+methylene iodide	132
Stearic acid	Cyclohexane+ethanol, or propan-2-ol; trichloroethylene, or tetrachloroethylene+propan-2-ol	
Arachidic acid	Cyclohexane+ethanol, or propan-1-ol; trichloroethylene, or 1,1,1-trichloroethane+ethanol, or propan-2-ol, or butan-1-ol	70
Behenic acid	Cyclohexane+ethanol, or propan-2-ol, or ethyl acetate; trichloroethylene+propan-2-ol	70
Ethylene glycol- monoarachidic acid ester	Cyclohexane, or 1,1,1-trichloroethane+ethanol, or propan-2-ol, or butan-1-ol; trichloroethylene+propan-2-ol, or butan-1-ol	73
Hexadecan-1-ol, octadecan-1-ol, eicosan-1-ol	Cyclohexane+propan-2-ol	128
5-Methyl-4-nitrophenol	Methanol, or ethanol, or propan-1-ol, or butan-1-ol, or pentan-1-ol, or hexan-1-ol, or octan-1-ol+ethyl acetate; ethanol, or pentan-1-ol+butyl acetate; ethanol, or pentan-1-ol+amyl acetate	131
3,4-Dimethylphenol	Methanol, or butan-1-ol, or hexan-1-ol+ethyl acetate	130

Table 4 *Systems showing negative synergistic effects*

Substance	*Binary solvent*	*Ref.*
Stearic acid, arachidic acid, behenic acid	1,4-dioxane or tetrahydrofuran+chloroform	136

deviations from linearity in the dilute solutions of the solute in the binary solvent ($\Delta\log x_1$) depends on the non-ideality of the binary solvent mixture ($G^E_{AB} > 0$), and in all cases there are positive deviations from the Raoult's law.

$$\Delta\log x_1 = G^E/2.303\ RT \tag{1}$$

$$\Delta\log x_1 = \log x_{1AB} - (x^0_A \log x_{1A} + x^0_A \log x_{1B}) \tag{2}$$

The use of mixed solvents to show the phenomenon of enhanced solubility (positive synergistic effect), was the subject of numerous investigations in a variety of ternary systems. The solutes belonging to the first group (compounds with stable intramolecular hydrogen bindings, non-polar compounds, solutes forming stable aggregates in solutions) showed the synergistic effect in binary solvents consisting of (hydrocarbon+alcohol), or (halohydrocarbon+alcohol), or (cyclohexane+methylene iodide). The synergistic effect was observed (see Table 3) for 5-methyl-2-nitrophenol, (which exhibit intramolecular hydrogen bonding, which is stable even in polar solvents,[131]) for naphthalene,[132] for phenanthrene,[133] for 2-acethyl-1-naphthol,[125] for 2-benzoyl-1-naphthol[129] and for compounds forming stable complexes in solutions, such as cyclic dimers of alkanoic and benzoic acids,[70–72] alkanoic acid-ethylene glycol monoesters[73] and long-chain alkan-1-ols.[128]

The second group (II) includes compounds forming intermolecular hydrogen bonding with the solvent, *i.e.* 5-methyl-4-nitrophenol,[131] 2,5-dimethylphenol and 3,4-dimethylphenol.[130] For this group the synergistic effect was observed in (ester+alcohol) mixtures (see Table 3). The results illustrate the role played by intra- and inter-molecular hydrogen bonding in the dissolution process.

It has been established that a positive synergistic effect is observed in binary solvent exhibiting a positive excess Gibbs free energy of mixing ($G^E \gg 0$). Furthermore, in the first group of solutes, the enhancement of solubility, in mixtures forming positive azeotropes, *i.e.* (cyclohexane+propan-2-ol) or (hexane+propan-2-ol) is observed. In many systems, a maximum in the isotherm "3" type was observed for the same concentration, which was noted for the azeotropic point.

One can expect a decrease in solubility ("negative synergistic effect" – see isotherm "5") in two component mixtures exhibiting a negative excess free energy of mixing ($G^E \ll 0$) and forming negative azeotropes. This is true for (tetrahydrofuran+ chloroform).[134] A negative synergistic effect can also be find in the solvents exhibiting strong *A–B* interactions as a result of intermolecular complexes in the solvent mixture, (these being the remains of a solid intermolecular compound in the solid phase at low temperatures). Such a solvent mixture is (1,4-dioxane+chloroform).[135] In both the above binary solvent systems, the negative synergistic effect was found for the alkanoic acids.[136] The minimum solubility on isotherm "5" corresponds to that composition of a binary system which shows a maximum negative deviation from ideality (azeotrope's point, solid compound *C* of solvents in the solid phase).

Azeotropic mixtures are used in industry, where azeotropes can be conveniently regenerated by distillation. For example, the solubility of stearic acid was measured in azeotropic mixtures of binary solvent systems of freons with alcohols, ketones and halogen derivatives.[137] All the solvent mixtures exhibited positive homo-azeotropes. The essential synergistic effect was observed in the (freon 1,1,3+ethanol or propan-2-ol) binary solvent system.[137]

The prediction of the synergistic effect for a given system follows from Hildebrand's theory of regular solutions[138] according to the condition of solubility parameters: $\delta_A < \delta_1 < \delta_B$. The solubility in binary solvent mixtures can be higher than in the pure solvents with maximum for $\delta_1 = \delta_{AB}$, where the solubility parameter for the binary solvent is:

$$\delta_{AB} = \frac{\phi_A \delta_A + \phi_B \delta_B}{\phi_A + \phi_B} \tag{3}$$

Excellent agreement was obtained for many mixtures. It is important, however, to remember that the Scatchard–Hildebrand theory predicts positive deviations from ideality, $\gamma_1 > 1$ for non-polar substances in non-polar solvents. Using the values of the solubility parameters presented by Barton,[139] it is possible to obtain a qualitative picture.

The solubility date for a number of polycyclic aromatic hydrocarbons (*i.e.* anthracene, naphthalene, pyrene) and hetero-atom polynuclear aromatics (*i.e.* carbazole, dibenzothiophene and xanthene) in binary solvent mixtures at one temperature (298.15 K) have been measured in the laboratory of Acree.[140–142]

8.3.2 Solubility of Mixtures of Two Solids in a Solvent

Unfortunately, real systems rarely contain just one solute and one solvent, rendering binary phase diagrams a somewhat inadequate way to represent crystallization systems. However, they are useful as building blocks to construct phase diagrams for systems with more than two components. The features of multi-component phase diagrams are discussed below, along with examples illustrating applications in the development of crystallization processes. Phase diagram plays a central role in the development of the crystallization processes, because it offers useful insights in the limitations of separations and the possible ways to overcome them.

For high-melting compounds with low solubility, a more practical approach is to measure the double saturation point directly, by preparing a mixture of solutes *A* and *B*, followed by the addition of small amounts of solvent to dissolve some solids. So long as both *A* and *B* exists in the solid phase, the liquid composition always corresponds to the double saturation composition.

There are very few data available for ternary aromatic systems consisting of solvent-solid–solid at different temperatures. Some results of this kind are reported in refs. 50 and 143–146. The easiest system includes three pairs of binary eutectics, *e.g.* (*cis*-decalin+naphthalene+biphenyl) or (tetralin+biphenyl+dibenzofuran).[146] The ternary phase diagrams at constant temperature and pressure shown ternary eutectic points. Other examples include systems with a solid solution on one side of the phase diagram. For the systems: (benzene+fluorine+dibenzofuran), or (benzene+fluorine+dibenzothiophene), or (benzene+dibenzothiophene+dibenzofuran), the fluorene – dibenzofuran binary forms a solid solution at any composition, while the fluorine-dibenzothiophene and dibenzothiophene – dibenzofuran systems have solid-phase immiscibility gaps.[50] Usually the different crystal structures of a component (solid–solid phase transition) and the solid immiscibility are seen in the ternary diagrams as characteristic inflections on the isotherms between the liquid end of the binary (liquid+solid) areas.[50]

Although, the SLE of multi-component systems are essential in separating organic melt mixtures, there have been only a few SLE studies involving the more complicated-ternary solid compounds. This is largely due to experimental difficulties. One reported example is of the ternary system of *ortho*-, *meta*- and *para*-nitroaniline. It was studied by a DSC method and was find to form a simple ternary eutectic mixture (from three binary eutectic mixtures).[147]

8.3.3 Liquid–Liquid Equilibria in Ternary Systems

An enormous amount of research has been conducted on the ternary LLE of the organic compounds as separation method for the industries use.[148] The binodal curves have usually been determined at certain temperature and pressure using the cloud point method either using a visual method, or an optical device involving a photo resistor with light passing through the solution. The equilibrium tie-lines have been determined using glc or some other analytical method such as refractive index or density.[149]

It is apparent that the magnitude of the heterogeneous region, its shape, the distribution coefficient values and the slopes of the tie-lines are determined primarily by the properties of the binary subsystems, although ternary interactions cannot be neglected, especially when two associating compounds are in the ternary mixture. Much discussion about the influence of the non-ideality and homogeneous binary systems on the size of the heterogeneous region and the slope of the tie-lines in ternary systems was presented earlier.[10] A typical phase diagram of a ternary organic system is given in Figure 4.

Hundreds of ternary systems LLE data are to be found in the Dortmund Data Bank[116] and at NIST.[117,118]

Because of the important industrial applications of *N*-methyl-2-pyrrolidone and sulfolane, several investigators have studied the LLE for ternary systems

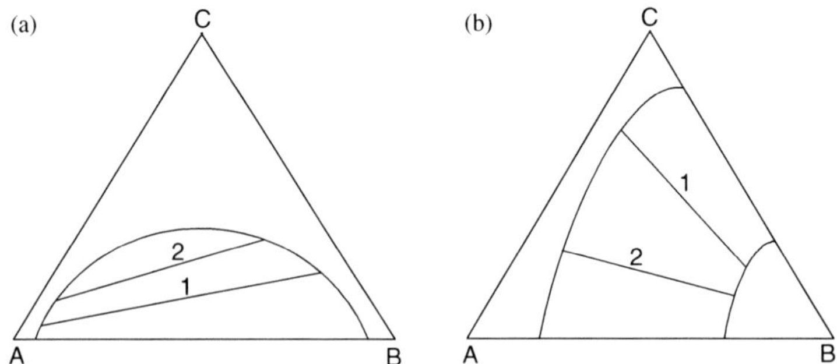

Figure 4 *Ternary LLE most common equilibrium type: (a) system with one heterogeneous binary subsystem (A+B); 1,2-represent typical tie-lines; and (b) system with two heterogeneous binary subsystems (A+B) and (B+C), 1,2-represent typical tie-lines.*

containing [*n*-alkane+aromatic hydrocarbon+*N*-methyl-2-pyrrolidone (NMP), or sulfolane (S)]. The separation of aromatic and saturated hydrocarbons is performed on a large scale in industry for the recovery of pure aromatic hydrocarbons from petroleum fractions. These separations are achieved by various techniques, which often require the addition of a solvent (extractive distillation, liquid–liquid extraction). The design of separation processes requires experimental information on phase equilibria and related thermodynamic properties. A typical system [heptane (*A*)+NMP (*B*)+benzene (*C*)],[150] of the type presented in Figure 4(a) was investigated 30 years ago together with [monoethanolamine (*A*)+heptane (*B*)+NMP (*C*)][150] ternary mixture of the type presented in Figure 4(b). Recent data deals with the LLE for mixtures of (hexane, or nonane, or tetradecane, or hexadecane+NMP+toluene) and mixtures of (hexadecane+NMP+*o*-, or *m*-, or *p*-xylene, or mesitylene, or ethylbenzene) [type presented in Figure 4(a)].[151] Better selectivity for the separation of aliphatic/aromatic mixtures was, however, shown by the binary solvent (NMP+water), or glycol, or glycerol.[152] Pseudo-ternary mixtures of [hexane+(NMP+co-solvent)+toluene] have highlighted the importance of glycerol as a co-solvent.[152]

Many ternary systems involving sulfolane have been measured. Examples are [hexane, or heptane, or decane, or cyclohexane+S+toluene],[153–157] and (dodecane+S+butylbenzene or hexylbenzene)[157] – all showing good selectivity. In multistage, countercurrent extraction, using sulfolane for the separation of toluene from heptane, the extract purity can evidently be increased by using water as a co-solvent.[154]

Other examples of ternary systems are: (heptane+methanol+propan-1-ol, or butan-1-ol, or butan-2-ol);[158] (cyclohexane+methanol+chlorobenzene, or benzene, or toluene, or methyl acetate, or tetrachloromethane, or tetrahydrofuran).[159–162] Examples of mixtures containing water are: (water+benzene+ethanol),[160] or (water+chloroform+acetone),[163] or (water+butyl acetate+acetic acid).[163] All the systems cited above are of the type given in Figure 4(a). More complicated organic mixtures such as (aniline+cyclohexane, or heptane, or hexane+benzene, or toluene)[164] are also of the same type. The related system (aniline+heptane+cyclohexane),[164] however, represents a mixture of the type Figure 4(b).

Recently, furfuryl alcohol, an inexpensive solvent formed as a byproduct in the manufacture of sugar, was used as a solvent for the separation of aromatic and aliphatic hydrocarbons in solvent extraction with good results.[165]

8.4 Correlation Methods

For practical purposes, solubilities of solids in pure or mixed solvents are of interest in chemical process design, especially when process conditions must be specified to prevent the precipitation of a solid. The thermodynamic description of SLE follows from the thermodynamic principle of equilibrium which requires that the fugacities of each component in each of the coexisting phases are equal. For the solubility of one solid component in a solvent, the fugacity may be expressed as in Equation (4), where f_1 is the fugacity of the solute in solid (s) and liquid (l) phases, respectively.

$$f_1^s = f_1^l \tag{4}$$

Introducing an activity coefficient (γ_1) and assuming that, there exists no solid solution in the solid phase; there is complete miscibility in the liquid phase, that T_t, the temperature of the triple point can be replaced by the melting-point temperature at normal pressure, $T_{fus,1}$, one can obtain Equation (5).[165,166] The solubility equation for temperatures below that of the phase transition must include the effect of the transition. The result for the first-order transition is:

$$\begin{aligned} -\ln x_1\gamma_1 = {} & \frac{\Delta_{fus}H_1}{R}\left(\frac{1}{T} - \frac{1}{T_{fus,1}}\right) + \frac{\Delta_{tr}H_1}{R}\left(\frac{1}{T} - \frac{1}{T_{tr,1}}\right) \\ & - \frac{\Delta_{fus}C_{p,1}}{R}\left(\ln\frac{T}{T_{fus,1}} + \frac{T_{fus,1}}{T} - 1\right) \end{aligned} \tag{5}$$

where x_1, γ_1, $\Delta_{fus}H_1$, $\Delta_{fus}C_{p,1}$, $T_{fus,1}$ and T refer to the mole fraction, activity coefficient, enthalpy of fusion, difference in solute heat capacity between the liquid and solid at the melting temperature, melting temperature of the solute (1) and equilibrium temperature, respectively. The $\Delta_{tr}H_1$ and $T_{tr,1}$ refer to the enthalpy of solid–solid phase transition and transition temperature of the solute, respectively. Equation (5) is valid for simple eutectic mixtures.

In many studies three simple methods are usually used to fit the solute activity coefficients, γ_1, to the so-called correlation equations that describe the Gibbs excess energy,[128] (G^E): the Wilson,[167] UNIQUAC[168]and NRTL[169]models.

The parameters of the equations can be fitted by an optimization technique. The objective function is:

$$F(A_1A_2) = \sum_{i=1}^{n} \omega_i^{-2}[\ln x_{1i}\gamma_{1i}(T_i, x_{1i}, A_1A_2) - \ln a_{1i}]^2 \tag{6}$$

where $\ln a_{1i}$ denotes an "experimental" value of the logarithm of the solute activity, taken as the left-hand side of the Equation (5), ω_i is the weight of an experimental point, A_1 and A_2 are the two adjustable parameters of the correlation equations, i denotes the ith experimental point and n the number of experimental data. The weights were calculated by means of the error propagation formula:

$$\omega_i^2 = \left(\frac{\partial \ln x_1\gamma_1 - \partial \ln a_i}{\partial T}\right)^2_{T=T_i} (\Delta T_i)^2 + \left(\frac{\partial \ln x_1\gamma_1}{\partial x_1}\right)^2_{x_1=x_{1i}} (\Delta x_{1i})^2 \tag{7}$$

where ΔT and Δx_1 are the estimated errors in T and x_{1i}, respectively.

According to the above formulation, the objective function was obtained by solving the non-linear Equation (5), using the Marquardt's method of minimization.[170] The root-mean-square deviation of temperature [σ_T defined by Equation (8)] was used as a measure of the goodness of the solubility correlation.

$$\sigma_T = \left(\sum_{i=1}^{n} \frac{((T_i)^{\exp} - (T_i)^{cal})^2}{n-2} \right)^{\frac{1}{2}} \tag{8}$$

where n is the number of experimental points (including the melting point) and 2 the number of adjustable parameters.

The pure component structural parameters r (volume parameter) and q (surface parameter) in the UNIQUAC equation were obtained by means of the following simple relationships:[171]

$$r_{\mathrm{i}} = 0.029281\ V_{\mathrm{m}} \tag{9}$$

$$q_i = \frac{(z-2)r_i}{z} + \frac{2(1-l_i)}{z} \tag{10}$$

where V_{m} is the molar volume of pure component i at 298.15 K, z the coordination number, assumed to be equal to 10 and l_{i} the bulk factor; it was usually assumed that $l_i = 1$ for cyclic molecules and that $l_i = 0$ for the linear molecules. The calculations were carried out for the many SLE data sets[20,29,32–34,36,50,63,70,72,73,81–84,90,101,102,104,113,121,125,126,128–130,172] and some examples are presented in Table 5.

For the polar compounds exhibited strong intermolecular interactions such as hydrogen bonding, good correlation results are obtained with the NRTL 1 or 2 equation[175] or using the UNIQUAC ASM (associated-solution model).[176]

The correlation of LLE using the same models as have been used for liquid-phase activity coefficients (UNIQUAC or NRTL) have been reviewed by Sørensen and co-authors.[177] When molecular addition compound melting congruently is observed in a SLE diagram, as is found in the systems (amine+alcohol),[13–16] or (NMP+phenol),[105] the simple Ott equation is usually used.[178]

8.5 Prediction Methods

Traditionally, models which are based on the thermodynamics of fluid-phase equilibria, and which describes the Gibbs energy as a function of composition and sometimes temperature, may be used for interpolation, extrapolation and prediction in ternary or multi-component systems using binary system's parameters. The correlation of SLE and LLE using Gibbs energy functions is used in most published data. Using parameters from VLE data it is possible to predict ternary or multi-component system properties. A systematic study of the correlation and prediction of SLE and LLE has been carried out in many laboratories of the world. Only the models based on the classical or molecular thermodynamics will be listed in this chapter. Three different types of models

Table 5 *Correlation of the solubility data (SLE) of [solute (1)+solvent (2)] mixtures by means of the Wilson, UNIQUAC and NRTL equations: Values of parameters and measures of deviations*

	Parameters			Deviations		
Binary system	Wilson Δg_{12} Δg_{21} $kJ\ mol^{-1}$	UNIQUAC Δu_{12} Δu_{21} $kJ\ mol^{-1}$	NRTL Δg_{12} Δg_{21} $kJ\ mol^{-1}$	Wilson $\sigma_T{}^a$ K	UNIQUAC $\sigma_T{}^a$ K	NRTL $\sigma_T{}^a$ K
Tetracosane (1)+cyclohexane (2)[b]	−0.7410	0.6827	—	0.16	0.18	—
	0.6926	−0.4538				
Octacosane (1)+heptane (2)[b]	−1.7175	1.3131	—	0.87	0.49	—
	1.5912	−0.8999				
2-benzoyl-1-naphthol (1)+hexane (2)[c]	5.192	−5.8081	0.7019[c]	1.44	1.16	1.16
	5.792	0.1716	7.4702			
4-benzoyl-1-naphthol (1)+butan-1-ol (2)[c]	1.2461	−1.2517	−3.5719[c]	0.96	0.59	0.59
	−0.8352	1.3599	3.4313			
Octadecanoic acid (1)+tetrahydrofuran (2)[d]	−0.4202	1.5775	−3.8819[c]	0.63	0.36	0.69
	0.3716	−1.1258	3.6612			
Eicosanoic acid (1)+chloroform (2)[d]	−3.4316	1.5181	−0.0571[c]	0.51	0.53	0.52
	3.4345	−0.6140	0.0107			
Hexadecan-1-ol (1)+propan-2-ol (2)[e]	9.9109	1.6050	−1.0944[c]	1.54	0.47	0.51
	1.4936	−0.3487	−0.4920			

[a] According to Eqation (9) in the text. [b] Ref. 173. [c] Ref. 129. Calculated with the third nonrandomness parameter $\alpha = 0.3$. [d] Ref. 174. [e] Ref. 128.

have been usually used: (i) models for activity coefficients or the excess Gibbs functions; (ii) group-contribution methods; and (iii) equation of state.

The Margules, Van Laar, Redlich–Kister equations have in common that they have all been useful in correlating SLE or LLE – often with good results. However, extrapolation to concentrations beyond the range of the data, or the prediction of ternary phase diagrams from only binary information should not be carried out with these models. Local composition models such as the Wilson's equation, UNIQUAC, NRTL, UNIQUAC ASM, NRTL 1 and 2 have proven superior to the older models, both for correlating binary and ternary systems and for predicting ternary phase diagrams.[20,50,125,126,128,130,158–164,174,177,179]

The more elaborated group contribution method models, used in the fluid-phase equilibrium calculations – the Modified UNIFAC,[180] DISQUAC[181] and the ASOG[182] – have been used in many predictions.[13–16,119–121,124,177,183–189]

The modified UNIFAC and/or DISQUAC models are perhaps the most use models today and each year the parameters describing the mixtures of (an organic substance+solvent) improves, and as a result so do the predictions of the thermodynamic properties of phase equilibria (VLE, LLE, SLE), activity coefficients and excess functions G_m^E and H_m^E of many mixtures. One advantage of the modified UNIFAC model is that it offers reasonable results using relatively few adjustable parameters. The biggest discrepancies between predicted and experimental results are with LLE systems. On the other hand, the DISQUAC model consistently and accurately, describes sets of thermodynamic properties including SLE and LLE. In many instances this is due to the additional fitting parameters, *e.g.* for the different alcohols, ethers, *etc.* The structural dependence of the DISQUAC parameters was taken into consideration for many mixtures and was found to be acceptable, provided that the mixtures showed regular properties.

The computation of SLE and LLE using equations of state (EOS) has only received limited attention. However, the few results available (see below) indicates that further work is warranted. One successful study was the application of the AEOS (Association+EOS) model for simultaneously predicting SLE, LLE and VLE of strongly non-ideal systems.[190] The equation is able to reasonably predict phase equilibria with the use of only pure component parameters and one adjustable binary parameter – see the results for SLE and LLE of the binary system (phenol+hexane).[190]

8.6 High-Pressure Solid–Liquid Equilibria

The SLE of alkanes systems has gained increasing interest in the recent decade. Most petroleum reservoir fluids contain heavy hydrocarbons which tend to precipitate as a waxy solid phase when conditions of temperature and pressure change. For petroleum the relative concentrations of alkanes with branch chain hydrocarbon, aromatic hydrocarbons or 1-alkynes appear to be very important, especially under conditions involving a wide range of temperature and pressure. Phase equilibrium data of mixtures including *n*-alkanes are of importance for the safe and efficient operation of chemical plants, not only oil refineries. They are also necessary for high-pressure polymerization processes and for the design of oil-recovery processes. Besides its importance for technological processes such as crystallization and purification at high pressure, SLE provides a good tool for examining the thermodynamic nature of many systems.

The modelling of (solid+liquid) equilibrium under high pressure has been approached from a number of angles: the Chain Delta Lattice Parameter Model,[191] the Sako–Wu–Prausnitz EOS (SWP)[192,193] or the van der Waals EOS.[194] Prediction of solid–fluid phase diagrams of light gases–heavy hydrocarbons systems up to 200 MPa using an equation of state, G^E model was developed by Pauly and co-workers.[195,196]

The fugacity of the solid phase at pressure P can be obtained through correcting the standard state fugacity with the Poynting correction by:

$$\ln f_i^s(P) = \ln f_i^s(P_0) + \frac{1}{RT}\int_{P_0}^{P} V_i^{s0}\,\mathrm{d}P \tag{11}$$

Since no equation of state for the solid phase is available, the solid-phase molar volume is taken as being proportional to the corresponding pure liquid molar volume:

$$V_i^{so} = \beta V_i^{lo} \tag{12}$$

In many works,[195–199] a constant value of β was used, with very good results for pressures up to 100 MPa. Preliminary calculations showed that this simplification would not hold above 250 MPa and a new approach to β was adopted.[200,201] Because experimental measurements indicate that $\mathrm{d}P/\mathrm{d}T$ is fairly constant over a broad range of pressures, and that the enthalpy of melting is reasonably pressure independent, some authors,[202] have used the Clapeyron equation accordingly

$$T\left(V_i^l - V_i^{so}\right) = \alpha \tag{13}$$

with α being a constant. Using Equation (13) we can write

$$\beta = \frac{V_i^l}{V_i^{so}} - \frac{\alpha}{TV_i^l} \cong 1 - \frac{\alpha}{TV_i^{lo}} \tag{14}$$

Here, β is pressure independent and the integration of the Poynting correction is affected.

The evaluation of liquid fugacities was performed using the Soave–Redlich–Kwong equation of state[203] corrected by the volume translation of Peneloux:[204]

$$P = \frac{RT}{(V' - b)} - \frac{a(T)}{V'(V' + b)} \tag{15}$$

with

$$V_i = V_i' + C_i \tag{16}$$

where V' was the molar volume calculated from the SRK-EOS. C_i was calculated from the GCVOL group contribution method[205] at atmospheric pressure by

$$C_i = V_i^{\mathrm{GCVOL}} - V_i^{\mathrm{EOS}}$$

For mixtures, the van der Waals one fluid mixing rules have been used:

$$a = \sum_i \sum_j x_i x_j \sqrt{a_i a_j}\left(1 - k_{ij}\right) \tag{17}$$

$$b = \sum_i x_i b_i \tag{18}$$

with the $k_{ij}'s$ obtained from the group contribution method.[206]

For crystallizing compounds the equilibrium constant for the solid phase is:[196]

$$K_i^s = \frac{x_i^s}{x_i^l} = \phi_i^l[P]\left(\phi_i^{lo}[P_o]\right)^{\beta-1}\left(\phi_i^{lo}[P]\right)^{-\beta}\left(\frac{P}{P_0}\right)^{1-\beta}$$
$$\exp\left\{\frac{(1-\beta)C_i(P-P_0)}{RT} + \frac{\Delta_{\text{fus}}H_i}{RT_{\text{fus},i}}\left(\frac{T_{\text{fus},i}}{T}-1\right)\right.$$
$$\left. + \frac{\Delta_{\text{tr}}H_i}{RT_{\text{tr},i}}\left(\frac{T_{\text{tr},i}}{T}-1\right) - \frac{\Delta_s^l Cp_{\text{m}}}{R}\left(\ln\frac{T}{T_{\text{fus},i}} + \frac{T_{\text{fus},i}}{T} - 1\right)\right\} \tag{19}$$

with the liquid-phase fugacity coefficients, ϕ_i^l, calculated from the equation of state. P_o is the reference pressure taken as atmospheric pressure.

This pressure–temperature-composition relation for high-pressure SLE has been successfully used for many *n*-alkane systems.

8.7 Polymers Solubility

Solid polymers have extremely wide application in industry, and so they should not be ignored when problems in solubilities are studied. It often happens, however, that they are neglected because solid polymers fit poorly within the conventional theory of phase equilibria and the statistical physics of solids. A solid polymer can be crystalline to a certain degree. Sometimes its structure approaches a high-degree crystalline, but in other cases it should be considered as an amorphous body. The individual polymer molecules within a given polymeric substance are not uniform. Nevertheless, the structural features, or phase equilibria of polymer molecules enables one to treat solid polymers as organic solids. The analogies between the behaviour of crystalline polymers and low-molecular weight organic substances have inspired new experiments and the extension of theoretical concepts. Details, relating to the physics of polymers and the principles of their structure are well documented (see the fundamental monograph[207]) and only a few problems, directly related to the subject, is given here.

Since Flory (1970) developed the well-known equation of state for polymer solutions, much work has been carried out in order to establish a model for the accurate predictions of solvent activities in polymer solutions. Several free-volume expressions have been proposed[208–210] and the activity coefficients at infinite dilution for athermal polymer solutions with hydrocarbons have been satisfactorily described by the UNIFAC FV model).[211–213] Acceptable results were also obtained for the LLE and SLE/LLE phase diagrams for polymers of different molecular weights by entropic FV and modified Flory–Huggins models.[214–218] Simultaneous VLE and LLE was estimated for (acetone+PS) at low pressures by the van der Waals equation of state.[219]

The SAFT (statistical association fluid theory) equation of state was found to represent phase transitions of *n*-alkanes, or polymers in supercritical methane, or *n*-alkanes, or alkynes.[220–224] Also more complicated systems such as (naphthalene+benzene, or carbon tetrachloride, or cyclohexane), or (benzene+cyclohexane) SLE diagrams have been described by SAFT.[221] Recently good results for the calculation of (polymer+solvent) systems was obtained by the simplified PC-SAFT[224] equation of state. Good representation of the LLE was obtained for the polymers showing UCST and LCST of mixing.

Using the Hansen solubility parameters for polar or hydrogen-bonded solvents the solubility of polymers can be calculated by Flory–Huggins/Hansen model which includes a universal correction factor.[225,226] This model gives comparable results to group contribution models.

8.8 Ionic Liquids Solubility

In principle, ionic liquids present some highly intriguing possibilities, giving rise to liquids with unusual force fields between the charged species, and to ions whose shape and size can be modified in a readily controllable way by skilful organic synthesis. Two main subdivisions can be identified with solubility of ionic liquids, ILs: those whose cations are charged organic molecules, such as phosphonium, or ammonium salts, *e.g.* $[NR_1R_2R_3R_4]^+X^-$ in which the anions X^- are familiar inorganic species, such as Cl^-, NO_3^-, *etc.*, and those salts in which both anion and cation are organic.

To design any process involving ionic liquids on an industrial scale it is necessary not only to know a range of thermo-physical properties including viscosity, density, heat capacity, activity coefficients at infinite dilution and other thermodynamic properties including phase equilibria as SLE[227–229] and LLE.[230–241] Recent works include phase equilibrium between alcohols and ILs and the partitioning of alcohols between ionic liquids and water.[232,233] In general, (IL+an alcohol) binary mixtures show LLE with upper critical solution points near $x_{IL} = 0.15$. An increase in the alkyl chain length of the *n*-alkanols results in an increase in the UCST (see Figure 5). Branching of the alcohol results in a higher solubility of the alcohol in the IL-rich phase. By increasing the alkyl chain length on the imidazolium ring, the UCST decreased (see Figure 6). The replacement of the hydrogen at C2 position of the ring with the methyl group resulted in an increase in the UCST.[233]

The solubilities of 1-alkyl-3-methylimidazolium hexafluorophosphate,[235] or methylsulfate[240] in aromatic hydrocarbons, in *n*-alkanes and in cyclohydrocarbons have shown that the solubilities of these ILs in aromatic hydrocarbons decreases with an increase of the molecular weight of the solvent, or with an increase of the alkyl chain attached to the benzene ring. The differences on the solubility in *o*-, *m*- *p*-xylene were not significant. The intermolecular solute – solvent interactions were found to be small.

The LLE measurements in ternary mixtures of (IL+aromatic hydrocarbon+*n*-alkane)[236] have shown that the selectivity of the extraction of benzene from *n*-alkane increases with increasing carbon number of the *n*-alkane.

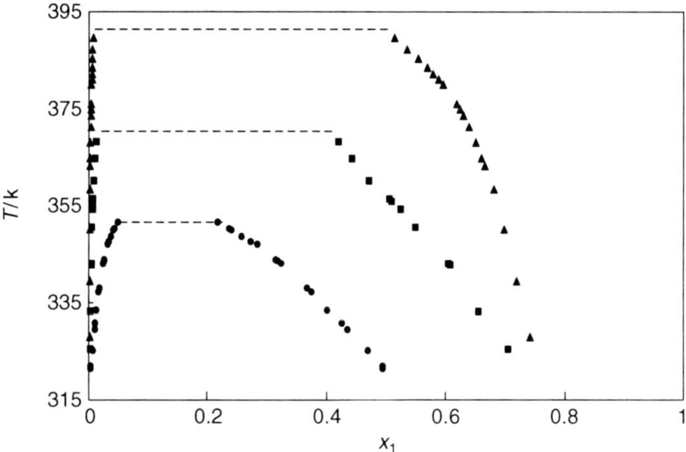

Figure 5 *LLE of* $\{x_1[emim][PF_6]+(1-x_1)$ *an alcohol} binary systems. Experimental points:*[235] *(●) ethanol, (■) propan-1-ol, (▲) butan-1-ol; (– –) boiling temperature of a solvent. (1-ethyl-3-methylimidazolium hexafluorophosphate,* $[emim][PF_6]$*)*

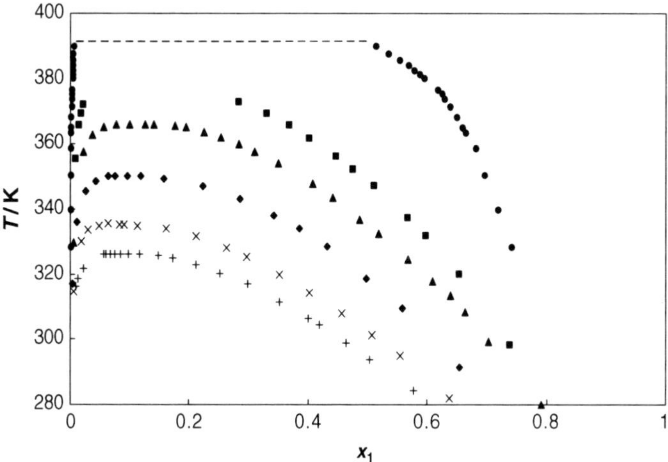

Figure 6 *LLE of* $\{x_1[Rmim][PF_6]+(1-x_1)$ *butan-1-ol} binary systems. Experimental points: (●)* $[emim][PF_6]$,[235] *(■)* $[bmim][PF_6]$,[230] *(▲)* $[pmim][PF_6]$,[234] *(◆)* $[hmim][PF_6]$,[234] *(x)* $[C_7mim][PF_6]$,[234] *(+)* $[omim][PF_6]$;[234] *(– –) boiling temperature of a solvent. (1-alkyl-3-methylimidazolium hexafluorophosphate,* $[Rmim][PF_6]$*: e-ethyl, b-butyl, p-penthyl, h-hexyl,* C_7*-heptyl, o-octyl).*

For many mixtures the traditional approaches using the excess Gibbs energy models (NRTL, UNIQUAC) for correlating the properties, *i.e.*, VLE, LLE, SLE or excess molar enthalpy of mixing in binary[227–229,235,237,239] and ternary mixtures[236] were used with acceptable results. There are still few measurements available to apply prediction methods to IL mixtures (*i.e.* the group contribution methods such as Modified UNIFAC or DISQUAC). The COSMO-RS model, a uni-molecular quantum chemical method has been successfully used to predict LLE in (IL+alcohol or hydrocarbon) binary mixtures.[230,234,240,241] This method has also predicted the correct trends for the variation of UCST with alkyl chain length of the alcohol.

Current studies focus on the solubilities of ILs, such as 1-hexyloxymethyl-3-methyl-imidazolium tetrafluoroborate [$C_6H_{13}OCH_2$mim][BF_4] or bis(trifluoromethylsulfonyl)-imide [$C_6H_{13}OCH_2$mim][Tf_2N], in *n*-alkanes and aromatic hydrocarbons with the aim of investigating the use of ILs in separating organic liquids, [239,242]

Other possible applications of ILs, which have a bearing on solubility issues, include lubricants, thermo fluids, plasticizers and electrically conductive liquids in electrochemistry.

References

1. G. Parkinson, *Chem. Eng. Progr.*, 2005, **101**, 15.
2. (*a*) L. O'Young, *Chem. Eng. Progr.*, 2005, **101**, 18; (*b*) C. Wibowo and L. O'Young, *Chem. Eng. Progr.*, 2004, **100**, 30.
3. M.J. Gentilcore, *Chem. Eng. Progr.*, 2004, **100**, 38.
4. C. Wibowo and K.M. Ng, *Ind. Eng. Chem. Res.*, 2002, **41**, 3839.
5. V. Norton and W. Wilkie, *Chem. Eng. Progr.*, 2004, **100**, 34.
6. W.J. Gennck, *Chem. Eng. Progr.*, 2004, **100**, 26.
7. L. Brunet, L. Prat, R. Wongkittipong, C. Gourdon, G. Casamatta and S. Damronglerd, *Chem. Eng. Technol.*, 2005, **28**, 110.
8. J. Nývlt, *Solid-Liquid Phase Equilibria*, Academia Publishing House of the Czechoslovak Academy of Sciences, Prague, 1977.
9. H. Knapp, M. Teller and R. Langhorst, *Solid–Liquid Equilibrium Data Collection, Binary Systems*, DECHEMA, Chemistry Data Series, Vol VIII, Part 1, Chemishe Technik und Biotechnologie e.v., Frankfurt am Main, 1987.
10. J.P. Novák, J. Matouš and J. Pick, *Liquid–Liquid Equilibria*, Czechoslovak Academy of Sciences, Prague, 1987 (in co-edition Elsevier Sciences Publishers, Amsterdam).
11. R. Haase and H. Schönert, *Solid–Liquid Equilibrium*, The International Encyclopedia of Physical Chemistry and Chemical Physics, Topic 13. Mixtures, Solutions, Chemical and Phase Equilibria, Vol 1, Pergamon Press, Oxford, London, Edinburgh, NY, Toronto, Sydney, Paris, Braunschweig, 1969.
12. A.I. Kitaigorodsky, *Mixed Crystals*, Springer, Berlin, Heidelberg, NY, Tokyo, 1984.
13. U. Domańska and M. Głoskowska, *Fluid Phase Equilib.*, 2004, **216**, 135.
14. U. Domańska and M. Głoskowska, *J. Chem. Eng. Data*, 2004, **49**, 101.
15. U. Domańska and M. Marciniak, *Ind. Eng. Chem. Res.*, 2004, **43**, 7647.
16. U. Domańska and M. Marciniak, *Fluid Phase Equilib.*, 2005, **235**, 30.

17. A.R. Ubbelohde, *The Molten State of Matter, Melting and Crystal Structure*, Wiley, Bristol, 1978.
18. A. Mączyński, A. Skrzecz, U. Domańska, H. Szatyłowicz and A. Bok, *NIST TRC SLE Floopy Book Database*, Boulder, CO, 1995, www.nist.gov/srd/nist95.htm.
19. Dortmund Data Bank, *Solid–Liquid Equilibria*, DDBST GmbH, Oldenburg, 2005, www.ddbst.de.
20. U. Domańska and J. Rolińska, Solid–Liquid Equilibrium Data Collection. Vol 1. Organic Compounds, *Monocarboxylic Acids*, PWN, IChF PAN, Warsaw, 1988.
21. G.I. Asbach, H.-G. Kilian and E. Stracke, *Colloid Polym. Sci.*, 1982 **260**, 151.
22. O.Yu. Batalin and M. Yu. Zakharov, *Russ. J. Phys. Chem.*, 1990, **64**, 444.
23. W.M. Mazee, *Erdoel und Kohle*, 1960, **13**, 88.
24. M. Dirand, M. Bouroukba, A.-J. Briard, V. Chevallier, D. Petitjean and J.-P. Corriou, *J. Chem Thermodyn.*, 2002, **34**, 1255.
25. U. Domańska and K. Kniaź, *Int. Data Ser., A, Sel. Data Mixtures*, 1990, **2**, 83–93; 1990, **3**, 194–206.
26. K. Kniaź, *Fluid Phase Equilib.*, 1991, **68**, 35.
27. U. Domańska, J. Rolińska and A.M. Szafrański, *Int. Data Ser., A, Sel. Data Mixtures*, 1987, **4**, 269–276.
28. U. Domańska, *Int. Data Ser., A, Sel. Data Mixtures*, 1989, **3**, 188–196.
29. U. Domańska, T. Hofman and J. Rolińska, *Fluid Phase Equilib.*, 1987 **32**, 273.
30. R.L. Snow, J.B. Ott, J.R. Goates, K.N. Marsh, S. O'Shea and R.H. Stokes, *J. Chem Thermodyn.*, 1986, **18**, 107.
31. S.-S. Chang, J.R. Maurey and W.J. Pummer, *J. Chem. Eng. Data*, 1983, **28**, 187.
32. Z. Plesnar, P. Gierycz and A. Bylicki, *J. Chem. Thermodyn.*, 1990, **22**, 403.
33. Z. Plesnar, P. Gierycz and A. Bylicki, *J. Chem Thermodyn.*, 1990, **22**, 393.
34. Z. Plesnar and A. Bylicki, *J. Chem Thermodyn.*, 1993, **25**, 1301.
35. Z. Plesnar, P. Gierycz, J. Gregorowicz and A. Bylicki, *Thermochim. Acta*, 1989, **150**, 101.
36. U. Domańska, K. Domański, C. Klofutar and Š. Paljk, *Fluid Phase Equilib.*, 1989, **46**, 25.
37. U. Domańska, *Int. Data Ser., A, Sel. Data Mixtures*, 1989, **2**, 108–115.
38. G. Gioia Lobbia, G. Vitali and R. Ruffini, *Thermochim. Acta*, 1982 **59**, 205.
39. G. Gioia Lobbia, G. Berchiesi and G. Vitali, *Thermochim. Acta*, 1983 **65**, 29.
40. G. Gioia Lobbia and G. Vitali, *J. Chem. Eng. Data*, 1984, **29**, 16.
41. H.S. Frank and M.J. Evens, *J. Chem. Phys.*, 1945, **11**, 507.
42. S.J. Gill, N.F. Nichols and I. Wadsö, *J. Chem. Thermodyn.*, 1976 **8**, 445.
43. T.R. Rettich, Y.P. Handa, R. Battino and E. Wilhelm, *J. Phys. Chem.*, 1981, **85**, 3230.

44. W.E. May, S.P. Wasik, M.M. Miller, Y.B. Tewari, J.M. Brown-Thomas and R.N. Golberg, *J. Chem. Eng. Data*, 1983, **28**, 197.
45. A. Mączyński, M. Góral, B. Wiśniewska-Gocłowska, A. Skrzecz and D. Shaw, *Monatshefte für Chemie*, 2003, **134**, 633.
46. A. Mączyński, B. Wiśniewska–Gocłowska and M. Góral, *J. Phys. Chem. Ref. Data*, 2004, **33**, 549.
47. P.B. Choi and E. McLaughlin, *Ind. Eng. Chem. Fundam.*, 1983, **22**, 46.
48. J.E. Coon, W.B. Sediawan, J.E. Auwaerter and E. McLaughlin, *J. Solution Chem.*, 1988, **17**, 519.
49. J.E. Coon, J.E. Auwaerter and E. McLaughlin, *Fluid Phase Equilib.*, 1989, **44**, 305.
50. U. Domańska, F.R. Groves Jr. and E. McLaughlin, *J. Chem. Eng. Data*, 1993, **38**, 88.
51. W.B. Sediawan, S. Gupta and E. McLaughlin, *J. Chem. Eng. Data*, 1989, **34**, 223.
52. H. Buchowski, U. Domańska and W. Ufnalski, *Polish J. Chem.*, 1979, **53**, 679.
53. H. Buchowski and U. Domańska, *Polish J. Chem.*, 1980, **55**, 97.
54. A. Jakob, R. Joh, C. Rose and J. Gmehling, *Fluid Phase Equilib.*, 1995, **113**, 117.
55. M. Teodorescu, M. Wilken, R. Witting and J. Gmehling, *Fluid Phase Equilib.*, 2003, **204**, 267.
56. M.E. Jamróz, M. Palczewska-Tulińska, D. Wyrzykowska-Stankiewicz, A.M. Szafrański, J. Polaczek, J.Cz. Dobrowolski, M.H. Jamróz and A.P. Mazurek, *Fluid Phase Equilib.*, 1998, **152**, 307.
57. D.G. Shaw and A. Mączyński, (eds), IUPAC-NIST solubility data series, hydrocarbons with water and seawater-revised and updated, "Part 8. C_9 hydrocarbons with water", *J. Phys. Chem. Ref. Data*, 2005, **4**, 2299.
58. Ch.L. Judy, N.M. Pontikos and W. E. Acree Jr., *J. Chem. Eng. Data*, 1987, **32**, 60.
59. M. Bissell, Ch.E. Chittick and W.E. Acree Jr., *Fluid Phase Equilib.*, 1988, **41**, 187.
60. J.W. McCargar and W.E. Acree Jr., *J. Solution Chem.*, 1989, **18**, 151.
61. U. Domańska and A.M. Szafrański, *Int. Data Ser., A, Sel. Data Mixtures*, 1987, **4**, 277–285.
62. U. Domańska and J. Rolińska, *Int. Data Ser., A, Sel. Data Mixtures*, 1991, **1**, 1–8; 1991, **2**, 90–97.
63. U. Domańska, *Fluid Phase Equilib.*, 1996, **114**, 175.
64. J.A. González, I. Garcia de la Fuente, J.C. Cobos, C. Casanova and U. Domańska, *Ber. Bunsenges. Phys. Chem.*, 1994, **98**, 955.
65. U. Domańska and J.A. González, *Fluid Phase Equilib.*, 1996, **119**, 131.
66. G. Gioia Lobbia, G. Berchiesi and M.A. Berchiesi, *J. Therm. Anal.*, 1976, **10**, 205.
67. C.W. Hoerr, H.J. Harwood and A.W. Ralston, *J. Org. Chem.*, 1944, **9**, 267.
68. A.W. Ralston and C.W. Hoerr, *J. Org. Chem.*, 1945, **10**, 170.

69. D.K. Kolb, *Chem., Dissert. Abstr.*, 1959, **20**, 82.
70. U. Domańska and T. Hofman, *J. Solution Chem.*, 1985, **14**, 531.
71. U. Domańska, *Fluid Phase Equilib.*, 1986, **26**, 201.
72. U. Domańska, *Ind. Eng. Chem. Res.*, 1987, **26**, 1153.
73. U. Domańska, *J. Solution Chem.*, 1989, **18**, 173.
74. R.S. Sedgwick, C.W. Hoerr and H.J. Harwood, *J. Org. Chem.*, 1952, **17**, 327.
75. W.E. Garner and A.M. King, *J. Chem. Soc.*, 1929, 1849.
76. W.O. Baker and Ch.P. Smyth, *J. Am. Chem. Soc.*, 1938, **60**, 1229.
77. E. Stenhagen and E. von Sydow, *Arkiv För Kemi*, 1952, **6**, 309.
78. K.N. Goswami, *J. Indian Chem. Soc.*, 1982, **20**, 968.
79. A. Martin, P.L. Wu and A. Beerbower, *J. Pharm. Sci.*, 1984, **73**, 188.
80. U. Domańska, *Polish J. Chem.*, 1986, **60**, 847.
81. U. Domańska, M.K. Kozłowska and M. Rogalski, *J. Chem. Eng. Data*, 2002, **47**, 8.
82. U. Domańska and M.K. Kozłowska, *J. Chem. Eng. Data*, 2003, **48**, 557.
83. U. Domańska and M.K. Kozłowska, *Fluid Phase Equilib.*, 2003, **206**, 253.
84. U. Domańska and A. Pobudkowska, *Fluid Phase Equilib.*, 2003, **206**, 341.
85. M. Della Monica, L. Janelli and U. Lamanna, *J. Phys. Chem.*, 1969, **72**, 1068.
86. L. Jannelli and A. Sacco, *J. Chem. Thermodyn.*, 1972, **4**, 715.
87. L. Jannelli, A. Inglese, A. Sacco and P. Ciani, *Z. Naturforsch.*, 1975, **30a**, 87.
88. L. Jannelli, A. Lopez, R. Jalent and L. Silvestri, *J. Chem. Eng. Data*, 1982, **27**, 28.
89. A. Sacco, A. Inglese, P. Ciani, A. Dell'Atti and A.K. Rakshit, *Thermochim. Acta*, 1976, **15**, 71.
90. U. Domańska, W.C. Moollan and T.M. Letcher, *J. Chem. Eng. Data*, 1996, **41**, 261.
91. J.A. González and U. Domańska, *Phys. Chem. Chem. Phys.*, 2001, **3**, 1034.
92. G. Che, J.B. Ott and J.R. Goates, *J. Chem. Thermodyn.*, 1986, **18**, 323.
93. Z. Bugajewski and A. Bylicki, *J. Chem. Thermodyn.*, 1988, **20**, 1191.
94. Z. Bugajewski and A. Bylicki, *J. Chem. Thermodyn.*, 1989, **21**, 37.
95. Z. Bugajewski and A. Bylicki, *J. Chem. Thermodyn.*, 1989, **21**, 847.
96. Z. Bugajewski and A. Bylicki, *J. Chem. Thermodyn.*, 1990, **22**, 573.
97. Z. Bugajewski and A. Bylicki, *J. Chem. Thermodyn.*, 1991, **23**, 901.
98. Z. Bugajewski and A. Bylicki, *J. Chem. Thermodyn.*, 1992, **24**, 29.
99. G. Che, Z. Huang, D. Li, X. Gu and S. Luo, *J. Chem. Thermodyn.*, 1996, **28**, 159.
100. U. Domańska and T.M. Letcher, *J. Chem. Thermodyn.*, 2000, **32**, 1635.
101. U. Domańska and J. Łachwa, *Fluid Phase Equilib.*, 2002, **198**, 1.
102. U. Domańska and J. Łachwa, *J. Chem. Thermodyn.*, 2003, **35**, 1215.
103. U. Domańska and J. Łachwa, *J. Chem. Thermodyn.*, 2005, **37**, 692.
104. U. Domańska and J. Łachwa, *Fluid Phase Equilib.*, 2005, **227**, 135.
105. U. Domańska and J. Łachwa, *Fluid Phase Equilib.*, 2005, **232**, 214.

106. A. El Basyony, J. Klimes, A. Knoechnel, J. Ochler and G. Rudolph, *Z. Naturforsch. B*, 1976, **31B**, 1192.
107. C.M. Grossel, R.M. Goldspink, R. Mark, A.J. Hriljacad and C.S. Weston, *Organometallics*, 1991, **10**, 851.
108. G.S. Heo, E.P. Hillman and A.R. Bartsch, *J. Heterocycl. Chem.*, 1982, **19**, 1099.
109. U. Domańska and J.A. González, *Fluid Phase Equilib.*, 2003, **205**, 317.
110. U. Domańska and K. Kniaź, *Int. Data Ser., A, Sel. Data Mixtures*, 1994, **22**, 37.
111. U. Domańska and K. Domański, *Int. Data Ser., A, Sel. Data Mixtures*, 1994, **22**, 43.
112. U. Domańska and J. Rolińska, *Int. Data Ser., A, Sel. Data Mixtures*, 1994, **22**, 50.
113. U. Domańska, *Polish J. Chem.*, 1998, **72**, 925.
114. A. Seidell, *Solubilities of Organic Compounds*, van Nostrand, New York, 1941.
115. J.M. Sørensen and W. Arlt, *Liquid–Liquid Equilibrium Data Collection, Dechema Chemistry Data Series*, Vol V, Chemishe Technik und Biotechnologie e.v., Frankfurt 1979, 1980.
116. Dortmund Data Bank, *Liquid-Liquid Equilibria*, DDBST GmbH, Oldenburg, 2005, www.ddbst.de.
117. A. Mączyński, A. Skrzecz and A. Bok, NIST/TRC LLE Floopy Book Database, Boulder, CO, 1994, www.nist.gov/srd/nist94.htm.
118. A. Mączyński, A. Skrzecz and A. Bok, NIST/TRC IDS Floopy Book Database, International Data Series/Selected data on Mixtures 1973–2004 (Version 2005), Boulder, CO, 2005, www.nist.gov/srd/nist92.htm.
119. J. Gmehling, K. Fisher, J. Li and M. Schiller, *Pure & Appl. Chem.*, 1993, **65**, 919.
120. J.A. González, I. Garcia, J.C. Cobos and C. Casanova, *J. Chem. Eng. Data*, 1991, **36**, 162.
121. U. Domańska, M. Szurgocińska and J.A. González, *Ind. Eng. Chem. Res.*, 2002, **41**, 3253.
122. U. Domańska and S. Malanowski, *Fluid Phase Equilib.*, 1999, **158–160**, 991.
123. T. Hauschild and H. Knapp, *J. Solution Chem.*, 1991, **20**, 125.
124. J.A. González and U. Domańska, *Phys. Chem. Chem. Phys.*, 2001, **3**, 1034.
125. U. Domańska, *Fluid Phase Equilib.*, 1990, **55**, 125.
126. U. Domańska, *Fluid Phase Equilib.*, 1987, **35**, 217.
127. H. Buchowski, U. Domańska, A. Książczak and A. Mączyński, *Roczniki Chem.*, 1975, **49**, 1889.
128. U. Domańska, *Fluid Phase Equilib.*, 1989, **46**, 223.
129. U. Domańska, *Ind. Eng. Chem. Res.*, 1990, **29**, 470.
130. U. Domańska, *Fluid Phase Equilib.*, 1988, **40**, 259.
131. H. Buchowski, U. Domańska and A. Książczak, *Polish J. Chem.*, 1979, **53**, 1127.

132. U. Domańska, *Polish J. Chem.*, 1981, **55**, 1715.
133. L.J. Gordon and R.L. Scott, *J. Am. Chem. Soc.*, 1952, **74**, 4138.
134. S.M. Byer, R.E. Gibbs and H.C. van Ness, *AIChE J.*, 1973, **19**, 245.
135. J.R. Goates, J.B. Ott and N.F. Mangelson, *J. Am. Chem. Soc.*, 1963, **67**, 2874.
136. U. Domańska and K. Kniaź, *Fluid Phase Equilib.*, 1990, **58**, 211.
137. D.A. Brandreth and R.E. Johnson, *J. Chem. Eng. Data*, 1971, **16**, 325.
138. I.H. Hildebrand, I.M. Prausnitz and R.L. Scott, *Regular and Related Solutions*, van Nostrand Reinhold Comp., New York, 1970.
139. A.F.M. Barton, *CRC Handbook of Solubility Parameters and other Cohesion Parameters*, CRC Press Inc., Boca Raton, FL, 1985.
140. J.R. Wallach, S.A. Tucker, B.M. Oswalt, D.J. Murral and W.E. Acree Jr., *J. Chem. Eng. Data*, 1989, **34**, 70.
141. J.W. McCargar and W.E. Acree, Jr., *J. Solution Chem.*, 1989, **18**, 151.
142. S.A. Tucker and W.E. Acree, Jr., *Phys. Chem. Liq.*, 1989, **19**, 73.
143. G.H. Dale, *Encyclopedia of Chemical Processing and Design*, Vol 13, Dekker, NewYork, 1981, 461.
144. A. Gomes, L. Serrano and F. Farelo, *J. Chem. Eng. Data*, 1988, **33**, 194.
145. V.K. Jadhav, R. Chivate and N.W. Tavare, *J. Chem. Eng. Data*, 1991, **36**, 249.
146. A. Gupta, U. Domańska, R.F. Groves Jr. and E. McLaughlin, *J. Chem. Eng. Data*, 1994, **39**, 175.
147. R. Ozawa and M. Matsuoka, *J. Cryst. Growth*, 1989, **98**, 411.
148. J.M. Muller and G. Hoefeld, *7th World Pet. Congr.*, 1967, **4**, 13.
149. T.M. Letcher and P.M. Siswana, *Fluid Phase Equilib.*, 1992, **72**, 203.
150. J.F. Fabries, J.-L. Gustin and H. Renon, *J. Chem. Eng. Data*, 1977, **22**, 303.
151. T.M. Letcher and P.K. Naicker, *J. Chem. Eng. Data*, 1998, **43**, 1034.
152. T.M. Letcher, D. Ramjugernath and R.D. Naidoo, *J. Chem. Eng. Data*, 2001, **46**, 1375.
153. G.W. Cassell, M.M. Hassan and A.L. Hines, *Chem. Eng. Comm.*, 1989, **85**, 233.
154. S.J. Ashcroft, A.D. Clayton and R.B. Shearn, *J. Chem. Eng. Data*, 1982, **27**, 148.
155. R.P. Tripathi, A.R. Ram and P.B. Rao, *J. Chem. Eng. Data*, 1975, **20**, 261.
156. S. Lee and H. Kim, *J. Chem. Eng. Data*, 1995, **40**, 499.
157. A. Masohan, S.M. Nanoti, K.G. Sharma, S.N. Puri, P. Gupta and B.S. Rawat, *Fluid Phase Equilib.*, 1990, **61**, 89.
158. I. Nagata, *Thermochim. Acta*, 1988, **127**, 337.
159. I. Nagata, *J. Chem Thermodyn.*, 1984, **16**, 737.
160. I. Nagata, *Thermochim. Acta*, 1982, **56**, 43.
161. I. Nagata, K. Miyamoto, P. Alessi and I. Kikic, *Thermochim. Acta*, 1987, **120**, 63.
162. I. Nagata and Y. Kawamura, *Chem. Eng. Sci.*, 1979, **34**, 601.
163. I. Nagata, *Fluid Phase Equilib.*, 1989, **51**, 53.

164. I. Nagata and K. Ohtsubo, *Thermochim. Acta*, 1986, **97**, 37.
165. R.F. Weimar and J.M. Prausnitz, *J. Chem. Phys.*, 1965, **42**, 3643.
166. P.B. Choi and E. McLaughlin, *AIChE J.*, 1983, **29**, 150.
167. G.M. Wilson, *J. Am. Chem. Soc.*, 1964, **86**, 127.
168. D.S. Abrams and J.M. Prausnitz, *AIChE J.*, 1975, **21**, 116.
169. H. Renon and J.M. Prausnitz, *AIChE J.*, 1968, **14**, 135.
170. D. Marquardt, *J. Soc. Appl. Math.*, 1963, **11**, 431.
171. T. Hofman and I. Nagata, *Fluid Phase Equilib.*, 1986, **25**, 113.
172. U. Domańska and T. Hofman, *Ind. Eng. Chem. Process Des. Dev.*, 1986, **25**, 996.
173. U. Domańska and J. Rolińska, *Fluid Phase Equilib.*, 1989, **45**, 25.
174. U. Domańska and K. Kniaź, *Fluid Phase Equilib.*, 1990, **58**, 211.
175. I. Nagata, Y. Nakamiya, K. Katoh and J. Koyabu, *Thermochim. Acta*, 1981, **45**, 153.
176. I. Nagata, *Fluid Phase Equilib.*, 1985, **19**, 153.
177. J.M. Sørensen, T. Magnussen, P. Rasmussen and A. Fredenslund, *Fluid Phase Equilib.*, 1979, **3**, 47.
178. J.B. Ott and J.R. Goates, *J. Chem Thermodyn.*, 1983, **15**, 267.
179. I. Nagata and K. Ohtsubo, *Thermochim. Acta*, 1986, **102**, 185.
180. J. Gmehling, J. Li and M. Schiller, *Ind. Eng. Chem. Res.*, 1993, **32**, 178.
181. H.V. Kehiaian, *Pure. Appl. Chem.*, 1985, **57**, 15.
182. K. Kojima and K. Tochigi, *Prediction of Vapor–Liquid Equilibria by the ASOG Method*, Elsevier, Amsterdam, 1979.
183. T. Magnussen, P. Rasmussen and A. Fredenslund, *Ind. Eng. Chem. Process Des. Dev.*, 1981, **20**, 331.
184. T. Hofman and U. Domańska, *J. Solution Chem.*, 1988, **17**, 237.
185. J.A. González, I. García de la Fuente, J.C. Cobos and U. Domańska, *Fluid Phase Equilib.*, 1996, **119**, 81.
186. C. Fiege, R. Joh, M. Petri and J. Gmehling, *J. Chem. Eng. Data*, 1996, **41**, 1431.
187. W.C. Moolan, U. Domańska and T.M. Letcher, *Fluid Phase Equilib.*, 1997, **128**, 137.
188. U. Domańska and J.A. González, *Fluid Phase Equilib.*, 2003, **205**, 317.
189. U. Domańska and J. Łachwa, *J. Chem Thermodyn.*, 2005, **37**, 692.
190. A. Anderko and S. Malanowski, *Fluid Phase Equilib.*, 1989, **48**, 223.
191. J.A.P. Coutinho, S.I. Andersen and E.H. Stenby, *Fluid Phase Equilib.*, 1996, **117**, 138.
192. P. Hemmingsen, *Doctoral Thesis, Department of Chemical Engineering*, Norwegian University of Science and Technology, Norway, 2000.
193. G. Sadowski, W. Arlt, A. De Haan and G. Krooshof, *Fluid Phase Equilib.*, 1999, **163**, 79.
194. C. Yokoyama, S. Moriya and T. Ebina, *Sekiyu Gakkaishi-J. Jpn. Petrol. Inst.*, 1998, **41**, 125.
195. J. Pauly, J.-L. Daridon, J.A.P. Coutinho, N. Lindeloff and S.I. Andersen, *Fluid Phase Equilib.*, 2000, **167**, 145.

196. J. Pauly, J.-L. Daridon and J.A.P. Coutinho, *Fluid Phase Equilib.*, 2001, **187–188**, 71.
197. J. Pauly, J.-L. Daridon, J.A.P. Coutinho and F. Montel, *Energy and Fuels*, 2001, **15**, 730.
198. J. Pauly, J.-L. Daridon, J.M. Sansot and J.A.P. Coutinho, *Fuel*, 2003 **82**, 595.
199. J. Pauly, J.-L. Daridon, J.A.P. Coutinho and M. Dirand, *Fuel*, 2005 **84**, 453.
200. P. Morawski, J.A.P. Coutinho and U. Domańska, *Fluid Phase Equilib.*, 2005, **230**, 72.
201. U. Domańska and P. Morawski, *J. Chem. Thermodyn.*, 2005, **37**, 1276.
202. G.W.H. Hohne and K. Blankenhorn, *Thermochim. Acta*, 1994 **238**, 351.
203. G. Soave, *Chem. Eng. Sci.*, 1972, **27**, 1197.
204. A. Peneloux, E. Rauzy and R.A. Frèze, *Fluid Phase Equilib.*, 1982, **8**, 7.
205. H.S. Elbro, A. Fredenslund and P. Rasmussen, *Ind. Eng. Chem. Res.*, 1991, **30**, 2576.
206. J.N. Jaubert and F. Mutelet, *Fluid Phase Equilib.*, 2005, **224**, 285.
207. D.W. Van Krevelen, *Properties of Polymers*, Elsevier Science Publishers, New York, 1990.
208. T. Oishi and J.M. Prausnitz, *Ind. Eng. Chem. Process Des. Dev.*, 1978 **17**, 333.
209. Y. Iwai and Y. Arai, *J. Chem. Eng. Jpn.*, 1989, **22**, 155.
210. H.S. Elbro, A. Fredenslund and P. Rasmussen, *Macromolecules*, 1990, **23**, 4707.
211. G.M. Kontogeorgis, A. Fredenslund and D.P. Tassios, *Ind. Eng. Chem. Res.*, 1993, **32**, 362.
212. G. Bogdanic and A. Fredenslund, *Ind. Eng. Chem. Res.*, 1994, **33**, 1331.
213. G.M. Kontogeorgis, P. Coutsikos, D. Tassios and A. Fredenslund, *Fluid Phase Equilib.*, 1994, **92**, 35.
214. G.M. Kontogeorgis, A. Saraiva, A. Fredenslund and D.P. Tassios, *Ind. Eng. Chem. Res.*, 1995, **34**, 1823.
215. V.I. Harismiadis, A.R.D. van Bergen, A. Saraiva, G.M. Kontogeorgis, A. Fredenslund and D.P. Tassios, *AIChE J.*, 1996, **42**, 3170.
216. V.I. Harismiadis and D.P. Tassios, *Ind. Eng. Chem. Res.*, 1996, **35**, 4667.
217. M.K. Kozłowska, U. Domańska, D. Dudek and M. Rogalski, *Fluid Phase Equilib.*, 2005, **236**, 184.
218. U. Domańska and M.K. Kozłowska, *Chem. Eur. J.*, 2005, **11**, 776.
219. A. Saraiva, G.M. Kontogeorgis, V.I. Harismiadis, A. Fredenslund and D.P. Tassios, *Fluid Phase Equilib.*, 1996, **115**, 73.
220. Ch. Pan and M. Radosz, *Ind. Eng. Chem. Res.*, 1998, **37**, 3169.
221. S.J. Han, D.J. Lohse, M. Radosz and L.H. Sperling, *Macromolecules*, 1998, **31**, 5407.
222. A.-Q. Chen and M. Radosz, *J. Chem. Eng. Data*, 1999, **44**, 854.
223. Ch. Pan and M. Radosz, *Fluid Phase Equilib.*, 1999, **155**, 57.
224. Ch. Pan and M. Radosz, *Ind. Eng. Chem. Res.*, 1999, **38**, 2842.

225. N. von Solms, I.A. Kouskoumvekaki, T. Lindvig, M.L. Michelsen and G.M. Kontogeorgis, *Fluid Phase Equilib.*, 2004, **222–223**, 87.
226. T. Lindvig, M.M. Michelsen and G.M. Kontogeorgis, *Fluid Phase Equilib.*, 2002, **203**, 247.
227. U. Domańska, E. Bogel-Łukasik and R. Bogel-Łukasik, *J. Phys. Chem. B*, 2003, **107**, 1858.
228. U. Domańska and E. Bogel-Łukasik, *Ind. Eng. Chem. Res.*, 2003, **42**, 6986.
229. U. Domańska and L. Mazurowska, *Fluid Phase Equilib.*, 2004, **221**, 73.
230. K.N. Marsh, A. Deev and A.C.-T. Wu, E. Tran and A. Klamt, *Kor. J. Chem. Eng.*, 2002, **19**, 357.
231. V. Najdanovic-Visak, J.M.S.S. Esperanca, L.P.N. Rebelo, M.N. da Ponte, H.J.R. Guedes, K.R. Seddon, R.F. de Souza and J. Szydłowski, *J. Phys. Chem. B*, 2003, **107**, 12797.
232. A. Heintz, J.K. Lehmann and C. Wertz, *J. Chem Eng. Data*, 2003, **48**, 472.
233. J.M. Crosthwaite, S.N.V. Akai, E.J. Maginn and J.F. Brennecke, *J. Phys. Chem. B*, 2004, **108**, 5113.
234. C.-T. Wu, K.N. Marsh, A.V. Deev and J.A. Boxall, *J. Chem Eng. Data*, 2003, **48**, 486.
235. U. Domańska and A. Marciniak, *J. Chem. Eng. Data*, 2003, **48**, 451.
236. U. Domańska and A. Marciniak, *J. Phys. Chem. B*, 2004, **108**, 2376.
237. T.M. Letcher and N. Deenadayalu, *J. Chem. Thermodyn.*, 2003, **35**, 67.
238. J.M. Crosthwaite, S.N.V. Aki, E.J. Maginn and J.F. Brennecke, *Fluid Phase Equilib.*, 2005, **228–229**, 303.
239. U. Domańska and A. Marciniak, *J. Chem. Thermodyn.*, 2005, **37**, 577.
240. U. Domańska, A. Pobudkowska and F. Eckert, *Green Chem.*, 2006 **8**, 268.
241. U. Domańska, A. Pobudkowska and F. Eckert, *J. Chem. Thermodyn.*, 2006, doi:10.1016/j.jct.2005.07.024. **38**, 685.
242. G.W. Meindersma, A.J.G. Podt and A.B. de Haan, *Fuel Processing Technol.*, 2005, **87**, 59.

CHAPTER 9

CO_2 Solubility in Alkylimidazolium-Based Ionic Liquids

ALIREZA SHARIATI,[1] SONA RAEISSI[1] AND COR J. PETERS[2]

[1] *Department of Oil, Gas, and Chemical Engineering, School of Engineering, Shiraz University, Shiraz, Iran*

[2] *Physical Chemistry and Molecular Thermodynamics, Faculty of Applied Sciences,, Delft University of Technology, Julianalaan 136, 2628 BL Delft, The Netherlands*

9.1 Introduction

Product purification and separation is among the costliest factors in the process industries and many of the conventional separation techniques have problems of solvent contamination. Both of these dilemmas arise because the solvents/extractants solubilize to some extent in all the phases. The current eruption of excitement over ionic liquids (ILs) as solvents is due to their great solvent power for polar and non-polar substances, while at the same time, they are generally believed to be insignificantly soluble themselves in the vapor phase, thus allowing for easy and clean separations.

ILs are organic salts that are liquid at room temperature. They consist of an organic cation such as quaternary ammonium, imidazolium, pyridinium, or pyrrolidinium ions combined with either an organic or an inorganic anion of usually smaller size and more symmetrical shape such as Cl^-, Br^-, I^-, $AlCl_4^-$, BF_4^-, PF_6^-, $ROSO_3^-$, or Tf_2N^-. It is the Coulombic attraction between these ions that is responsible for the recognized lack of IL vapor pressures even up to their thermal decomposition temperatures. In fact, they have been called "liquid solids" because they incorporate some of the most useful physical properties of both phases.[1] In addition to facilitating separations and minimizing expenses resulting from solvent loss, the seemingly insignificant vapor pressures of ILs minimize environmental pollution problems and highly reduce working exposure hazards in comparison with the conventional organic solvents being used in industries today. Aside from this, ILs posses a range of other outstanding characteristics which suggest a promising industrial future for them. Most ILs are in the liquid state at room temperature and will remain liquid till high temperatures, so processes such as heterogeneous reactions can instead be performed in a single liquid phase within a wide temperature range. ILs are thermally stable, are non-flammable, and have high ionic conductivity. Unlike conventional organic solvents, it is possible to adjust the properties of ILs to produce task-specific solvents by combining different anions and cations or their functional groups such as the alkyl chain length. Based on these properties, there is an extensive perspective of potential industrial applications for ILs in catalytic reactions, gas drying and separations, liquid–liquid extractions, electrolyte/fuel cells, and as lubricants, heat transfer fluids, plasticizers, and solvents for cleaning operations.[2]

Mixtures of CO_2 and imidazolium-based ILs are, in particular, interesting for a variety of applications, partly because CO_2 has shown remarkable solubilities in these ILs. Biphasic CO_2+IL mixtures have been used for a range of homogeneously catalyzed reactions. Supercritical CO_2 can be used as an environmentally benign solvent to extract organic products or contaminants from ILs.[3,4] It has also been shown that CO_2 can be used to separate organic liquids and water from ILs by inducing a liquid-liquid phase split.[5,6] Owing to the remarkable solubility of CO_2 in imidazolium-based ILs, there is an emerging interest in applying ILs for gas separations.[7] It is obvious that a fundamental understanding of the phase behavior of CO_2 with ILs is important for all these applications and for future developments.

Symbolism

In what is to follow, the following abbreviations are used for various cations and anions: 1-ethyl-3-methylimidazolium ([emim]), 1-butyl-3-methylimidazolium ([bmim]), 1-hexyl-3-methylimidazolium ([hmim]), 1-octyl-3-methylimidazolium ([omim]), hexafluorophosphate ([PF_6]), tetrafluoroborate ([BF_4]) bis(trifluoromethylsulfonyl)imide ([Tf_2N]), dicyanamide ([DCA]), nitrate ([NO_3]), trifluoromethanesulfonate ([TfO]), and (trifluoromethylsulfonyl) methide ([methide]).

9.2 Phase Behaviour

The experimental solubilities of the binary system CO_2+[bmim][PF_6] is presented in Figure 1, in the form of bubble point pressures at different temperatures for several isopleths.[8] As expected for a gas dissolving in a liquid, CO_2 solubility decreases with increasing temperature and increases with increasing pressure. However, it is the large quantities of CO_2 that can be dissolved in the IL, for example reaching up to about 65 mol% at 60 MPa and 330 K, which makes such binary systems particularly interesting. This can be better observed

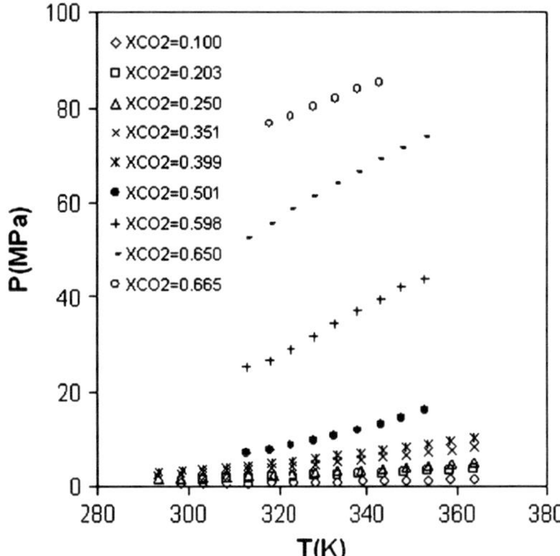

Figure 1 *P–T diagram of the binary system of CO_2+[bmim][PF_6] at different molar concentrations.*[8]

on P-x coordinates, as shown for example, in Figure 2 at three different temperatures.[8] Carbon dioxide shows very high solubilities at lower pressures, while with almost a sudden sharp break in slope the curves steep upwards, indicating that very little CO_2 can be further dissolved with further increase of pressure. There is a chance that the curve will bend back down at extremely high pressures to meet the dew point curve but it is more likely that it will simply continue upward to infinitely high pressures. Attempts to measure dew point curves, *i.e.*, the solubility of [bmim][PF_6] in CO_2 have indicated immeasurably small IL concentrations. For example, Blanchard and co-workers[9] indicated that the solubility is less than 5×10^{-7} in mole fraction at 40°C and 138 bars. A binary mixture of [bmim][PF_6] with 97 mol% CO_2 showed the existence of two phases even up to a pressure of 3100 bars at 40°C. Blanchard and co-workers[9] state that such diverging behavior, with a large immiscibility gap even up to extremely high pressures, is very unusual for a mixture of CO_2 in a liquid, *i.e.*, normally when a large amount of CO_2 dissolves in the liquid phase at low pressures, the system shows a simple phase envelope with a mixture critical point at moderate pressures. However, in what is to follow, we will show that according to phase behavior principles, this kind of behavior is not unusual, although indeed, it is not so commonly observed. If this uncommon behavior is not intriguing enough on its own, the realization of the very extreme phase behaviors that can occur in these systems in pressure and temperature regions outside those observed experimentally, can be flabbergasting. Fortunately, the wide variety of phase behavior that can occur when operating at high pressures has been classified by Scott and Van Konynenburg[10] into a limited number of phase diagram types.

Trying to understand the one type of phase diagram that is most likely the behavior of the systems of concern in this chapter, out of context from the other

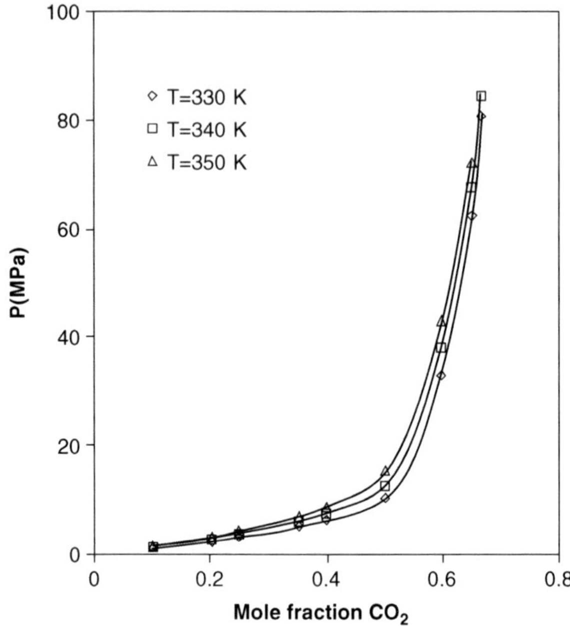

Figure 2 *P-x diagram[8] of the binary system CO_2+[bmim][PF_6] at 330, 340, and 350 K.*

evolutionary transforming phase types, is a difficult task, especially for the unfamiliar reader. But including all the types of phase diagrams requires a full chapter on its own. However, due to the valuable (qualitatively) extrapolating information that Scott and van Konynenburg's phase descriptions can reveal, we will try our best to explain the one type of phase behavior of interest to this work. The interested reader is, however, encouraged to read further[10,11] for a better understanding.

To be able to determine the type of fluid phase behavior, the occurrence of a second liquid phase was investigated in a mixture of 98 mol% CO_2 and 2 mol% [bmim][PF_6], but this time within a temperature and pressure range close to the critical point of CO_2.[8] Within the uncertainty of the experimental data, a three-phase line L_1L_2V was found, which turned out to be almost indistinguishable compared to the location of the vapor pressure curve of pure CO_2. In addition, a critical point of the nature $L_1 = V + L_2$ was found.[8] This is the point at which L_1 and V are critical in the presence of L_2. According to the classification of Scott and Van Konynenburg,[10] this system could have Type III, IV, or V fluid phase behavior. However, because no binary CO_2 systems are known to show Type V behavior in literature, and because the occurrence of Type IV systems are rare, the system CO_2 + [bmim][PF_6] most likely has Type III fluid-phase behavior.

Type III behavior occurs in systems where the difference in size, structure, and/or strength of the intermolecular forces between the mixture constituents are very large. The following explanations of Type III behavior are mostly taken directly from the book of McHugh and Krukonis.[11] Figure 3 shows this type of behavior in a simplified form by using a two-dimensional P-T *projection* of critical mixture curves and three phase equilibrium lines from the three-dimensional P–T-x diagram of Figure 4. The formation of (multiple) solid phases and multiple liquid phases at cryogenic temperatures are not discussed here. In Figure 3, the solid curves are the pure component vapor pressure curves and the dash-dot curve is the projection of the three-phase equilibrium surface. The dashed curves are projections of the critical mixture curves (the locus of the mixture critical points of mixtures of varying compositions). The filled circles are pure component critical points and the open circle is a critical endpoint. (A critical endpoint is the limiting point at which two or three coexisting phases become identical).

Figure 3 also shows that one branch of the critical locus ($L_1 = V$) connects the critical point of pure component A (the more volatile component) to the critical endpoint of the three-phase equilibrium L_1L_2V. A second branch of the critical locus ($L_2 = V$) originates in the critical point of pure component B and extends to higher pressures while gradually changing its nature from $L_2 = V$ into $L_1 = L_2$.

For a better understanding of Figure 3, the full P–T-x diagram and some isothermal intersections of interest are shown in Figure 4. Figure 4(b) shows that at low pressures a single vapor phase exists (This is a general statement for Type III behavior; ILs most probably cannot exist as a single vapor phase, no matter how low the pressure[15]). At higher pressure, the dew point curve is intersected and liquid and vapor phases now coexist. As the pressure is increased still further, the three-phase LLV line is intersected. If the pressure is increased still further, and if the overall mixture composition is greater than

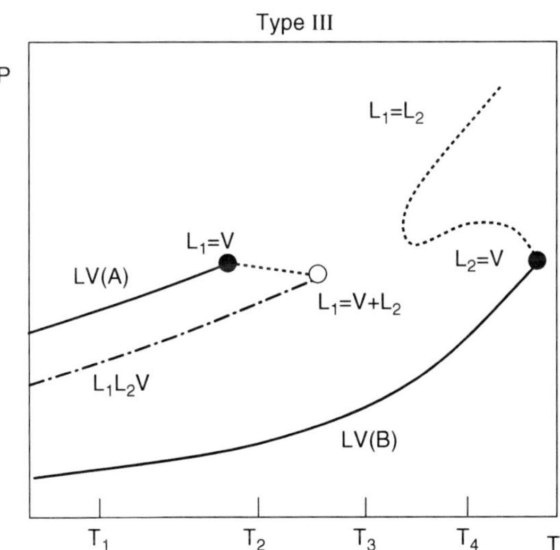

Figure 3 *Schematic diagram for Type III phase behaviour (P–T projection).*

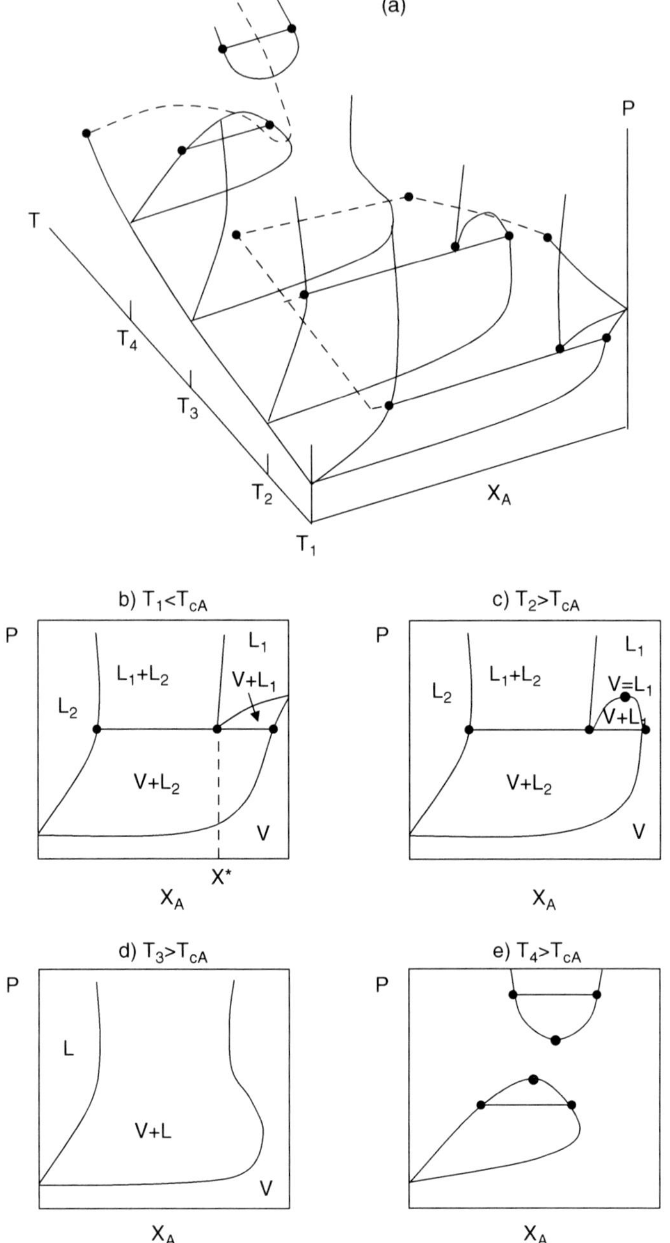

Figure 4 *The P–T-x (a) and isothermal P-x diagrams (b–e) for Type III binary mixtures.* Tc_A *is the critical temperature of the more volatile component.*

x*, a vapor–liquid envelope is observed, which intersects the pressure axis at P_A^{vap}, corresponding to the intersection of the vapor pressure curve of the more volatile component at T_1 in Figure 3. However, the liquid–liquid envelope at overall mixture concentrations less than x* does not exhibit a closed dome with a mixture critical point. Both branches of the liquid–liquid envelope rise steeply with increasing pressure; in fact, they can diverge at very high pressures. This type of liquid–liquid phase behavior is representative of mixtures in which the components have a strong "dislike" for each other, as for instance with hydrocarbon-water mixtures at modest temperatures.

If the next P-x diagram is constructed at a slightly higher temperature, T_2, the LLV line is intersected but the vapor pressure curve of the more volatile component is not intersected (Figure 3). At pressures higher than the three-phase LLV pressure, the left hand side of the vapor–liquid envelope exhibits a closed dome with a maximum in pressure equal to the pressure of the isothermal intersection of the branch of the critical mixture curve closest to the critical point of the volatile component, *A*. But, again, the liquid–liquid envelope at lower concentrations does not exhibit a closed dome with a mixture critical point [see Figure 4(c)].

If the temperature is raised to T_3, the phase behavior shown in Figure 4(d) occurs. This temperature is higher than the upper critical endpoint temperature and, therefore, two phases exist as the pressure is increased as long as the critical mixture curve is not intersected. The two branches of the vapor–liquid phase envelope approach each other in composition at an intermediate pressure and it appears that a mixture critical point may occur. But as the pressure is further increased, a mixture critical point is not observed and the two curves begin to diverge.

An interesting type of phase behavior occurs if the temperature of the system is increased to T_4 [see Figure 4(e)]. In this case, the vapor–liquid envelope does exhibit a closed dome with a mixture critical point at a moderate pressure equal to the intersection of the mixture critical curve at this temperature. A single fluid phase now exists at this temperature for pressures greater than the mixture critical pressure. But if the pressure is increased much beyond the mixture critical pressure, the single fluid phase splits into two phases. Two representative tie lines are shown in the two-phase regions of this temperature. Figure 4(e) shows that two mixture critical points occur at this temperature, depending on the overall composition of the mixture. One critical point occurs at the maximum of the vapor–liquid envelope as the pressure is isothermally raised from a low to a moderate value. The other mixture critical point occurs at the minimum of the fluid–liquid envelope, which exists at higher pressures.

When the locus of mixture critical points is connected, the P–T diagram shown in Figure 3 is generated. The reader is cautioned that the critical mixture curve shown in Figure 3 is only a schematic representation of Type III fluid phase behavior as the curve does not necessarily have to exhibit a minimum in pressure. In fact, the branch of the mixture critical curve that starts at the critical point of the less volatile component can have many shapes, as shown schematically in Figure 5(a).

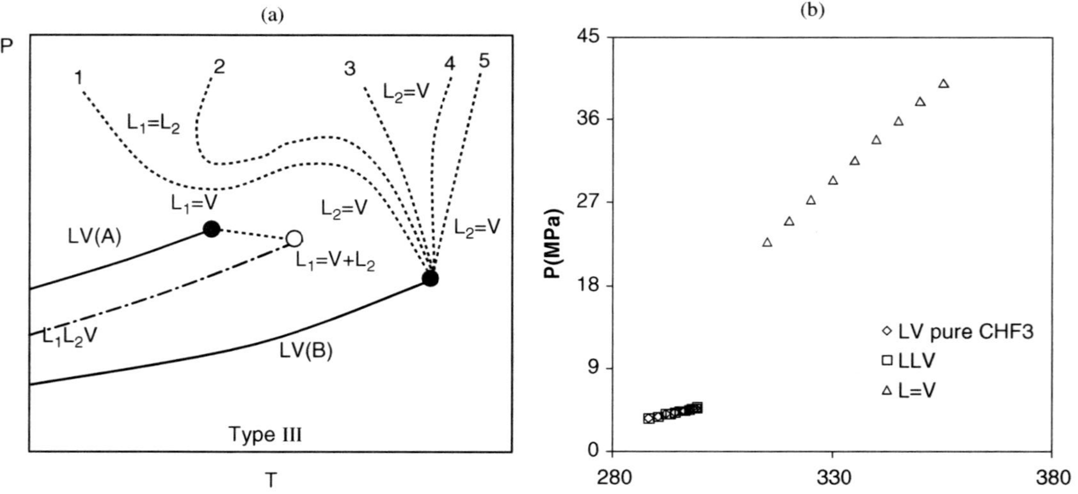

Figure 5 *(a) Schematic diagram for Type III phase behaviour showing the different shapes and trends that the mixture critical curve can take (b) P–T projection of the system* CHF_3*+[bmim][*PF_6*].*

For example, the critical curve can also show a temperature minimum combined with a pressure minimum and a pressure maximum (Branch 2), or it can have a negative slope of $(dP/dT)_c$ (Branch 3) at the critical point of component B, or only a temperature minimum (Branch 4), or it can have a positive slope of $(dP/dT)_c$ at the critical point of component B (Branch 5).[12] Molecules showing a critical line of the shape of Branch 5 have more "dislike" than molecules showing a critical line of the shape of Branch 1.

Unlike what is shown in Figure 3, if the above-mentioned critical branch extends to temperatures lower than the critical pressure of pure component *A* (as shown for branch 1 of Figure 5(a)), then the diverging curves (immiscibility gaps) of Figure 4(b–e) will instead come together to meet at a common critical point, and in doing so, would result in the more commonly recognized closed loop phase envelopes.

Going back to Figure 2, we realize that the CO_2+[bmim][PF_6] system shows a behavior similar to Figure 4(d) at the temperatures shown. In addition, considering the L_1L_2V behavior that it exhibited[8] at temperatures below the critical temperature of CO_2, we believe it to have a behavior similar to Figure 4(c) within a lower temperature range. So, concluding that CO_2+[bmim][PF_6] most likely has Type III behavior (although Types IV and V should not be excluded as possibilities), we now know the fascinating kinds of behaviors that may be expected of this system *outside* the regions measured so far (any of the schematic behaviors of Figs. 4(b) to 4(e) are possible).

Also worth paying special attention to is the shape of the dew point curve of Figure 4(d). The region just above the "nose" on the right side of this diagram, where the slope of the two-phase boundary is negative, illustrates a type of behavior sometimes associated with enhanced solubility of a non-volatile solute (IL) in a supercritical solvent (CO_2)[13]. With increasing pressure, there is a dramatic increase in the solubility of the non-volatile phase in the gas phase, shown by negative slope of the gas boundary in this region. A statistical mechanics-based study by Kroon *et al.*[14] has in fact predicted the presence of some IL in the supercritical CO_2 phase at pressures higher than 10 MPa. Also, a recent experimental study[15] indicated rather high solubilities of phosphonium-based ILs in CO_2, for example, a solubility of up to about 7 mass% (equivalent to 6.6×10^{-3} mol%) trihexyltetradecylphosphonium chloride in CO_2. Paulaitis and co-workers[13] mention that at much higher pressures, the trend toward enhanced solubility is usually reversed and the gas phase boundary again takes on a positive slope as shown in Figure 4(d). Although such dew point behavior has not yet been observed in any CO_2+IL systems, it is well advised to be aware of its possibility. Especially when the major advantage of ILs is considered to be their lack of solubility in CO_2, such a behavior can pose a serious drawback.

The specific type of behavior, with high CO_2 solubility at low pressures but a steep P-x curve at higher CO_2 concentrations is not only limited to CO_2+[bmim][PF_6], but seems to be the predominant type of behavior with most binary CO_2+IL mixtures. Aki and co-workers[16] have compiled a number of CO_2+[bmim]-based IL systems, all showing such solubility trends. The different anions of their study included [NO_3], [DCA], [BF_4], [TfO], [PF_6],

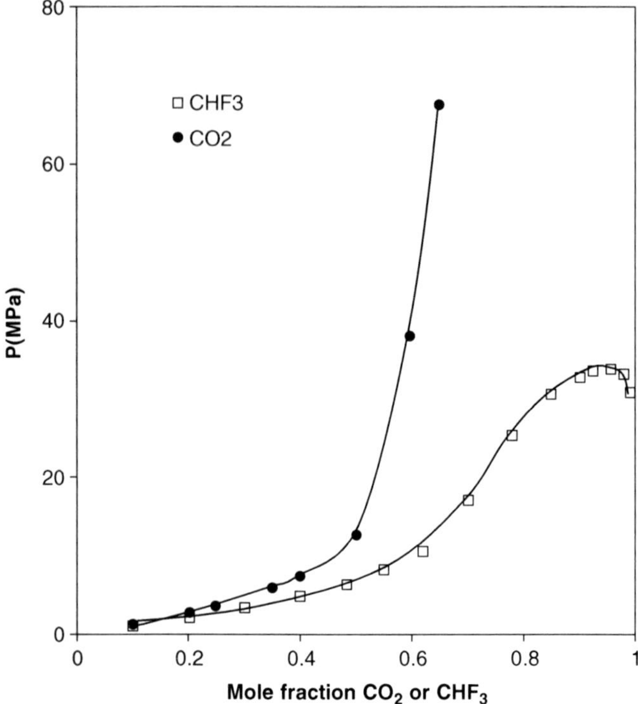

Figure 6 *Comparison between the phase behaviors of the systems CHF_3+[bmim][PF_6] and CO_2+[bmim][PF_6] at 340 K.*

[Tf_2N], and [methide]. Although it is tempting to generalize, one should be aware that not all gas+IL systems will necessarily behave this way. For example, a completely different solubility curve, with a precautionary message warning against the general belief of IL immiscibility in gases, was discovered by Shariati and Peters[17] and Shariati *et al.*[8] when they measured the phase behaviors of the systems CHF_3+[emim][PF_6] and CHF_3+[bmim][PF_6]. Their experimental results illustrated that both of these systems exhibit closed-dome P-x-y solubility curves, in contrast to the previously measured gas+IL systems. Figure 6 compares, for example, the P-x diagrams of the systems of CO_2+[bmim][PF_6] and CHF_3+[bmim][PF_6] at 340 K. The completely different phase behaviors are evident in this diagram. The closed loop of the system CHF_3+[bmim][PF_6] evidences the occurrence of a critical point at the maximum of the curve, in addition to the noticeable solubility of [bmim][PF_6] in CHF_3, whereas CO_2+[bmim][PF_6] binary systems have immiscibility gaps between their supercritical phases and the IL-rich phases even up to very high pressures, with no measurable [bmim][PF_6] solubility in CO_2. This can be due to stronger molecular interactions between CHF_3 and the ILs compared to CO_2 and the ILs. CHF_3 has a permanent dipole moment (=1.65 debye[18]) while CO_2 has no dipole moment. In addition, CHF_3+[bmim][PF_6] also showed[8] a three phase L_1L_2V region close to the vapor pressure curve of pure CHF_3 with a critical endpoint of the nature $L_1 = V+L_2$. Therefore the system CHF_3+[bmim][PF_6] may also have any of Type III, IV, or V of the Scott and van Konynenburg[10] phase behavior, but as explained more by Shariati *et al.*[8], Type III is the most probable type of behavior for this system as well.

The P–T projection of the critical locus of CHF_3+[bmim][PF_6] has a positive slope, as shown in Figure 5(b). This critical line is located between the critical temperatures of CHF_3 and that of the IL. Therefore we can expect[8] the critical loci of the system CHF_3+[bmim][PF_6] to have the shape of either Branch 1 or 2 in Figure 5(a). As discussed extensively by Levelt Sengers,[12] binary mixtures of a strongly interacting solvent and a volatile component can have critical lines that run to much lower temperatures and pressures than binary systems of the same solvent with a less interacting volatile molecule. Therefore, the critical locus of the less interactive CO_2 with [bmim][PF_6] may be expected[8] to have the shape of either Branch 3 or 4 in Figure 5. These branch shape speculations also confirm with the closed loop shape of CHF_3+[bmim][PF_6] and the immiscibility gap of CO_2+[bmim][PF_6] in Type III phase behavior, as discussed earlier.

And yet another example of differing phase behavior is the recent unpublished experimental data from our group, which indicate that the temperature dependence of hydrogen solubility in several IL families is the reverse of that with CO_2, *i.e.*, whereas CO_2 solubilities in ILs decrease with increasing temperature, H_2 shows better solubility in the investigated systems as temperature is increased.

9.3 Molecular Interactions

Understanding the nature of CO_2–IL interactions at a molecular level is also vital for further developments concerning such mixtures. In an IL, the anions and cations form ion pairs due to strong Coloumbic interactions that keep them closely associated, even in systems diluted with CO_2. Therefore, IL molecules are considered to be highly asymmetric neutral ion pairs with large dipole moments as a result of the charge distribution over the ion pair.[14] On the other hand, CO_2 molecules have quadruple moments. So it is expected that the interactions between CO_2 molecules and IL anions should be of primary importance in solubility.

Using ATR-IR spectroscopy for CO_2+[bmim][PF_6] and CO_2+[bmim][BF_4], Kazarian and co-workers[19] suggested that the high solubility of CO_2 in ILs results from weak Lewis acid–base complexation between CO_2 (the electron-pair acceptor) and the IL anion (the electron pair donor). They did not find spectroscopic evidence of specific interactions of CO_2 with the cation. They noted that the CO_2-anion interaction is stronger in [bmim][BF_4], while the solubility of CO_2 in [bmim][BF_4] is less than in [bmim][PF_6]. This being consistent with their experimental solubility results, Blanchard and co-workers[9] concluded that the relatively high CO_2 solubility in the ILs with fluorinated anions is at least in part due to weak Lewis acid–base complex formations, but that additional factors such as free volume, are important in determining the ultimate solubility. In fact, by noticing the roughly linear correlation between the liquid molar volume of the IL and CO_2 solubility, as well as the almost linear CO_2 solubility as a function of pressure, Blanchard and co-workers[9] suggested a "space-filling" mechanism. Once the void space within the IL is saturated, very little CO_2 can be further dissolved in the IL, even under very high pressures.

Cadena and co-workers[20] used a combined experimental and molecular simulation approach to understand the factors governing the high solubility of CO_2 in alkylimidazolium-based ILs. In agreement with previous studies, they found that the anion dominates the interactions with CO_2, with the cation playing a secondary role. Their simulations indicated that CO_2 organizes strongly about the [PF_6] anion in a "tangent-like" configuration that maximizes favorable interactions, but is more diffusely distributed about the imidazolium ring. Their results suggested that the best indicator of CO_2 solubility in alkylimidazolium-based ILs is the association of CO_2 with the anion.

Huang *et al.*[21] interpreted CO_2 solubility in [bmim][PF_6] in terms of spontaneously forming cavities in the IL phase, and they proposed that CO_2 occupies extremely well-defined locations in the IL. In fact, they claimed that this dissolution is characterized by a process very similar to percolation through a "quasi-static" glassy material. This view is consistent with observations by Hu and Margulis[22] and by Popolo and Voth[23] describing the non-Gaussian characteristics of ILs. This semirigid and sticky glassy structure is a result of the strong Coulombic attractions between the ions. Through their simulations, Huang and co-workers[21] found that most of the space occupied by CO_2 in the IL phase consists of very localized cavities of larger size than those spontaneously forming in the neat IL. They further suggested that the cavities, which are occupied by CO_2, are for the most part not generated by expansion of the [bmim][PF_6] phase, but instead they are formed by small angular rearrangements of the anions. With these small angular rearrangements that do not significantly change radial distribution functions in the liquid, CO_2 is able to fit above and below the imidazolium ring. CO_2 is also typically found close to the long alkyl tail of the imidazolium ring. The partial molar volume of CO_2 is so low that CO_2 molecules dissolving in the IL phase occupy a space that is nearly equivalent to the sum of their van der Waals volume and the liquid structure of [bmim][PF_6] in the presence of CO_2 is nearly identical to that of the neat IL, even at fairly large mole fractions of CO_2. Huang and co-workers[21] believe that the sudden change of slope on P-x diagrams, above which little CO_2 can be further dissolved even by considerable increases in pressure, occurs at a "maximum" concentration. The liquid structure of the ions would have to change significantly in order to accommodate more CO_2.

Recently, Kroon *et al.*[14] used the Perturbed Chain Polar Statistical Associating Fluid Theory equation of state to successfully model the "double-slope" behavior of CO_2 in various 1-alkyl-3-methylimidazolium-based ILs. This equation accounts explicitly for the microscopic characteristics of ILs and CO_2. Their study indicated that the dominant interactions in the nearly flat portion of the solubility curve are the polar and dispersive interactions between the IL molecules themselves, while it is the Lewis acid–base association between the CO_2 molecules and the anions of the ILs that contribute the most in calculating the nearly vertical portion of the curve.

Blanchard and co-workers[9] have observed that while large amounts of CO_2 dissolve into the IL phase, the normal volume expansion of the liquid is not observed. They hypothesized that this was due to strong Coulombic forces between the ions such that separation of those ions would result in too large a thermodynamic penalty. Since the liquid phase does not noticeably expand upon the solution of CO_2, the two phases will never become identical; i.e., a mixture critical point will never be reached. Because of this phenomenon, CO_2–IL systems remain two-phase even at extremely high pressures[9].

9.4 Effect of Anions

Figure 7 compares the P-x diagrams of three commonly studied binary systems of CO_2 in ILs differing only in their anions, namely, [bmim][BF_4],[24] [bmim][PF_6],[8] and [bmim][Tf_2N][25] at 333.15 K. The same trend is also observed at other temperatures. Although all three systems show the same type of behavior, it is immediately clear that the choice of anion has a dramatic effect on the extent of CO_2 solubility. As mentioned previously in Section 3, CO_2 solubility depends primarily on the strength of interactions of CO_2 with the anion. In Figure 7, it is evident that solubility increases in the ILs in the following order of anions [BF_4]<[PF_6]<[Tf_2N]. Aki and co-workers[16] have made similar comparisons for a wider range of anions and their results indicate that the solubility of CO_2 in [bmim] cation-based ILs increase in the following

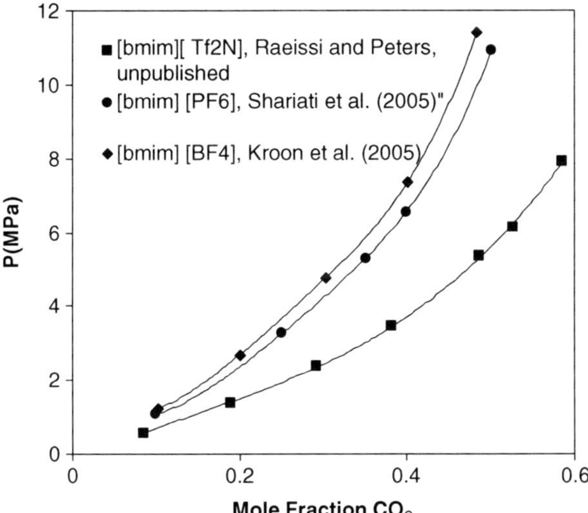

Figure 7 *Comparison of binary systems of CO_2 with ILs having the same cation but different anions at 333.15 K.*[8,24,25]

order $[NO_3]<[DCA]<[BF_4]\sim[PF_6]<[TfO]<[Tf_2N]<$[methide]. They explained the very high solubility in [TfO] and $[Tf_2N]$ to be attributed to the CO_2-philic nature of the fluoroalkyl groups. In fact, CO_2 solubility increases with increasing number of CF_3 groups in the anion. For instance, Aki and co-workers[16] mentioned that at 20 bars, the solubility of CO_2 in [bmim][methide] is 40% greater than in [bmim][TfO]. In addition, they showed that the CO_2 solubility of the investigated samples does not correlate with the measure of basicity or hydrogen-bond strength. In view of this, Aki and co-workers[16] concluded that acid/base interactions of CO_2 with anions are only one mechanism of interaction. Carbon dioxide interactions with fluorous alkyl chains and the molar volumes of ILs may also be important, such that ILs with larger molar volumes will contribute less to the solubility parameter, and hence to higher CO_2 solubility.

9.5 Effect of Cation Alkyl Chain Length

Figure 8 shows the effect of the length of the alkyl chain group on gas solubility at 333.15 K in the systems of CO_2+[1-alkyl-3-methylimidazolium]-based cations together with the following anions: $[BF_4]$,[24,26,27] $[PF_6]$,[8,28,29] and $[Tf_2N]$.[25] It is evident that the size of the alkyl side chain of the cation does indeed affect solubility, however, the effect is not as pronounced as the substitution of the anion discussed in the previous section. In all three systems shown, gas solubility increases with increasing alkyl chain length at all pressures, for instance, at a pressure of 40 MPa, the solubilities of the [bmim], [hmim] and [omim] members of the $[BF_4]$ family are approximately 57, 65, and 70 mol%, respectively. Aki and co-workers[16] explained this based on the decreasing densities of imidazolium-based ILs with increasing alkyl chain length. The greater free volume in ILs with longer alkyl chains allows for more CO_2 to dissolve. As seen in Figure 8, the differences are more distinct at higher CO_2 concentrations and higher pressures. Similar trends are observed at other temperatures as well. There may exist an almost linear relationship between the alkyl chain length and the solubility of CO_2 in such ILs. This is seen for example in Figure 9, for the 1-alkyl-3-methylimidazolium hexafluorophosphates at 333.15 K and 10 MPa.[30]

9.6 Substitution at the C2 Position

Aki and co-workers[16] indicated that replacement of the acidic hydrogen on the C2 carbon of the alkylimidazolium cation with a CH_3 group decreases the solubility marginally at low pressures, but this decrease becomes more apparent at high pressures. In a combined experimental and molecular simulation approach, Cadena and co-workers[20] suggested that replacing the C2 carbon of the [bmim] cation with a methyl group leads to a reduction of the experimental enthalpy of absorption by about 1–3 kJ mol^{-1} and a modest loss of

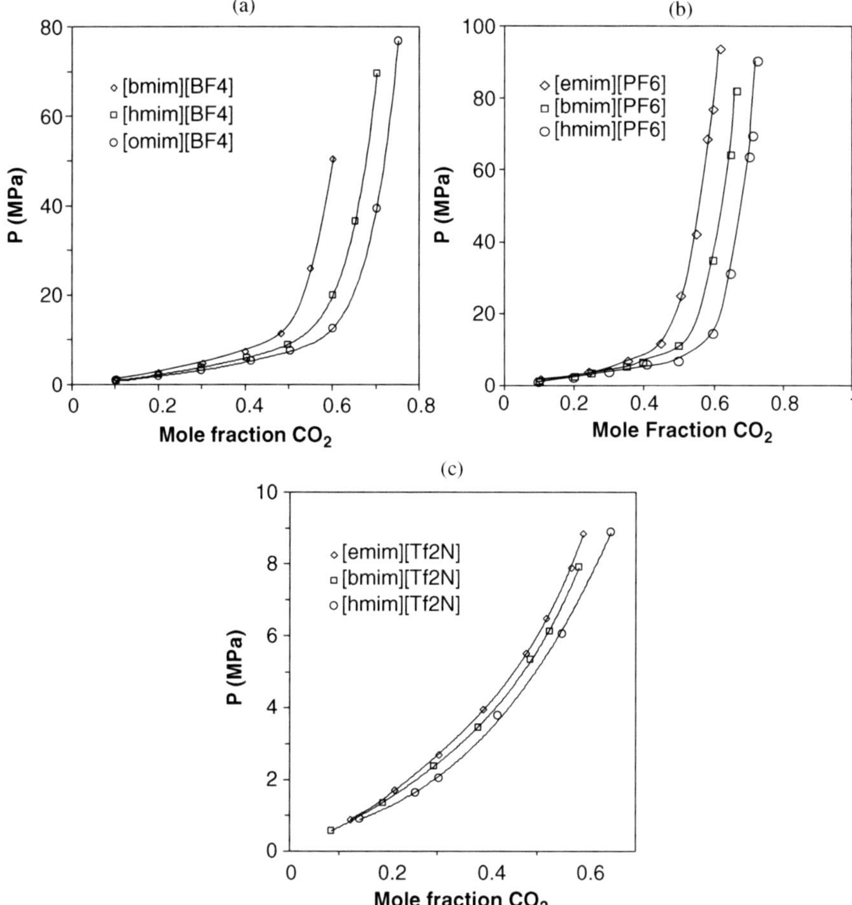

Figure 8 *Comparison of effect of alkyl side chain in (a) CO_2+[1-alkyl-3-methylimidazolium][BF_4],[24,26,27] (b) CO_2+[1-alkyl-3-methylimidazolium][PF_6],[8,28,29] and (c) CO_2+[1-alkyl-3-methylimidazolium][Tf_2N][25] at 333.15 K.*

organization of the anion and CO_2 about the cation. These changes should have a small effect on overall solubility. This is because most of the CO_2 is located relatively far away from the cations (ca. 5Å or more). Subtle differences of 0.2–0.4 Å in interaction distance will not result in large enough energetic differences to influence solubility to a great amount.

9.7 Effects of Impurities

Brennecke and co-workers[9,31,32] compared the phase behavior of CO_2 in both dry and water-saturated samples of [bmim][PF_6]. They reported[9] a dramatic difference of solubility between the two samples, for example, the CO_2

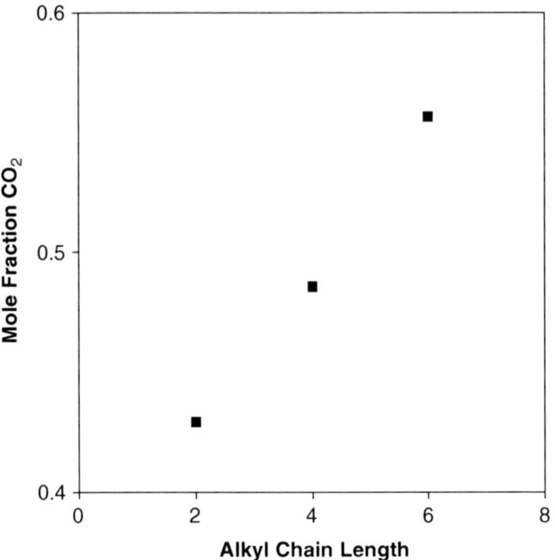

Figure 9 *The effect of alkyl chain length on the solubility of CO_2 in 1-alkyl-3-methyl-imidazolium hexafluorophosphate at 333.15 K and 10 MPa.*[30]

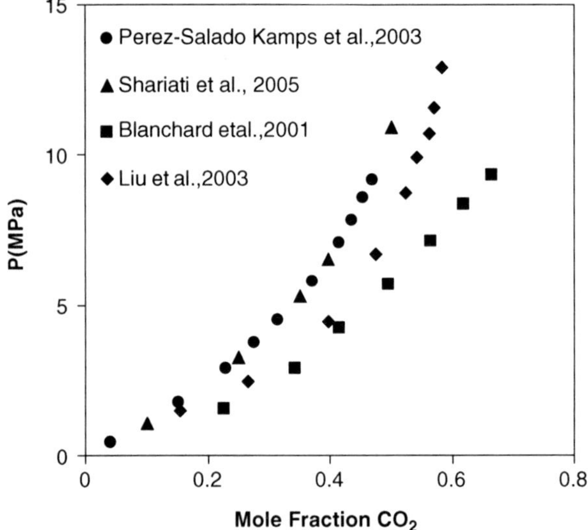

Figure 10 *Comparison of experimental literature data on the system CO_2+[bmim][PF_6] at 333.15 K.*

solubility at 313.15 K and 5.7 MPa was 0.54 in the mole fraction of CO_2 in the dried sample (having approximately 0.15 wt% water), compared to only 0.13 in the water-saturated sample (having up to 2.3 wt% water). The solubility of CO_2 in [bmim][PF_6] at a temperature of 333.15 K, as measured by several different research groups is collectively shown in Figure 10. The differences in these results are probably due not only to the different experimental techniques used for measurement, but also partly to differing sample purities.

However, when the same group[16] carried out similar comparative experiments with [Tf_2N]-based ILs, they noticed that the presence of water had essentially no effect on the solubility of CO_2. In this case, the two samples were dried [bmim][Tf_2N] with 450 ppm water content *vs* water-saturated [bmim][Tf_2N] with 13500 ppm water, equivalent to 1.35 wt% or 24.2 mol%.

These results indicate that impurities in particular (even in small amounts), and multiple components in general, may or may not have profound effects on solubilities in ILs. In any case, it is important to be aware and take such effects into consideration in any IL solubility study.

9.8 Conclusions and Summary

Because of the often-complicated nature of different phase behavior diagrams, the more practically-minded scientists mostly neglect or avoid such fundamental issues. In this chapter, we have shown that lack of such knowledge can lead to seriously incorrect generalizations. Reading the introductory paragraph of almost any IL-related article, it is nearly impossible not to find a statement about the lack of IL vapor pressure or the negligible solubility of ILs in gases. However, we have shown through a detailed review of Scott and van Konynenburg's Type III phase behavior that, in principle, it is indeed possible to have regions where there is a significant increase of IL solubility in the gaseous phase. In addition to a theoretical explanation, we have also given actual examples of experimentally measured binary systems of gas+IL in which there is an appreciable amount of gas "contamination" by the IL. A statistical mechanics-based study has also predicted such presence of IL in supercritical CO_2. So the miraculously green ILs may be too good to be true after all, at least at some particular conditions of pressure and temperature.

Further experimental evidence within wider ranges of temperatures and pressures and for a greater variety of ILs is vital for revealing more about possible phase behaviors. The nearly non-existent solubility data of a decade ago on the phase behavior of supercritical gases in ILs is fortunately growing now.[8,9,17,24–30,33–37] Although data on tens of other systems than those presented in this work are currently available in literature, it was not our intention to make an inventory. Rather, this chapter is focused on giving a general *understanding* of the phase behavior of such systems. For this purpose, we have limited the systems to three analogous families for the sake of simplicity. The solubility of CO_2 in nine imidazolium-based ILs with the [BF_4], [PF_6], and [Tf_2N] anions were discussed in order to compare the effects of the anion and the alkyl chain size of the cation. All these systems showed a similar type of phase behavior. Carbon dioxide showed very good solubilities in every system at lower pressures while solubilities of the corresponding ILs in CO_2 were immeasurable. Solubility increased with increasing pressure in all the nine systems, however after surpassing a certain CO_2 concentration, there was very little gain in solubility with pressure, indicative of immiscibility gaps, at least up to the pressures investigated. This behavior, however, should in no way be taken as a generalization because closed loop phase behaviors have been detected for at least a two gas+IL systems. In every system, the solubility of CO_2 decreased with increasing temperature, but this should also not be generalized for all supercritical gases since hydrogen gas seems to show the reverse temperature trend. Comparison of the different analogous systems, in addition to spectroscopic studies, molecular simulations, and statistical mechanics-based predictions, suggests the strong dependence of CO_2 solubility on the choice of the anion, being much higher for the systems with Tf_2N. It seems that solubility increases with increasing fluoroalkyl groups in the anion. The size of the cation side-chain also affects solubility but the effect is not as dramatic as with the choice of anion. There seems to be a linear relationship between the increase of the alkyl chain length and the increase of solubility of CO_2 in, at least, the first few members of the 1-alkyl-3-methylimidazolium hexafluorophosphate ILs. So, as suggested by Aki and co-workers,[16] one can increase CO_2 solubility in ILs by increasing alkyl chain length on the cation, as well as by adding expensive CF_3 groups to the anion. Replacement of the C2 hydrogen of the cation with a CH_3 group also has a minor CO_2 solubility decreasing effect.

References

1. E.D. Bates, R.D. Mayton, I. Ntai and J.H. Davis, Jr., *J. Am. Chem. Soc.*, 2002, **124**(6), 926.
2. J.F. Brennecke and E.J. Maginn, *AIChE J.*, 2001, **47**, 2384.
3. L.A. Blanchard and J.F. Brennecke, *Ind. Eng. Chem. Res.*, 2001, **40**, 287.
4. P. Scovazzo, A.E. Visser, J.H. Davis, R.D. Rogers, C.A. Koval, D.L. Dubois and R.D. Noble, *ACS Symp. Ser.*, 2002, **818**, 69.
5. A.M. Scurto, S.N.V.K. Aki and J.F. Brennecke, *J. Am. Chem. Soc.*, 2002, **124**, 10276.
6. A.M. Scurto, S.N.V.K. Aki and J.F. Brennecke, *Chem. Comm.*, 2003, Issue 5, 572.
7. P. Scovazzo, Jesse Kieft, D.A. Finan, C. Koval, D. Dubois and R. Noble, *J. Membrane Sci.*, 2004, **238**, 57.
8. Shariati, K. Gutkowski and C.J. Peters, *AIChE J.* 2005, 51, 1532.
9. L.A. Blanchard, Z. Gu and J.F. Brennecke, *J. Phys. Chem. B*, 2001, **105**, 2437.
10. R.L. Scott and P.H. Van Konynenburg, *Discuss Faraday Soc.*, 1970, **49**, 87.
11. M.A. McHugh and V.J. Krukonis, *Supercritical Fluid Extraction; Principles and Practice*, 2nd edn., Butterworth–Heinemann series in chemical engineering, Stoneham, MA, 1994.
12. J.M.H. Levelt Sengers, *J. Supercritical Fluids*, 1991, **4**, 215.
13. M.E. Paulaitis, J.M.L. Penninger, R.D. Gray Jr., and P. Davidson, *Chemical Engineering at Supercritical Fluid Conditions*, Ann Arbor Science Publishers, Ann Arbor, Michigan, 1983.

14. M.C. Kroon, E.K. Karakatsani, I.G. Economou, G.J. Witkamp and C.J. Peters, *J. Phys. Chem. B*, 2006, **110**, 9262.
15. J.W. Hutchings, K.L. Fuller, M.P. Heitz and M.M. Hoffmann, *Green Chem.*, 2005, **7**, 475.
16. S.N.V.K. Aki, B.R. Mellein, E.M. Saurer and J.F. Brennecke, *J. Phys. Chem. B*, 2004, **108**, 20355.
17. A. Shariati and C.J. Peters, *J. Supercritical Fluids*, 2003, **25**, 109.
18. H. Yuan and S.V. Olesik, *Anal. Chem.*, 1998, **70**, 1595.
19. S.G. Kazarian, B.J. Briscoe and T. Welton, *Chem. Commun.*, 2000, **20**, 2047.
20. C. Cadena, J.L. Anthony, J.K. Shah, T.I. Morrow, J.F. Brennecke and E.J. Maginn, *J. Am. Chem. Soc.*, 2004, **126**, 5300.
21. X. Huang, C.J. Margulis, Y. Li and B.J. Berne, *J. Am. Chem. Soc.*, 2005, **127**, 17842.
22. Z.H. Hu and C.J. Margulis, *Proceedings of the National Academy of Sciences of the United States of America*, 2006, **103**(4), 831.
23. M.G.D. Popolo and G.A. Voth, *J. Phys. Chem. B*, 2004, **108**, 1744.
24. M.C. Kroon, A. Shariati, M. Costantini, J. Van Spronsen, G.J. Witkamp, R.A. Sheldon and C.J. Peters, *J. Chem. Eng. Data*, 2005, **50**, 173.
25. unpublished data from our group (to be published).
26. M. Constantini, V.A. Toussaint, A. Shariati, C.J. Peters and I. Kikic, *J. Chem. Eng. Data*, 2005, **50**, 52.
27. K. Gutkowski, A. Shariati and C.J. Peters, *J. Supercrit. Fluids*, 2006, in press.
28. A. Shariati and C.J. Peters, *J. Supercrit. Fluids*, 2004, **29**, 43.
29. A. Shariati and C.J. Peters, *J. Supercrit. Fluids*, 2004, **30**, 139.
30. A. Shariati and C.J. Peters, *J. Supercrit. Fluids*, 2005, **34**, 171.
31. L.A. Blanchard, D. Hancu, E.J. Beckman and J.F. Brennecke, *Nature*, 1999, **399**, 28.
32. J.L. Anthony, E.J. Maginn and J.F. Brennecke, *J. Phys. Chem. B*, 2001, **105**, 10942.
33. J.L. Anthony, J.L. Anderson, E.J. Maginn and J.F. Brennecke, *J. Phys. Chem. B*, 2005, **109**, 6366.
34. P. Husson-Borg, V. Majer and M.F. Costa Gomes, *J. Chem. Eng. Data*, 2003, **48**, 480.
35. A.P.S. Kamps, D. Tuma, J. Xia and G. Maurer, *J. Chem. Eng. Data*, 2003, **48**, 746.
36. F. Liu, M.B. Abrams, R.T. Baker and W. Tumas, *Chem. Commun.*, 2001, Issue 5, 433.
37. D. Camper, P. Scovazzo, C. Koval and R. Noble, *Ind. Eng. Chem. Res.*, 2004, **43**, 3049.

Modelling and Simulation

CHAPTER 10

Solubility and Molecular Modelling

MARGARIDA F. COSTA GOMES AND AGÍLIO A.H. PÁDUA

Laboratoire de Thermodynamique des Solutions et des Polymères, CNRS/ Université Blaise Pascal, Clermont-Ferrand, 24 avenue des Landais, 63177, Aubière, France

10.1 Introduction

Hildebrand and Scott have stressed more than 40 years ago[1] that "solubility has a much wider scope than the term itself" because at the fundamental level it concerns the nature and strength of the intermolecular forces, affecting in this way many other physical properties and phenomena of scientific and practical interest. Besides providing a useful way to understand the interactions in solution, solubility can also bring significant information on the microscopic structure of the solutions. These are presently active research domains in physical and biophysical chemistry. Concerning the applications, solubility data are vital to the calculation of phase equilibria in problems of technological and industrial interest namely in the design of chemical engineering processes and, of course, solubility is one of the key properties determining the fate of substances in the environment. The scientific and technological aspects are often closely associated, for example, in the search for new strategies to improve the choice of novel solvents or separation media for chemical reactions or industrial processes. In a society demanding less hazardous and more efficient chemical industries, the choice of alternative solvents (acceptable both from an economic and an environmental point of view) for reactions or separations is regarded as one of the promising ways to advance.[2]

The purpose of the present chapter is to show how concepts of classical thermodynamics and macroscopic experimental information can be combined with molecular modelling tools to provide a better understanding of solubility phenomena. This two-way approach brings mutual benefits, since experimental data are still essential to set up the best molecular models that, in turn, can offer detailed microscopic-level insights through the use of atomistic simulation methods. Solubility is not just simply determined by the energies between molecules, but also by configurational aspects resulting from the particular organization of the solvent molecules around the solute, or by conformational changes in the solute itself. This entropic contribution may even be the predominant factor and cannot be ignored if a full understanding and prediction is sought. Consideration of such structural features requires modelling techniques where the bulk solvent is explicitly represented and the theoretical methods appropriate to handle condensed phases are based on statistical mechanics.

This chapter is organized in three parts. Some aspects of the classical thermodynamics of solutions are introduced first, particularly those that establish the connection to quantities that are directly accessible to molecular modelling techniques. Although the thermodynamic formalism of solution processes is an old and meticulously worked out subject, that link between experimental thermodynamics and calculation by molecular models is not always straightforward to establish. In the second part the fundamental tools used in the molecular modelling of solubility are introduced, together with an explanation of atomistic simulation techniques and free energy routes that enable the calculation of chemical potentials. This property that is only indirectly accessible to experimental thermodynamics is readily available by simulation. In both fields the chemical potential is the key to study solubility. In the third part, an application example is given involving a novel type of solvent with promising applications as reaction of separation medium. The solvation of carbon dioxide by several ionic liquids sharing the same cation is analysed in terms of solute–solvent interactions and microscopic structure.

10.2 Thermodynamics of Solution

The chemical potential of component i in a mixture[3,5] is

$$\mu_i \equiv \mu_i^{\text{ref}} + RT \ln (\gamma_i x_i) \tag{1}$$

Equation (1) defines the activity coefficient γ_i in terms of the choice of the chemical potential in a reference state μ_i^{ref}. It gives the freedom to specify different combinations of reference state chemical potential and activity coefficient, according to our convenience. For example, both quantities become unambiguously determined when the situation in which activity coefficient becomes unity is specified. Two conventions[3] are usually adopted, based on the assumptions that a component of a real mixture approaches ideal behaviour either when its mole fraction approaches unity or zero. When that mixture is seen as a solution, that is, one of the components is present in a much larger quantity, which convention is adopted in practice, depends largely on the state of the pure components at the thermodynamic conditions of the mixture, but this is also a matter of tradition among certain communities of researchers. For example, if in a binary system, both components are liquids then the symmetric convention is more often used; if one component is a gas or a solid (the solute) and the other is a liquid (the solvent) then the asymmetric convention is more natural.

In the symmetric convention the activity coefficient of each component (solute or solvent) approaches unity as its mole fraction approaches unity: $\gamma_i^{\mathrm{R}} \to 1$ as $x_i \to 1$ for all i. In these limits the mixture approaches ideal behaviour in the sense implied by Raoult's law.[3] This convention is usually adopted for mixtures where all the pure components are in the same physical state as the mixture at the same temperature and pressure, for example, a liquid solute dissolved in a liquid solvent to form a liquid solution. The reference chemical potential in this convention, $\mu_i^{*\mathrm{R}}$, is equal to the molar Gibbs free energy of the pure substance i at the same temperature, pressure and physical state as the mixture, and the activity coefficient reflects this choice

$$\mu_i = \mu_i^{*\mathrm{R}} + RT \ln (\gamma_i^{\mathrm{R}}\, x_i) \quad x_i \to 1 \Rightarrow \gamma_i^{\mathrm{R}} \to 1 \tag{2}$$

This activity coefficient accounts for deviations from ideal behaviour owing to the unlike interactions between the different components not being similar to the interactions found in the pure substances. If the symmetric convention is applied to a system where a pure substance is not in the same physical state as the mixture, then the reference state of that component will not be the real pure substance, but instead a hypothetical pure substance having same physical state as the mixture.

The asymmetric convention is preferably applied to solutions, understood as mixtures where some of the components are not in the same physical state as the solution at a given temperature and pressure, for example, a gas or a solid dissolved in a liquid solvent to form a liquid solution. In these cases it is convenient to distinguish between the solvent, which in a binary system is generally the component present in large excess, and the solute, present in a smaller concentration. The solvent is not necessarily a pure substance and there may also be several solutes in a multi-component system. For a solvent, the activity coefficient approaches unity when its mole fraction is approximately unity (the situation is the same as in the symmetric convention). But for a solute, $\gamma_i^{\mathrm{H}} \to 1$ when $x_i \to 0$, meaning that its activity coefficient becomes unity in the limit of infinite dilution: the solution approaches ideal behaviour in the sense of Henry's law.[3,4] In this convention, the reference chemical potential, $\mu_i^{*\mathrm{H}}$, no longer refers to the pure solute but instead stands for the chemical potential of the solute in a hypothetical reference state related to the condition of infinite-dilution. This reference state is obtained by extrapolation of the infinite-dilution limit: the solute would remain infinitely dilute (no solute-solute interactions are present) but its mole fraction would be unity (like if it would become "pure while infinitely dilute").[†]

$$\mu_i = \mu_i^{*\mathrm{H}} + RT \ln (\gamma_i^{\mathrm{H}}\, x_i) \quad x_i \to 0 \Rightarrow \gamma_i^{\mathrm{H}} \to 1 \tag{3}$$

The activity coefficient in the asymmetric convention accounts for the presence of solute–solute interactions, since solute–solvent and solvent–solvent interactions are already present in the reference state.

[†] The reference state of the solute in the asymmetric convention should not be mistaken with the chemical potential "at infinite dilution" since, according to Equation (2), the value of the later is necessarily $-\infty$.

The state of infinite dilution may be expressed equally well using the symmetric conventions, only here the activity coefficient will take its value at infinite dilution: $\gamma_i^{\mathrm{R}} \to \gamma_i^{\mathrm{R}\infty}$ when $x_i \to 0$. The reference states and the activity coefficients in the two conventions can be related: suppose that the quantity $RT \ln \gamma_i^{\mathrm{R}\infty}$ is added and subtracted from Equation (2), then

$$\mu_i = \mu_i^{*\mathrm{R}} + RT \ln \gamma_i^{\mathrm{R}\infty} + RT \ln\left(\frac{\gamma_i^{\mathrm{R}}}{\gamma_i^{\mathrm{R}\infty}} x_i\right) \tag{4}$$

and when $x_i \to 0$, and accordingly $\gamma_i^{\mathrm{R}}/\gamma_i^{\mathrm{R}\infty} \to 1 = \gamma_i^{\mathrm{H}}$, the identity

$$\mu_i^{*\mathrm{H}} = \mu_i^{*\mathrm{R}} + RT \ln \gamma_i^{\mathrm{R}\infty} \tag{5}$$

is obtained by comparing Equations (3) and (4).

The Gibbs energy of solution $\Delta_{\mathrm{sol}}\, G_i$ is defined as being the difference in chemical potential when transferring the solute, at constant pressure and temperature, from its pure state into the infinitely dilute solution but retaining a mole fraction of unity, that is, into the reference state of the asymmetric convention:

$$\Delta_{\mathrm{sol}}\, G_i \equiv \mu_i^{*\mathrm{H}} - \mu_i^{*} \tag{6}$$

In the departure state the pure solute may be a solid, a liquid or a gas. Let us suppose that solubility equilibrium is realized in practice by allowing phase coexistence between the solute in its stable physical state and the solution. If the solute is a solid it normally remains pure. But if it is liquid, then there may be mutual solubility of the "solvent" in the "solute"-rich phase. If it is a gas, then there may be some evaporation of the solvent into the headspace, depending on its vapour pressure. In these last two situations a complication arises when trying to relate the Gibbs energy of solution with experimentally determined solubilities: the free-energy difference between the pure solute and the actual solute-rich phase that exists in equilibrium with the solution has to be computed. If the solute remains pure at equilibrium with the solution and its measured solubility x_i^{sol} is low enough that $\gamma_i^{\mathrm{H}} \approx 1$, then from Equation (3) and (6) the following approximate, practical relation can be derived:

$$\Delta_{\mathrm{sol}}\, G_i \approx -RT \ln x_i^{\mathrm{sol}} \tag{7}$$

The Gibbs energy of solution may also be related to quantities expressed in the symmetric convention. If the pure solute is in the same physical state as the solution, then from Equation (5),

$$\Delta_{\mathrm{sol}}\, G_i = \mu_i^{*\mathrm{H}} - \mu_i^{*\mathrm{R}} = RT \ln \gamma_i^{\mathrm{R}\infty} \tag{8}$$

Comparing this exact result with Equation (7) yields $\gamma_i^{\mathrm{R}\infty} \approx 1/x_{\mathrm{i}}^{\mathrm{sol}}$ to the same levels of approximation as assumed in Equation (7), provided that the pure solute is in the same physical state as the solution. The Gibbs energy of solution expresses the difference between the solute–solute interactions in the pure species, which may be a condensed phase, and the solute–solvent interactions in an infinitely dilute solution. It would be interesting to isolate the role of the solute–solvent interactions in the process of dissolution, by defining a thermodynamic transformation called solvation.[5] The Gibbs energy of solvation is defined as the difference in chemical potential when the solute is transferred from an ideal gas at standard pressure p^0, into the reference state at infinite dilution:

$$\Delta_{\text{solv}}\, G_i \equiv \mu_i^{*\text{H}}\ \mu^{\text{ig},0} \tag{9}$$

where $\mu^{\text{ig},0}$ is the chemical potential in the ideal gas state at the standard conditions. (This classical thermodynamic definition differs from the often quoted Ben–Naim's definition of solvation,[6,7] which is based on statistical mechanics. The solvation process defined here corresponds to the "x-process" in Ben–Naim's terms[8]). If the chemical potential of the solute is expressed in terms of its fugacity,[4] f_i,

$$\mu_i = \mu_i^{\text{ig},0} + RT\,\ln\left(\frac{f_i}{p^0}\right) \tag{10}$$

then Equations (3) and (10) lead to

$$\Delta_{\text{solv}} G_i = RT\,\ln\left(\frac{K_{\text{H},i}}{p^0}\right) \tag{11}$$

in which Henry's law constant is defined as $K_{H,i} \equiv \lim_{x_i \to 0}(f_i/x_i)$. This expression of Henry's law constant is general and not restricted to gaseous solutes.

It is possible to relate the Gibbs energy of solution with that of solvation by comparing Equation (6) and (9):

$$\Delta_{\text{solv}}\, G_i = \Delta_{\text{sol}}\, G_i + (\mu_i^{*\text{R}} - \mu_i^{\text{ig},0}) \tag{12}$$

The difference between reference-state chemical potentials in Equation (12) is called the residual chemical potential. It approaches zero for gaseous solutes at low pressure, situation in which the thermodynamic properties of solution are approximately equal to the thermodynamic properties of solvation. The value of the residual chemical potential becomes important for liquid or solid solutes.[9]

From the behaviour with temperature or with pressure of the Gibbs energy of solvation it is possible to calculate the other thermodynamic properties,[10]

$$\begin{aligned}
\Delta_{\text{solv}}\text{H}_i &= -\,T^2 \frac{\partial}{\partial T}\left(\frac{\Delta_{\text{solv}} G_i}{T}\right)_p \\
\Delta_{\text{solv}} S_i &= -\left(\frac{\partial \Delta_{\text{solv}} G_i}{\partial T}\right)_p \\
\Delta_{\text{solv}} V_i &= \left(\frac{\partial \Delta_{\text{solv}} G_i}{\partial p}\right)_T
\end{aligned} \tag{13}$$

The enthalpic and entropic contributions to the Gibbs energy of solvation provide insights into the molecular mechanisms involved; the enthalpic term reflects the energy of molecular interactions between solute and solvent, while the entropic contribution is more directly related to structural organization in the solution.

The conceptual thermodynamic transformations underlying the Gibbs energy of solution and of solvation, Equations (6) and (9), can be related to experimental accessible quantities, namely through measurements of solubility, limiting activity coefficient and Henry's law constant.[9] The same transformations can be realized using the methods of molecular modelling based on statistical thermodynamics. However, in many situations the most natural way of performing such calculations is not at constant pressure and temperature, but at constant volume (density) and temperature. Connecting both paths requires simply the evaluation of the difference in chemical potential between two ideal gas states, for example one at the standard pressure and the other at the same density as the solution.[‡]

10.3 Modelling Solubility

Atomistic simulation is the tool of choice to interpret the solubility and its derived thermodynamic properties explicitly in the light of structure and interactions at a molecular scale. Different molecular methods of calculating solvation properties are available, ranging from first-principles quantum theories to classical models. Although at present the atomistic simulation of condensed phases is possible using a purely quantum-mechanical Hamiltonian,[11] these techniques are still computationally expensive and therefore restricted to small systems over very short timescales.[12] In general, quantum mechanical calculations applied to solvation still represent the solvent by a dielectric continuum[13,14] responding to the electrostatic field created by the solute, hosted in a cavity with appropriate size and shape. The calculated solvation energies agree well with experimental data.[15] The main drawback of these implicit solvent models is their lack of information about the structure of the solvent and how its molecules organize themselves around the solute.

Mixed quantum mechanics/molecular mechanics methods[16,17] are fully discrete methods that overcome the cost of representing all the atoms in the system using quantum mechanics. The basic idea[18] is to retain a quantum mechanical description of the solute whereas the solvent is represented discretely by a classical force field. A quantum treatment of the solute means that polarization effects are taken into account and even chemical transformations on the solute can be considered. A discrete representation of the solvent allows entropic and energetic contributions due to the reorganization of its molecules to be considered. Several variations exist, adapted to specific problems. If the solute is a biological macromolecule such as an enzyme, the subsystem treated by quantum methods may just be an active site, with the remaining solute atoms and the entire solvent treated by classical methods.[18] Conversely, for small solutes that exert a strong polarization influence on the solvent, it may be interesting to treat the solute and the first solvation shells using quantum mechanics and the remote solvent molecules using classical models.[19]

‡ This difference would be $\mu_i^{ig}(\rho) - \mu_i^{ig,0} = RT \ln(\rho RT/p^0)$.

Finally, the entire system can be represented by a classical force field, which is a mathematical description within the scope of classical mechanics of the structure and interactions of the molecules. These are the most efficient of the discrete methods and so a large number of configurations can be generated for systems consisting of thousands of atoms according to statistical ensemble theory: here we are in the field of numeric statistical mechanics.[20] The parameters in a force field can be derived from quantum chemical calculations on isolated molecules and from experimental spectroscopic or thermodynamic data. Although no electronic-structure calculations are performed during a simulation, strategies to include polarization effects explicitly have been adopted in the latest force-field models[21,22] introducing a further degree of realism.

Still larger timescales can be achieved if the all-atom representation is abandoned in favour of a united-atom or even a meso-scale model,[23] where interaction sites no longer correspond to an atom but instead may translate a group of atoms or some repeating unit in a macromolecule, respectively.

The traditional scheme to study solubility phenomena by molecular simulation requires two elements: simulation of the solvent, which is not necessarily a pure liquid, in order to sample sufficiently well its configurational space, and then determination of the chemical potential of the solute in that solvent. The first step is accomplished by the Monte Carlo or molecular-dynamics methods and the second through free-energy routes such as the free-energy perturbation (FEP) or thermodynamic-integration (TI) techniques.

In spite of the finite (actually very small) size of systems that are simulated in molecular modelling, solvation properties can be obtained even in the limit of infinite dilution. If the simulated system is sufficiently large, then the chemical potential, which is a partial molar quantity, can be evaluated by adding one solute molecule to the initial system containing N solvent molecules:

$$\mu_i = \left(\frac{\partial A}{\partial N_i}\right)_{N_j VT} \approx A_{N+1} - A_N \tag{14}$$

where A is the Helmholtz free energy. The connection to statistical mechanics is made through the relation between the Helmholtz free energy and the canonical partition function Q_{NVT} expressed in its classical limit[24]

$$A = -kT \ln Q_{NVT} = -kT \ln\left(\frac{1}{\Lambda^{3N} N!}\int \cdots \int \mathrm{e}^{-U/kT} \mathrm{d}\mathbf{r}^N\right) \tag{15}$$

where $1/\Lambda^{3N} N!$, with $\Lambda = (h^2/2\pi mkT)^{1/2}$, accounts for the translational part and the configurational integral[24] contains the contribution of the potential energy of the system U. The chemical potential can therefore be expressed as

$$\mu_i = -kT \ln \frac{Q_{N+1,VT}}{Q_{NVT}} \tag{16}$$

For simplicity we have taken here, as example, the Helmholtz free energy for a system at constant NVT but other cases, such as the Gibbs free energy G at constant NpT, can be obtain in analogous ways.[8,24] When the simulated solvent is a pure liquid, or even if it is a mixture not containing the solute species, then the free energy difference calculated corresponds to the transfer, at constant volume and temperature, of the solute molecule from an ideal gas state into a state of infinite dilution, since no solute–solute interactions are present. The free energy difference obtained can be directly related to Henry's law constant, as given in Equation (11). In any case, the system considered explicitly needs to be large enough to allow the influence of the solute on the structure of the solvent to vanish before the boundaries of the simulation box are met. For example, an intermolecular potential cut-off distance of 12 Å in water at room temperature requires a simulation box containing about 500 molecules, a small system for today's computers.

10.3.1 Molecular Force Fields

Many different molecular force-field models have been developed and reported in literature,[25–29] each containing parameters that describe several families of organic and biochemical compounds. Some force fields are more specifically targeted at the intramolecular features, their aim being to reproduce accurately geometries, vibration modes, and conformational energies;[30] others are more dedicated to the intermolecular aspects and are suited to calculate properties of condensed phases[26] including phase equilibria.[31]

The functional form of these force fields contains in general four kinds of potential energy: stretching of covalent bonds (between every two bonded atoms), bending of valence angles (between every three atoms connected by two bonds), torsion around dihedral angles (between every four atoms connected by three bonds), and non-bonded interactions. The latter are exerted between atoms of different molecules and also between atoms of the same molecule that are separated by more than three bonds. The potential energy associated with bonds and angles is in the simplest way described by harmonic terms or sometimes by rigid constraints. Dihedral torsion energy profiles are usually translated by series of cosines. Nonbonded terms may be given by the Lennard–Jones 12–6 repulsive-dispersive potential and by electrostatic interactions between partial-point charges placed on the atomic series. The potential energy of the system is expressed by a function like:

$$
\begin{aligned}
U = & \sum_{ij}^{\text{bonds}} \frac{k_{r,ij}}{2}\left(r_{ij} - r_{0,ij}\right)^2 \\
& + \sum_{ijk}^{\text{angles}} \frac{k_{\theta,ijk}}{2}\left(\theta_{ijk} - \theta_{0,ijk}\right)^2 \\
& + \sum_{ijkl}^{\text{torsions}} \sum_{n=1}^{4} \frac{V_{ijkl,n}}{2}\left[1 - (-1)^n \cos(n\phi_{ijkl})\right] \\
& + \sum_{ij}^{\text{nonbonded}} \left\{ 4\varepsilon_{ij}\left[\left(\frac{\sigma_{ij}}{r_{ij}}\right)^{12} - \left(\frac{\sigma_{ij}}{r_{ij}}\right)^{6}\right] + \frac{q_i q_j}{r_{ij}} \frac{e^2}{4\pi\varepsilon_0} \right\}
\end{aligned}
\tag{17}
$$

A specification of the force field requires that bond equilibrium distances and force constants be parameterized, as well as equilibrium valence angles and the respective force constants; that the coefficients of the torsions around dihedral angles be calculated; finally, that van der Waals diameters and well-depths, and partial charges be defined for all sites. Just as an example, a molecule like butan-1-ol contains 15 sites, 14 bonds (1 O–H, 1 C–O, 3 C–C, and 9 C–H), 25 angles (6 H–C–H, 13 C–C–H, 2 O–C–H, 2 C–C–C, 1 C–C–O, and 1 C–O–H), and 30 dihedrals (14 H–C–C–H, 9 C–C–C–H, 2 H–C–C–O, 2 H–C–O–H, 1 C–C–C–C, 1 C–C–C–O, 1 C–C–O–H). The parameters for bonds, angles and torsions can be obtained from *ab initio* quantum–chemical calculations on isolated molecules, and compared to spectroscopic data to validate the geometry, vibration frequencies and internal rotation barriers. The relatively small energy scale and the nature of the non-bonded interactions are such that it is better to obtain them empirically from adjustment to experimental thermodynamic properties such as densities and heats of vaporization. Partial charges can be calculated *ab initio* by algorithms that adjust the charge distribution in order to reproduce the electrostatic field created by the molecule.[32,33] Comparison is possible with the experimental multi-pole moments to verify the results.

The literature on force fields forms today a large body but more often than not some terms necessary to describe certain functional groups, or arrangements of such groups within molecules, are missing not allowing immediate application of the simulation approach to the system under study. Development of some force-field terms, most frequently torsion energy profiles and partial charges, may be necessary in these cases.[34] Nowadays such tasks of force-field development can be accomplished using *ab initio* quantum chemical tools available in the major software packages.

10.3.2 Free Energy Routes

The traditional routes to evaluate the free energy of a solute in solution are the FEP and TI techniques,[35] each presenting advantages and disadvantages.[36] Other methods of calculating free energies exist[20] but they will not be discussed here. Both FEP and TI rely on the introduction of an activation parameter[37] λ in the molecular potential energy U such that the interval $0 \leq \lambda \leq 1$ connects the initial system, the pure solvent, with potential energy U_0 to the final system, the solvent plus one solute molecule, with potential energy U_1:

$$U_\lambda = U_0 + \lambda(U_1 - U_0) \tag{18}$$

A thermodynamic path can now be built over which to calculate the free-energy difference. Depending on the situation, that path is decomposed in a succession of steps, smaller or larger according to the requirements of good statistical sampling. The TI route is based on the integration:

$$\Delta A = \int_0^1 \left(\frac{\partial A}{\partial \lambda}\right) \mathrm{d}\lambda \tag{19}$$

and the derivative in the integrand is developed using Equation (15):

$$\left(\frac{\partial A}{\partial \lambda}\right)_{NVT} = -\frac{kT}{Q_{NVT}}\frac{\partial Q_{NVT}}{\partial \lambda} = \frac{\int \dots \int \frac{\partial U}{\partial \lambda} \mathrm{e}^{-U/kT} \mathrm{d}\mathbf{r}^N}{\int \dots \int \mathrm{e}^{-U/kT} \mathrm{d}\mathbf{r}^N} = \left\langle \frac{\partial U}{\partial \lambda} \right\rangle_{NVT} \tag{20}$$

where $\langle \dots \rangle_{\mathrm{NVT}}$ is an ensemble average.[§] The free energy difference is evaluated numerically:

$$\Delta A = \int_0^1 \left\langle \frac{\partial U_\lambda}{\partial \lambda} \right\rangle \mathrm{d}\lambda \approx \sum_{i=0}^{n} w_i \left\langle \frac{\partial U_\lambda}{\partial \lambda} \right\rangle_{\lambda_i} \Delta\lambda \tag{21}$$

The necessary number n and distribution of intermediate points along the interval depends on the smoothness of the integrand and on the quadrature rule chosen (trapezes, Simpson, Gaussian, *etc.*) that also prescribes the weights w_i.

At each value of the activation parameter an independent simulation run is performed with the solute partially activated to a different extent in order to sample the ensemble average in Equation (21). Let us take as a rule of thumb that the step $\Delta\lambda$ is of the order of 0.1, but in reality it has to be carefully adapted to each application. The independence of the simulations is an advantage, since more can be added if some part of the integrand needs a finer mesh. On the other hand, the TI method is not sensitive to changes in the solute-solvent interactions in a dilute system; the effect of changing the solute–solvent interactions by a small amount as 0.1 may be concealed by the fluctuations of the total energy of the system, since there are many more solvent–solvent pairs (order N^2) than there are solute–solvent pairs (order N).

The alternative route, FEP,[38] is also derived from the ensemble partition function via Equation (16):

$$\Delta A = -kT \ln \frac{V}{\Lambda^3 (N+1)} - kT \ln \frac{\int \dots \int \mathrm{e}^{-U_{N+1}/kT} \mathrm{d}\mathbf{r}^N}{\int \dots \int \mathrm{e}^{-U_N/kT} \mathrm{d}\mathbf{r}^N} \tag{22}$$

where the free energy difference between the N-molecule system (the solvent) and the $(N + 1)$-molecule system (solvent+solute) was split into the sum of a translational part and a configurational contribution arising from the interactions. The first part can be interpreted as the translational free energy acquired by the solute when it is created in the system.[6] It corresponds strictly to an ideal gas term only if the solute is monatomic; in the general case the intramolecular components of the solute's potential energy are comprised in the configurational part. Since we can write $U_{N+1} = U_N + \Delta U$, with ΔU corresponding to the solute–solvent interactions, Equation (22) becomes

$$\Delta A = \Delta A^{\mathrm{trans}} - kT \ln \langle \mathrm{e}^{-\Delta U/kT} \rangle_{NVT} \tag{23}$$

[§] To obtain an expression for the Gibbs free energy difference an ensemble average in the NpT ensemble would have to be performed. In this ensemble the volume fluctuates. If the pressure imposed is chosen so that the average volume observed $\langle V \rangle_{\mathrm{NpT}}$ is identical to the volume in the NVT ensemble, then the Gibbs free energy difference and the Helmholtz free energy difference will be the same.[8]

the ensemble average being performed over configurations of the N-molecule system. This one-step insertion of the solute is called the Widom[39] test-particle insertion method, carried out in practice by simulating configurations of the pure solvent and then repeatedly inserting (and then removing) the solute as a ghost molecule in different positions (and eventually orientations) in each one of the solvent configurations, to evaluate the average in Equation (23). If the solute overlaps with a solvent molecule then the potential energy is repulsive, $\Delta U \gg 0$ and that configuration weights very little in the average of Boltzmann factors in Equation (23). If, conversely, the solute fits in a spontaneously present cavity and interacts favourably with the solvent molecules, then $\Delta U < 0$ and that configuration will count significantly.

The test-particle method is used for solutes treated as rigid molecules, and in this case yields a solvation free energy, analogous to Equation (19), corresponding to the transfer at constant temperature and density of the solute from the ideal gas state into the reference state of the solution at infinite dilution. Widom's method is very simple to apply and efficient but limited to cases where the solute is a small molecule that does not interact strongly with the solvent. The solute has to be small enough to fit into the cavities spontaneously present in the solvent, otherwise sampling will be very poor. Moreover, if there are directional or strongly associating interactions with the solvent then the method will not give good results even if the solute is small: because the solute will induce a significant and specific reorientation of the solvent molecules, the most likely configurations of the solvation shell will not be adequately sampled. The solvent does not "feel" the presence of the solute.

The FEP route may also be formulated with a multi-step activation of the solute, allowing the molecules of the solvent to reorganize themselves around the solute and thus overcoming the major difficulties of the Widom technique:

$$\Delta A^{\text{conf}} = -\sum_{i=0}^{n-1} kT \ln\langle \exp[-(U_{\lambda_i+\Delta\lambda} - U_{\lambda_i})/kT]\rangle_{\lambda_i} \tag{24}$$

The solute–solvent system is simulated at a series of values of the coupling parameter λ_i (the reference system of each step) and, using the same configurations, the energy is recalculated at $\lambda_{i+1} = \lambda_i + \Delta\lambda$ (the perturbed system at each step). It is obvious that the configurations obtained for the reference system have to be also representative of the perturbed system: the step has to be sufficiently small. Otherwise, poor sampling will result and hysteresis would be observed if the perturbations were performed first increasing and then decreasing the values of the coupling parameter between 0 and 1. This possible hysteresis is a disadvantage of the technique. On the other hand, the FEP method has a fine sensitivity in dilute systems because only the solute–solvent interactions that are perturbed appear in the potential energy difference. All the solvent–solvent interactions remain identical between the reference and perturbed systems.

A combined technique that assembles the best features of TI and FEP exists; it is the finite-difference thermodynamic integration[40] (FDTI). The basic equation is the same as that of TI, Equation (19), but now the derivative in the integrand is evaluated numerically, for example using a centred 3-point formula around each value of λ_i,

$$\frac{\partial A}{\partial \lambda} \approx \frac{(A_{\lambda+\delta\lambda} - A_\lambda) - (A_{\lambda-\delta\lambda} - A_\lambda)}{2\delta\lambda} = \frac{\delta_+ A - \delta_- A}{2\delta\lambda} \tag{25}$$

The finite differences in this numerical derivative are obtained using the FEP formalism, Equation (24):

$$\delta_\pm A = -kT \ln\langle \exp[-(U_{\lambda_i \pm \delta\lambda} - U_\lambda)/kT] \rangle_{\lambda_i} \tag{26}$$

where $\delta\lambda$ can be of the order of 0.001. In this way the hysteresis problem of FEP is avoided while its sensitivity is retained since only solute–solvent interactions are concerned.

Free-energy methods are general, allowing the calculation of free-energy differences in a multitude of situations,[35] and are not circumscribed to solvation problems, still less to the calculation of the chemical potential of gases in liquids. Even a procedure like the test-particle insertion, *a priori* suited to gas-solubility calculations only, can be used to obtain partition coefficients of a solute between liquid phases if a thermodynamic cycle like that of Figure 1 is constructed.

The solvation free energies in Figure 1 can be obtained by the Widom method or by a stepwise free energy route, complemented by the necessary ideal gas terms (to connect the ideal gas states that have the same density of the solutions, starting points of Widom's method, to the standard ideal gas state). If one of the liquids is the same chemical species as the solute, the free energy

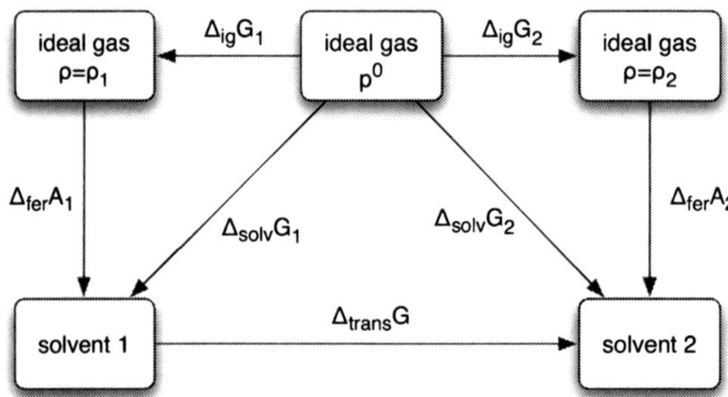

Figure 1 *Isothermal thermodynamic cycle used to calculate the difference in solvation free energies of a solute in two solvents. The quantities $\Delta_{fer}A_i$ refer to the free energies obtained using free energy routes at constant NVT. The $\Delta_{ig}G_i$ (numerically identical to $\Delta_{ig}A_i$ for ideal gas isotherms) connect the ideal gas at the standard pressure to the ideal gas states at the same densities as the solutions. The $\Delta_{solv}G_i$ are Gibbs energies of solvation and $\Delta_{trans}G$ is the free-energy difference upon transfer of the solute from solvent 1 to 2, between reference states at infinite dilution. The partition coefficient is defined as $\ln P = -\Delta_{trans}G/RT$.*

difference $\Delta_{trans}G$ is identical to the Gibbs energy of solution $\Delta_{sol}G_i$ and is related to the activity coefficient at infinite dilution of a solute that is liquid in the conditions of the solution, Equation (8).

A difference in solvation free energies between two solutes in the same solvent can equally well be calculated using a thermodynamic cycle, as shown in Figure 2. Now the solute 1 (for example benzene) is mutated into the solute 2 (for example, toluene) by means of an activation parameter that causes creation, disappearance or mutation of certain atoms (one H becomes a C and three new H atoms are created, together with the necessary intramolecular bonds, angles, and torsions). The transformation must be carried out both in the ideal-gas phase and in solution. The free energy difference from the ideal gas transformation depends only on the intramolecular part, whereas the free energy difference in solution contains the effect of solute–solvent interactions as well as the structural and energetic effects on the solvent caused by the presence of the solutes.

A cycle close to the one in Figure 2 can lead to the free energy of solvation of a complex solute, one that has a molecular skeleton with many intramolecular degrees of freedom. Assume that solute 2 corresponds to the same molecule as solute 1 but in which all the intermolecular interactions have been switched off. Now solute 2 does not interact with the solvent, but all of the intramolecular terms are still present. Going around the cycle, we can transfer solute 1 from the ideal gas state into the solution by gradually deactivating its interactions ($\Delta_{ig}G$), then inserting it into the solvent ($\Delta_{solv}G_2$) with no difficulty even if no large cavities are present, and then gradually activating the intermolecular terms ($\Delta\Delta G$). Numerous applications exist of this kind of thermodynamic cycle to the study of solvation and association in solution.

Solvation processes are in many instances understood as the result of two steps:[41] one is the formation of a cavity in the solvent capable of hosting the solute, and a second is associated with the activation of solute–solvent interactions. Simulation methods, in particular free energy routes, can also be successfully used to access directly the free energy of cavity formation and the distribution of cavity sizes, for example using the test-particle method to

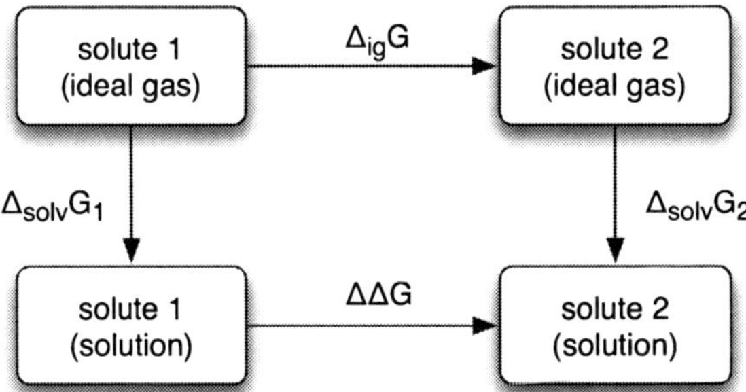

Figure 2 *Thermodynamic cycle used to calculate the difference in solvation free energies between two different solutes in the same solvent.*

determine the probability of insertion of hard spheres that do not overlap with the atoms of the solvent.[42]

10.4 Solute–Solvent Interactions in Ionic Liquids

Ionic liquids are organic salts that are molten at room temperature.[43] The most common are composed of a large organic cation and of a smaller anion. At least the cation has an asymmetric molecular shape that, together with the diffuse electrostatic charge, hinders crystallization resulting in a wide liquid range. Ionic liquids are one of the newest types of solvent in industry, with vast and still unexplored possibilities, and whose properties are still not well understood. They can be synthesized in numerous combinations of cation and anion, generating a variety that poses difficulties to the use of macroscopic thermodynamic models for the prediction of these properties. The approach based on molecular force fields is more promising in this case, since it relies on a physico-chemical model whose parameters are defined on a mode fundamental level than in macroscopic solution theories, and are thus more transferable. The traditional molecular models that have been used in the theoretical treatment of molten salts, such as charged hard bodies, cannot capture the sophisticated molecular shape of ionic liquids neither the subtle conformational features that are typical of ionic liquids.

We present here, as an example of the molecular approach described above to understand solubility phenomena, the study of the solvation of carbon dioxide in three ionic liquids sharing the same cation, 1–butyl-3–methyl-imidazolium: [bmim][BF_4], [bmim][PF_6], and [bmim][tf_2N], where tf_2N^- is the *bis*(trifluoromethanesulfonyl)amide anion $(CF_3SO_2)_2N^-$. The effect of changing the anion in the solubility of CO_2 is related to the microscopic structure and interactions in the pure solvents. The nature of the solute–solvent interactions is investigated by selectively modifying certain terms in the force field of the solute and observing the changes in the chemical potential.

It is observed experimentally[44–46] that the solubility of CO_2 in the ionic liquids increases when the anion is changed from BF_4^- to PF_6^- to tf_2N^-, being significantly larger in the last solvent as seen in Figure 3. Molecular simulation associated with the test particle insertion method was used to calculate the solubility of carbon dioxide in these three ionic liquids.[47,48] The relative order of the solubility in the different ionic liquids is correctly predicted, even for [bmim][BF_4] and [bmim][PF_6] for which the difference in solubility is very low. The temperature dependence of the solubility is correctly reproduced. Predictions are qualitative, validating both the molecular force fields describing the solvents[49,50] and the solute,[51] and the free energy route.

The cavities spontaneously present in the three ionic liquids were studied by calculating the distribution of cavity sizes and hence the free energy of cavity formation. This was done by testing the probability of inserting hard spheres using Widom's method.[52] The free energy of cavity formation is represented in Figure 4. It shows that more cavities of a given size are found in [bmim][tf_2N],

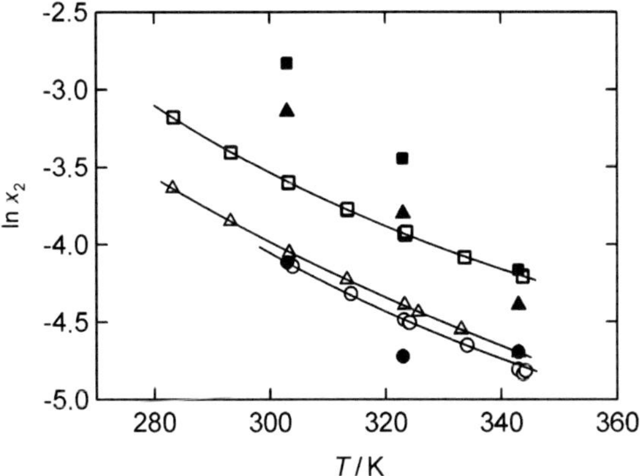

Figure 3 *Experimental (open symbols) and simulated (filled symbols) solubilities of carbon dioxide in: (○,●), [bmim][BF$_4$], (△,▲), [bmim][PF$_6$], and (□,■) [bmim][tf$_2$N].*

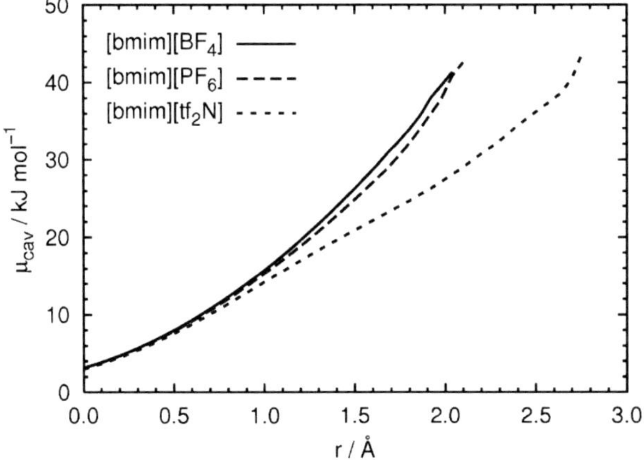

Figure 4 *Free energy of cavity formation in three ionic liquids.*

followed by [bmim][PF$_6$] and [bmim][BF$_4$], the last two being close. This observation agrees with the experimental relative solubility of carbon dioxide in the three ionic liquids. The trend is compatible with a strong dependence of solubility on the structure of the pure solvent.

Because carbon dioxide is a quadrupolar molecule, it would be interesting to inquire if the role of electrostatic forces in the gas-solvent interactions dominates its solvation by ionic liquids. Calculations were repeated using a modified intermolecular potential model for carbon dioxide, this time removing the partial electrostatic charges. Solubility suffers a significant decrease, as can be observed in Table 1, proving that the electrostatic components play a major role in the interactions between carbon dioxide and the ionic liquids.

Table 1 *Effect of the quadrupole on the solubility of carbon dioxide in three ionic liquids calculated at 303 K. Mole fraction solubility is given at a partial pressure of 1 bar*

	Full electrostatics	*No electrostatics*	*Experimental $x_2/10^{-3}$*
[bmim][BF_4]	16	1.3	17.2[45]
[bmim][PF_6]	43	2.8	18.7[44]
[bmim][tf_2N]	59	22	27.5[46]

These energetic aspects can be complemented with structural details of the solutions, in the form of site–site radial distribution functions of the solvent atoms around those of the solute.[53] Such radial distribution functions permit to identify the preferential solvation sites and so understand the effect of the change in the molecular structure of the solvent, on solubility.

10.5 Conclusion

As a general conclusion, it can be retained that molecular modelling tools used in a predictive manner, with no adjusted parameters to the property or to the multi-component system in question, provide qualitative answers when thermodynamic properties such as solubility are calculated. Quantitative results can be attained using specific molecular models, tailored for certain properties and systems, or by adjusting parameters to an appropriate property of the mixture. The importance of qualitative information should not be underestimated; the results obtained from molecular models allows a fundamental interpretation of the major trends underlying the behaviour of solutes and solvents, providing a guide when scanning for new compounds in pure and applied research.

Quantum chemistry and molecular-simulation packages, and the required computer performance, are currently available to non-specialist researchers. These tools certainly change the approach to experimental thermodynamics, since fewer but more carefully chosen data are required. A rigorous analysis of the classical thermodynamic formalism is indispensable to establish the connection between the properties accessible to experiment and the quantities calculated through molecular models and simulation methods based on statistical mechanics.

Acknowledgments

The authors are grateful to Prof. Vladimir Majer for the excellent discussions about solution thermodynamics.

References

1. J.H. Hildebrand and R.L. Scott, *The Solubility of Nonelectrolytes*, 3rd edn, Dover Publications, NY, 1964.

2. P.T. Anastas and J.C. Warner, *Green Chemistry, Theory and Practice*, Oxford University Press, Oxford, 1998.
3. K. Denbigh, *The Principles of Chemical Equilibrium*, 4th edn, CUP, Cambridge, 1981.
4. J.M. Prausnitz, R.N. Lichtenthaler and E. Gomes de Azevedo, *Molecular Thermodynamics of Fluid-Phase Equilibria*, 3rd edn, Prentice-Hall, NJ, 1999.
5. J.M. Smith, H.C. Van Ness and M.M. Abbott, *Introduction to Chemical Engineering Thermodynamics*, 5th edn, McGraw Hill, NY, 1996.
6. A. Ben-Naim, *J. Phys. Chem.*, 1978, **82**, 792.
7. A. Ben-Naim and Y. Marcus, *J. Chem. Phys.*, 1984, **81**, 2016.
8. A. Ben-Naim, *Statistical Thermodynamics for Chemists and Biochemists*, Plenum Press, New York, 1992.
9. V. Majer, J. Sedlbauer and R.H. Wood, in *Aqueous Systems at Elevated Temperatures and Pressures: Physical Chemistry, in Water, Steam and Hydrothermal Solutions*, D.A. Palmer, R. Fernandez-Prini and A.H. Harvey, (eds),Elsevier, Amsterdam, 2004.
10. B.B. Benson and D. Krause, Jr., *J. Solution Chem.*, 1989, **18**, 803.
11. R. Car and M. Parrinello, *Phys. Rev. Lett.*, 1985, **55**, 2471.
12. I.-F.W. Kuo, C.J. Mundy, M.J. McGrath, J.I. Siepmann, J.V. Vondele, M. Sprik, J. Hutter, M.L. Klein, F. Mohamed, M. Krack and M. Parrinello, *J. Phys. Chem. B*, 2004, **108**, 12990.
13. J. Tomasi, B. Mennucci and R. Cammi, *Chem. Rev.*, 2005, **105**, 2999.
14. C.J. Cramer and D.G. Truhlar, *Chem. Rev.*, 1999, **99**, 2161.
15. G. Alagona, C. Ghio and P.I. Nagy, *Int. J. Quantum Chem.*, 2004, **99**, 161.
16. J. Gao, *Acc. Chem. Res.*, 1996, **29**, 298.
17. M. Orozco and F.J. Luque, *Chem. Rev.*, 2000, **100**, 4187.
18. A. Warshel and M. Levitt, *J. Mol. Biol.*, 1976, **103**, 227.
19. W. Liu, S. Sakane, R.H. Wood and D.J. Doren, *J. Phys. Chem. A.*, 2002, **106**, 1409.
20. D. Frenkel and B. Smit, *Understanding Molecular Simulation*, 2nd edn, Academic Press, San Diego, 2002.
21. S.W. Rick, S.J. Stuart and B.J. Berne, *J. Chem. Phys.*, 1994, **101**, 6141.
22. S. Patel and C.L. Brooks, III, *J. Comp. Chem.*, 2003, **25**, 1.
23. S.C. Glotzer and W. Paul, *Annu. Rev. Mater. Res.*, 2002, **32**, 401.
24. D.A. McQuarrie, *Statistical Mechanics*, Harper Collins, New York, 1976.
25. W.D. Cornell, P. Cieplak, C.I. Bayly, I.R. Gould, K.M. Merz, Jr., D.M. Ferguson, D.C. Spellmeyer, T. Fox, J.W. Caldwell and P.A. Kollman, *J. Am. Chem. Soc.*, 1995, **117**, 5179.
26. W.L. Jorgensen, D.S. Maxwell and J. Tirado-Rives, *J. Am. Chem. Soc.*, 1996, **118**, 11225.
27. A.D. MacKerell, D. Bashford, M. Bellott, R.L. Dunbrack, J.D. Evanseck, M.J. Field, S. Fischer, J. Gao, H. Guo, S. Ha, D. Joseph-McCarthy, L. Kuchnir, K. Kuczera, F.T.K. Lau, C. Mattos, S. Michnick, T. Ngo, D.T. Nguyen, B. Prodhom, W.E. Reiher, B. Roux, M. Schlenkrich, J.C. Smith, R. Stote, J. Straub, M. Watanabe, J. Wiorkiewicz-Kuczera, D. Yin and M. Karplus, *J. Phys. Chem. B*, 1998, **102**, 3586.

28. T.A. Halgren, *J. Comp. Chem.*, 1996, **17**, 490.
29. J. Hermans, H.J.C. Berendsen and van Gunsteren, *Biopolymers*, 1984, **23**, 1.
30. N.L. Allinger, K. Chen and J.-H. Lii, *J. Comp. Chem.*, 1996, **17**, 643.
31. M.G. Martin and J.I. Siepmann, *J. Phys. Chem. B*, 1998, **102**, 2569.
32. C.M. Brenneman and J. Wiberg, *J. Comput. Chem.*, 1987, **8**, 894.
33. W.D. Cornell, P. Cieplak, C.I. Bayly and P.A. Kollmann, *J. Am. Chem. Soc.*, 1993, **115**, 9620.
34. A.A.H. Padua, *J. Phys. Chem. A*, 2002, **106**, 10116.
35. P. Kollman, *Chem. Rev.*, 1993, **93**, 2395.
36. D.A. Kofke and P.T. Cummings, *Mol. Phys.*, 1997, **92**, 6.
37. J.G. Kirkwood, *J. Chem. Phys.*, 1935, **3**, 300.
38. R.W. Zwanzig, *J. Chem. Phys.*, 1954, **22**, 1420.
39. B. Widom, *J. Chem. Phys.*, 1963, **39**, 2802.
40. M. Mezei, *J. Chem. Phys.*, 1987, **86**, 7084.
41. G.L. Pollack, *Science*, 1991, **251**, 1323.
42. L.R. Pratt and A. Pohorille, *Proc. Nat. Acad. Sci. USA*, 1992, **82**, 2995.
43. P. Wasserscheid and T. Welton (Eds), *Ionic Liquids in Synthesis*, Wiley-VCH, Weinheim, 2003.
44. J. Jacquemin, P. Husson, V. Majer and M.F. Costa Gomes, *Fluid Phase Equilib.*, 2006, **240**, 87.
45. J. Jacquemin, M.F. Costa Gomes, P. Husson and V. Majer, *J. Chem. Thermodyn.*, 2006, **38**, 490.
46. J. Jacquemin, P. Husson, V. Majer and M.F. Costa Gomes, *J. Solution Chem.*, 2006, to be submitted.
47. J. Deschamps, M.F. Costa Gomes and A.A.H. Padua, *Chem. Phys. Chem.*, 2004, **5**, 1049.
48. J. Deschamps and A.A.H. Padua, in *Ionic Liquids IIIA: Fundamentals, Progress, Challenges, and Opportunities, Properties and Structure*, R.D. Rogers and K.R. Seddon (eds), ACS Symposium Series 901, ACS, Washington, DC, 2005.
49. J.N. Canongia Lopes, J. Deschamps and A.A.H. Padua, *J. Phys. Chem. B.*, 2004, 108, 2038. Corrections 11250.
50. J.N. Canongia Lopes and A.A.H. Padua, *J. Phys. Chem. B.*, 2004, **108**, 16893.
51. J.G. Harris and K.H. Yung, *J. Phys. Chem.*, 1995, **99**, 12021.
52. M.F. Costa Gomes and A.A.H. Padua, *J. Phys. Chem. B*, 2003, **107**, 14020.
53. M.F. Costa Gomes and A.A.H. Padua, *Pure Appl. Chem.*, 2005, **77**, 653.

CHAPTER 11

Molecular Simulation Approaches to Solubility

KELLY E. ANDERSON AND J. ILJA SIEPMANN

Departments of Chemistry and of Chemical Engineering and Materials Science and Minnesota Supercomputing Institute, University of Minnesota, 207 Pleasant Street S.E., Minneapolis MN 55455-0431, USA

11.1 Introduction

Over the past 50 years, molecular simulation has emerged as a useful tool for providing microscopic insight into a variety of chemical systems and processes. There are two main particle-based simulation techniques – molecular dynamics, which uses deterministic equations of motion integrated over small time steps to evolve the system, and Monte Carlo sampling, which uses a Markov chain to construct a stochastic sequence of system configurations. Molecular dynamics can provide insight into time-dependent and time-averaged properties of the system, whereas Monte Carlo simulations provide ensemble-averaged properties. Both techniques have been used extensively to investigate solubility in different chemical systems.

11.2 Solubility

The solubility of a solute in a solvent, or more generally the transfer of a solute between two solvation environments, is usually expressed either in terms of the infinite-dilution limit or the solubility limit, where the latter often involves the mutual solubility limit in the sense that both "solute" and "solvent" can transfer between the two environments. Commonly, these environments are homogeneous bulk phases, but in principle one might also be interested in the transfer involving a micro-phase region, such as a surfactant micelle or an octadecyl-bonded phase in reversed-phase liquid chromatography. Most commonly studied are the transfer of the solute between the gas and liquid phases (free energy of solvation), gas and water phases (free energy of hydration), or two liquid phases.

Phase coexistence for a system consisting of two phases α and β requires that

$$\begin{aligned} \mu_i^\alpha &= \mu_i^\beta \\ p^\alpha &= p^\beta \\ T^\alpha &= T^\beta \end{aligned} \tag{1}$$

where μ_i is the chemical potential of species i (and the equality needs to be satisfied for all independent components), p the pressure, and T the absolute temperature of a given phase.[1–3] The solubility is often examined in terms of the Gibbs free energy of transfer of solute i from phase α to phase β, and is defined by the following expression[4,5]

$$\begin{aligned} \Delta G_i^{\alpha\to\beta}(T,p,X^\alpha,X^\beta) &= -RT\ln K_i^{\alpha\to\beta}(T,p,X^\alpha,X^\beta) \\ &= -RT\ln\left(\frac{\rho_i^\beta(T,p,X^\alpha,X^\beta)}{\rho_i^\alpha(T,p,X^\alpha,X^\beta)}\right) \end{aligned} \tag{2}$$

where X denotes the equilibrium composition, R the universal gas constant, and ρ the number density of i in a given phase. For solute i, the partition coefficient K is defined by the ratio of the concentrations of i in the two phases. When evaluating solubility experimentally, K is the observable property of interest. The specific value of K depends on the concentration units with molarity (or number density) being preferred on statistical mechanical grounds (because the translational entropy depends on volume), but molality often used on practical grounds (because it avoids the need to measure the volume).

Because K depends on temperature, pressure, and concentration, it is useful to relate this property to the Gibbs free energy of the standard state, ΔG°. In the gas phase, the standard state is defined as the pure gas at one atmosphere for a given temperature. The standard state can be defined in different ways for liquids and solutions. For a pure liquid, the standard state is the pure liquid at one atmosphere for a given temperature. In solution, it may be defined as a (hypothetical) ideal solution with unit molarity or molality, or with respect to infinite dilution properties.[1] It is naturally of great importance to clarify the definitions used for the standard states before comparing different experimental or simulation results.

The most direct comparison between experiment and simulation is the comparison of K under the same conditions. Until recently, this was not possible, since there was no method to directly determine K from simulation. Therefore, techniques were developed to determine the solubility of a given species in terms of the standard Gibbs free energy of transfer. Defined for the transfer between phases

$$\Delta G^\circ = G^{\circ,\beta} - G^{\circ,\alpha} \tag{3}$$

This becomes equal to the difference in the standard chemical potential when phase α is the N particle system and β is the $N+1$ particle system (or the $N-1$ and N particle systems, respectively). The chemical potential of the solute, μ_s, is the partial derivative of the Gibbs free energy with respect to the number of solute molecules present at constant temperature, and pressure. The standard chemical potential, μ_s°, is defined as

$$\mu_s^\circ (T, p) = \mu_s (T, p, x_s) - k_B T \ln a_s \tag{4}$$

where $a_s = x_s\gamma_s$ is the activity of the solute, x_s the mole fraction of the solute, γ_s the activity coefficient of the solute, and k_B the Boltzmann's constant. A key point here is that when using the standard state to determine solubility, one is evaluating the difference in the standard chemical potential between two phases, while the actual chemical potential difference is zero (see Equation (1). Because of this condition of phase coexistence, the difference in the standard chemical potential is related to the ratio of the activity of the solute in the two phases

$$\Delta\mu_s^\circ = \mu_s^{\circ,\beta} - \mu_s^{\circ,\alpha} = -k_B T \ln(a_s^\beta / a_s^\alpha) \tag{5}$$

For infinitely dilute systems, the activity coefficient of each phase is assumed to be unity, and the ratio of activities is simply the ratio of the mole fraction of the solute between the phases. Most simulation techniques seek to determine the chemical potential at infinite dilution and relate this to the solubility coefficient. While this is a reasonable approach in many situations, it is important to keep the system size of the simulation and the properties of the solute in mind. Due to computational limitations, many "infinitely dilute" simulation systems consist of only a few hundred solvent molecules. For many solutes, one solute molecule in the presence of a few hundred solvent molecules is already above the solubility limit.

More recently, the development of the Gibbs ensemble Monte Carlo (GEMC) technique[6,7] has allowed for the direct determination of K by simulating two phases simultaneously. In the Gibbs ensemble, molecules can transfer directly between the phases over the course of a simulation and the system properties, such as the solute number densities, are evaluated as ensemble averages. This does not pose a restriction on the number of solute molecules in a specific phase and together with the direct evaluation of solute number densities leads to very precise determinations of solubility coefficients.[8]

11.3 Computing Solubility for the Infinite Dilution Limit

The infinitely dilute solution is of particular interest because it provides a means to examine solute–solvent interactions without the presence of solute–solute interactions.[4] There are many different standards used to measure solubility based on different criteria.[1] The most common measure of the solubility of a gas in a liquid at infinite dilution is Henry's law constant,

$$H_s = \lim_{x \to \infty} \frac{p_s^{\text{gas}}}{x_s^{\text{liq}}}, \tag{6}$$

where p_s is the partial pressure of the solute in the gas and x_s the concentration of the solute in the solvent. Henry's law states that at infinite dilution the partial pressure of a solute is directly proportional to the concentration of the solute in the solvent phase. H_s is related to the excess, or residual, chemical potential, μ_s^E, by[9]

$$\frac{\mu_s^E}{k_B T} = \ln \frac{H_s}{\rho k_B T} \tag{7}$$

where ρ is the density of the solvent. The excess chemical potential is the difference between the chemical potential of the real system and that of an ideal system at the same temperature and pressure. According to the potential distribution theorem,[10] the change in free energy in the canonical ensemble (constant: N,V,T) upon the addition of a particle (*i.e.* the chemical potential) can be calculated from an ensemble average over configurations of the original system (containing N–1 particles)

$$\mu_s^E = -k_B T \ln \left\langle \exp\left(\frac{U_{test}}{k_B T}\right)\right\rangle_{N-1,V,T} \tag{8}$$

where U_{test} is the interaction energy of a test particle (the Nth particle) with the system, *i.e.* $U_{test} = U_N - U_{N-1}$. Most simulation techniques seek to use this expression to determine the excess chemical potential of the system, and from there, determine Henry's law constant. It should be noted that Equation (8) is only valid in the canonical ensemble and different expressions need to be used in the isobaric–isothermal and Gibbs ensembles.[11]

11.3.1 Thermodynamic Integration

The most widespread methods used to calculate free energy differences are thermodynamic integration (TI) and free energy perturbation (FEP). Both methods can be applied equally well in the canonical and isobaric–isothermal ensembles to obtain Helmholtz and Gibbs free energies, respectively. TI seeks to determine the free energy of the system of interest by constructing a reversible pathway between a reference system of known free energy and the system of interest.[11] TI is based on the equation

$$\Delta F = F_1 - F_0 = \int_0^1 \frac{\partial F(\lambda)}{\partial \lambda} d\lambda = \int_0^1 \left\langle \frac{\partial H(\mathbf{q},\mathbf{p},\lambda)}{\partial \lambda} \right\rangle_\lambda d\lambda \tag{9}$$

where ΔF is the free energy difference between state 1 and state 0, $F(\lambda)$ the free energy of the system as a function of the coupling parameter λ, which can run from zero to unity, and $H(\mathbf{q},\mathbf{p},\lambda)$ the Hamiltonian of the system as a function of λ and the set of generalized coordinates and momenta.[12] $F(\lambda)$ is chosen such that when $\lambda = 0$, $F(\lambda) = F_0$, the free energy of the reference or initial system, and when $\lambda = 1$, $F(\lambda) = F_1$, the free energy of the target system. In contrast to experimental techniques, a simulation is not constrained to follow a physical thermodynamic integration pathway. For example, if the reference system is defined as the $N - 1$ particle system and the N particle system is the target system, one technique is to use the strength of the interactions between the Nth

particle and the rest of the system as the coupling parameter λ. A series of simulations, then, proceeds along this pathway, progressively inserting a particle into the system, by turning on its full interaction potential as λ moves from 0 to 1. Using a finite number of values for λ, the integration over λ is performed numerically to determine the free energy difference.[11]

While reliable and applicable to a wide variety of systems, including difficult systems such as solids, TI is subject to some drawbacks. As summarized by Kofke and Cummings,[13] one main drawback to TI is its "perceived inefficiency." This perception arises primarily from the reliance of TI on sampling a series of intermediate and thermodynamically uninteresting states as the system moves from state 0 to state 1. Traversing these intermediate states is not inherently less efficient than other methods used in the calculation of the chemical potential, according to Kofke and Cummings. Additionally, care must be used when choosing an integration pathway so as not to cross a phase boundary.[13]

Thermodynamic integration has been used alone and in combination with other techniques to model the solubility properties of a variety of systems. These include non-polar solutes, such as *n*-alkanes in water, where differences in the solvation free energies between alkane species agreed with experimental results while the statistical error within the calculated values of $\langle \partial H(\lambda)/\partial \lambda \rangle$ for each value of λ was at most $\pm 5\%$, as estimated by block averaging.[14,15] The free energies of hydration for *p*-substituted benzamidine derivatives were used to predict solvation order, with errors around 30% for the free energy values, based on standard deviations in the exponential term (Equation (8)).[16] In addition, the solubility of charged species in water has also been examined using TI.[17] Straatsma and Berendsen provide many good points to be taken into consideration when modeling such systems, but conclude that TI is generally a reliable technique.[17] In one novel application, TI was used to predict the solubility of atomic species such as silicon and sulfur in liquid and solid iron under conditions of extreme pressures and temperatures, with statistical errors around 5% in the free energy calculations.[18]

11.3.2 Free Energy Perturbation

The second common class of methods used to calculate chemical potential includes those based on FEP. These methods are primarily concerned with the relationship[13]

$$\Delta F = F_1 - F_0 = -k_B T \ln \langle \exp [\Delta H_{10}(\mathbf{q},\mathbf{p})] \rangle_0 \tag{10}$$

The subscript 0 on the braces indicates an ensemble average at state 0. In most applications, state 0 is taken as the reference state and state 1 is the perturbed state, although in principle these labels are interchangeable. FEP and TI, thus, are very similar in theory; both are primarily concerned with the construction of a thermodynamic pathway between two states of interest. The distinction is how this pathway is constructed. TI is concerned with the integral in Equation (9). It is not necessary for intermediate states to overlap, but the accuracy of the calculation depends on the smoothness of a plot of $\partial H(\lambda)/\partial\lambda$ *vs.* λ.[14] In FEP, ΔF is determined by simulating one state and perturbing the interaction parameters at each point to sample the second state. This requires that the two states of interest be close in energy so that they overlap substantially in phase space. The total free energy difference between the first state and the final state is the sum of the free energy differences between the intermediate pairs of states.[14] A series of different FEP techniques have been proposed; these can be loosely arranged into single-stage and multiple-stage methods.

11.3.2.1 Single-Stage Techniques

The most common single-stage FEP method is that of Widom's test particle insertion.[10] This method follows directly from Equation (10). Choosing the reference system to have $N - 1$ particles and the perturbed system as that with N particles, ΔF is equal to the excess chemical potential given by Equation (8) (for the canonical ensemble). In practice, the ensemble average is measured by inserting the test particle into the system, calculating the potential energy of the test particle with the system, and removing the particle before continuing the simulation. This means that as the simulation progresses, the system itself is not disturbed; in fact, the calculation of the residual chemical potential could be performed simply on the configuration file from a simulation of the $N - 1$ particle system.

The primary drawback of this method is its increasing inability to adequately sample the N particle system as the density of the fluid increases, particularly as one moves beyond single atom insertions to insertions of whole molecules. Methods, such as configuration-biased insertions,[19–24] are available to overcome or lessen this restriction. In principle, it is possible to proceed in the opposite direction, *i.e.* from the N system to the $N - 1$ system, through particle removal, but in practice this method is not as reliable or as straightforward and can suffer from inadequate sampling of the $N - 1$ particle system.[13]

Widom's test particle insertion technique has been used to probe the solubility of small molecules in a variety of systems. The simulation results for the solubility of oxygen and carbon dioxide in fluorocarbons show good agreement with experimental results, and statistical errors of, at most, 7% in the calculation of μ^{E}.[25,26] Others have used the technique to examine the solubility of alkanes in a rubber polymer; over the progression from methane to hexane, the uncertainty in the calculated solubility coefficient ranged from 10 to 69% based on data from eight different starting configurations.[27] Widom's insertion technique is also often used to compute Henry's law constant. Recently, Boutard *et al.* calculated the Henry's law constants for carbon dioxide, methane, oxygen, and nitrogen in ethanol with a statistical uncertainty of about 5%.[28] Widom's method may also be used to incrementally insert long-chain molecules into a fluid.[29] For a homopolymer of length l in the solvent system, the chemical potential of the $l + 1$ chain is determined by inserting an additional bead at the end of the l chain. Kumar *et al.*[29] show that μ^{E} for each step along the growth of the chain may be computed and, through a building up process, used to determine the chemical potential of the whole chain. This provides a useful technique for inserting larger molecules into solutions.

11.3.2.2 Multiple-Stage Techniques

FEP methods may also proceed through a series of stages, where intermediate states may be joined together in the course of calculating the chemical potential. It is possible to use multiple intermediates, but for the ease of discussion, the examples presented here will use only one, with M indicating the intermediate state. Just as before, the reference and perturbed states must be defined, and it is this definition that separates the available multistage FEP methods. Among the most common techniques are umbrella sampling and the overlapping distribution method. In umbrella sampling, the M state is defined as the reference state, with the $N-1$ and N states as perturbations.[30] In a single simulation, a sampling distribution is constructed to sample both inserting and removing a single particle. A bias is used in the selection of the sampling distribution to ensure that regions important to both the $N-1$ and N "perturbations" are sampled evenly. The excess chemical potential is calculated according to

$$\mu^{\mathrm{E}} = -k_{\mathrm{B}}T \ln\left(\frac{\langle \exp(-U_N/k_{\mathrm{B}}T)/\pi\rangle_\pi}{\langle \exp(-U_{N-1}/k_{\mathrm{B}}T)/\pi\rangle_\pi}\right) \tag{11}$$

where U_m is the potential energy of the N or $N-1$ particle system and π the distribution weighting function.[30] Care must be taken while using this method to select an appropriate weighting function so as not to bias the system toward either perturbation. A variation of umbrella sampling known as "double-wide sampling"[31] (but care is required to ensure a subset relationship between the perturbed and sampled systems[13]) has been used to study the partition coefficients of small organic molecules in water and chloroform. The trends in solubility predicted by simulation followed those known from experiment, although the range of free energies was larger for the simulation results and the statistical error in the calculated free energies was on the order of 10%.[32] This technique was recently used to examine solvent effects on partition coefficients for a wider range of solvents, including tetrahydrofuran and acetonitrile.[33]

Alternatively, in the overlapping distribution method, both the $N-1$ and N states are considered reference states, while M is a perturbation to each. As the name implies, the perturbation state is chosen so that the phase space of M overlaps that of both the $N-1$ and N systems. The calculation of the chemical potential requires two separate simulations, one for each reference system[34]

$$\mu^{\mathrm{E}} = -k_{\mathrm{B}}Tln\left(\frac{\langle exp(-U_{(N-1)\to M}/k_{\mathrm{B}}T)\rangle}{\langle exp(-U_{N\to M}/k_{\mathrm{B}}T)\rangle}\right) \tag{12}$$

Although this method has not been widely used, Kofke and co-workers have shown recently that it may be more efficient than some methods that are more popular.[35,36] A separate study by Shirts and Pande[37] came to a similar conclusion, showing that Bennett's method,[38] a type of overlap sampling, is more efficient than TI in many situations. One example of successful application is the calculation of the chemical potential of flexible chain molecules of up to 14 hard spheres, where the overlapping distribution method was shown to be as reliable as Widom's technique for certain systems with errors in the residual chemical potential on the order of a few percent.[39]

11.3.3 Expanded Ensembles

The next two techniques used to determine excess chemical potential are also related. The first is the expanded ensemble method.[40–42] This method determines the free energy difference due to the insertion of a particle by constructing a series of intermediate states between the $N-1$ and N systems and tracking the frequency with which these subsystems are sampled. The initial subsystem is that of $N-1$ interacting particles and 1 non-interacting "solute" particle, which may or may not be of the same chemical species as the solvent molecules. A series of m intermediates is constructed with the interactions, α_m, between the solute and the solvent slowly increasing until the Mth subsystem, where the solute fully interacts with the solvent system ($\alpha_M = 1$), which corresponds to the true N particle ensemble. A Monte Carlo walk through the expanded system samples each state between $\alpha_0 = 0$ and $\alpha_M = 1$ to produce a probability distribution, p_m, over the subensembles. The excess chemical potential is related to the ratio of the probability distributions of the two end states,

$$\mu^{\mathrm{E}} = -\ln\left(\frac{p_M}{p_0}\right) + \pi_M - \pi_0 \tag{13}$$

where π_m is the weighting factor of the mth state.[40] These weighting factors are necessary to ensure that all the subsystems are sampled with roughly equal probability; without the weighting factors, those subsystems with larger chemical potentials would be less frequently sampled, which introduces statistical errors. In general, π_m is chosen to be proportional to the relative free energy of the subsystem m, but this is the quantity of interest to be determined *via* simulation.[43] Therefore, a trial run must be used to determine an approximate value for the free energy, and this initial weight can then be iteratively updated over the course of a simulation based on the sampling at the previous weight. One major drawback is that this iterative approach means that the data collected using previous weighting functions must be discarded because the excess chemical potential is dependent upon the weighting factor. This method has been used to determine solvation free energies for molecules from methane to benzylamine in water, and for aqueous ionic solutions with standard deviations of the residual chemical potential generally less than 5% of the value.[44] The solubility of a series of drug-related compounds has also been examined using expanded ensembles. Solvation free energies were calculated with a precision of about 2 kJ mol^{-1} (11%).[45]

11.3.4 Transition Matrix Monte Carlo

The transition matrix Monte Carlo (TMMC) method used by Cichowski *et al.*[43] to determine Henry's law constants seeks to improve upon the expanded ensemble approach by integrating transition matrix techniques and the expanded ensemble technique. Beginning with the same expanded ensemble framework, Cichowski *et al.* use the transition matrix approach to determine the weighting factors. Instead of tracking only the times each subsystem is visited and then discarding the accumulated statistics each time the weighting factor is updated, TMMC monitors the transition probabilities between the ensembles. For each attempt to transition between microstates of two subensembles, a collection matrix is updated regardless of whether or not the transition is completed. From this matrix, the transition probabilities between the subensembles are determined. The key is that although biasing is still used during sampling to ensure even sampling of all subsystems, the bias is removed before the collection matrix is updated. Thus the statistics accumulated over the whole simulation may be used, not only those of a given biasing factor. This makes the technique more efficient than the expanded ensemble approach. Cichowski *et al.* determined Henry's law constants for a few small solutes in ethanol with no more than 4% standard deviation as determined from four independent simulations.

11.3.5 Gibbs Ensemble

The methods described so far, all require the use of standard states to compute Gibbs free energies of transfer at infinite dilution *via* Equation (3) from separate simulations of the two phases. Rather different in spirit, the GEMC[6,7] approach may also be used to explore infinitely dilute systems, but the calculation of Gibbs free energies proceeds analogously as experiments *via* Equation (2). The Gibbs ensemble uses two (or more) separate simulation boxes representing distinct bulk phases. A series of specialized moves are used in addition to the standard translation, rotation, and conformation moves to bring the bulk phases into equilibrium. At constant pressure and temperature, this includes volume exchange moves with a pressure bath and particle swap moves between the phases to equilibrate the pressure and the chemical potential of the phases, respectively.[8] Owing to the dependence on swapping particles, the original GEMC method is not very efficient for dense systems or articulated solutes where the acceptance rate for swap moves can become very small (*i.e.* similar problems are encountered as with Widom's insertion method). For flexible molecules (and to a lesser extent for rigid molecules), the use of configurational-bias growth procedures can dramatically enhance the swap acceptance rates.[24,46–49] In addition, expanded ensembles can also be used that allow for a gradual transfer of the solute[50,51] and pre-weighting factors can be employed to improve the statistical precision by ensuring roughly equal number of solute molecules in both phases.[52] For liquid–liquid equilibria, the transfer rate can be greatly enhanced by using one (or multiple) intermediate phase, such as a vapor phase.[53,54]

The primary advantage of the Gibbs ensemble is that the Gibbs free energy of transfer for particles allowed to swap between the two phases may be calculated directly from the number densities of the particles in the two phases[8]

$$\Delta G_i = -k_{\mathrm{B}}T\ln\left(\frac{\rho_i^{\alpha}}{\rho_i^{\beta}}\right) \tag{14}$$

The direct computation of a ratio of mechanical properties greatly improves the statistical precision compared to indirect methods, such as thermodynamic integration and FEP, and it avoids the need to specify a standard state for the solution with the usual assumption of ideal solution behavior. Similarly, the Henry's law constant can be directly computed *via* Equation (6), where the property $x_{\mathrm{s}}^{\mathrm{liq}}$ is obtained from its ensemble average. Dalton's law of partial pressures may be used to determine $p_{\mathrm{s}}^{\mathrm{gas}}$ from the mole fraction of the solute in the gas phase and the ensemble-averaged saturation pressure. Using this technique, it is possible to determine Henry's law constant in a manner that more closely mimics experimental conditions. Another advantage of the Gibbs ensemble approach is that it avoids the systematic errors that can be encountered in FEP calculations depending on the phase space relationship between the reference and the perturbed system.[13,55]

Recently, GEMC simulations were used to compute Henry's law constants for methane, carbon dioxide, nitrogen, and oxygen in ethanol with a precision of about 3% based on the standard error of the mean taken over four independent simulations.[56] GEMC has also been used to examine partitioning of alkanes and alcohols in neat and water-saturated 1-octanol as well as into water to examine the effects of multiple solvents on partitioning.[52,54] In the field of chromatography, GEMC has been used to examine the partitioning of solutes between stationary and mobiles phases in both gas–liquid and liquid–liquid chromatography.[57–62] Using GEMC, it is possible to simulate systems with multiple organic solutes present and to calculate free energies of transfer and partition coefficients, *K*, for each solute. The *K* may be determined directly from the ensemble-averaged number densities with great precision (often less than 3% uncertainty) so that differences between isomers, such as 2-methylpentane and 3-methylpentane, may be resolved.[57] Furthermore, GEMC allows investigations of systems where one or more solutes are present in larger concentration,[58,60] where one or multiple interfaces are present, such as the liquid–air interface in gas-liquid chromatography[59] or the silica–bonded phase and bonded phase/mobile phase interfaces in reversed-phase liquid chromatography.[62] Finally, the precision of GEMC simulations for Gibbs free energies of transfer is also sufficient to explore deviation from van't Hoff behavior caused by the heat capacity of transfer.[63]

11.3.6 Continuum Solvation Models

Although not usually considered in the context of molecular simulation, continuum solvation models are discussed here briefly because these methods are very popular for the computation of solubilities. Instead of using explicit molecular models to examine the interactions of solute and solvent, as is done in particle-based simulations, continuum solvation models use a simplified representation of the solvent, such as a continuum characterized by properties of the bulk solvent.[64] The continuum models are generally parameterized for specific solvents to reproduce the interactions of these solvent with a variety of solutes. The continuum representation of the solvent results in significant gains in computational efficiency not only because one does not need to compute the interactions with a large number of solvent molecules but more so because continuum solvation models do not require sampling of the phase space of the solvent molecules, *i.e.* one does not need to move solvent molecules to obtain free energies. However, the neglect of sampling the solvent degrees of freedom also implies that continuum solvation models cannot be simply applied over a range of state points (temperatures and pressures).

The free energy of solvation is determined by placing the solute into a cavity in the continuum solvent, which becomes polarized. The key concept in understanding the polarization effects on the solvent and solute is the reaction field.[65] The reaction field is the electric field exerted by the solvent on the solute once the solvent has been polarized. The reaction field polarizes the solute, which in turn changes the polarization of the solvent. Out of this process comes the self-consistent reaction field method.[66] The first continuum model is the Born–Kirkwood–Onsager approach,[67–69] where the solvent is usually modeled as a homogeneous, isotropic continuum characterized by its dielectric constant.[66] While it is widely implemented, this method has two strict limitations. First, the solute must be nearly spherical and second, multipole effects, beyond monopole and dipole interactions, must be negligible.[66] A second approach is the polarizable continuum model[70] (PCM) which divides the surface of the solute into discrete elements each containing a point charge. The overall interaction of the solute and the continuum is then a summation of the point charge interactions. Thirdly, the conductor-like screening model[71] (COSMO) uses a similar approach as PCM but instead of a dielectric outside of the solute cavity, a conductor is used with corrections added for dielectric behavior. Klamt and co-workers[72–74] have used COSMO, and its statistical mechanical extension, COSMO-RS, to study the solubility of drugs, pesticides, and other molecules in water and organic solvents. A final example of a popular continuum solvation models is the SMx family.[75] These models use a generalized Born approach where Born electrostatics are combined with a quantum mechanical description of the solute and the standard state free energy of solvation is calculated from the solvent free energy and the gas phase electronic energy contributions. The current model, SM6, was standardized using aqueous solvation free energy calculations for a range of 273 neutral solutes, as well as 112 ions.[76] The addition of one or more explicit water molecules showed significant improvement in free energy calculations for solutes that interact strongly with the solvent.

11.4 Computing the Solubility Limit

The second problem in solubility where simulation is used to probe behavior is that of the solubility limit. This is the maximum amount of a solute soluble in a given volume of solvent. It should be noted that in particular for cases where both the "solute" and the "solvent" exist in the liquid state, one should discuss the mutual solubility in the sense that both components have the ability to influence the solubility characteristics (and other properties) of the phase in which they are the minor component.

The solubility limit has been explicitly explored far less frequently than the infinite dilution limit and fewer simulation techniques have been used to probe this behavior. The most common is the use of GEMC simulations, particularly through modeling binary mixtures. By creating two bulk phases and propagating the system to reach equilibrium, the solubility of molecule A in the bulk phase that is rich in molecule B, and vice versa, may be determined. Various research groups have simulated the binary vapor–liquid phase behavior of hydrocarbon mixtures using GEMC (for example, see Refs. 77-82). The statistical uncertainty of the mole fraction of A in the bulk phase of B is usually less than 10%, often much less than this. Simulations of systems with supercritical carbon dioxide as one of the phases give similar uncertainties and provide insight into partitioning and fluid structure for these systems.[80,82] GEMC has also been used to study the solubility of gas molecules in aqueous electrolyte solutions.[83] Additionally, systems with three or more components and more than two phases are accessible with GEMC.[84–86] In one recent application, GEMC was used to examine *n*-decane/*n*-perfluorohexane/carbon dioxide and *n*-hexane/*n*-perfluorodecane/carbon dioxide systems.[86] The results show how an increase in pressure leads to a swelling of the two immiscible liquid phases and ultimately leads to an upper critical solution pressure (being the endpoint of the liquid–liquid–vapor three-phase region). These simulations closely followed experimental results and the phase compositions were calculated within a few percent of statistical uncertainty. The latter is one of the few examples of computational investigations of mutual solubilities involving two (condensed) liquid phases. Another example is a recent investigation of the mutual solubility of water and 1-butanol that also explored the surface tension behavior at the solubility limit.[87]

Other Monte Carlo methods have also been used to examine binary phase behavior. One recent example uses TMMC to determine the phase behavior of monatomic systems.[88] Important features of the phase diagrams, such as triple points and azeotropes, were determined with less than 1% uncertainty based on four independent simulations. Monte Carlo simulations in the grand canonical ensemble (GCMC) have also been used extensively to study fluid phase behavior. The thermodynamic constraints for the grand canonical ensemble are fixed temperature, volume, and chemical potential, *i.e.* the number of particles and energy can fluctuate.[11] Over the course of the simulation, the number of particles and energy of a given configuration may be stored and the probability distribution may be extracted after a simulation. Histogram-reweighting is often used to connect the different simulations. In order to allow for an exploration of the two-phase region throughout the course of a single simulation, the initial run is often carried out near the critical point with additional simulations and their histograms used to separately sample the liquid and vapor regions further away from the critical point.[89] Thus, although GCMC is an excellent method to map large regions in pressure–temperature space, it is less well suited to explore solubility at a single-state point (in particular, if this state point is far away from the critical region). Potoff *et al.* used GCMC to examine the phase behavior of a variety of non-polar and polar binary systems.[89] The average uncertainty in the pressures and compositions of the diagrams was a few percent.

Another method to probe the solubility limit was suggested by MacCullum and Tieleman.[90] In their study of water/1-octanol binary phase behavior, they simulated a series of systems with varying mole fractions of solute (in this case, water) and determined the excess chemical potential of water at each composition using a variant of Widom's particle insertion technique. Using the relationship that the chemical potential of a given molecular species must be equal between two phases that are in contact and at equilibrium, they determined the solubility limit of water in 1-octanol by comparing the chemical potential of water/octanol systems to that of pure water. Similarly, Ferrario *et al.*[91] used TI to determine the solubility limit of potassium fluoride in water by simulating a series of different systems with increasing potassium fluoride concentrations. This technique, while more intensive than GEMC for many systems, has potential applications for systems for which GEMC is untenable.

Phase behavior may also be probed with continuum solvation models. Thompson *et al.* used SM5.42 to predict solubility based on solvation free energy calculations.[92] Using a series of liquid and solid solutes (90 total solutes), the mean unsigned error of the logarithm of the solubility was about 0.4 compared to experimental solubility. For these calculations, the activity coefficient is unity for the saturated system. Based on the results for these 90 solutes, spanning seven orders of magnitude in solubility, this solvation model parameterized for infinite dilution can be used to predict saturated solubility as well. The authors postulate that these findings may be extended to the validity of other implicit and explicit solvation models.[92]

11.5 Finite Size Effects

One concern for all particle-based simulations is the necessary system size limitation due to computational efficiency and memory constraints. This restricts most simulations to no more than a few thousand molecules, often much less. In order to determine bulk properties, larger systems are necessary. The use of periodic boundaries to replicate the system in each dimension is the most common means to simulate bulk properties using a finite system.[15] Nevertheless, the finite size of the system can introduce systematic errors for computations of the solubility and, in particular, for calculations of the chemical potential.[8,93] It is always best to perform test simulations with systems of varying size to estimate the finite size corrections for the system of interest. It should be noted here that an expression for this correction has been derived for chemical potential calculations.[93]

11.6 Acknowledgments

We thank Dave Kofke for helpful comments. Financial support by the National Science Foundation and the 3M Graduate Fellowship (KEA) is gratefully acknowledged.

References

1. S.I. Sandler, *Chemical and Engineering Thermodynamics*, Wiley, New York, NY. 1989.
2. H.T. Davis, *Statistical Mechanics of Phases, Interfaces, and Thin Films*, VCH Publishers, New York, NY. 1996.
3. W.R. Fawcett, *Liquids Solutions, and Interfaces*, Oxford University Press, Oxford, 2004.
4. A. Ben-Naim, *Statistical Thermodynamics for Chemists and Biochemists*, Plenum Press, NY. 1992.
5. A. Ben-Naim, *J. Phys. Chem.*, 1978, **82**, 792.
6. A.Z. Panagiotopoulos, *Mol. Phys.*, 1987, **61**, 813.
7. A.Z. Panagiotopoulos, N. Quirke, M. Stapleton and D.J. Tildesley, *Mol. Phys.*, 1988, **63**, 527.
8. M.G. Martin and J.I. Siepmann, *Theor. Chem. Acc.*, 1998, **99**, 347.
9. K.S. Shing, K.E. Gubbins and K. Lucas, *Mol. Phys.*, 1988, **65**, 1235.
10. B. Widom, *J. Chem. Phys.*, 1963, **39**, 2808.
11. D. Frenkel and B. Smit, *Understanding Molecular Simulation*, Academic Press, San Diego, CA, 2nd edn, 2002.
12. R.J. Radmer and P.A. Kollman, *J. Comput. Chem.*, 1997, **18**, 902.
13. D.A. Kofke and P.T. Cummings, *Mol Phys.*, 1997, **92**, 973.
14. J.T. Wescott, L.R. Fisher and S. Hanna, *J. Chem. Phys.*, 2002, **116**, 2361.
15. M.P. Allen and D.J. Tildesley, *Computer Simulation of Liquids*, Oxford University Press, New York. 1996.
16. C.R.W. Guimarães and R.B. de Alencastro, *Int. J. Quant. Chem.*, 2001, **85**, 713.
17. T.P. Straatsma and H.J.C. Berendsen, *J. Chem. Phys.*, 1988, **89**, 5876.
18. D. Alfè, M.J. Gillan and G. D. Price, *J. Chem. Phys.*, 2002, **116**, 7127.
19. J. Harris and S.A. Rice, *J. Chem. Phys.*, 1988, **88**, 1298.
20. J.I. Siepmann, *Mol. Phys*, 1990, **70**, 1145.
21. J.I. Siepmann and D. Frenkel, *Mol. Phys.*, 1992, **75**, 59.
22. J.J. de Pablo, M. Laso and U.W. Suter, *J. Chem. Phys.*, 1992, **96**, 6157.
23. D. Frenkel, G.C.A.M. Mooij and B. Smit, *J. Phys.: Cond. Mat.*, 1992, **4**, 3053.

24. J.I. Siepmann, in *Monte Carlo Methods in Chemical Physics*, vol 105, D.M. Ferguson, J.I. Siepmann and D.G. Truhlar (eds), Wiley, New York, 1999, 443.
25. A.M.A. Dias, R.P. Bonifácio, I.M. Marrucho, A.A.H. Pádua and M.F. Costa Gomes, *Phys. Chem. Chem. Phys.*, 2003, **5**, 543.
26. M.F. Costa Gomes and A.A.H. Pádua, *J. Phys. Chem. B*, 2003, **107**, 14020.
27. V.E. Raptis, I.G. Economou, D.N. Theodorou, J. Petrou and J.H. Petropoulos, *Macromolecules*, 2004, **37**, 1102.
28. Y. Boutard, Ph. Ungerer, J.M. Teuler, M.G. Ahunbay, S.F. Sabater, J. Pérez-Pellitero, A.D. Mackie and E. Bourasseau, *Fluid Phase Equil.*, 2005, **236**, 25.
29. S.K. Kumar, I. Szleifer and A.Z. Panagiotopoulos, *Phys. Rev. Lett.*, 1991, **66**, 2935.
30. K. Ding and J.P. Valleau, *J. Chem. Phys.*, 1993, **98**, 3306.
31. W.L. Jorgensen and C. Ravimohan, *J. Chem. Phys.*, 1985, **83**, 3050.
32. W.L. Jorgensen, J.M. Briggs and M.L. Contreras, *J. Phys. Chem.*, 1990, **94**, 1683.
33. H.S. Kim, *Chem. Phys. Lett.*, 2000, **317**, 553.
34. D. Wu and D.A. Kofke, *J. Chem. Phys.*, 2005, **123**, 084109.
35. N. Lu, J.K. Singh and D.A. Kofke, *J. Chem. Phys.*, 2003, **118**, 2977.
36. N. Lu, D.A. Kofke and T.B. Woolf, *J. Comput. Chem.*, 2004, **25**, 28.
37. M.R. Shirts and V.S. Pande, *J. Chem. Phys.*, 2005, **122**, 144107.
38. C.H. Bennett, *J. Comput. Phys.*, 1976, **22**, 245.
39. G.C. A. M. Mooij and D. Frenkel, *J. Phys.: Cond. Mat.*, 1994, **6**, 3879.
40. A.P. Lyubartsev, A.A. Martsinovski, S.V. Shevkunov and P.N. Vorontsov-Velyaminov, *J. Chem. Phys.*, 1992, **96**, 1776.
41. A.P. Lyubartsev, A. Laaksonen and P.N. Vorontsov-Velyaminov, *Mol. Phys.*, 1996, **28**, 455.
42. A.P. Lyubartsev, O.K. Førrisdahl and A. Laaksonen, *J. Chem. Phys.*, 1998, **108**, 227.
43. E.C. Cichowski, T.R. Schmidt and J.R. Errington, *Fluid Phase Equil.*, 2005, **236**, 58.
44. K.M. Åberg, A.P. Lyubartsev, S.P. Jacobsson and A. Laaksonen, *J. Chem. Phys.*, 2004, **120**, 3770.
45. A.P. Lyubartsev, S.P. Jacobsson, G. Sundholm and A. Laaksonen, *J. Phys. Chem. B*, 2001, **105**, 7775.
46. G.C.A.M. Mooij, D. Frenkel and B. Smit, *J. Phys.: Cond. Matt.*, 1992, **4**, L255.
47. M. Laso, J.J. de Pablo and U.W. Suter, *J. Chem. Phys.*, 1992, **97**, 2817.
48. K. Esselink, L.D.J.C. Loyens and B. Smit, *Phys. Rev. E*, 1995, **51**, 1560.
49. A.D. Mackie, B. Tavitian, A. Boutin and A.H. Fuchs, *Mol. Simul.*, 1997, **19**, 1.
50. F.A. Escobedo and J. J. de Pablo, *J. Chem. Phys.*, 1996, **105**, 4391.
51. F.A. Escobedo, *J. Chem. Phys.*, 1998, **108**, 8761.
52. B. Chen and J.I. Siepmann, *J. Phys. Chem. B.*, 2006, **110**, 3555.
53. J.N.C. Lopez and D.J. Tildesley, *Mol. Phys.*, 1997, **92**, 187.

54. B. Chen and J.I. Siepmann, *J. Am. Chem. Soc.*, 2000, **122**, 6464.
55. D.A. Kofke, *Mol. Phys.*, 2004, **102**, 405.
56. L. Zhang and J.I. Siepmann, *Theor. Chem. Acc.*, 2006, **115**, 391.
57. M.G. Martin, J.I. Siepmann and M.R. Schure, *J. Phys. Chem. B*, 1999, **103**, 11191.
58. C.D. Wick, J.I. Siepmann and M.R. Schure, *Anal. Chem.*, 2002, **74**, 37.
59. C.D. Wick, J.I. Siepmann and M.R. Schure, *Anal. Chem.*, 2002, **74**, 3518.
60. C.D. Wick, J.I. Siepmann and M.R. Schure, *Anal. Chem.*, 2004, **76**, 2886.
61. L. Sun, C.D. Wick, J.I. Siepmann and M.R. Schure, *J. Phys. Chem. B*, 2005, **109**, 15118.
62. J.L. Rafferty, L. Zhang, J.I. Siepmann and M.R. Schure, *Nature*, submitted for publication.
63. C.D. Wick, J.I. Siepmann and M.R. Schure, *J. Phys. Chem. B*, 2003, **107**, 10623.
64. C.J. Cramer and D.G. Truhlar, in *Quantitative Treatments of Solute/Solvent Interactions*, vol 1, P. Politzer and J.S. Murray (eds), Elsevier, Amsterdam, 1994, 9.
65. G.B. Bacskay and J.R. Reimers, in *Encyclopedia of Computational Chemistry*, vol 4, P.v.R. Schleyer, Wiley, Chichester, 1998, 2620.
66. C.J. Cramer and D.G. Truhlar, *Chem. Rev.*, 1999, **99**, 2161.
67. M.Z. Born, *Physik*, 1920, **1**, 45.
68. J.G. Kirkwood, *J. Chem. Phys.*, 1939, **7**, 911.
69. L. Onsager, *J. Am. Chem. Soc.*, 1936, **58**, 1486.
70. J. Tomasi and M. Persico, *Chem. Rev.*, 1994, **94**, 2027.
71. A. Klamt and G. Schüürmann, *J. Chem. Soc., Perkins Trans.*, 1993, **2**, 799.
72. A. Klamt, F. Eckert, M. Hornig, M.E. Beck and T. Bürger, *J. Comput. Chem.*, 2002, **23**, 275.
73. R. Putnam, R. Taylor, A. Klamt, F. Eckert and M. Schiller, *Ind. Eng. Chem. Res.*, 2003, **42**, 3635.
74. M. Diedenhofen, F. Eckert and A. Klamt, *J. Chem. Eng. Data.*, 2003, **48**, 475.
75. C.J. Cramer and D.J. Truhlar, *Science*, 1992, **256**, 213.
76. C.P. Kelly, C.J. Cramer and D.G. Truhlar, *J. Chem. Theory Comput.*, 2005, **1**, 1133.
77. J.J. de Pablo, M. Bonnin and J.M. Prausnitz, *Fluid Phase Equil.*, 1992, **73**, 187.
78. I. Yu. Gotlib, E.M. Piotrovskaya and S.W. de Leeuw, *Fluid Phase Equil.*, 1997, **129**, 1.
79. M.G. Martin and J.I. Siepmann, *J. Am. Chem. Soc.*, 1997, **119**, 8921.
80. M.G. Martin, B. Chen and J. I. Siepmann, *J. Phys. Chem. B*, 2000, **104**, 2415.
81. J. Carrero-Mantilla and M. Llano-Reptrepo, *Mol. Simul.*, 2003, **29**, 549.
82. A. van 'T Hoff, S.W. de Leeuw, C.K. Hall and C.J. Peters, *Mol. Phys.*, 2004, **102**, 301.
83. J. Vorholz, V.I. Harismiadis, A.Z. Panagiotooulos, B. Rumpf and G. Maurer, *Fluid Phase Equil.*, 2004, **226**, 237.

84. P.C. Tsang, O.N. White, B.Y. Perigard, L.F. Vega and A.Z. Panagiotopoulos, *Fluid Phase Equil.*, 1995, **107**, 31.
85. B. Neubauer, B. Tavitian, A. Boutin and P. Ungerer, *Fluid Phase Equil.*, 1999, **161**, 45.
86. L. Zhang and J. I. Siepmann, *J. Phys. Chem. B*, 2005, **109**, 2911.
87. B. Chen, J.I. Siepmann and M.L. Klein, *J. Am. Chem. Soc.*, 2002, **124**, 12232.
88. V.K. Shen and J.R. Errington, *J. Chem. Phys.*, 2005, **122**, 064508.
89. J.J. Potoff, J.R. Errington and A.Z. Panagiotopoulos, *Mol. Phys.*, 1999, **97**, 1073.
90. J.L. MacCullum and D.P. Tieleman, *J. Am. Chem. Soc.*, 2002, **124**, 15085.
91. M. Ferrario, G. Ciccotti, E. Spohr, T. Cartailler and P. Turq, *J. Chem. Phys.*, 2002, **117**, 4947.
92. J.D. Thompson, C.J. Cramer and D.G. Truhlar, *J. Chem. Phys.*, 2003, **119**, 1661.
93. J.I. Siepmann, I.R. McDonald and D. Frenkel, *J. Phys.: Cond. Mat.*, 1992, **4**, 679.

CHAPTER 12

Prediction of Solubility with COSMO-RS

FRANK ECKERT

COSMOlogic, GmbH & Co KG, Burscheider Str. 515, D-51381 Leverkusen, Germany

12.1 Introduction

Solubility is one of the most fundamental processes in chemistry and biology. The understanding and the prediction of the thermodynamic properties of neutral compounds and of salts in water and organic solvents are of crucial importance in many areas of chemistry and biochemistry. Most physiological and technical processes occur in solution, and the choice of solvent or solvent mixture is very important to the rates and outcome of a process.[1] When considered at thermodynamic equilibrium, solubility can be expressed in terms of the free energy (or chemical potential) difference of a compound X dissolved in a solvent phase S and the pure compound X. The theoretical calculation of a chemical potential is complicated, because we do not only have to calculate the interaction energy of a solute X in a solvent S, but also have to take into account the change in the entropy and in the interactions of the solvent molecules caused by the solute molecule X.

Molecular dynamics (MD) or Monte Carlo (MC) methods are the most straight-forward procedures to compute the change in free energy of an ensemble of solvent molecules S by insertion of a solute X. To get reasonably accurate values one has to consider a very large ensemble of solvent molecules. Nowadays such calculations can be done routinely based on force-field pair-potentials,[2] but one should be aware that all interactions, which are of quantum-chemical nature, are described by a classical force-field approximation. Unfortunately these calculations are rather time-consuming. Jorgensen and Duffy[3] introduced a shortcut of the MD/MC approach, which is based on averaged interaction descriptors derived from rapid simulations in reference solvents and combined with a quantitative-structure-property-relationship (QSPR) analysis with respect to the solubility.

In contrast, the computationally fastest, but chemically least detailed approach to the estimation of partition coefficients is the fragment- or group-based increment approach, also known as linear free energy relationship (LFER) approach. The basic assumption of these methods is that the change in free energy of a solute *X* between solvents *S* and S′ can be split into independent contributions of the chemical groups of *X*. Thus the logarithmic solubility becomes a sum of group contributions, which can be fitted by linear regression from a sufficiently large set of experimental data. The LFER approach has been applied successfully to the prediction of partition properties such as octanol-water partition coefficients.[4] A disadvantage however, is that a data set of several thousand partition coefficients is required in order to fit the large number of group-parameters. Moreover, while the chemical potential of a solute *X* diluted in a solvent S can be approximated quite well by a sum of fragment contributions, the chemical potential of a pure compound *X* that acts as both solute and solvent, is not readily available. In the pure compound the assumption of linear additivity of free energy contributions of structural fragments to a target molecule is not valid, because the addition of a certain fragment can either increase or decrease the solubility of a compound, depending on the remainder of the compound *X*. Thus solubility, being a strongly non-linear property, can hardly be expressed by a linear regression analysis, unless one of the descriptors does include most of the non-linear behaviour.

A different type of group contribution models (GCMs) such as UNIFAC,[5] in which solute and solvent are represented by groups, is widely used in chemical engineering applications. In such models the chemical potentials of the compounds are derived from an approximate statistical thermodynamic treatment of pair-wise interacting surface pieces. For each pair of functional groups an interaction parameter has to be fitted to experimental thermodynamic data in a – non-linear fitting procedure. The advantage of this approach, compared with the LFER approach, is its ability to treat any solvent or solvent mixture as well as complete binary phase diagrams, provided the interaction parameters for all groups involved in the system are known. A disadvantage is the fact that in GCMs the number of required parameters involved, is related to the square of the number of groups. For the LFER method, however, the number of parameters is linearly related to the number of groups. Therefore, UNIFAC can afford a much less detailed definition of groups compared to typical LFER methods. Nevertheless, GCMs are of limited applicability to typical solid drug-like compounds, because they are not capable of treating heterocycles or multifunctional aromatic rings.

A different approach is the explicit representation of the solute combined with a continuum representation of the solvent. Most of these continuum solvation models,[6,7] (CSMs) concentrate on the electrostatic behaviour of the solvent. Either by solution of the dielectric boundary conditions or by solution of the Poisson–Boltzmann equations (both of which represent the same physics in non-ionic solvents) the solute is treated as if it is embedded in a dielectric medium. Usually the macroscopic dielectric constant of the solvent is used. The conductor-like screening model (COSMO)[8] is a model of this type, which by a slight approximation achieves superior efficiency and robustness compared with others. The advantage of CSMs is that the solute can be treated with great rigor, typically at a quantum chemical level. If supplemented with surface specific descriptors characterizing dispersive interactions, cavitation energies, and other non-electrostatic contributions, the results of such CSMs appear to be capable of describing liquid partition properties and solubility.[6]

To summarize, there are several approaches to the calculation of free energies of molecules in solution, each of them covering different aspects of the problem. Despite the usually assumed picture of pair-wise and distance-dependent interactions of atoms, the relative success of MD/MC derived interaction parameters, of group interaction models, and of surface parameter supplemented CSMs suggests that many aspects of free energies of molecules in solution can be as well or even better described by a model of surface interactions without explicit knowledge of the atom positions of the solvent. The COSMO-RS method is a model based on such an assumption, combining the advantages of quantum chemically based CSMs and sound statistical thermodynamics. The method is described in the following sections and is applied to problems of solubility predictions.

12.2 COSMO-RS

COSMO-RS is a predictive method for the thermodynamic properties of fluids and liquid mixtures that combines a statistical thermodynamic approach with a quantum chemistry method. The theory of COSMO-RS has been described in detail in several articles.[9–13] Therefore only a short survey of the basic concept will be given here.

The starting point for any COSMO-RS calculation is a molecule X in its ideally screened state. This state can be calculated with reasonable effort by a quantum chemical method, the COSMO,[8] which is an efficient variant of dielectric continuum solvation method (DCSM).[6,7] In the COSMO model, a solute molecule is calculated in a virtual conductor environment. In such an environment, the solute molecule induces a polarization charge density σ on the interface between the molecule and the conductor, that is, on the molecular surface. These charges act back on the solute and thus generate a polarized electron density. During the quantum chemical self-consistency algorithm, the solute molecule is thus converged to its energetically optimal state in a conductor with respect to its electron density, including geometry optimization. The quantum chemical calculation has to be performed once for each molecule of interest. The resulting polarization charge densities on the molecular surface σ are – good local descriptors of the molecular surface polarity and can be stored in a database. The σ values from the COSMO calculation allow one to extend the model towards "Real Solvents", which results in the COSMO-RS method. In COSMO-RS, the deviations of a real solvent, compared to an ideal conductor are taken into account in a model of pair-wise interacting molecular surfaces. For this purpose, the three-dimensional polarization density distribution on the surface of each molecule X is converted into a distribution function, the so-called σ-profile $p^X(\sigma)$, which gives the relative amount of surface with polarity σ on the surface of the molecule. The σ -profile for the entire solvent of interest S, which might be a mixture of several compounds, $p_S(\sigma)$, can be built by adding the $p^X(\sigma)$ values of the components weighted by their mole fractions x_i in the mixture. Now, electrostatic energy differences and hydrogen bonding energies are quantified as functions of the local COSMO polarization charge densities σ and σ' of the two interacting surface pieces. The chemical potential differences arising from these interactions are evaluated using an exact statistical thermodynamic algorithm for independently pair-wise interacting surfaces, which takes into account local deviations from dielectric behaviour as well as hydrogen bonding. In this approach all information about solutes and solvents is extracted from initial QC-COSMO calculations, and only very few parameters have been adjusted to experimental values of partition coefficients and vapour pressures of a wide range of neutral organic compounds. COSMO-RS is capable of predicting partition coefficients, vapour pressures, and solvation free energies of neutral compounds with an error of 0.3 log-units (rms) and better. This corresponds to an accuracy of about 1.5 kJ mol^{-1} for large chemical-potential differences like those typically involved in octanol–water partition coefficients or in water solubility and slightly less than a factor 2 for equilibrium constants at room temperature. A lot of experience has been gathered during the past years about COSMO-RS' surprising ability to predict thermodynamic properties of mixtures.[13]

12.3 Computational Details

For all molecular species involved, the standard two step procedure for COSMO-RS calculations has been applied:

(i) Quantum chemical COSMO calculations. These involve setting up initial molecular geometries and determination of the lowest energy conformations. For each compound a MD calculation has been done with the molecular modelling program package Alchemy.[14] The MM3 force field[15] has been used to obtain the potential energy during the MD calculation, using an overall MD run time of 5 ps, a time step of 0.001 ps and a initial temperature of 293 K. From the geometries created by the MD calculation up to five significant lowest energy conformations have been picked for each molecule. Special care has been taken in choosing conformations of molecules which are able to build internal hydrogen bonds, since the polarization-charge densities σ computed in the subsequent QC-COSMO calculations (and thus also COSMO-RS' chemical potentials $\mu_S{}^X$) critically depend upon the correct representation of such hydrogen bonds. Subsequently the geometry of the chosen conformations has been optimized with the Turbomole quantum chemistry program package[16–18] using the B–P density functional[19,20] with TZVP quality basis set and the RI approximation.[21,22] During these calculations the COSMO CSM was applied in the conductor limit

($\varepsilon = \infty$). Element-specific default radii from the COSMO-RS parameterizations have been used for the COSMO cavity construction.[12,13] Such calculations end up with the self-consistent state of the solute in the presence of a virtual conductor that surrounds the solute outside the cavity.

(ii) COSMO-RS calculations have been done using the COSMO*therm* program.[23] If more than one conformations were considered to be potentially relevant for the neutral or ionic form of a compound, several conformations have been calculated in step 1 and a thermodynamic Boltzmann average over the total Gibbs free energies of the conformers was consistently calculated by the COSMO*therm* program in step 2. Details on the COSMO-RS calculation method and all COSMO-RS parameters used are given in Ref. 9.

12.4 Solubility

Considering solubility in thermodynamic equilibrium, the quantity required for its calculation is the chemical potential $\mu_S{}^X$ of a compound X in a solvent S, at a given temperature T and dilution. Using the pseudo-chemical potential $\mu^*{}_S{}^X$ according to Ben Naim:[24]

$$\mu_S^{*X} = \mu_S^X - kT \ln x_S^X \quad (1)$$

where $x_S{}^X$ is the molar concentration of compound X in solvent S, the equilibrium condition of equal chemical potentials of X in two phases S and S' reads:

$$\mu_S^{*X} - kT \ln x_S^X = \mu_{S'}^{*X} - kT \ln x_{S'}^X \quad (2)$$

Thus the solubility $S_S{}^X$ of a liquid compound X in a solvent S is related to the difference $\Delta_S{}^X = \mu^*{}_S{}^X - \mu^*{}_X{}^X$ of the pseudo-chemical potentials of X in solvent S and of pure compound X. If $S_S{}^X$ is sufficiently small so that the solvent behaviour of the X-saturated solvent S is not significantly influenced by the solute X (infinite dilution of X in S), then the decadic logarithm of the solubility is given by:

$$\log S_S^X = \log\left(\frac{MW_X \rho_S}{MW_S}\right) - \frac{1}{kT \ln(10)} \Delta_S^X \quad (3)$$

Note that in the case of high solubility (solubility greater than 10 mol%), Equation (3) becomes approximate and the true solubility would have to be derived from a detailed search for a thermodynamic equilibrium of a solvent-rich and a solute-rich phase. However, if the zeroth order $S_S{}^{X(0)}$ as initially provided by Equation (3) using infinite dilution of X in S, is re-substituted into the solubility calculation *via* $\Delta_S{}^{X(1)} = \mu^*{}_{S[x(0)]}{}^X - \mu^*{}_X{}^X$, a better approximation for

S_S^X is achieved. In other words, the solute pseudo-chemical potential $\mu^*{}_{S[x(0)]}{}^X$ is computed for the solvent-solute mixture with the finite mole fraction of X in S that was predicted by the zeroth order $S_S^{X(0)}$. Now, using Equation (3) with the new $\mu^*{}_{S[x(0)]}{}^X$ and the resulting $\Delta_S^{X(1)}$ values, an improved solubility $S_S^{X(1)}$ is computed. This value again can be used to compute further refined $\mu^*{}_{S[x(1)]}{}^X$ and again re-substituted into Equation (3), a better guess for S_S^X can be achieved. This procedure can be repeated until the computed value of S_S^X is constant. This iterative-refinement procedure is implemented in the COSMO*therm* program[23] and allows the accurate prediction of solubility values even for cases of high solubility (solubility up to 50 mol%). Thus except for rare cases of very high solubility, a complicated search for a multiphase thermodynamic equilibrium of a solvent-rich and a solute-rich phase is not necessary, but instead Equation (3) and the iterative refinement procedure can be used. All of the following examples were calculated with an iteratively refined Equation (3).

Since the molecular weights (MW) and the solvent density usually ρ are known, Equation (3) is sufficient for the prediction of the solubility of compounds which are liquid at room-temperature. Unfortunately most drugs are solid at room temperature. Since the solid state of a compound X is related to its liquid state by the free energy difference $\Delta G_{\text{fus}}{}^X$ which is negative in the case of solids, a more general expression for solubility reads:

$$\log S_S^X = \log\left(\frac{MW_X \rho_S}{MW_S}\right) + \frac{1}{kT \ln(10)}\left[-\Delta_S^X + \min(0, \Delta G_{\text{fus}}^X)\right] \tag{4}$$

Since for liquids $\Delta G_{\text{fus}}{}^X$ is positive, Equation (4) reduces to Equation (3) in this case. For the precise calculation of $\Delta G_{\text{fus}}{}^X$ it is necessary to evaluate the free energy of a molecule of compound X in its crystal, *i.e.* the crystal structure has to be known. In general, crystal-structure prediction for complex molecules has to be considered as an unsolved problem.[25] Thus there is no viable way to a fundamental model. However, typically $\Delta G_{\text{fus}}{}^X$ is small compared to Δ_S^X. Hence it is reasonable to use Δ_S^X of the liquid as a fundamental input for the calculation of log S_S^X and to search for some plausible-empirical approximation for $\Delta G_{\text{fus}}{}^X$.

In a study on the aqueous solubility of 150 drug-like organic molecules taken from Ref. 3, we found that a simple correlation of log S_S^X *vs.* Δ_S^X (as computed by COSMO-RS) yields a correlation coefficient of $r^2 = 0.65$ and a rms-deviation of 1.2 log-units. The slope in this regression is close to the theoretical expectation. This clearly show the great significance of the pseudo-chemical potentials as calculated by COSMO-RS. In a second step the theoretical liquid solubility values of $\Delta_S^X\ kT^{-1}\ \ln(10)$ were subtracted from the experimental values of solubility log S_S^X in order to obtain reasonable data values for $\Delta G_{\text{fus}}{}^X$. Now it was possible to find a simple linear QSPR expression for $\Delta G_{\text{fus}}{}^X$ based on molecular descriptors provided by COSMO-RS.[26] Thus we derived the equation:

$$\Delta G^X_{\text{fus}} = 12.2\ V^X + 0.54\,\mu^{*X}_{\text{water}} - 0.76\ N^X_{\text{ringatom}} \quad (5)$$

where the units for $\Delta G_{\text{fus}}{}^X$ and $\mu^*{}_{\text{water}}{}^X$ are (kJ mol^{-1}) and for V^X it is (nm^3). The descriptor V^X is the COSMO volume. Descriptor $\mu^*{}_{\text{water}}{}^X$, the pseudo-chemical potential of solute X in solvent water at infinite dilution, is a combined measure of the solutes polarity and hydrogen bonding properties. Hence this descriptor also appears in the solubility calculation of arbitrary non-aqueous solvents S. The number of ring atoms $N^X{}_{\text{ringatom}}$ acts as a descriptor of molecular rigidity. In Equation (5) the regression constant c_0 is omitted, because it was found to be insignificant for this regression. Equation (5) is applicable to room-temperature solubilities of a wide range of solid-organic solutes in arbitrary solvents S. The correlation of the experimental-aqueous solubilities log $S_{\text{water}}{}^X$ of the 150 compounds taken from Ref. 3 with COSMO-RS properties *via* Equations (4) and (5) yielded a correlation coefficient of $r^2 = 0.90$ and a rms deviation of 0.66 log ($x_{\text{water}}{}^X$) units.[26] The COSMO-RSol solubility model thus defined was verified using a data set of aqueous solubility values for 548 pesticide compounds, yielding a standard deviation of $s = 0.61$ log ($x_{\text{water}}{}^X$) units.[26] Originally, COSMO-RSol was developed as a tool for the prediction of the solubility of pure, neutral, and undissociated drug-like compounds in water.[26] But its applicability is not restricted to these classes of solutes and solvents. Corrections for dissociation or protonation can be trivially made for compounds with known pK-values. If the dissociation constant is not known from experiment, it can be calculated routinely by COSMO-RS, for acids[27] as well as for bases.[28] The application of COSMO-RSol to the solubility of non-neutral compounds is demonstrated in Section 5. Although some applications of COSMO-RSol have been reported,[29,30] the method has not been systematically verified, using a large data set of solubilities in non-aqueous solvents. In the recent years Acree[31] has built up a large data source for experimental room temperature solubilities of complex organic compounds in various solvents. Acree's set of 706 room temperature solubilities of 21 solutes in a variety of solvents ranging from non-polar alkanes to strongly polar alcohols, amides, carbon acids, and water has been predicted by the COSMO-RSol method as outlined above and is given in Figure 1. Note that a value of log ($x_S{}^X$) = 0 means that arbitrary miscibility of solute and solvent was predicted. The 706 solubilities were predicted with an overall rms deviation of 0.74 log ($x_S{}^X$) units.[32] If experimental data for $\Delta G^X{}_{\text{fus}}$ as recommended by NIST[33] are used in Equation (4), the rms error of the predictions reduces to 0.43 log ($x_S{}^X$) units.[32] This is well within the expected error ratio of COSMO-RSol and corroborates the broad scope and general applicability of the method. Note, that a major part of the error of the full COSMO-RSol predictions (using the $\Delta G^X{}_{\text{fus}}$ estimate) is caused by a single solute, 4-nitrobenzoic acid, where the $\Delta G^X{}_{\text{fus}}$ value is underestimated strongly by COSMO-RSol-QSPR. If the 29 data values for 4-nitrobenzoic acid are removed from the data set, the rms error of the full COSMO-RSol predictions reduces to 0.64 log

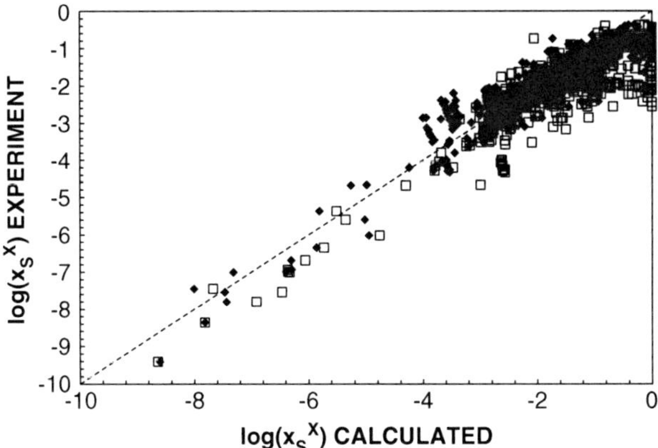

Figure 1 *Experimental[31] vs. calculated solubility log (x_S^X) of organic compounds in different solvents.[32] Filled rhombus: predictions using ΔG^X_{fus} data estimated by COSMO-RSol. Open squares: predictions using experimental ΔG^X_{fus} data.*

(x_S^X) units, which is even below the rms error obtained for the COSMO-RSol fitted data set.[26]

Cinchona alkaloids are an example for compounds that display an interesting solubility behaviour their solubilities in solvents of different polarity vary by 5–6 orders of magnitude and they show non-trivial behaviour in mixed solvents composed of water and organic solvents.[34] For solute cinchonidine (CAS-RN: 485-71-2) high solubility in solvent 1,4-dioxane is reported, which increases if a small amount of water is added to the solvent.[34] If the water fraction in the solvent mixture is increased further, the solubility decreases and finally drops to −5.91 log (x_S^X) units for pure water. This non-trivial behaviour is predicted by COSMO-RSol, qualitatively as well as quantitatively (see Figure 2) achieving an rms error of 0.40 log (x_S^X) units, if the experimental cinchonidine ΔG^X_{fus} was used.[32]

12.5 Salt Solubility

The prediction of the solubility of salts involves a few complications. First, in COSMO-RS a salt A^-C^+ is always treated by means of its anion A^- and cation C^+ separately. To obtain a salts solubility, the pseudo-chemical potentials and the free energy of fusion ΔG_{fus} have to be determined for the individual anion A^- and cation C^+. Now, the salts solubility is determined from the mean pseudo-chemical potentials and the sum of the free energy of fusion of the ions. Thus for salts, Equation (4) is modified to:

$$\log S_S^{\text{AC}} = \log\left(\frac{MW_{\text{AC}}\rho_S}{MW_S}\right) + \frac{1}{kT\ln(10)}\left[-\Delta_S^{\text{AC}}/2 + \min(0, \Delta G_{\text{fus}}^{\text{AC}})\right] \qquad (6)$$

wherein $\Delta_S{}^{\text{AC}} = \mu^*{}_S{}^{\text{AC}} - \mu^*{}_{\text{AC}}{}^{\text{AC}}$. The pseudo-chemical potential of the pure salt $\mu^*{}_{\text{AC}}{}^{\text{AC}}$ is the sum of the pseudo-chemical potentials of the anion A^- and

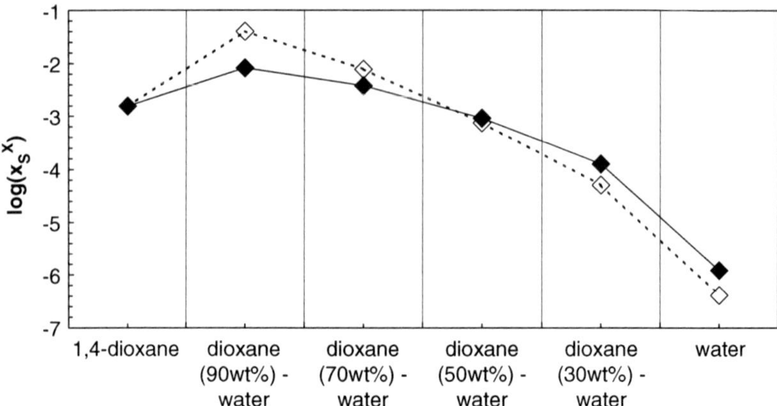

Figure 2 *Experimental vs. calculated solubility log* $(x_S{}^X)$ *of cinchonidine in dioxane – water solvent mixtures at* $T = 25°C$.[32] *Filled rhombus: Experimental data of Ma and Zaera.*[34] *Open rhombus: COSMO-RSol predictions using experimental cinchonidine* $\Delta G^X{}_{fus}$ *data.*

cation C^+ as determined in an equimolar (50:50) mix of A^- and cation C^+

$$\mu^{*\mathrm{AC}}_{\mathrm{AC}} = \mu^{*A^-}_{(50:50)} + \mu^{*\mathrm{C}^+}_{(50:50)} \tag{7}$$

The chemical potential of the salt in solution $\mu^*{}_S{}^{\mathrm{AC}}$ is the sum of the chemical potentials of the A^- and C^+ computed in infinite dilution in solvent *S*:

$$\mu^{*\mathrm{AC}}_S = \mu^{*A^-}_S + \mu^{*C^+}_S \tag{8}$$

In a study on the aqueous solubility of 22 *para*-substituted benzoic acid salts of benzylamine at $T = 37°C$,[35] it was found that a simple correlation of log $S_S{}^{\mathrm{AC}}$ *vs.* $\Delta_S{}^{\mathrm{AC}}$ as computed from Equations (7) and (8), yielded a correlation coefficient of $r^2 = 0.69$ and a rms-deviation of 1.46 log-units. Again, it turned out that the chemical potential difference between the ions in aqueous solution and the virtual super-cooled melt of the salt, *i.e.* the virtual ionic liquid, is the most important and significant contribution to the logarithmic solubility. To be able to obtain a QSPR expression for the salts free energy of fusion, the theoretical liquid solubility values of $\Delta_S{}^{\mathrm{AC}}\ \mathrm{kT}^{-1}\ \ln(10)$ were subtracted from the experimental values of solubility log $S_S{}^{\mathrm{AC}}$ to obtain pseudo-experimental data values for $\Delta G_{\mathrm{fus}}{}^{\mathrm{AC}}$. The performance of different combinations of molecular descriptors provided by COSMO-RS in multilinear regression has been tested. The following descriptors of potential significance for $\Delta G_{\mathrm{fus}}{}^{\mathrm{AC}}$ have been tested in different combinations: molecular volume V or area A as a measure of the molecules size, the number of ringatoms N_{ringatom} as a measure of molecular rigidity, the dielectric COSMO energy E_{diel} as a descriptor of polarity, the pseudo chemical potential $\mu^*{}_{\mathrm{water}}$ of the salt in water as a combined measure of polarity and hydrogen bonding capacity and the generic

COSMO-RS descriptors known as σ-moments[13] M_i. Finally, it turned out, that the descriptor combination V^{AC}, $N^{AC}_{ringatom}$, and $\mu^*_{water}{}^{AC}$ is best suited for the regression of $\Delta G_{fus}{}^{AC}$. The polarity descriptor $E_{diel}{}^{AC}$ did not achieve any significance. The improvement achieved by introducing the COSMO-RS σ-moments into the regression was negligible. Thus, the QSPR equation for salt free energy of fusion has the same functional form as the equation for neutral compounds, Equation (5). The QSPR parameters however, have to be readjusted for the computation of the salt's free energy of fusion. The QSPR descriptors of the salt are just the sums of the QSPR descriptors of the anion A^- and the cation C^+. Thus, for salts, Equation (5) translates to:

$$\Delta G_{fus}^{AC} = c_0 + c_V V^{AC} + c_\mu \mu_{water}^{*AC} + c_N\ N_{ringatom}^{AC} \tag{9}$$

wherein $\mu^*_{water}{}^{AC} = \mu^*_{water}{}^{A-} + \mu^*_{water}{}^{C+}$, $V^{AC} = V^{A-} + V^{C+}$ and $N^{AC}{}_{ringatom} = N^{A-}{}_{ringatom} + N^{C+}{}_{ringatom}$. The correlation of the experimental aqueous solubilities $\log S_{water}{}^{AC}$ of a high quality data set[35] of 22 *para*-substituted benzoic acid salts of benzylamine at $T = 37°C$ with COSMO-RS properties via Equations (6)–(9) yielded a correlation coefficient of $r^2 = 0.56$ and a rms deviation of only 0.30 log ($x_{water}{}^{AC}$) units.[32] Regression coefficients $c_V = -36.9$, $c_\mu = 0.046$, $c_N = -0.14$, and a regression constant $c_0 = 7.83$ were determined, where the units for $\Delta G_{fus}{}^{AC}$ and $\mu^*_{water}{}^{AC}$ are (kJ mol^{-1}) and for V^{AC} it is (nm^3). The fitted data set is given in Figure 3. The fitted coefficients were tested on a set of 8 aromatic carbon acid salts of the local anaesthetic agent bupivacaine (CAS-RN: 2180-92-9) at $T = 37°C$. This data was measured with the same experimental methodology as the fit data set.[36,37] Since the test data set is consistent with the measurement method and temperature of the training data set, but chemically much more complex than the fit data set, it can provide an indication for the extrapolative quality of the COSMO-RSol method for salts. The results for the test data set is also given in Figure 3. The experimental solubility values are predicted with an rms deviation of 0.37 log ($x_{water}{}^{AC}$) units and a mean deviation of 0.15 log ($x_{water}{}^{AC}$) units.[32]

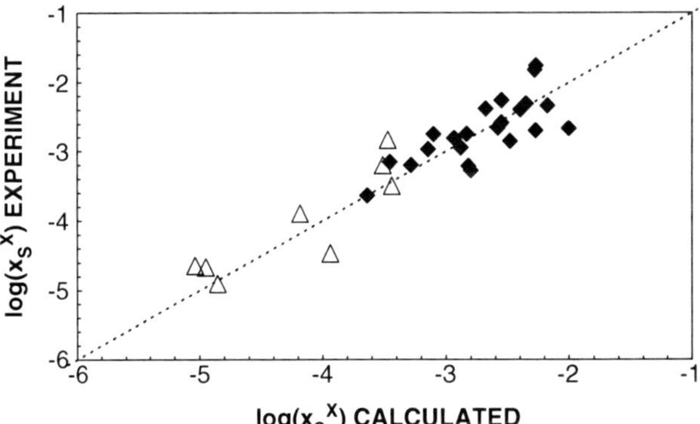

Figure 3 *Experimental vs. calculated aqueous solubility log ($x_{water}{}^{AC}$) of organic salts at $T = 37°C$. Filled rhombus: fitted data set of benzylamine salts. Open triangles: test data set of bupivacaine salts.*

Considering the increased chemical complexity of the test data set, it can be concluded that the methodology is predictive well beyond the boundary of similar or identical chemical functionality and structure of the organic salts. It should be noted, however, that the fitted data set presented is too small and not diverse enough to be accounted for as a general and transferable prediction method for salt solubility by COSMO-RSol. Furthermore, the adjusted QSPR parameters are valid for $T = 37°C$, the temperature of the solubility measurements, only. This data set was used to demonstrate the principle functionality and practical workability of the method. The generalization of the method to the whole of organic chemistry is straightforward, but ultimately depends on an appropriate set of experimental salt solubilities that allows the fitting of the QSPR parameters. Thus, currently the main problem remaining is the collection of a reliable and validated set of experimental room temperature solubilities of organic salts where the anions and cations show a broad distribution of chemical functionality and structure.

12.6 Summary and Conclusions

The applicability and capacity of the novel COSMO-RSol method as a prediction tool for the solubility of neutral solid compounds and salts in water and non-aqueous solvent media has been demonstrated. It was found that same empirical QSPR formula for the estimation of ΔG_{fus} can be used for salts and for neutral compounds. The coefficients of the QSPR model, however, have to be readjusted for salts. The readjustment of the QSPR coefficients for salts has been demonstrated at a coherent but small-data set of experimental salt solubilities. The data set presented is not sufficiently large and diverse enough to provide general and transferable COSMO-RSol QSPR parameters for salts. Unfortunately the lack of reliable and diverse experimental solubilities for organic salt compounds may be the major drawback for further developments in this area. As a result it can be concluded that despite the empirical character of the expression for the free energy of fusion, COSMO-RSol has a rather sound physico-chemical basis compared to all presently available prediction methods for solution phenomena and thus provides the most fundamental and transferable-prediction tool for the solubility of organic compounds. Another advantage of this new method is that based on the same COSMO calculations used for aqueous solubility many other physico-chemical properties, such as solubility in non-aqueous solvents, partition coefficients, vapour pressures, Henry constants, *etc.* are easily available by COSMO-RS.[9] Even physiological-partition behaviour can be calculated based on COSMO-RS.[13]

References

1. P. Kolar, J.-W. Shen, A. Tsuboi and T. Ishikawa, *Fluid Phase Equil.*, 2002, **194–197**, 771.
2. W.L. Jorgensen, in *Encyclopedia of Computational Chemistry*, Vol 2, P.V.R. Schleyer and L. Allinger (eds.), Wiley, New York, NY, 1998.

3. E.M. Duffy and W.L. Jorgensen, *J. Am. Chem. Soc.*, 2000, **122**, 2878.
4. C. Hansch and A.J. Leo, *Substituent Parameters for Correlation Analysis in Chemistry and Biology*, Wiley, New York, NY, 1979.
5. A. Fredenslund, J. Gmehling and P. Rasmussen, *Vapor Liquid Equilibria Using UNIFAC*, Elsevier, Amsterdam, 1977.
6. C.J. Cramer, and D.G. Truhlar, in *Reviews in Computational Chemistry*, Vol 6, K.B. Lipkowitz and D.B. Boyd, (eds), VCH Publishers, New York, NY, 1995.
7. J. Tomasi, B. Mennucci and R. Cammi, *Chem. Rev.*, 2005, **105**, 2999.
8. A. Klamt and G. Schüürmann, *J. Chem. Soc. Perkins Trans.*, 1993, **2**, 799.
9. F. Eckert and A. Klamt, *AIChE J.*, 2002, **48**, 369.
10. A. Klamt and F. Eckert, *Fluid Phase Equilibria*, 2000, **172**, 43.
11. A. Klamt, V. Jonas, T. Buerger and J. C. W. Lohrenz, *J. Phys. Chem.*, 1998, **102**, 5074.
12. A. Klamt, *J. Phys. Chem.*, 1995, **99**, 2224.
13. A. Klamt, *COSMO-RS From Quantum Chemistry to Fluid Phase Thermodynamics and Drug Design*, Elsevier, Amsterdam, 2005.
14. Alchemy32, Version 2.0.5, Tripos, Inc., St. Louis, MO, 1998.
15. N.L. Allinger, Y.H. Yuh and J.-H. Lii, *J. Am. Chem. Soc.*, 1989 **111**, 8551.
16. A. Schäfer, A. Klamt, D. Sattel, J.C.W. Lohrenz and F. Eckert, *Phys. Chem. Chem. Phys.*, 2000, **2**, 2187.
17. R. Ahlrichs, M. Bär, M. Häser, H. Horn and C. Kölmel, *Chem. Phys. Lett.*, 1989, **162**, 165.
18. Turbomole, Version 5.7, Universität Karlsruhe, Germany, 2004.
19. A.D. Becke, *Phys. Rev. A*, 1988, **38**, 3098.
20. J.P. Perdew, *Phys. Rev. B*, 1986, **33**, 8822.
21. K. Eichkorn, O. Treutler, H. Öhm, M. Häser and R. Ahlrichs, *Chem. Phys. Lett.*, 1995, **242**, 652.
22. K. Eichkorn, F. Weigend, O. Treutler and R. Ahlrichs, *Theor. Chem. Acc.*, 1997, **97**, 119.
23. F. Eckert and A. Klamt, COSMO*therm*, Version C2.1-Revision 01.05, COSMO*logic* GmbH & Co KG, Leverkusen, Germany, 2005.
24. A.B. Naim, *Solvation Thermodynamics*, Plenum Press, New York, NY, 1987.
25. P. Verwer and F. Leusen, in *Reviews in Computational Chemistry*, Vol 12, K.B. Lipkowitz and D.B. Boyd (eds), Wiley-VCH, New York, 1998.
26. A. Klamt, F. Eckert, M. Hornig, M. Beck and T. Bürger, *J. Comp. Chem.*, 2002, **23**, 275.
27. A. Klamt, F. Eckert, M. Diedenhofen and M. Beck, *J. Chem. Phys. A.*, 2003, **107**, 9380.
28. F. Eckert and A. Klamt, *J. Comp. Chem.*, 2006, **27**, 11.
29. H. Ikeda, K. Chiba, A. Kanou and N. Hirayama, *Chem. Pharm. Bull.*, 2005, **53**, 253.
30. S. Oleszek-Kudlak, M. Grabda, E. Shibata, F. Eckert and T. Nakamura, *Env. Tox. Chem.*, 2005, **24**, 1368.

31. A.K. Charlton, C.R. Daniels, R.M. Wold, E. Pustejovsky, W.E. Acree, Jr., and M.H. Abraham *J. Mol. Liquids*, 2005, **116**, 19, and references therein.
32. Supporting material with additional information and calculational details is available free of charge from the web-address: http://www.cosmologic.de/IUPAC-Solubility.html.
33. NIST Standard Reference Database Number 69, June 2005 Release, http://webbook.nist.gov/chemistry/.
34. Z. Ma and F. Zaera, *J. Phys. Chem. B*, 2005, **109**, 406.
35. H. Parshad, K. Frydenvang, T. Liljefors and C.S. Larsen, *Int. J. Pharm.*, 2002, **237**, 193.
36. J. Østergaard, S.W. Larsen, H. Parshad and C. Larsen, *Eur. J. Pharm. Sci.*, 2005, **26**, 280.
37. H. Parshad, *Design of Poorly Soluble Drug Salts*, Ph.D. Thesis, The Danish University of Pharmaceutical Sciences, Copenhagen, 2003.

Industrial Applications

CHAPTER 13

Solubility of Impurities in Cryogenic Liquids

VANIA DE STEFANI[1] AND DOMINIQUE RICHON[2]

[1] *Department of Chemical Engineering and Chemical Technology, Imperial College, London SW7 2AZ, United Kingdom*

[2] *Laboratoire de Thermodynamique, CNRS FRE 2861, ENSMP, CEP/TEP 35, rue Saint Honoré, 77305, Fontainebleau, France*

13.1 Introduction

The accurate knowledge of solid solubility in condensed gases is of great importance in designing safe cryogenic processes involving separation of gas mixtures into their components.

The great interest in the solid solubility is closely associated with the problem of accumulation of solid impurities in process equipment and storage tanks. Solid formations can hinder the passage of the liquid behind the plug, resulting in either an unexpected, rapid release of gas as the line warms, or the catastrophic failure of the line as the liquid warms behind the plug. In a cryogenic plant, such accumulations in liquid oxygen (LOX) may cause fouling and blockage in heat exchangers and piping and it may eventually cause serious explosions, for instance, the serious accident that occurred in Bintulu, Malaysia, in 1997.[1]

Flammability, high-pressure gases, and materials of construction are the principal areas of hazard, related to processing cryogenic liquids. These categories of hazards are usually present and must be carefully considered in air separation plants or processes, in order to reduce the probability of incidents to acceptable values. Fire and explosion events may occur when gases such as hydrogen, methane, and acetylene are involved in the process. Moreover, the presence of highly concentrated oxygen mixtures could lead to very high reactivity of ordinary combustibles, and may even cause some non-combustible materials like carbon steel to burn readily under the appropriate conditions.

At the state of the art, one of the main unsolved problems in air separation industries is the accurate understanding of the solid deposition rate of air contaminants in cryogenic units. Air contaminants, *i.e.*, hydrocarbons and carbon dioxide, have melting temperatures higher than the LOX liquefaction temperature. Therefore, they can solidify and cause plugs during air separation process.

This has been the cause of serious safety problems in many plants throughout the world,[2] especially when the solid deposits are flammable.

Accurate data about phase equilibria are indispensable for both improvement of existing processes and design of new ones. In particular, solubility data of solid flammable hydrocarbons in LOX is fundamental in reaching safe design and working conditions, for example, in the reboiler-condenser of an air distillation tower, where LOX evaporates.

13.2 Loss of Prevention in Cryogenic Plants

The fractional distillation of liquid air is one of the most widely used processes in the chemical industry. Cryogenic air separation plants are used to generate pure oxygen, pure argon, and pure nitrogen for use in steel, metallurgical, petrochemical, and semiconductor manufacturing processes.

As oxygen and nitrogen are key requirements for many advanced energy technologies, air separation units (ASU) are commonly used in a variety of processes.

Most commercial air separation plants are based on Linde's double distillation column process. Many variations on the above concept are known. These include separation of air into gaseous products, liquid products, and all kind of combinations thereof.

Hydrocarbons in combination with pure oxygen might create safety problems. Therefore, when large amounts of hydrocarbons are present in the feed air, these hydrocarbons should be removed. Also, when extremely high-purity oxygen and/or nitrogen have to be produced, *e.g.*, for use in the production of micro-electronics components, hydrocarbons are removed even more carefully from the feed air. However, it is extremely difficult to remove all impurities and hydrocarbons from the feed air.

Usually, the higher hydrocarbons, especially C5+, are removed at the same time as water and/or carbon dioxide are filtered off to avoid plugging of the cold process lines. This is often done by refrigeration purification and/or (low temperature) solid adsorbent purification. This will also remove higher hydrocarbons. Furthermore, acetylene is often also removed using adsorbent beds. Provided that the amounts of Cl-4 hydrocarbons in the feed air (especially unsaturated C1-4 hydrocarbons, and in particular ethane) is relatively low, (*e.g.*, less than 100 ppm for each hydrocarbon and often less than 40 ppm) it is generally accepted that these hydrocarbons do not create any safety problems in any gaseous oxygen streams or in any of the LOX streams or reservoirs. In this respect, it is observed that these lower hydrocarbons are soluble in LOX, and thus are expected not to accumulate in the system. Even when higher concentrations of hydrocarbons occur somewhere in the system, especially the distillation section, it would not be expected that this would result in any safety problem, especially in view of the low temperature. Furthermore, a simple drain from a specific part in the distillation unit would be sufficient to reduce any increasing concentrations. Thus, it is general practice not to apply any special processes to remove any lower hydrocarbons from cryogenic ASUs producing large amounts of oxygen, more particularly oxygen of industrial quality. This is only done for small-scale units when there is a clear need to do so.

Under specific conditions, there might be a large and unwarranted accumulation of air impurities in the cryogenic section of an ASU. Without being bound by any theory, it is acknowledged that insoluble compounds may be formed, especially on the inside walls of the cryogenic distillation units. These solid compounds are potentially hazardous and could result in a disaster. A possible ignition source might be internal friction, or the falling down of material, *e.g.*, blocks of ice or other solid material, or the (spontaneous) falling down of grown insoluble compound at a higher stage in the reactor. Also chemical excitation might be possible, for instance, initiated by a higher olefin molecule or by a transition metal. The build up of the insoluble material is presumably slow, and will depend on the presence of unsaturated hydrocarbons, especially ethene, and radicals in the feed air. In the case where there is no build up, or the distillation unit has been derimed (which is usually done once every two or three years), there are no real problems. However, sometimes ignition does occur, resulting in (small) damage to the internal structure, but sometimes larger explosions take place.

The world's first middle distillate synthesis plant (i.e. Shell Middle Distillate Plant, SMDP) was built in 1993 in Bintulu (Malaysia) and was the only integrated low-temperature Fischer–Tropsch plant in existence. The plant produced 517,000 tons per years of products such as kerosene, naphtha, paraffin, and gas oil from locally supplied natural gas. The oxygen required for the shell gasification process was taken from an adjacent ASU. On December 1997, the SMDP suffered an explosion in the ASU, which was caused by forest fires which brought about prolonged local hazy air conditions. This was unrelated to the Shell Middle Distillate Synthesis, SMDS technology. Lessons learned from this event were shared industry-wide to avoid re-occurrence in ASUs elsewhere. The Bintulu plant was subsequently rebuilt, and the improved performance during the second start-up validated the value of prior experience. Following reconstruction and upgrading the plant went back on stream only in May 2000, leading to extensive production losses.

A small explosion, it would later be learned, had occurred in an ASU of the complex which supplied oxygen for the production of the synthesis gas. The findings of the investigation and the conclusions lead to an understanding of the problems of accumulation of impurities in process plants.

13.3 Experimental Methods

The composition of solid impurities in cryogenic liquids must be determined with great precision. The main challenges, which are encountered during the measurements, are temperature control, components interaction, and measurement of low compositions.

Even if there are a substantial number of solubility data of air contaminants in cryogenic liquids, their accuracy is sometimes doubtful.

The accurate determination of solid–liquid equibrium data at very low temperatures remains difficult. Nevertheless, over the years many different apparatuses have been developed and used. Since 1950, solubility measurements have been carried out using techniques, such as synthetic optical method, evaporation method, and static-analytic method.

Synthetic methods visually analyse the solidification process of a solution of known composition and do not require sampling devices. Consequently, the experimental apparatus is relatively simple and the main difficulty is the preparation of the solution to be studied. However, the solubility is measured at a fixed temperature and in order to obtain a complete data set for multi-solute solutions, it is necessary to link this technique with an analytical one.

Static-analytical methods require the establishment of the thermodynamic equilibrium in the cell followed by the analysis of solution samples. The limitations of the static-analytical methods are those of the applied analysis technique, *e.g.*, spectroscopy or chromatography. Furthermore, it is essential to develop a reliable sampling technique, which allows withdrawing samples small enough not to perturb the established equilibrium.

Table 1 summarizes the range of applicability of each of the method and the solubility uncertainties.

The following paragraphs detail some of the most common experimental technique used to determine solid–liquid equilibrium.

13.3.1 Synthetic-Optical Method

The optical method is a synthetic method, which involves the preparation of solute–solvent mixture of the desired concentration in a loading reserve connected to the equilibrium cell. The amount of each component is found out by weighing or by volume pressure measurement. Weighing is the most accurate method, but it cannot be applied to highly diluted solutions (less than 500 ppm) with great accuracy. In addition, it is essential for the reserve to weigh considerably less than the mixture itself, in order to minimize the errors accompanying the weighing procedure. Loading through a volume-measured

Table 1 *Summary of the performances of the experimental methods*

Technique	*Temperature (K)*	*Composition range (ppm)*	*Uncertainty on solubility (%)*
Synthetic-optical	> 70	> 500	±10
Evaporation	> 70	> 500	±10
Static-analytic: transmission cell	> 70	> 1	±25
Static-analytic: reflection cell	> 70	> 1	±10
Static-analytic: chromatographic analysis	> 70	> 1	±5

procedure is simple and accurate, but very precise knowledge of pure components PVT properties is required.

Once the solute–solvent mixture at a given composition has been prepared in the loading reserve, the solution is introduced into a transparent equilibrium cell. For a given composition, the melting process inside the equilibrium cell is visually observed and the temperature, at which the solid phase just disappears, is determined. As a result when the solute is a mixture, this method of noting the appearance and disappearance of crystals in a solution as the temperature is alternately lowered and the raised, cannot be used.

Jakob A. *et al.*,[3] developed an apparatus composed of a 160 cm^3 pyrex cell, A (see Figure 1) surrounded by a three-chamber envelope: in the first chamber, 6, the vacuum ensures there is not condensation of atmospheric vapour that should disturb the phenomena observation. In the chamber, 8, circulates the refrigerant, which regulates the temperature of thermostatic bath circulating in the chamber, 9. The temperature inside equilibrium cell is measured by means of a platinum probe, 3.

Similar equipment has been utilized by McKinley and Himmelberger.[4]

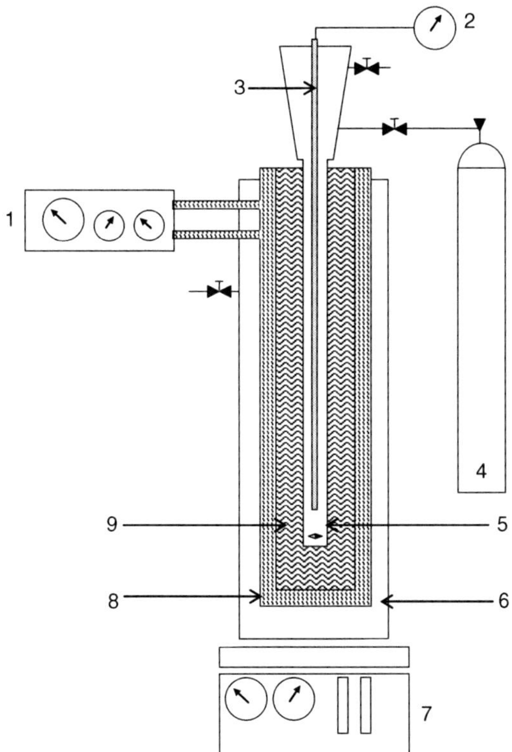

Figure 1 *Optical method for measuring solid–liquid equilibrium*[3]*: 1, cryostat; 2, temperature indicator; 3, platinum probe; 4, nitrogen; 5, equilibrium cell; 6, vacuum chamber; 7, magnetic stirring system; 8, thermostatic bath; 9, refrigerant.*

13.3.2 Evaporation Method

The evaporation method is a static method based on the indirect determination of the solute. In 1959, Din and Goldman[5] developed the equipment presented in Figure 2.

The equilibrium cell, 6, is placed in a thermostatic bath of LOX. When the thermodynamic equilibrium is reached, a sample of filtered saturated solution is removed by opening valve. The sample vaporizes in 4 at room temperature and the pressure is measured by a mercury manometer, 7. The total number of moles in the sample is determined by the ideal gas law:

$$n_{\text{tot}} = \frac{\text{P}_{\text{tot}} V_{\text{E}}}{RT} \tag{1}$$

where:

P_{tot} = total pressure in E (Pa)
V_{E} = volume of ampoule E (m^3)
T = temperature (K)
R = universal ideal gas constant (m^3 Pa mol^{-1} K^{-1}]

Subsequently, the contents of 6 flows through the condenser, H, which is maintained at low temperature by liquid nitrogen: the solvent is removed by a vacuum pump, 9, and the solute is solidified in the bottom of 8. Then the valve is closed and the solid solute restrained in 8 evaporates at room temperature. The pressure (P_i) is measured and the number of moles of component *i*

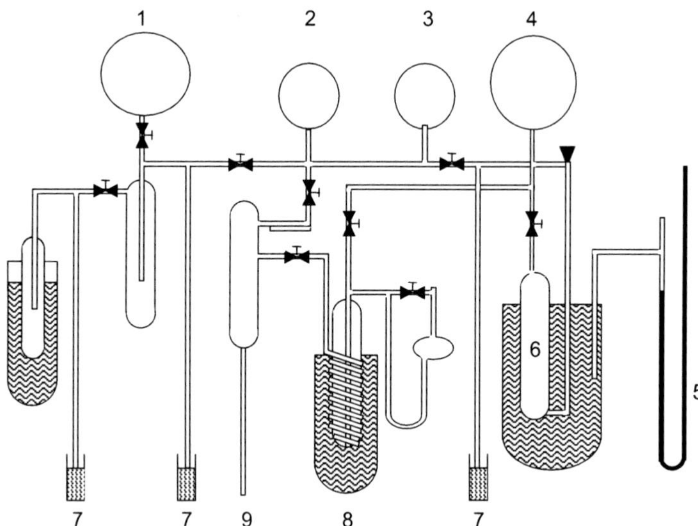

Figure 2 *Evaporation method of Din and Goldman*[5]*: 1, 2, 3, 4, ampoules; 5, oxygen vapour manometer; 6, equilibrium cell; 7, manometer; 8, condenser; 9, vacuum pump.*

contained in the sample is determined:

$$n_i = \frac{P_i V_H}{RT} \tag{2}$$

where:

P_i = pressure in H (Pa)

V_H = volume of ampoule H (m^3)

The molar fraction of solute present in the sample withdrawn from equilibrium cell 6 is:

$$x_i = \text{const}\frac{P_i}{P_{tot}} \tag{3}$$

where:

const = V_H/V_E

To obtain accurate data, it is necessary to carefully measure the volumes of ampoules 8 and 4. The uncertainty from this method, based on estimated molar composition is ±10%. This technique is applicable to the measurements of solute mixtures.

13.3.3 The Static-Analytical Methods: Spectroscopic Analysis

Static-analytic methods are very simple: the solute–solvent mixture is enclosed in a cell where the thermodynamic equilibrium is to be achieved (see Figure 3).

The differences between the various methods are usually associated with the analytical technique adopted. *In situ* analyses of the sample composition are carried out through spectroscopic techniques, which have the advantage of using relatively simple equilibrium cells, as sampling is not necessary.

For cryo-systems, the most used spectroscopic technique is infrared spectroscopy. This technique has the disadvantage of not being able to make a distinction between solubilized particles and suspended ones. For example, the colloids that might not be eliminated by filtration are detected by the infrared spectroscopic analysis.

We can classify the cells present in literature in two different classes: transmission cells and reflection cells.

13.3.3.1 Reflection Cell

In 1973, Bulanin[7] developed the apparatus shown in Figure 4. The equilibrium cell, 8, is placed in a cryostat filled of refrigerant, 9. At the bottom of the cell, there is a spherical mirror, 11, a unique window is sufficient to analyse the liquid phase with an infrared beam.

There are many applications of this cell, which allows the determination of co-solubility of a solid mixture in liquids.

Meneses *et al.*[8] measured the solid-vapour equilibrium for several mixtures. A gaseous mixture of known composition is injected into a cryostat in front of a

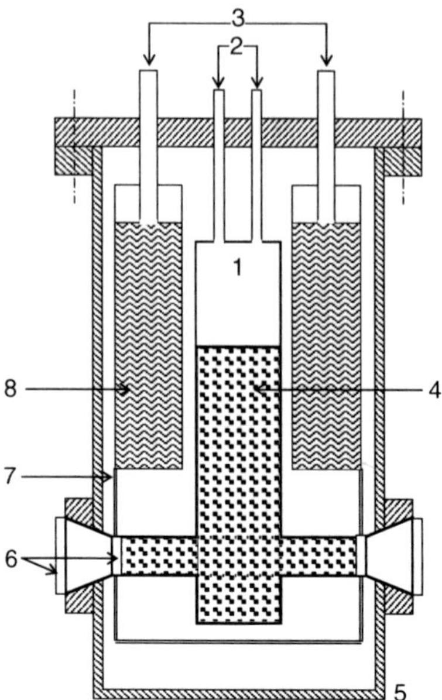

Figure 3 *Transmission cell of Rest et al.*[6]*: 1, sampling chamber; 2, inlet-line sample; 3, inlet and outlet lines of refrigerant; 4, sample; 5, Dewar; 6, optic window; 7, thermal insulator; 8, refrigerant fluid.*

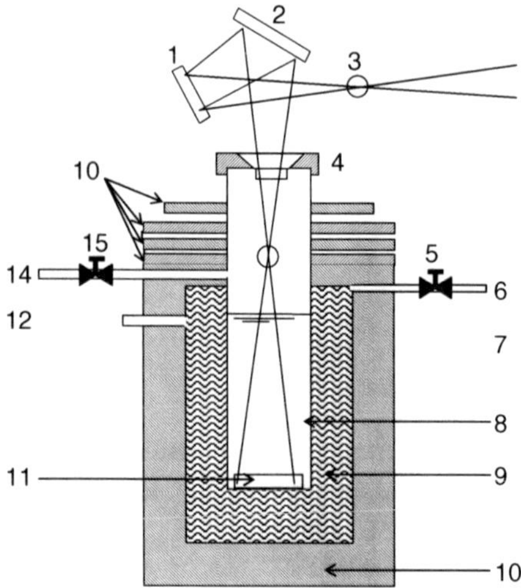

Figure 4 *Reflection cell of Bulanin*[7]*: 1, 2, spherical mirrors; 3, focus; 4, optical window; 5, filling of refrigerant; 6, inlet-line of the refrigerant; 8, equilibrium cell; 9, refrigerant; 10, thermal insulator; 11, plan mirror; 13, refrigerant vast; 14, inlet-line of the sample; 15, inlet valve of the sample.*

polished aluminum mirror and then the temperature of the mirror surface is decreased to the target value. For each mixture, the total pressure is measured by an optical method. When the system is stabilized, the pressure is recorded and the composition of the deposited solid is measured by infrared spectroscopy. From the results obtained for the solid–vapour equilibrium, it is possible to extrapolate the solid–liquid equilibrium by making the following assumption: the diluted components are close to infinite dilution, therefore the liquid can be considered as an ideal mixture; the gaseous phase is a mixture of ideal gases; the solvent solubility in the solid is negligible. The accuracy of the Meneses's method on estimated molar composition is ±15% for liquid phases and ±10% for solid ones.

13.3.3.2 Transmission Cell

The cell developed by Rest *et al.*[6] (see Figure 3) includes an analysis chamber, 1, fitted with two optical windows, 6, transparent in the spectrum region of interest. A refrigerant bath, 8, maintains the cell at low temperature.

The solute–solvent solution is prepared outside the cell and, after filtering, is introduced in 1. The mixture is efficiently stirred by a magnetic rod. Temperature inside 1, is measured by two thermocouples. When the thermodynamic equilibrium is reached, an infrared beam pass through the optical windows and the liquid phase is analysed. The uncertainty of this method on estimated molar composition is ±25%.

13.3.4 The Static-Analytical Methods: Chromatographic Analysis

The static methods using chromatographic analysis are characterized by the use of a sampling device to send the sample to the chromatograph. It is essential that the sampling procedure does not perturb the equilibrium, and the sample is representative of the phase under study. For this reasons, the sampling devices must be reliable, and allow withdrawing of samples small enough not to disturb the equilibrium reached inside the equilibrium cell.

Miller *et al.*[9] developed the apparatus presented in Figure 5. It consists of a stainless equilibrium cell, 1, enclosed into a cryostat, 2. The temperature of the cell is measured by a platinum probe. The experimental procedure is the following: at room temperature, a solute mixture is introduced into the equilibrium cell and then diluted with a cryogenic fluid, *e.g.*, oxygen. Afterwards, the temperature is lowered and helium is loaded to pressurize the mixture in order to allow sampling. At thermodynamic equilibrium, samples of liquid phase are withdrawn using a pneumatic valve.

In previous papers, we demonstrated that this loading procedure can lead to a non-homogeneous solid system. In fact, because of their different melting points, the two mixture components can condense and deposit in the equilibrium cell independently, forming two independent solid layers. Then the measurements of liquid phase would be representative of the solid phase of the component having the lowest melting point (*i.e.*, the component producing the layer in contact with the liquid phase), altering the measured solubility data.

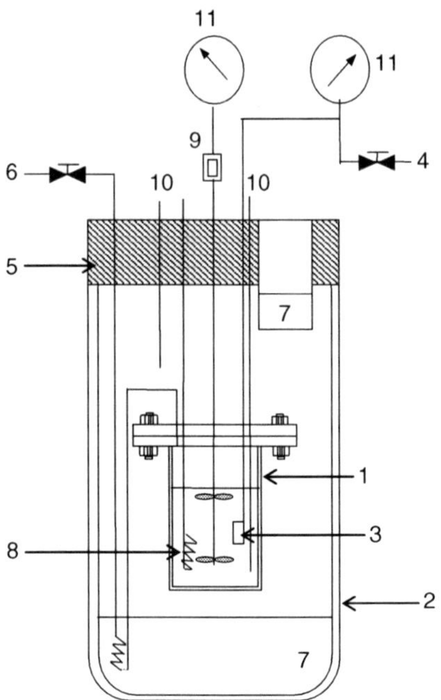

Figure 5 *Static-analytic method of Miller et al.*[7]*: 1, equilibrium cell; 2, Dewar; 3, filter; 4, line to the chromatograph; 5, thermal insulator; 6, loading circuit; 7, liquid nitrogen; 8, temperature regulator; 9, stirring assembly; 10, platinum probe; 11, pressure transducer.*

To overcome this just described problem, De Stefani *et al.*[10] designed and developed an innovative apparatus adapted to measure the co-solubility in cryogenic liquids, see Figure 6.

This apparatus is based on a static-analytic method: it consists of an equilibrium cell suspended into a Dewar partially filled with liquid nitrogen. A brass envelope maintains a fine local temperature regulation in the equilibrium cell by means of an electrical resistance. A new device, named "atomiser – injector", 5, has been developed in order to introduce, into the equilibrium cell, fine particles of a homogeneously dispersed solid mixture. This new equipment represents a significant step forward over the previous traditional apparatuses. The pneumatic sampler rapid on line sampler and injector[11,12] (ROLSI) allows the on line sampling and analyses of the withdrawn samples; their mass is in the range of 0.01 mg to a few milligram and is small enough to ensure that the thermodynamic equilibrium inside the cell is not perturbed.

13.4 Review of Literature Data

A selection of solubility data of solid impurities in LOX and liquid nitrogen is summarized in Figures 7–13.

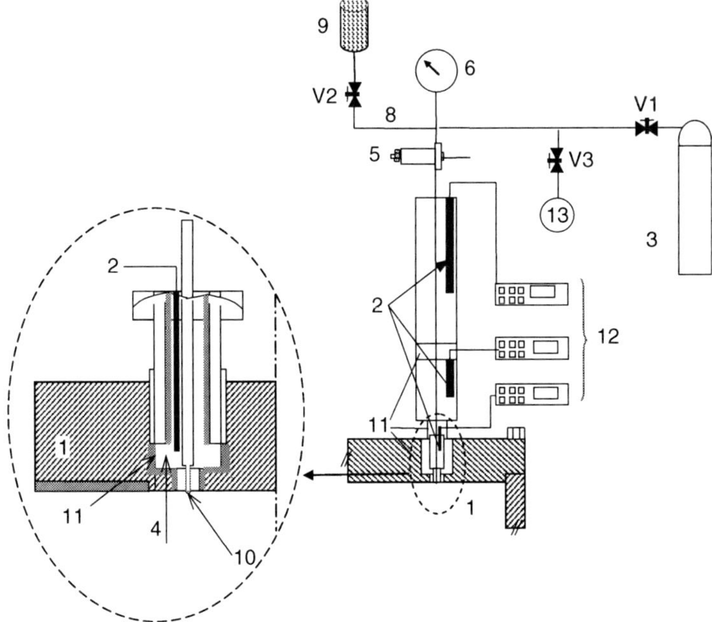

Figure 6 *Overview of the atomiser-injector system[10]: 1, equilibrium cell; 2, heating cartridge; 3, helium; 4, heat exchanger; 5, injector; 6, pressure transducer; 7, screw plug; 8, solute loading circuit; 9, solvent reservoir; 10, spray tip; 11, thermal insulator; 12, PID temperature regulator; Vi, Valve i; 13, vacuum pump.*

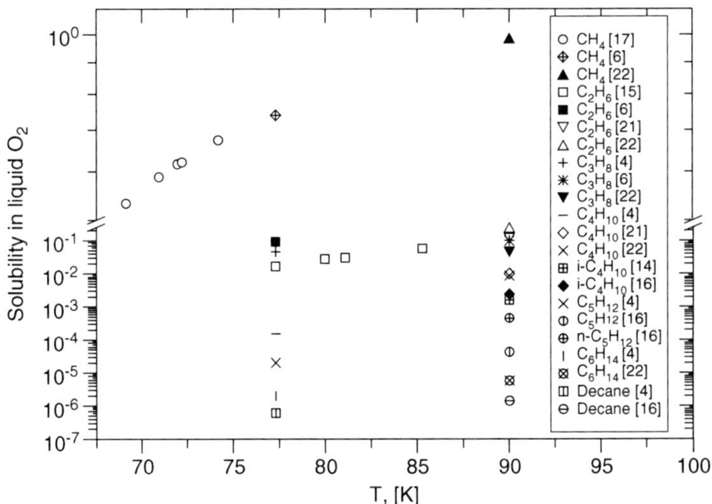

Figure 7 *Measured solubility of alkanes in LOX.*

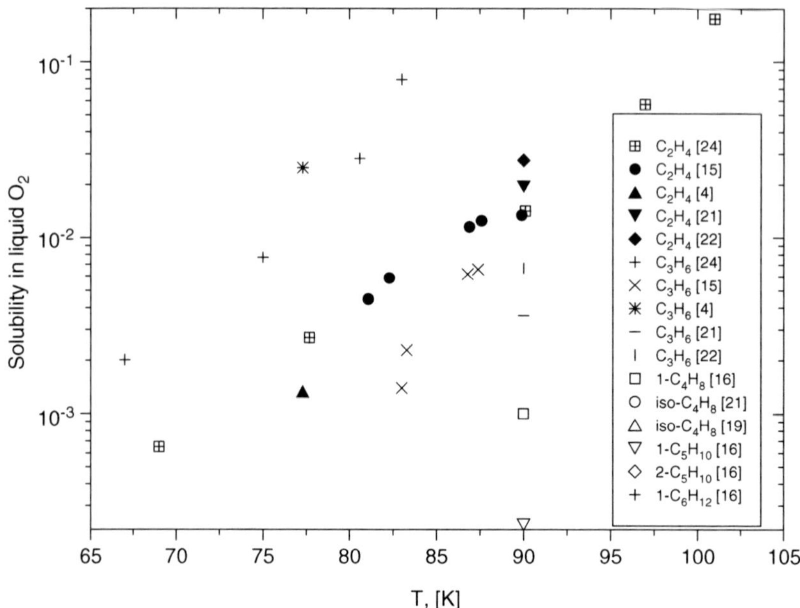

Figure 8 *Measured solubility of alkenes in LOX.*

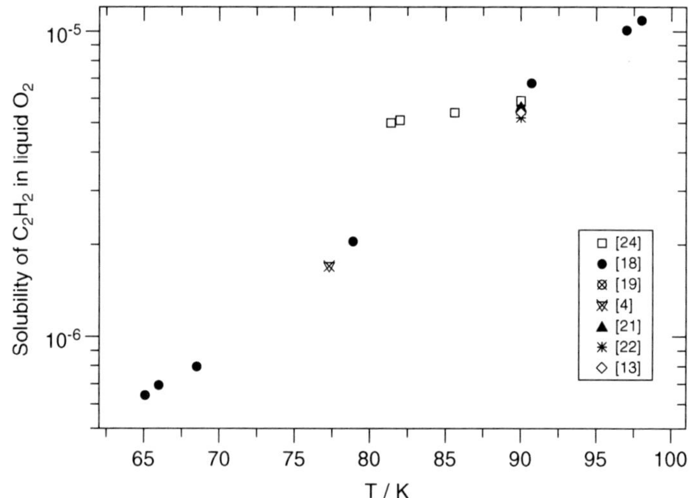

Figure 9 *Measured solubility of acetylene in LOX.*

Very few data exist for long-chain hydrocarbons and in the most cases solubility was measured at a fixed temperature (90 or 70 K), *i.e.*, the working temperature of air distillation column. As a reasonable amount of experimental points is not available, serious data correlation is not possible.

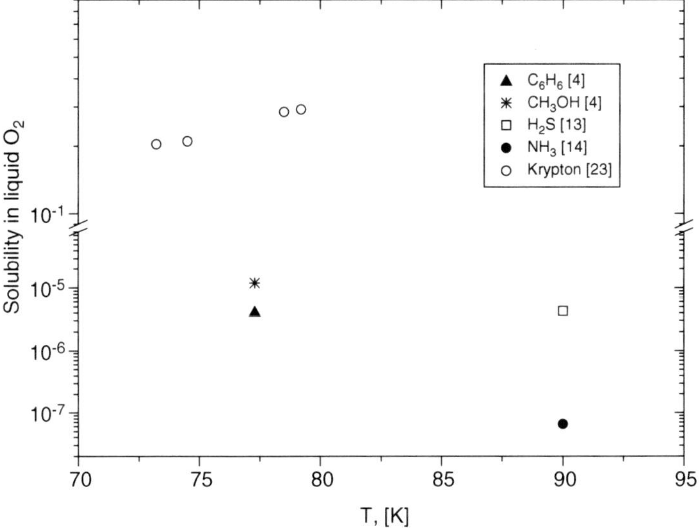

Figure 10 *Measured solubility of different substances in LOX.*

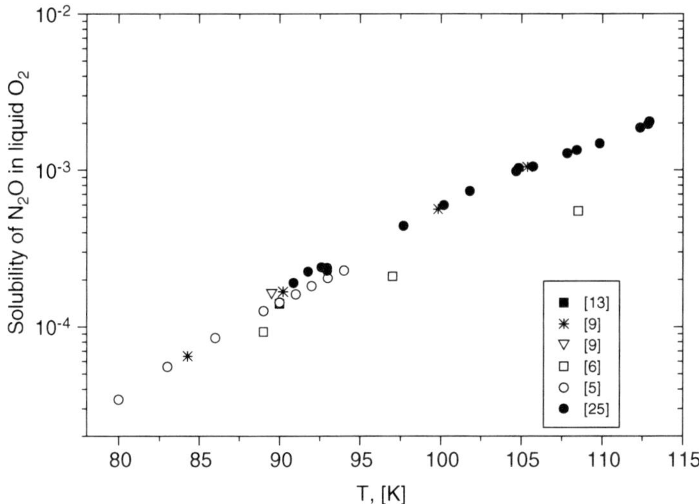

Figure 11 *Measured solubility of nitrous oxide in LOX.*

Some alkanes and alkenes (C1–C5) are characterized by having relatively high solubility (between 10^{-1} and 10^{-4} in mol fraction). Therefore, all methods, including the optical method or evaporation method, are suitable to determine the solubility of these compounds.

Substances with low solubility (lower than 10^{-5} in mol fraction) are long-chain hydrocarbons, acetylene, carbon dioxide, nitrous oxide, methanol,

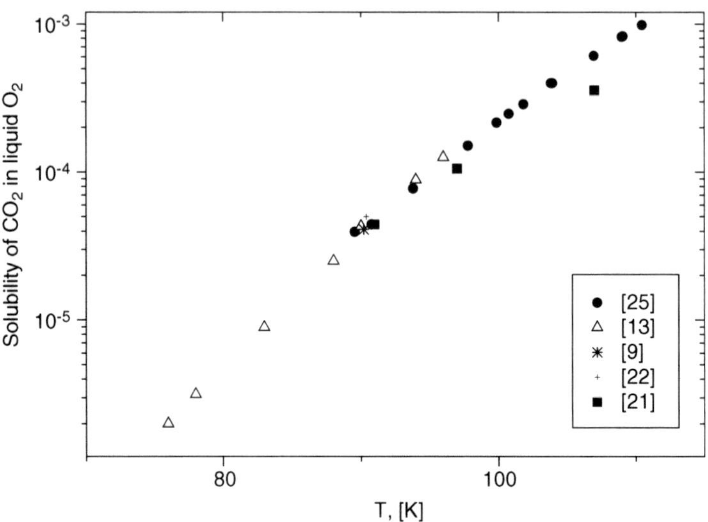

Figure 12 *Measured solubility of carbon dioxide in LOX.*

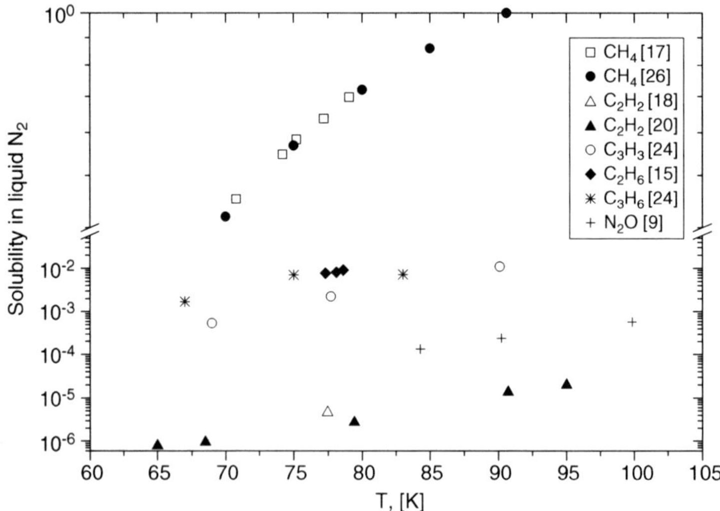

Figure 13 *Measured solubility of different substances in liquid nitrogen as a function of temperature.*

ammonia, and sulfuric acid. To measure such low compositions, an analytical method is required.

As expected, the solubility of the hydrocarbons in LOX decreases in the homologous series together with the increase of the number of carbon atoms in the carbon chain.

The solubility of solid substances in liquid nitrogen is represented in Figure 13, similar conclusions can be made of those for the solubility in LOX.

13.5 Conclusion

Several instruments available to measure the solubility data in cryogenic fluids have been reviewed here. Synthetic methods allow seeing, directly, solidification phenomena and do not necessarily require sampling devices. Analytic methods are excellent to determine phase equilibria provided analytical instrument is available with good calibration and reliable phase samplers are also available (ROLSI is good choice for samplings in larges ranges of temperatures and pressures, see http://www.rolsi.com/English.htm).

References

1. R.M. van Hardeveld, M.J. Groeneveld, J.-Y. Lehman and D.C. Bull, *J. Loss Prevent. Process Ind.*, 2001, **14**, 167.
2. F.G. van Dongen, J.D. Graaf, R.M. Groeneveld and R.M. van Hardeveld, *Proceeding of the 12th Intersociety Cryogenic Symposium*, AIChE 2000 Spring National Meeting, Atlanta, 2000, p. 3.
3. A. Jakob, R. Jho and J. Gmehling, *Fluid Phase Equilib.*, 1995, **113**, 117.
4. C. McKinley and F. Himmelberger, *Chem. Eng. Progr.*, 1957, **53**, 112.
5. F. Din and K. Goldman, *Trans. Faraday Soc.*, 1958, **55**, 239.
6. A. Rest, R. Scurlock and M. Fai Wu, *Chem. Eng. J.*, 1990, **43**, 25.
7. M.O. Bulanin, *J. Molecu. Struct.*, 1973, **19**, 59.
8. D. Meneses, J.-Y. Thonnelier, C. Szulman and E. Werlen, *Proceeding of Cryogenics 2000*, October 2000, p. 109.
9. E.J. Miller, S.R. Auvil, N.F. Giles and G.M. Wilson, *Proceedings of the 12th Intersociety Cryogenic Symposium*, AIChE 2000 Spring National Meeting, Atlanta, 2000, p. 18.
10. V. De Stefani, A. Baba-Ahmed and D. Richon, *J. Fluid Phase Equilib.*, 2003, **207**, 131.
11. Automatic pressurized fluid microsampling and injection device. US patent from Armines: 4488436.
12. Procédé et dispositif pour prélever des micro-échantillons d'un fluide sous pression contenu dans un container. French patent from Armines: 98 10708.
13. R.G. Amamchyan, V.V. Bertsev and M.O. Bulanin, *Zavodskaya Lab.*, 1973, **4**, 432.
14. W.L. Ball, *Safety Air Ammonia Plant*, 1966, **8**, 12.
15. A.L. Cox and T. de Vries, *J. Phys. Colloid. Eng.*, 1960, **4**, 11.
16. G.F. Densenko and W.I., Fajnsztein, Technika bezopastnosti pri poizwodstwie kisloroda, Moska, 1958.
17. V.G. Fastovskii and I.A. Krestinskii, *Zh. Fiz. Khim.*, 1941, **15**, 525.
18. M.F. Federova, *Zh. Fiz. Khim.*, 1940, **14**, 422.
19. P. Ishkin, P.Z. Burbo and L.T. Pashkovskaia, *Zh. Khim. Prom.*, 1937, **8**, 560.
20. J.P. Ishkin and P.Z. Burbo, *Zh. Fiz. Khim.*, 1939, **13**, 1137.
21. E. Karwat, *Chem. Eng. Prog.*, 1958, **54**, 10.

22. C. McKinley and E.S. Wang, *Advan. Cryog. Progr.*, 1960, **53**, 11.
23. M. Stackelberg, *Z. Phys. Chem.*, 1934, **170**, 262.
24. N.M. Tsin, *Zh. Fiz. Chim.*, 1940, **14**, 418.
25. V. De Stefani, A. Baba-Ahmed, A. Valtz, D. Meneses and D. Richon, *Fluid Phase Equilib.*, 2002, **200**, 19.
26. B. Dabrowska, *Cryogenics*, 1996, **36**, 985.

CHAPTER 14

Solubility of BTEX and Acid Gases in Alkanolamine Solutions in Relation to the Environment

CHRISTOPHE COQUELET AND DOMINIQUE RICHON

CNRS FRE2861 CEP/TEP, Ecole Des Mines de Paris, 35 rue Saint Honoré, 77305, Fontainebleau, France

14.1 Introduction

Recent observations clearly show a dramatic increase of greenhouse and acid gas concentrations with great damages to the environment. This leads to increasing rain acidity. Acid rain are highly responsible for forest destruction and the impoverishment of nutrients. The last point affects particularly the animals living in corresponding regions. The main pollutants, which acidify the rain, are carbon dioxide, sulfur dioxide and nitrogen oxide (NOx). Energy supply industries and transports are the main acid gas producers. Natural gas and petroleum industries have to deal with raw materials containing very high concentrations of both hydrogen sulfide and carbon dioxide. Other sulfur species normally occur in much smaller amounts. Treating processes are needed in order to remove not only H_2S and CO_2 but also all other sulfur species and prohibited compounds such as aromatics.

In 1930, R.R. Bottoms originally applied aqueous alkanolamine processes to gas treating. Figure 1 presents a typical gas treating process. Alkanolamine processes remain the most attractive ones for large treating units whatever the acid gas or aromatic initial content of fluids to be treated, because of the large number of degrees of freedom allowed. For these reasons our attention in this chapter is focused on absorption processes with alcanolamine solutions.

These processes are commonly used where acid gas partial pressures are low and/or low levels of acid gas are desired in the residue gas (or purified gas).[1] Problems usually encountered with amine chemical absorption are related to corrosion especially in the presence of oxygen. As a result the choice of amine and of its concentration must be considered together with the choice of reactor materials, in order to avoid corrosion problems and to allow adequate operating

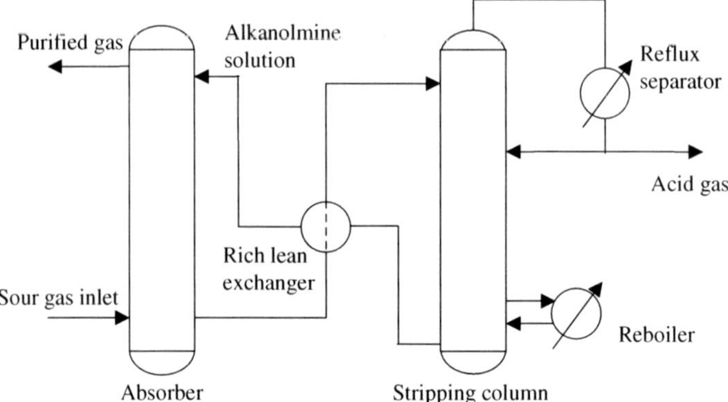

Figure 1 *Principle of gas treating by chemical reaction with aqueous alkanolamine solution.*

conditions. The amine-treating unit absorbs an appreciable quantity of heavy hydrocarbons and aromatics from the feed gas. This quantity is a function of several parameters and in particular, the amine content in the aqueous solution. A substantial portion of the heavy hydrocarbons and especially the aromatics in the inlet gas, is absorbed by the circulating amine, and is released to the acid-gas stream during amine regeneration. Hydrocarbons (HC) are desorbed from the solvent during the stripping step. If the overhead of the regenerator (stripping unit) is vented to the atmosphere, these HC emissions are more closely regulated. The clean air Act has specified the limiting amount of heavy HCs which may be emitted from facilities or power plants to 250 $t.year^{-1}$.

If the overhead of the regenerator is fed to a burner such as in a Claus plant, the BTEX components are more difficult to destroy, relative to other HCs, and can cause overheating in the catalyst bed. In effect, heavy HCs and particularly aromatics have harmful effects[2] in a sulfur plant feed; aromatics absorbed in an amine unit tend to deactivate the catalyst in the downstream Claus unit. Knowledge of aromatics (BTEX) and acid-gas solubility is essential in order to develop models, predict absorption, desorption and to optimize the amine recovery processes.

14.2 Choice of Amine

There are several types of amines. Industry employs mainly 6 amines to extract, preferentially in a selective way, CO_2 and H_2S: primary amine i.e. Monoethanolamine (MEA), secondary amine i.e. Diethanolamine (DEA), and tertiary amines: Methyldiethanolamine (MDEA), Diglycolamine® (DGA), Triethanolamine (TEA) and Diisopropanolamine (DIPA). In processing, acid gases are removed from gaseous streams using reactive absorption with aqueous alkanolamine solution. A desorption step follows where the reactions are reversed by increasing temperature and/or decreasing pressure. The reactions of acid gases with amines depend on the type of amine: CO_2 reactions with primary or secondary alkanolamines produce carbamates (fast reactions), whereas CO_2 reactions with tertiary alkanolamines produce bicarbonates and carbonates (the corresponding rate of reaction is slower then the previous one). Reactions with H_2S involving a proton exchange, are very fast whatever the amine. As a consequence, tertiary amines can be considered as selective with respect to H_2S compared to CO_2. For example, MDEA is selective towards H_2S in H_2S–CO_2 containing mixtures while DEA or MEA are not. Carbamate formation is responsible for the high heat of reaction and so significant energy is required for regeneration of either primary or secondary amine during desorption steps. DEA is often preferred to MEA because less heat is required to strip the amine solution[1] and reactions with COS and CS_2 are not irreversible. MEA has the highest vapour pressure. This is a disadvantage, as solution is lost through vapourization from the contactor and stripper can be very high. DGA® is a primary amine capable of removing not only H_2S and CO_2, but also COS and mercaptans[1] from gas and liquid streams reversibly contrary to MEA. The concentration of alkanolamine is in 15–70 mass% range. It depends on choice of amine and its choice is usually made on the basis of operating experience.[2] Finally, the last criterion concerns the corrosive aspect: primary amines are more corrosive than secondary or tertiary amines. This is due to the degradation products of primary amines which are more acidic than those of the other amines.

14.3 Experimental Techniques

The relationships between the solubility of hydrocarbons and acid gases in amine solution (mole of gas per mole of amine) and their partial pressures in the gas phases at equilibrium are required if one wants to be able to design and optimize efficient removal units. A number of parameters have a significant influence on phase equilibria: temperature, total pressure, composition of aqueous phase, nature of amine, respective partial pressures of solutes, *etc.* Consequently, the choice of the experimental technique, allowing accurate and relevant determinations, proves to be very important. Analytical methods are generally preferred to determine solubilities of solutes in multicomponent mixtures. They can be divided into two classes dynamic and static methods.

14.3.1 Dynamics Methods (or Flow Methods)

One phase (or more) circulates through the equilibrium cell. This technique has been widely used for the determination of acid-gas solubilities by Jou *et al.*[3–6] Their apparatus basically consists of an equilibrium cell with large windows (windowed Jerguson cell) fitted at the top with a 250 cm^3 reservoir for vapour phase. The vapour circulates from the top of the reservoir into the bottom of the Jerguson cell and thus through the liquid phase. Gas chromatographic techniques are used to analyse samples withdrawn from vapour phase by means of a sample loop set in the gas sample line extended from the reservoir. Liquid samples were withdrawn into a vessel and analysed through an acid–base reaction using aqueous NaOH solution. The aim of this step is to convert free dissolved acid gas into ionic species. Finally the carbonate is precipitated as $BaCO_3$, using a solution of $BaCl_2$, and is then titrated against an HCl solution.

H_2S in the liquid was determined by using the iodine-thiosulfate titration. The amine concentration was determined by direct titration with a solution of H_2SO_4. The main disadvantages of this method are complex shape of equilibrium cell, the liquid recirculation and the necessity of withdrawing large samples of both liquid and vapour phases.

Austgen *et al.*[7] use another experimental technique. The experimental apparatus is a continuous flow apparatus. It is composed of two cells: the first one is a water saturator and the second contains an alkanolamine solution. The aim is to find the right-gas flow, composed mainly of acid gas with the concentration of the feed gas being identical to the concentration in the effluent from the equilibrium cell. The acid gas partial pressure of the feed gas is then taken to be the acid gas partial pressure in the equilibrium with the alkanolamine solution.

Another dynamic method which can be used to determine solubility of a solute (activity coefficient or Henry's constant at infinite dilution) in a solvent is the gas-stripping technique. The principles and equipment have been fully described previously by Richon *et al.*[8] An equilibrium cell contains non-volatile solvent (alkanolamine aqueous solution) and infinitely diluted solutes. The solutes are continuously carried away in a stripping gas flow. The decrease of concentration of solute in the stripping gas is recorded from chromatographic analysis. Coquelet and Richon[9] have determined, using this technique, Henry's law coefficients of mercaptans in 50 wt% methyldiethanolamine-aqueous solution. Solute-partition coefficients at infinite dilution at pressures in the range 1–20 MPa and temperatures in the range 278–473 K have also been determined[10] after considering mass balance and equilibrium equations.

14.3.2 Static Methods

(a) Static-analytic methods: the components under investigation are loaded into an equilibrium cell. The mixture is stirred to promote contact between the phases and consequently rapidly reach equilibrium. At equilibrium, the constant temperature and pressure are recorded and samples of liquid or/and vapour are withdrawn from the equilibrium cell to determine their compositions, generally through gas chromatography.

Valtz *et al.*[12,13] used a static-analytic method with phases sampling through ROLSI™ samplers[11] (see http://www.rolsi.com/English.htm). The flow diagram of this apparatus is presented in Figure 2. Each phase is analysed by gas chromatography.

(b) Static-synthetic method: such a method was selected[12] and used to perform determination of the solubility of BTEX in aqueous-alkanolamine solutions. A solution of desired composition is prepared and transferred into an equilibrium cell. The quantity of solvent introduced into the equilibrium cell is known accurately through differential weighing. The vapour pressure of solution is recorded first. Then, small quantities of alkylbenzene are added progressively through a variable volume cell. After each introduction of solute, total pressure is measured again. The curve representing total pressure *vs.* injected alkylbenzene mass displays a break point corresponding to the saturation of solvent with alkylbenzene. This technique can bc conveniently used to determine acid-gas solubility data.

14.4 Experimental Results

Many physical parameters influence the solubility of acid gases and BTEX in aqueous amine solution. These parameters include: temperature, amine concentration, amine structure, acid gases loading, acid gas nature, aromatic structure, *etc.*

(a) Acid gases

Several experimental data are available in literature concerning the solubility of acid gases in amine-aqueous solutions. Studies concern aqueous solutions for different amines and different amine concentrations. Tables 1–3 present the more recent references concerning MEA, DEA and MDEA.

Few data exist for DGA® (Dingman *et al.*,[32]) and the other amines.

Figures 3 and 4, present, respectively, the H_2S and CO_2 partial pressures in aqueous-amine solutions as a function of acid-gas loadings at several temperatures.

Figures 5 and 6 give, at 313.15 K, acid gas partial pressure as a function of acid-gas loading for several initial concentrations of the complementary acid gas.

According to the Figures 3 and 4 the partial pressure of acid gas increases at constant loading, with temperature. The main reason is that absorptions and chemical reactions are exothermic phenomena.

H_2S is more acidic than CO_2 and reacts directly with the amine in aqueous solutions: $R_1R_2R_3N + H_2S = R_1R_2R_3NH^+ + HS^-$, whereas CO_2 reacts with amine following the acid–base reaction: $R_1R_2R_3N + H_2O + CO_2 = R_1R_2R_3NH^+ + HCO_3^-$. Moreover, CO_2 may also react directly with many primary and secondary amines to form stable carbamates: $R_1R_2HN + H_2O + CO_2 = R_1R_2NCOO^- + H_3O^+$.

It is worth noting that the reaction of CO_2 with MDEA, as a tertiary amine, does not form stable carbamates. According to the Figures 3 and 4, for a same acid-gas loading, partial H_2S pressure is lower than CO_2 partial pressure in aqueous MDEA solution.

CO_2 and H_2S often occur simultaneously in natural gas. The influence between these two gases is not negligible. Starting from a given loading of H_2S and CO_2 an increase in the loading of H_2S leads to a larger increase of CO_2

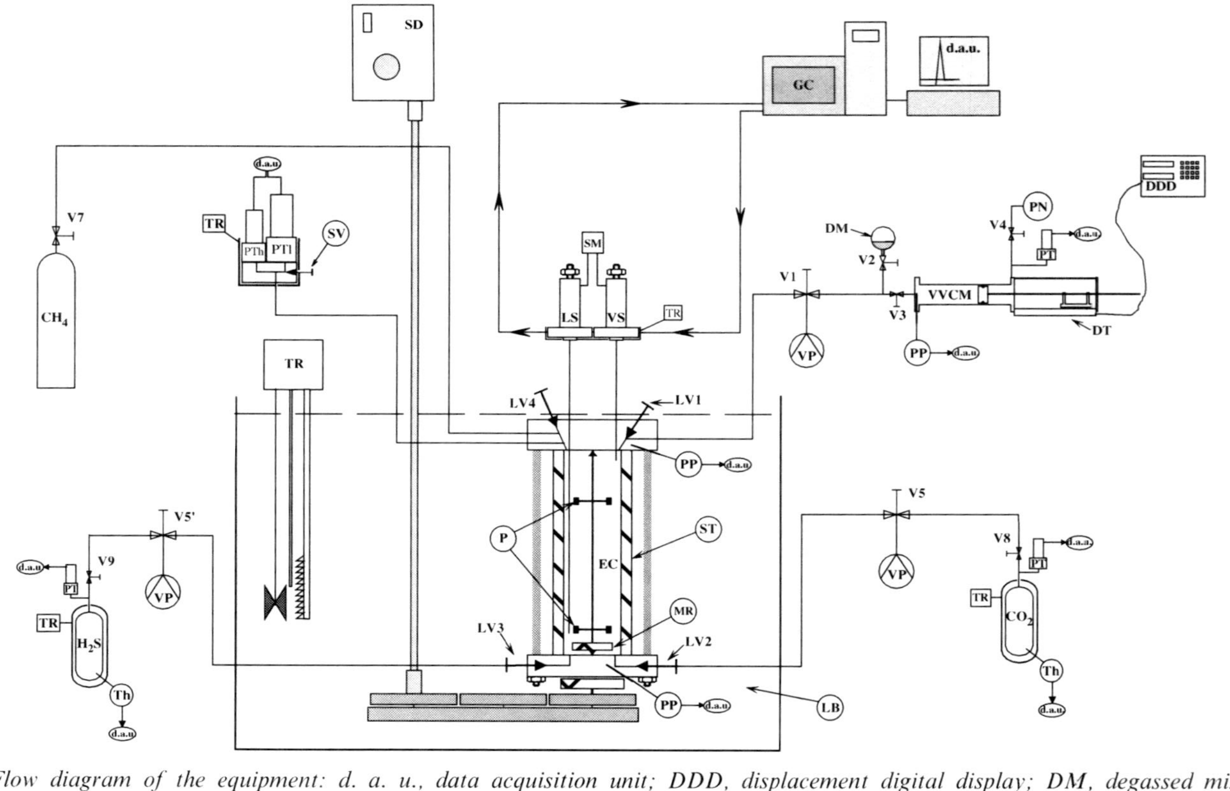

Figure 2 *Flow diagram of the equipment: d. a. u., data acquisition unit; DDD, displacement digital display; DM, degassed mixture; DT, displacement transducer; EC, equilibrium cell; GC, gas chromatograph; LB, liquid bath; LS, liquid sampler; LVi, loading valve; MR, magnetic rod; P, propeller; PP, platinum probe; PN, pressurized nitrogen; PT, pressure transducer; PTh, pressure transducer for high pressure values; PTl, pressure transducer for low pressure values; SD, stirring device; SM, sample monitoring; ST, sapphire tube; TR, thermal regulator; Th, thermocouple; Vi, valve; VP, vacuum pump; VS, vapour sampler; VVCM, variable volume cell for mixture.*

Table 1 *Solubility data for CO_2 and H_2S in aqueous MEA solutions. Mixtures are composed of $H_2S + CO_2$*

References	*MEA concentration kmol m^{-3}*	*Acid gas*
Lee *et al.*[14]	5.0	Mixtures
Lee *et al.*[15]	1.0, 2.5, 4.0, 5.0	CO_2
Lawson and Garst[16]	2.5, 5.0	Mixtures
Nasir and Mather[17]	2.5, 5.0	CO_2 and H_2S
Isaacs *et al.*[18]	2.5	Mixtures
Maddox *et al.*[19]	2.5	CO_2 and H_2S
Shen and Li[20]	2.5, 5.0	CO_2
Murrieta-Guevara *et al.*[21]	2.5, 5.0	CO_2 and H_2S
Jou *et al.*[6]	5.0	CO_2

Table 2 *Solubility data for CO_2 and H_2S in aqueous DEA solutions. Mixtures are composed by H_2S and CO_2*

References	*DEA concentration kmol m^{-3}*	*Acid gas*
Lee *et al.*[22]	0.5, 2.0, 3.5, 5.9	CO_2
Lee *et al.*[23]	0.5, 5.0	H_2S
Lee *et al.*[24]	0.5, 2.0, 3.5, 5.0	H_2S
Lee *et al.*[25]	2.0, 3.5	Mixtures
Lal *et al.*[26]	2.0	H_2S, CO_2 and mixtures
Kennard and Meisen[27]	1.0, 2.0, 3.0	CO_2

Table 3 *Solubility data for CO_2 and H_2S in aqueous MDEA solutions. Mixtures are composed by H_2S and CO_2*

References	*MDEA concentration kmol m^{-3}, (wt %)*	*Acid gas*
Jou *et al.*[3]	1.0, 2.0, 4.28 (11.8, 23.3, 48.8)	CO_2 and H_2S
Bhairi *et al.*[28]	1.0, 1.75, 2.0	CO_2 and H_2S
Ho and Eguren[29]	(23, 49)	Mixtures
MacGregor and Mather[30]	2.0	CO_2 and H_2S
Austgen *et al.*[7]	4.28	CO_2
Shen and Li[31]	2.6	CO_2
Jou *et al.*[4]	(35)	Mixtures
Jou *et al.*[3]		CO_2

partial pressure than of H_2S partial pressure. This is illustrated in Figures 5 and 6 (MDEA weight fraction in aqueous solution is 0.5).

Alkanolamines are generally used at several concentrations. The choice of amine and amine concentration is usually up to operating experience. Increasing amine concentration generally allows a reduction of the required solution circulation rates and consequently the plant costs.[2] Unfortunately this positive effect is not as advantageous as might be expected because the acid-gas vapour

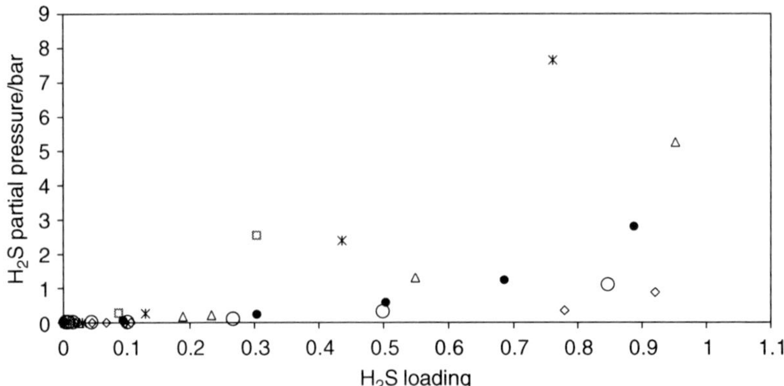

Figure 3 *H_2S Partial pressure as a function of H_2S loading in H_2O (1) + MDEA (2) solution (with 0.5 amine weight fraction in aqueous solution) at various temperatures: (◊) T = 298 K[6], (○) T = 313.15 K[6], (●) T = 323.15 K (unpublished data), (△) T = 343.15 K[6], (*) T = 373.15 K[6] and (□) T = 393.15 K[6].*

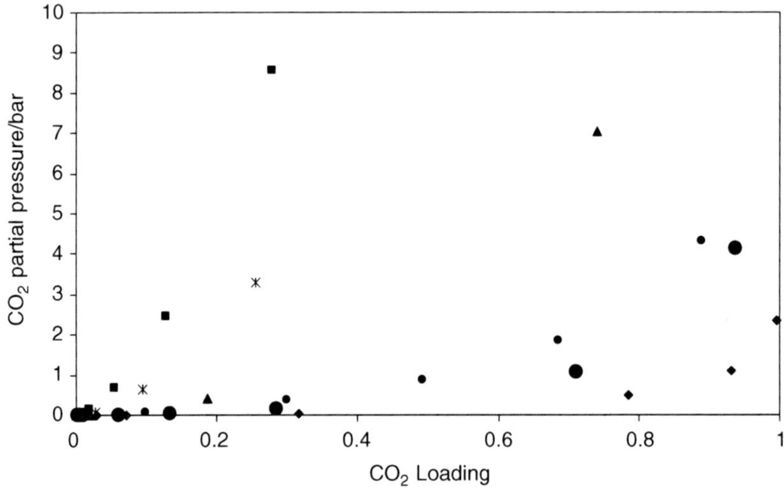

Figure 4 *CO_2 Partial pressure as a function of CO_2 loading in H_2O (1) + MDEA (2) solution (with 0.5 amine weight fraction in aqueous solution) at various temperatures: (◊) T = 298 K[6], (○) T = 313.15 K[6], (●) T = 323.15 K (unpublished data), (△) T = 343.15 K[6], (*) T = 373.15 K[6] and (□) T = 393.15 K[6].*

pressure is higher in more concentrated solutions at equivalent acid gas/amine mole ratios (see Figures 7 and 8). Another drawback of the increasing concentrations and decreasing solvent volumes is the large increase in temperatures for a given reaction, resulting in a large increase of acid-gas vapour pressure over the solution.

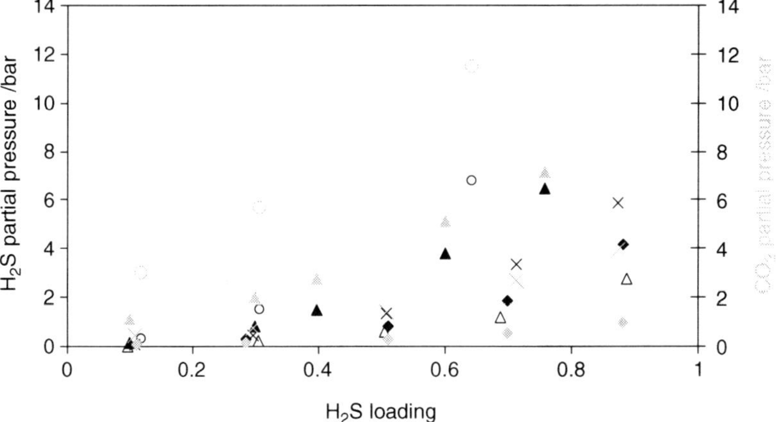

Figure 5 *H_2S and CO_2 partial pressures at 313.15 K as a function of H_2S loading for five CO_2 loadings: (△) 0, (◆) 0.09, (×) 0.31, (▲) 0.51, (○) 0.70.*

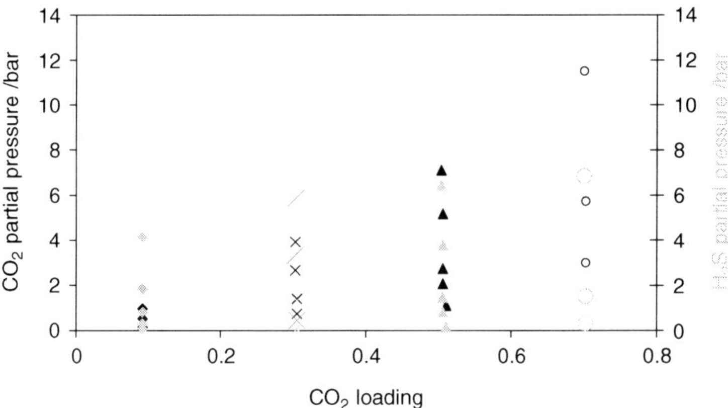

Figure 6 *CO_2 and H_2S partial pressures in function of CO_2 loading ((◆) 0.09, (×) 0.31, (▲) 0.51, (○) 0.70) at different H_2S loading. Grey symbols are related to H_2S partial pressure.*

In order to improve the selectivity of alkanolamine-aqueous solution in relation to acid-gas removal, aqueous mixtures of two alkanolamines are often used. The aim is to combine the desirable features of both amines. For example, solutions of MEA and MDEA may be used to remove CO_2 and H_2S: MDEA does not form carbamate and it has a smaller heat of reaction than MEA whereas MEA forms stable carbamate and needs more energy to remove CO_2 during the regenerating step. Moreover, reaction between MEA and H_2S seems to be irreversible contrary to the reaction between MDEA and H_2S. Association between MEA and MDEA can be advantageous and can lead to improve process design. Jou *et al.*[3] 1994 measured the solubility of CO_2 in

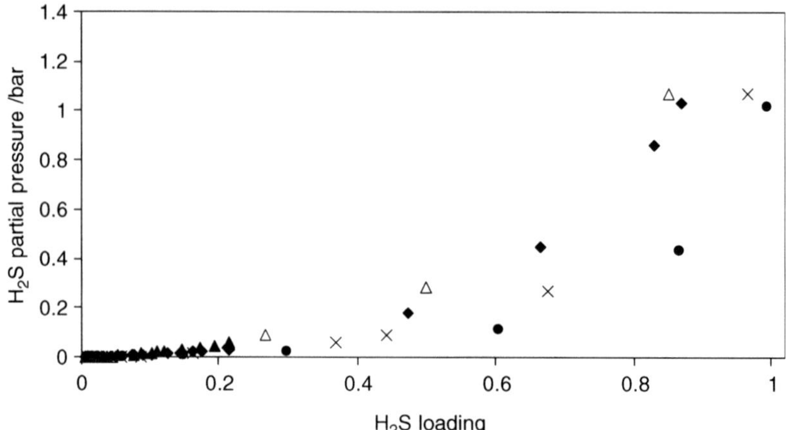

Figure 7 *H_2S partial pressure as a function of H_2S loading at 313.15 K for several amine weight fractions, w_2, in (H_2O (1) + MDEA (2)) system: (△) $w_2 = 0.5^3$, (▲) $w_2 = 0.5^5$, (◆) $w_2 = 0.35^5$, (×) $w_2 = 0.24^3$, (●) $w_2 = 0.12^3$.*

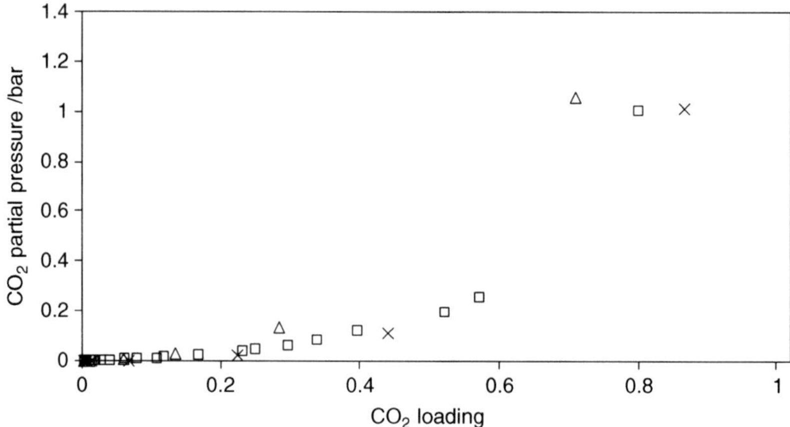

Figure 8 *H_2S partial pressure as a function of CO_2 loading at 313.15 K for several amine weight fractions, w_2, in (H_2O (1) + MDEA (2)) system. (△) $w_2 = 0.5^3$, (□) $w_2 = 0.35^5$, (×) $w_2 = 0.24^3$.*

aqueous mixtures of MEA and MDEA (the global amine weight fraction is 0.3). They conclude that the optimal MEA/MDEA ratio depends on the CO_2 residual specification and the stripping-energy cost. In effect, they show that the enthalpy of the solution of CO_2 was a strong function of the MEA concentration in the mixed-amine solution and so CO_2 removal is more difficult with the mixed amine than with MDEA solution. Shen and Li[31] measured CO_2

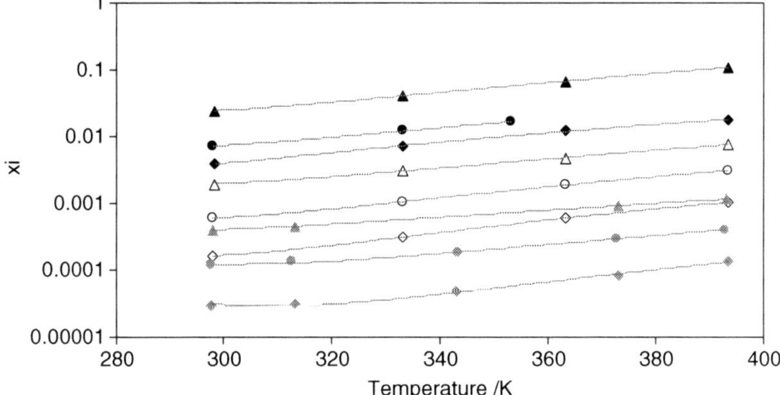

Figure 9 *Solubility of benzene, toluene and ethylbenzene in DGA aqueous solutions vs. temperature at 0.5 MPa total pressure. (▲) benzene in pure water, (△) benzene in 35 wt.% DGA, (▲) benzene in 70 wt.% DGA, (●) toluene in pure water, (○) toluene in 35 wt.% DGA, (●) toluene in 70 wt.% DGA, (◆) ethylbenzene in pure water, (◇) ethylbenzene in 35 wt.% DGA, (◆) ethylbenzene in 70 wt.% DGA.*

solubility in the same solvent. Austgen *et al.*[7] have measured CO_2 solubility in mixed MDEA and DEA-aqueous solution.

(b) BTEX

Hydrocarbons are partly dissolved by alkanolamine solution under gas treating process. Many experimental data concerning the solubility of light hydrocarbons (except BTEX) in aqueous alkanolamines are available (see Further reading). Only, Valtz *et al.*[12,13] have measured the BTEX solubility in alkanolamine solutions. Figure 9 presents the solubility of benzene, toluene and ethylbenzene in DGA-aqueous solution (weight fraction of DGA: 0.35 and 0.5) for several temperatures and at 0.5 MPa total pressure (pressure adjusted through methane addition).

The solubility of aromatics is found to increase logarithmically with temperature. However, addition of organic alkanolamine to water results in substantially higher solubility of the aromatic over that of water (see Figure 9). Moreover Valtz *et al.*[12] observed that the solubility of aromatics decreased with CO_2 loading because the reaction between the amine and CO_2 produces carbamates.

14.5 Thermodynamic Frameworks

(a) Acid gases

In 1981, Deshmukh and Mather[33] proposed a model to represent the solubility of acid gases (H_2S and CO_2) in alkanolamine solutions. This model is based on an equilibrium-reaction model coupled with thermodynamic-dissymmetric approach. In aqueous solution, H_2S and CO_2 react through an acid/base buffer mechanism with alkanolamine. The following chemical reactions can be considered where carbonate, bicarbonate or bisulfide ions are formed.

$$\begin{aligned} &2H_2O = OH^- + H_3O^+ \\ &H_2O + H_2S = H_3O^+ + HS^- \\ &H_2O + HS^- = H_3O^+ + S^{2-} \\ &2H_2O + CO_2 = H_3O^+ + HCO_3^- \\ &H_2O + HCO_3^- = H_3O^+ + CO_3^{2-} \\ &H_2O + R_1R_2R_3NH^+ = H_3O^+ + R_1R_2R_3N \end{aligned} \tag{1}$$

R_i represents an alkyl group, alkanol group or hydrogen depending on the alkanolamine. Furthermore, CO_2 can react directly with many primary and secondary amines to form stable carbamates:

$$\begin{aligned} &R_1R_2R_3N + CO_2 + H_2O = R_1R_2R_3NH^+ + HCO_3^- \\ &R_1R_2R_3N + H_2S = R_1R_2R_3NH^+ + HS^- \end{aligned} \tag{2}$$

The model requires equilibrium constants for the chemical reactions:

$$K = \Pi a_i^{\nu_i} \tag{3}$$

Activity-coefficient models are chosen to calculate equilibrium constants. The solubility of acid gases are expressed in terms Henry's law. The reference state chosen for molecular solute is the ideal, infinitely dilute aqueous solution at the system temperature and pressure. The same standard state is chosen for ionic solutes. Concerning the solvents, the standard state is the pure liquid at the system temperature and pressure. As a result, the convention adopted for the normalization of the activity coefficient is:

Solvents: $\gamma_S \rightarrow 1$ as $x_S \rightarrow 1$
Ionic and neutral solutes: $\gamma_i^* \rightarrow 1, x_i \rightarrow 0, x_{S \neq W} = 0.$
The thermodynamics equations are:

Ionic and neutral solutes:

$$y_i \Phi_i^V(T, P, y)P = x_i \gamma_i^* H_i^{P^0}(T, P^0) \exp\left(\frac{v_i^\infty (P - P^0)}{RT}\right) \tag{4}$$

Solvents:

$$y_s \Phi_s^V(T, P, y)P = x_s \gamma_s \Phi_s^0(T, P_s^0) \exp\left(\frac{v_s (P - P_s^0)}{RT}\right) \tag{5}$$

Φ is the fugacity coefficient, γ the activity coefficient, H the Henry's coefficient, v the molar volume, y and x are the vapour and liquid composition, respectively. P^0 is the pure component vapour pressure at the temperature T and (∞) referred to the infinite dilution state.

The originality in this approach comes from the model used to calculate activity coefficients for all species in the solution. Deshmukh and Mather used an extended Debye–Huckel expression. They considered excess Gibbs energy divided into two terms which take into account short-range van der Waals forces and long-range Debye–Huckel terms.

$$g^E = g^E_{\mathrm{SR}} + g^E_{\mathrm{DH}} \tag{6}$$

Deshmukh and Mather used the expression given by Guggenheim.[34] Li and Mather[35] used the same approach but they chose a Margules expansion for the short-range term and the original Pitzer equation[36] for the long-range term. In 1991, Austgen *et al.*[7] used the model described by Equation (6) with different expressions. They applied the Chen Local Composition model. This model[37] combines Pitzer's extended Debye–Hückel equation for long-range ion–ion interactions with the NRTL model[38] for short-range interactions in a local composition framework. The Born equation[39] is introduced to convert the reference condition of an ion at infinite dilution from water to the reference condition at infinite dilution in a solvent–water mixture.

$$g^E = g^E_{\mathrm{BORN}} + g^E_{\mathrm{PDH}} + g^E_{\mathrm{NRTL}} \tag{7}$$

Then, activity coefficient of any species is determined after derivation, with respect to mole number, of the excess Gibbs energy.

$$\ln \gamma_i = \left[\frac{\partial\left(\mathrm{ng^E}/\mathrm{RT}\right)}{\partial \mathrm{n_i}}\right]_{\mathrm{T,P,n_{j\neq i}}} \tag{8}$$

Another way is to consider an equation of state dedicated to electrolytes. Furst *et al.*[40] have developed an equation based on the following expression of free energy with two contributions:

$$\begin{aligned}\left[\frac{a-a^0}{RT}\right] = &\left\{\left[\frac{a-a^0}{RT}\right]_{ATT} + \left[\frac{a-a^0}{RT}\right]_{REP}\right\}_{\text{non electrolyte}} \\ &+ \left\{\left[\frac{a-a^0}{RT}\right]_{\mathrm{SOLV}} + \left[\frac{a-a^0}{RT}\right]_{LD}\right\}_{\text{electrolyte}}\end{aligned} \tag{9}$$

The non electrolyte contribution is composed of two terms: attractive and repulsive. The electrolyte contribution is also composed of two terms: a short–range interaction (SOLV) term and one long–range interaction (LD). This model can be applied to the vapour and to the liquid phase together. Moreover, influence of the pressure can be taken into account which is not possible in the liquid-phase model. Supercritical state can be easily described by this approach. This approach leads to a successful representation of systems H_2S-CO_2-aqueous DEA[41] and H_2S-CO_2-aqueous MDEA[42] and give similar results as the approach of Austgen *et al.*[7]

In 1996, Posey *et al.*[43] developed a simple model leading to good approximation of the acid gas partial pressure. By introducing the total loading $L_T\left(L_T = \frac{mol\, H_2S + mol\, CO_2}{mol\, A}\right)$ (A: amine) and the Henry's constant, the partial pressure of each species *i*, given by:

$$P_i = K_i H_i X_i \frac{L_T}{1 - L_T} \tag{10}$$

Where

$$X_i = \frac{mol\, i}{mol\, A + mol\, H_2S + mol\, CO_2 + mol\, H_2O} \tag{11}$$

The authors determined the value of K_iH_i by a correlation. The expression of this correlation is:

$$Ln(K_iH_i) = C + \frac{D}{T} + EL_TX_A^0 + F\sqrt{L_TX_A^0} \tag{12}$$

Where

$$X_A^0 = \frac{mol\, A}{mol\, A + mol\, H_2O} \tag{13}$$

Unfortunately, this model is no more accurate when the acid gases loading reach unity but nevertheless it can be used to get convenient approximations of acid gases partial pressures.

(b) BTEX

Owing to very low mutual solubility in aqueous solvent-hydrocarbon systems, it is very easy to reach liquid–liquid equilibrium (LLE) conditions. One liquid phase is mainly composed of alkylbenzene (organic phase). The second liquid phase is mainly composed of water and alkanolamine (aqueous phase).

Thus, for alkylbenzene, equilibrium-thermodynamic equations are:

$$f_i^o(T, P, x_i^o) = f_i^w(T, P, x_i^w), \quad \text{i.e.} \quad x_2^o\gamma_2^o = x_2^w\gamma_2^w \tag{14}$$

between the two liquid phases and

$$f_i^o(T, P, x_i^o) = f_i^V(T, P, y_i^V) \tag{15}$$

between organic and vapour phases

f_i is the fugactity of alkylbenzene in organic phase (*o*), aqueous phase (*w*) and vapour phase (*V*). Thanks to this approach, Valtz *et al.*[44] show clearly the influence of amine concentration on benzene-activity coefficient (Figure 10).

Benzene-activity coefficient decreases with temperature and amine concentration. The type of amine has no influence on the benzene solubility.

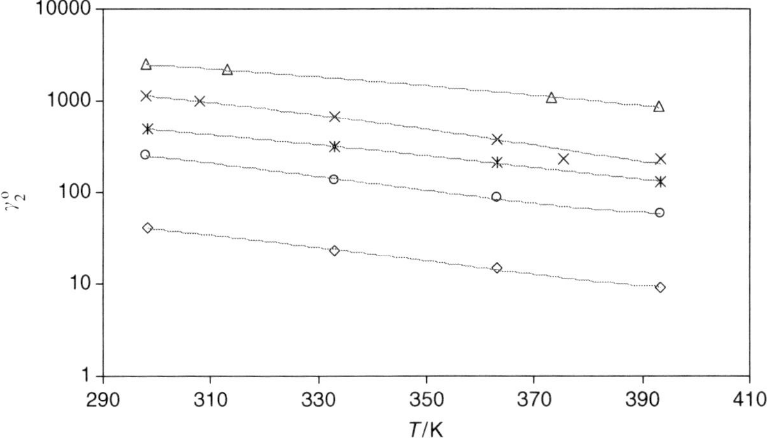

Figure 10 *Benzene-activity coefficient as a function of temperature: (△) Water, (×) MDEA water (25/75) wt% mixture, (*) DGA water (35/65) wt% mixture, (○) MDEA water (50/50) wt% mixture, (◇): DGA water (70/30) wt% mixture.*

14.6 Conclusion

Acid gas and BTEX solubility have been presented in this chapter. Many apparatus either based on static-analytic and dynamic-analytic methods have been used to get experimental data. From these data, influence of temperature, acid-gas loadings and of other parameters have been highlighted.

The modelling presented in this work takes into account the chemical reaction and the interactions between ions and molecule in the liquid phase. Three approaches are presented: empirical approach, dissymmetric approach (G^E model + equation of state) and symmetric approach (equation of state for all phases). The empirical approach is convenient for rapidly estimating solubility data. The other two approaches are more accurate, the symmetric one being preferred for high pressures.

Further Reading

J.J. Carroll, J. Maddocks and A.E. Mather, The solubility of hydrocarbons in amine solutions, *Laurence Reid Gas Conditioning Conference*, Norman, OK, 1998.

F.-Y. Jou, J.J. Carroll, A.E. Mather and F.D. Otto, The solubility of methane and ethane in aqueous solutions of methyldiethanolamine, *J. Chem. Eng. Data*, 1998, **43**, 781.

J.D. Lawson and A.W. Garst, Hydrocarbon gas solubility in sweetening solutions: methane and ethane in aqueous monoethanolamine and diethanolamine, *J. Chem. Eng. Data*, 1976, **21**, 30.

J.J. Carroll, F.-Y. Jou, A.E. Mather and F.D. Otto, Phase equilibria in the system water- methyldiethanolamine-propane, *AIChE J.*, 1992, **385**, 11.

F.-Y. Jou, H.-J. Ng and A.E. Mather, Solubility of propane in aqueous alkanolamine solutions, *Fluid Phase Equilib.*, 2002, **194**, 825.

F.-Y. Jou, J.J. Carroll, A.E. Mather and F.D. Otto, Phase equilibria in the system n-butane-water-methyldiethanolamine, *J. Chem. Eng. Data*, 1996, **116**, 407.

J. Critchfield, P. Holub, H.-J. Ng, A.E. Mather, F.-Y. Jou and T. Bacon, Solubility of hydrocarbons in aqueous solutions of gas treating amines, *Proceedings of the Laurance Reid Gas Conditioning Conference*, Norman, OK, 2001, 199–227.

References

1. GPSA (Gas Processors Suppliers Assoc.), Engineering Data book, SI Version, 11th edn, Section 21: hydrocarbon treating, Processors Suppliers Association, 6526 East 60th St., Tulsa, OK, 1998.
2. A. Kohl and R. Nielsen, in *Gas purification*, 5th ed, Gulf Publishing company, Houston, 1997.
3. F.Y. Jou, A.E. Mather and F.D. Otto, *Ind. Eng. Chem. Process Des. Dev.*, 1982, **21**, 539.
4. F.Y. Jou, J.J. Carroll, A.E. Mather and F.D. Otto, *J. Chem. Eng. Data*, 1993, **38**, 75.
5. F.Y. Jou, J.J. Caroll, A.E. Mather and F.D. Otto, *Can. J. Chem. Eng.*, 1993a, **71**, 264.
6. F.Y. Jou, A.E. Mather and F.D. Otto, *Can. J. Chem. Eng.*, 1995 A, **73**, 140.
7. D.M. Austgen, G.T. Rochelle and C.C. Chen, *Ind. Eng. Chem. Res.*, 1991, **30**, 543.
8. D. Richon, P. Antoine and H. Renon, *Ind. Eng. Chem. Process Des. Dev.*, 1980, **19**, 144.
9. C. Coquelet and D. Richon, *J. Chem. Eng. Data*, 2005, **50**, 2053.
10. D. Legret, J. Desteve, D. Richon and H. Renon, *AIChE J.*, 1983, **29**, 137.
11. P. Guilbot, A. Valtz, H. Legendre and D. Richon, *Analusis*, 2000, **28**, 426.
12. A. Valtz, P. Guilbot and D. Richon, in *Amine BTEX Solubility*, RR180, Gas processors Association, Tulsa, OK, 2002.
13. A. Valtz, M. Hegarty and D. Richon, *Fluid Phase Equilib.*, 2003, **210**, 257.
14. J.I. Lee, F.D. Otto and A.E. Mather, *Can. J. Chem. Eng.*, 1976, **54**, 214.
15. J.I. Lee, F.D. Otto and A.E. Mather, *J. Appl. Chem. Biotechnol.*, 1976 **26**, 541.
16. J.D. Lawson and A.W. Garst, *J. Chem. Eng. Data*, 1976, **21**, 20–30.
17. P. Nasir and A.E. Mather, *Can. J. Chem. Eng.*, 1977, **55**, 715–717.
18. E.E. Isaacs, F.D. Otto and A.E. Mather, *J. Chem. Eng. Data*, 1980, **25**, 118–120.
19. R.N. Maddox, A.H. Bhairi, J.R. Diers and P.A. Thomas, RR-104, Gas Processors Association, Tulsa, OK, 1987.
20. K.P. Shen and M.H. Li, *J. Chem. Eng. Data*, 1992, **37**, 96–100.

21. F. Murrieta-Guevara, E. Robolledo-Libreros and A. Trejo, *Fluid Phase Equilib.*, 1993, **86**, 225–231.
22. J.I. Lee, F.D. Otto and A.E. Mather, *J. Chem. Eng. Data*, 1972, **17**, 465.
23. J.I. Lee, F.D. Otto and A.E. Mather, *J. Chem. Eng. Data*, 1973, **18**, 71.
24. J.I. Lee, F.D. Otto and A.E. Mather, *J. Chem. Eng. Data*, 1973, **18**, 420.
25. J.I. Lee, F.D. Otto and A.E. Mather, *J. Chem. Eng. Data*, 1974, **52**, 125.
26. D. Lal, F.D. Otto and A.E. Mather, *Can. J. Chem.*, 1985, **63**, 681.
27. M.L. Kennard and A. Meisen, *J. Chem. Eng. Data*, 1984, **29**, 309.
28. A. Bhairi, G.J. Mains and R.N. Maddox, AIChE Spring National Meeting, Atlanta, GA, 1984.
29. B.S. Ho and R.R. Eguren, AIChE Spring National Meeting, New Orleans, LA, 1988.
30. R.J. MacGregor and A.E. Mather, *Can. J. Chem. Eng.*, 1991, **69**, 1357.
31. K.P. Shen and M.H. Li, *J. Chem. Eng. Data*, 1992, **37**, 96.
32. J.C. Dingman, J.L. Jackson, T.F. Moore and J.A. Branson, Equilibrium data for the H_2S-CO_2-Diglycolamine-Water system, paper presented at 62nd Annual Gas Processors Assoc., Convention San Fransisco, CA, March, 1983, 14.
33. R.D. Desmukh and A.E. Mather, *Chem. Eng. Sci.*, 1981, **36**, 355.
34. E.A. Guggenheim, *Phil. Mag.*, 1935, **19**, 588.
35. Y.G. Li and A.E. Mather, *Ind. Eng. Chem. Res.*, 1994, **33**, 2006.
36. K.S. Pitzer, *J. Phys. Chem.*, 1973, **77**, 268.
37. C.C. Chen and L.B. Evans, *AIChE J.*, 1986, **32**, 444.
38. H. Renon and J.M. Prausnitz, *AIChE J.*, 1968, **14**, 135.
39. J.W. Tester and M. Modell, in *Thermodynamics and its Applications*, 3rd ed, Prentice-Hall, Upper Saddle River, 1997.
40. W. Fürst and H. Renon, *AIChE J.*, 1993, **39**, 335.
41. G. Vallée, P. Mougin, S. Jullian and W. Fürst, *Ind. Eng. Chem. Res.*, 1999, **38**, 3473.
42. L. Chunxi and W. Fürst, *Chem. Eng. Sci.*, 2000, **55**, 2975.
43. M.L. Posey, K.G. Tapperson and G.T. Rochelle, *Gas. Sep. Purif.*, 1996 **10**, 181.
44. A. Valtz, C. Coquelet and D. Richon, *Thermochimica Acta*, 2006, **443**, 259.

CHAPTER 15

Solubility of Solids in Bayer Liquors

ERICH H. KÖNIGSBERGER, GLENN HEFTER AND PETER M. MAY

School of Chemical and Mathematical Sciences, Murdoch University, Murdoch WA 6150, Australia

15.1 Introduction

The Bayer process[1] involves the selective dissolution of $Al(OH)_3$ (gibbsite) and/or AlOOH (boehmite) in hot, concentrated caustic (NaOH) solution and the subsequent precipitation, on cooling, of purified gibbsite from the resulting supersaturated alkaline aluminate solutions ('Bayer liquors'). This is arguably the largest industrial recrystallisation process in the world being the only method used for the large-scale extraction of alumina from bauxitic ores. Global annual production of alumina *via* the Bayer process exceeds 60 million tonnes; in Australia, the world's largest producer, over 17 million tonnes are produced per annum.

The dissolution and precipitation of gibbsite are controlled by the solubility equilibrium

$$Al(OH)_3(s) + NaOH(aq) \rightleftharpoons \text{'}NaAl(OH)_4(aq)\text{'} \tag{1}$$

Quantitative knowledge of the solubility of gibbsite in Bayer liquors is, therefore, essential to optimise the design and engineering of plant equipment so as to maximise yields and product quality.

The industrial process is complicated by the variable composition of the bauxite ores, which include a variety of inorganic and organic contaminants. This results in complex, multi-component liquors that may become supersaturated with respect to undesirable solid phases, such as kogarkoite (Na_3FSO_4) which can precipitate from concentrated Bayer liquors (when both fluoride and sulfate are present) and form hard crusts in heat exchangers and pipes that consequently require costly clean up operations.[2] The organic material in bauxite typically includes high molecular weight compounds such as cellulose and lignin, which is decomposed under Bayer plant conditions and forms a great variety of organic sodium salts including formate, acetate, succinate and especially oxalate. The last has a very low solubility in caustic liquors[3] and interferes with gibbsite precipitation.[1,4] As the process stream is continuously recycled, these decomposition products build up, which alters the solubilities of solids and the thermodynamic and other properties of the liquor that are required for process engineering calculations. A physicochemical model that permits reliable simulations of Bayer-type solutions over a wide range of temperature, pressure and composition is therefore of great interest to the alumina industry.

There are currently no fundamental theories that permit the reliable *prediction* of the thermodynamic properties of concentrated binary (single salt plus solvent water), let alone multi-component, electrolyte solutions from first principles. Thus, it is necessary to use semi-empirical models parameterised with respect to experimental data to *correlate* such properties, particularly those of the binary solution subsystems. It transpires that mixing of these binary systems to form ternary and higher order ones is then governed by relatively simple, often linear, mixing rules so that in general only a few ternary interaction parameters are required.

An important issue in developing such models is their *thermodynamic consistency*. In practice, this means that the various properties and functions should be related to each other *via* the standard thermodynamic relationships such as partial differentiation or the Gibbs–Duhem equation[5] rather than be developed as 'stand-alone' models for each property, which has been common industrial practice until now. Within the thermodynamics of chemical systems, the Gibbs energy function, G, is an especially useful quantity, as other measurable properties of interest such as enthalpy, entropy, heat capacity, volume (density) and chemical potentials (activities) can be derived by appropriate differentiation thereby ensuring thermodynamic consistency. Moreover, at chemical or phase equilibrium in homogeneous (single-phase) or heterogeneous (multi-phase) systems respectively, the Gibbs energy assumes a minimum with respect to the compositional variables at constant temperature and pressure. Therefore, knowledge of the Gibbs energy of the system permits *all* thermodynamic properties of the liquor, together with solubilities, vapour pressures, boiling point elevations, *etc.* to be calculated.

Thermodynamically consistent electrolyte models, based on various kinds of data measured over wide ranges of temperatures (from *ca.* −60 to 300°C), pressures and concentrations up to saturation, have been reported for several binary and multi-component systems.[6] The semi-empirical ion-interaction model developed by Pitzer[7,8] is particularly useful as it is able to correlate, in a thermodynamically consistent way, the properties of electrolyte solutions within the experimental uncertainty of high-precision measurements over wide ranges of conditions. Parameterisation is performed on the binary and ternary subsystems and the model equations are then used to predict the properties of the multi-component system.

Until recently,[9] no comprehensive, thermodynamically consistent model for Bayer liquors had been reported in the literature. The purely empirical property functions almost invariably used industrially to describe the properties of multicomponent mixtures are generally thermodynamically inconsistent. In other words, independently derived functions, not related to each other by the laws of thermodynamics, are used to describe properties like densities, heat capacities or solubilities. Furthermore, such property functions are often valid only over limited concentration and temperature ranges and frequently fail to extrapolate correctly to the known properties of their binary solution subsystems or even to those of pure water. In such circumstances, it is difficult to perform process simulations in a coherent way. When the conditions fall outside the parameterisation range of the empirical model, it becomes virtually impossible to perform useful process simulations. An important issue here is the tendency of protagonists of various models to represent 'good fits' to limited sets of data as evidence of their preferred models' fundamental validity[10]: an approach which is scientifically unjustified and which will ultimately fail. Real progress in this area cannot occur without a more thermodynamically rigorous approach.

In various projects sponsored by the major Australian alumina producers, we have developed the first thermodynamically consistent model of synthetic Bayer liquors. This model employing the Pitzer equations consists of ten components: NaOH, '$NaAl(OH)_4$', Na_2CO_3, Na_2SO_4, NaCl, NaF, $Na_2C_2O_4$ (sodium oxalate), NaHCOO (sodium formate), $NaCH_3COO$ (sodium acetate) and H_2O, where '$NaAl(OH)_4$' is used to represent the hypothetical (pure) salt of this composition. With this Pitzer model, it is possible to generate, in good agreement with available experimental data, thermodynamic properties, such as heat capacities, enthalpies, vapour pressures, boiling-point elevations, volumes (densities) and activities of the electrolyte components and the solvent (water) over wide ranges of industrially relevant conditions. The model also predicts solubilities of gibbsite, $Al(OH)_3$, boehmite, AlOOH and the sodium salts of oxalate, fluoride, sulfate, carbonate, along with other relevant solid phases in synthetic Bayer liquors. It reproduces the large body of thermodynamic data reported in the literature or measured in our earlier projects. This synthetic liquor model has been delivered to our sponsors in the alumina refining industry in the form of a software product (BAYER.EXE).

The present review gives an overview of the solubility predictions that can be achieved with this model. Possible extensions of this knowledge to predict the properties of actual plant liquors and applications to Bayer process simulation will also be discussed.

15.2 Pitzer Equations

The set of equations developed by Pitzer[7,8] has been reviewed many times, so only the essential points are given here. The Pitzer equations can precisely represent the excess properties of even highly concentrated electrolyte solutions as a function of composition, temperature and pressure. The total excess Gibbs energy, G^E, of a binary solution is given by

$$G^{E} = w_{w} \nu m RT\,(1 - \phi + \ln \gamma_{\pm}), \qquad (2)$$

where R and T have their usual meanings, w_w is the mass of the solvent, $v = v_M + v_X$ where v_M, v_X are the stoichiometric coefficients of cations and anions respectively, m the molality of the salt (solute), ϕ the osmotic coefficient of the solvent (water) and $\gamma_{\pm}$ the mean activity coefficient of the solute. In the Pitzer model, G^E contains a Debye–Hückel term, a representation of the ionic-strength-dependent second virial coefficient B_{MX} and a third virial coefficient C_{MX} that is taken to be independent of ionic strength,

$$G^E/(w_w RT) = -A_\phi\,(4I\,b^{-1})\ln(1 + bI^{1/2}) + 2v_M v_X\,[m^2 B_{MX} + m^3 v_M z_M C_{MX}] \tag{3}$$

where

$$B_{MX} = \beta^{(0)}_{MX} + 2\beta^{(1)}_{MX}[1 - (1 + \alpha_1 I^{1/2})\exp(-\alpha_1 I^{1/2})]\,(\alpha_1{}^2 I)^{-1} + 2\beta^{(2)}_{MX}[1 - (1 + \alpha_2 I^{1/2})\exp(-\alpha_2 I^{1/2})]\,(\alpha_2{}^2 I)^{-1} \tag{4}$$

In Equation (3) and (4), I is the molality-based stoichiometric ionic strength, A_ϕ the Debye–Hückel coefficient for the osmotic function (at 25°C, $A_\phi = 0.3915$ kg$^{1/2}$ mol$^{-1/2}$)[11] and z_i the formal charge of the ion i. The constant b equals 1.2 for all solutes, $\alpha_1 = 2$ and $\alpha_2 = 0$ (*i.e.* $\beta^{(2)}{}_{MX}$ is not needed) unless $z_i \geq 2$ for both ions, in which case $\alpha_1 = 1.4$ and $\alpha_2 = 12$ and $\beta^{(2)}{}_{MX}$ must be included. For 3–2 and 4–2 electrolytes, $\alpha_1 = 2$ and $\alpha_2 = 50$ have been used.[8] It should be noted that b, α_1 and α_2 are taken as temperature independent. Thus at fixed pressure and temperature, G^E is expressed in terms of up to four adjustable parameters $\beta^{(0)}{}_{MX}$, $\beta^{(1)}{}_{MX}$, $\beta^{(2)}{}_{MX}$ and C_{MX} per electrolyte, which are generally determined by fitting the equations to osmotic and/or activity coefficient data. In most Pitzer models, the electrolytes are treated as fully dissociated and departures from ideal behaviour are considered to arise from specific ion interactions. However, values of $\beta^{(2)}{}_{MX}$ are relatively large and negative for solutes which are significantly ion-paired (associated). If the value of the ion-association constant exceeds some critical value, the resulting complexes are better taken into account as individual species.[12]

Equations for ϕ and $\ln \gamma_{\pm}$, which are related to the partial molar excess Gibbs energies of solvent and solute, are obtained by appropriate differentiation.[8] There are also general equations for G^E, ϕ and $\ln \gamma_{\pm}$ that are valid for multi-component solutions.[8] For the latter, the so-called higher-order electrostatic terms, ${}^E\theta(I)$, are usually taken into account. These can be derived from a theoretical treatment[13] of the effect of electrostatic forces arising from the unsymmetrical mixing of ions of the same charge type, *e.g.* between OH^- and $CO_3{}^{2-}$, and are functions of ionic strength. In addition, the parameters ${}^S\theta_{M,M'}$, ${}^S\theta_{X,X'}$, $\psi_{M,X,X'}$ and $\psi_{M,M',X}$ are used to describe specific interactions in ternary systems. Pitzer presents a convenient method to calculate ${}^E\theta(I)$ and reviews extensive tables of binary and ternary interaction parameters.[8] As the Pitzer model assumes that quaternary and higher-order interactions are negligible, predictions of thermodynamic properties can be made for multi-component electrolyte solutions once all relevant binary and ternary parameters are known.

Equations for apparent molar volumes, enthalpies and heat capacities can be derived from the excess Gibbs energy by appropriate differentiation with respect to pressure or temperature. The respective equations and a discussion of the strengths and limitations of the Pitzer model are given in our recent papers.[9,14]

15.3 Comprehensive Pitzer Model for Synthetic Bayer Liquors

It should be emphasised that in the present Pitzer model for synthetic Bayer liquors all solutes are treated as strong electrolytes, *i.e.* the model does not contain any species other than Na^+, OH^-, $Al(OH)_4^-$, CO_3^{2-}, SO_4^{2-}, Cl^-, F^-, $C_2O_4^{2-}$, CH_3COO^- and $HCOO^-$. All activity coefficients are calculated from the interactions among these species rather than, for instance, *via* formation of ion pairs or complexes. Redox equilibria are not considered in the present model.

When available in the literature, critically evaluated Pitzer models for binary electrolyte solutions, which are capable of calculating thermal (*e.g.* heat capacities) and volumetric (*e.g.* densities) properties over wide ranges of conditions, were employed (models a–d):

(a) NaOH,[15]
(b) NaCl,[16]
(c) Na_2SO_4[17,18] and
(d) Na_2CO_3[19]
(e) For hypothetical pure $NaAl(OH)_4$ solutions, two literature models were combined[15,20] and a volumetric Pitzer model was derived.[21]
(f) To incorporate fluoride into the model, thermodynamic data for aqueous and solid phases, including sodium fluoride and kogarkoite, Na_3FSO_4, were taken from various sources and adjusted with respect to experimental data.
(g) For $NaCH_3COO$, a literature model[22] has been tentatively adopted.
(h) Thermodynamic measurements for NaHCOO and $Na_2C_2O_4$ are sparse. For sodium formate, various approximations have been made, including the plausible assumption that the heat capacities of NaHCOO(aq) are temperature independent.[14,19]
(i) Sodium oxalate is modelled assuming that its Pitzer parameters are equal to those of Na_2SO_4.[17] This model is mainly intended for use in solubility calculations, as the very low oxalate concentrations in caustic solutions contribute negligibly to water activities, heat capacities, densities, *etc.*

The complete Bayer liquor model includes a few ternary interaction parameters selected from the literature.[6,14,20] Gibbs energies of solid phases, which are required for solubility calculations, were either adopted (occasionally slightly adjusted) from the literature (gibbsite, boehmite, sodium sulfate, carbonate, fluoride, kogarkoite) or optimised with respect to solubility data (*e.g.* sodium oxalate solubilities in water) using the ChemSage optimiser.[23] To calculate the properties of water, the IAPWS Industrial Formulation 1997 was used.[24]

The Bayer liquor model outlined above has been incorporated into a 'stand-alone' code that calculates the thermodynamic properties of aqueous solutions and two-phase equilibria between the aqueous phase and stoichiometric solid phases (or the water vapour phase). This code is based on the condition of equal chemical potentials of the components in the two phases at equilibrium, which is mathematically equivalent to the condition of minimal Gibbs energy.

After specifying the temperature, the concentrations of the synthetic Bayer liquor components are entered. Among other options this allows specification of concentrations, including 'industrial' units (*e.g.* [NaOH] expressed as grams Na_2CO_3 per litre of solution), that have been determined at room temperature for prediction of properties at other temperatures. All unit conversions are performed internally, using densities calculated from the various volumetric Pitzer models. The calculations can be performed either at 'saturation' pressure (vapour pressure of steam-saturated water at $t \geq 100°C$, 1 bar at $t < 100°C$) or at a pressure specified by the user.

As a result of the calculation, temperature, pressure as well as ionic strength and concentrations of the components (in various units) are displayed. Furthermore, boiling point elevations, vapour pressure, heat capacities, densities (and their associated apparent molar quantities), water activity and osmotic coefficient as well as conventional single-ion and mean ionic activity coefficients of solutes are shown. Also, saturation indices ($SI = a/a_{sat}$, which means that the solution is saturated when $SI = 1$, undersaturated when $SI < 1$ and supersaturated when $SI > 1$) and solubilities of selected solid phases are given.

15.4 Model Validation and Solubilities in Multi-Component Systems

The model has been validated for all binary and ternary subsystems over wide temperature and concentration ranges using experimental data (heat capacities, water activities, solubilities, densities) available in the literature or measured in our laboratories, as detailed in our recent publications.[9,25] It also performs well for more complicated solutions, *e.g.* the model calculations agree with experimental densities of twelve 7-component synthetic Bayer liquors to better than 0.1%.[21]

The following examples of solubility calculations for gibbsite, boehmite, sodium oxalate and other solid phases relevant to the Bayer process demonstrate the good performance of the present model. Sodium oxalate has a very low solubility in caustic liquors and frequently co-precipitates with gibbsite, which is detrimental to crystallisation kinetics and product quality. The experimental data[3,26] shown in Figure 1 were not used in the parameterisation of the model and a very good agreement is achieved without the use of any ternary Pitzer parameters. This suggests that the assumption to approximate

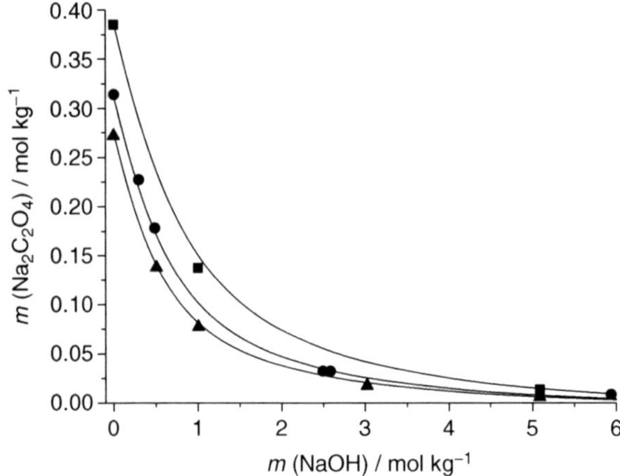

Figure 1 *Solubility of sodium oxalate in NaOH solutions. The Bayer liquor model (lines) is compared to experimental data: triangles, 25°C*[26]*; dots, 40°C*[3]*; squares, 75°C.*[26]

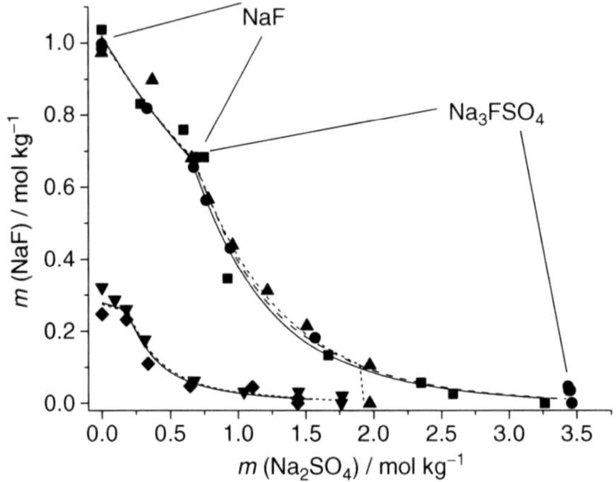

Figure 2 *Solubilities in the NaF, – Na_2SO_4 – H_2O and NaF – Na_2SO_4 – NaOH – H_2O systems. Experimental values, in water: triangles, 25°C*[28]*; dots, 35°C*[27]*; squares, 50°C*[28]*; in 3 mol kg^{-1} NaOH: inverted triangles, 25°C*[28]*; diamonds, 50°C.*[28] *Lines were calculated using the present Pitzer model: dotted, 25°C; dashed, 35°C; solid, 50°C.*

sodium oxalate activity coefficients by the sodium sulfate model of Holmes and Mesmer[17] is a good one.

Fluoride is a small but significant impurity in Bayer liquors. In some plants, sodium fluoride and/or kogarkoite, Na_3FSO_4, may become supersaturated, which leads to the deposition of hard scale in the evaporators and consequently to an interference with heat transference.[4] Figure 2 compares results from two of the very few NaF/kogarkoite solubility studies in the open literature[27,28] with our model. The agreement is well within the scatter of the experimental data. It should be noted that the solubilities of NaF and kogarkoite are significantly lower in NaOH solutions than in water; in both cases, however, they depend only slightly on temperature.

Figure 3 compares boehmite and gibbsite solubility data in NaOH solutions[29] with those calculated from the present model. As discussed recently,[25] the satisfactory agreement indicates thermodynamic consistency between the measured calorimetric and solubility data for both solid phases. Figure 3 also shows that, consistent with industry practice, much higher digestion temperatures are needed for boehmitic than gibbsitic bauxite ores: the solubility of boehmite at 150°C is only slightly higher than the solubility of gibbsite at 80°C.

The following figures show gibbsite solubilities in more complex electrolyte mixtures.[25] Figure 4 compares model predictions for gibbsite solubilities in NaOH/NaCl mixtures with experimental data by Lyapunov *et al.*,[30] which have not been used in the current parameterisation. The agreement is excellent.

The predictions of the present model also agree very well with gibbsite solubilities in NaOH/Na_2CO_3 solutions, again measured by Lyapunov *et al.*[30] (Figure 5). The model predicts gibbsite solubilities in synthetic Bayer liquors containing all of the common major inorganic 'impurities'[31]: Cl^-, $CO_3{}^{2-}$ and

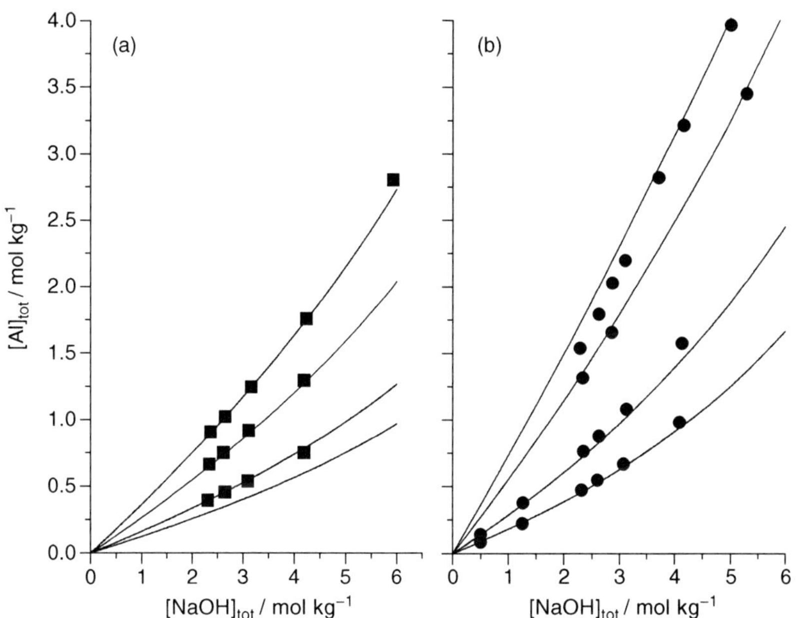

Figure 3 *Solubility of boehmite (a) and gibbsite (b) in NaOH solutions at 60, 80, 120 and 150°C (with isotherms increasing bottom to top). Experimental data[29]: squares, boehmite; dots, gibbsite; lines, present model. Square brackets denote concentrations (in the unit given) and the subscript 'tot' indicates total (analytical) concentrations (Reproduced from Königsberger et al.[25] by permission of Springer Verlag, Wien)*

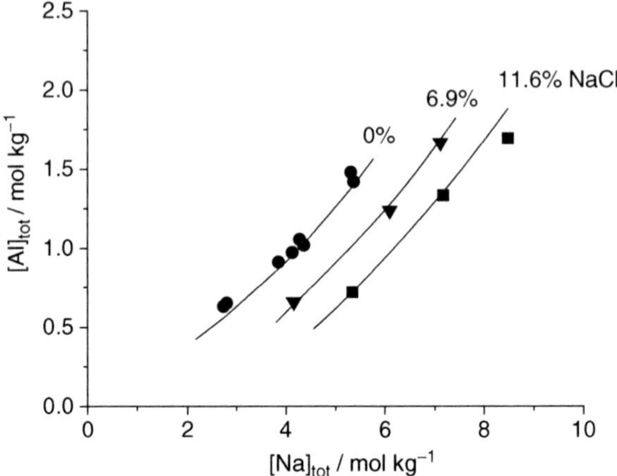

Figure 4 *Gibbsite solubilities[30] at 60°C in NaOH solutions containing various amounts of sodium chloride (in mass-%) compared with the present Pitzer model (lines) (Reproduced from Königsberger et al.[25] by permission of Springer Verlag, Wien)*

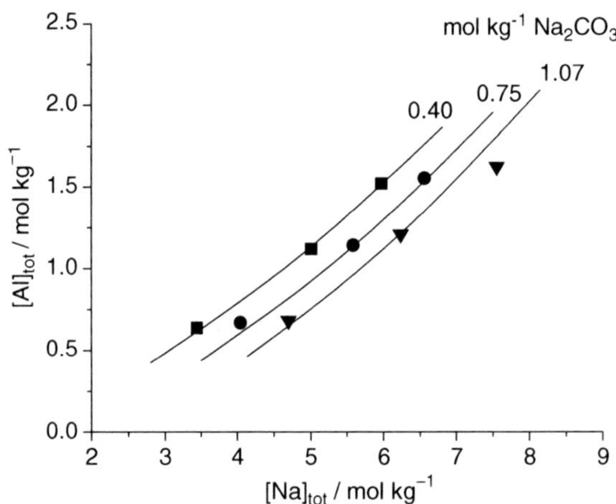

Figure 5 *Gibbsite solubilities[30] at 60°C in NaOH solutions containing varying amounts of Na_2CO_3 compared with the predictions of the present Pitzer model (lines) (Reproduced from Königsberger et al.[25] by permission of Springer Verlag, Wien)*

SO_4^{2-} (Figure 6). The agreement is very good at 50°C and reasonable at 60°C. In both cases, no additional ternary parameters were required.

No Bayer *plant* liquor data have been used for parameterisation of the present model. Predictions for gibbsite solubilities in plant liquors (containing all the components of the present model as well as many other minor ones) were compared with values carefully measured by Rosenberg and Healy[32] who also

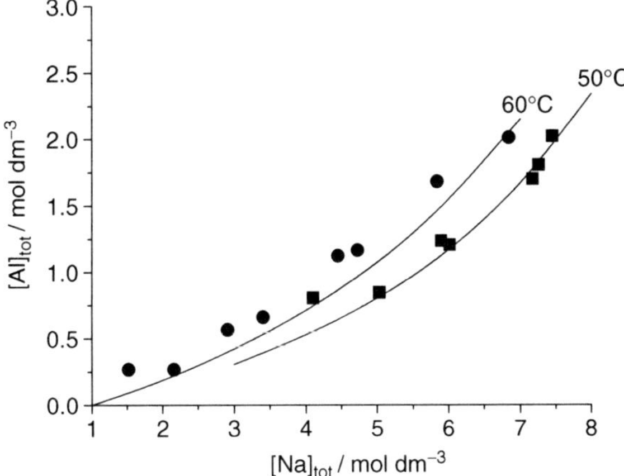

Figure 6 *Solubility of gibbsite in synthetic Bayer liquors*[31] *containing* $[Cl^-] = 0.24$ *mol* dm^{-3}, $[SO_4^{2-}] = 0.22$ *mol* dm^{-3} *and* $[CO_3^{2-}] = 0.22$ *mol* dm^{-3} *compared with the predictions of the present Pitzer model (lines). The calculated molarities refer to 25°C, since the analyses are usually performed close to this temperature. (Reproduced from Königsberger et al.*[25] *by permission of Springer Verlag, Wien)*

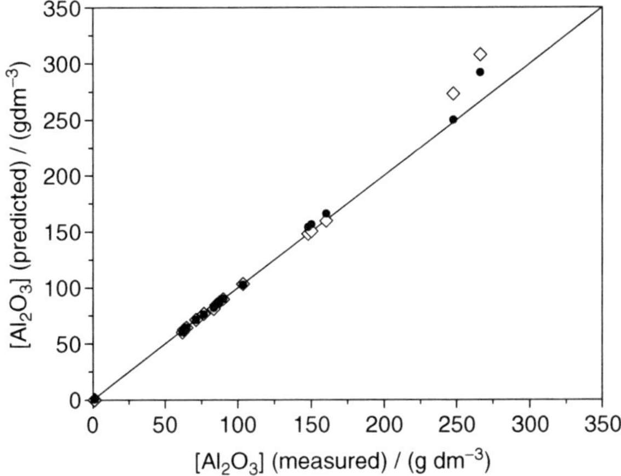

Figure 7 *Predicted vs. experimental*[32] *gibbsite solubilities in plant liquors. Open symbols, model of Rosenberg and Healy*[32]*; solid symbols, present Pitzer model for synthetic Bayer liquors. The line represents exact agreement between observed and calculated data.*

developed an empirical gibbsite solubility model that was fitted to their experimental data (Figure 7, open symbols). After a simplifying assumption was made regarding the modelling of organic species, an excellent agreement between our model predictions and experimental gibbsite solubility data was found (Figure 7, solid symbols).

15.5 Conclusion

In terms of number of components and ranges of temperature and concentration, which cover all aspects of the Bayer process from bauxite digestion to gibbsite precipitation, the present Pitzer model of *synthetic* Bayer liquors is one of the largest thermodynamically consistent electrolyte models ever developed for an industrial process. The model calculates, in a consistent manner, the thermodynamic properties (*e.g.* C_p, ϕ, ρ) of alkaline sodium aluminate solutions containing most of the major industrial 'impurities', both inorganic and organic, together with the solubilities of gibbsite, boehmite and other solid phases relevant to the Bayer process over concentration and temperature ranges of industrial interest.

Improvements in the mature Bayer process, used industrially for more than 100 years, require the development of sound and robust models of Bayer *plant* liquors. Such models must not only describe the thermodynamic and transport properties well, but also be capable of being incorporated in process simulators. The results discussed in this brief summary indicate that the present model can be applied favourably to plant liquors under industrially realistic conditions. However, accurate computer simulations of the Bayer process at conditions far outside current plant operating ranges are likely to be critical for future increases in productivity, to minimise refinery discharges, lower energy consumption and cut greenhouse gas emissions. This will require high-quality measurements of key thermodynamic properties such as heat capacities and densities up to high temperatures. Such data are currently being measured in our laboratory to 300°C to develop hitherto unavailable modelling capabilities for industrial Bayer liquors.

Acknowledgements

This work was funded through the Australian Mineral Industries Research Association (AMIRA) by the Australian alumina industry (Alcan Engineering, Alcoa World Alumina, Comalco/Rio Tinto, Queensland Alumina and Worsley/BHP-Billiton), and the Australian Government through SPIRT (No. C29940103) and Linkage (Nos. LP0349107 and LP0560477) Grants and under its Cooperative Research Centres program.

References

1. L.K. Hudson, Alumina production, In *Production of Aluminium and Alumina*, A.R. Burkin, (ed.), Critical Reports on Applied Chemistry, vol 20, Wiley, New York, 1987, 11–46.
2. D.J. Wilson, A.A. Aboagye, C.A. Heath, S.P. Rosenberg, W. Tichbon and C.R. Whitaker, *Proceedings of the 6th International Alumina Quality Workshop*, Brisbane, Australia, 2002, 281.
3. W.H.H. Norris, *J. Chem. Soc. Abstr.*, 1951, 1708.

4. T.G. Pearson, *The Chemical Background of the Aluminium Industry* Royal Institute of Chemistry, London, Lecture Monograph Report No. 3, 1955.
5. M.L. McGlashan, *Chemical Thermodynamics*, Academic Press, London, 1979.
6. R.T. Pabalan and K.S. Pitzer, Mineral solubilities in electrolyte solutions, In *Activity Coefficients in Electrolyte Solutions*, 2nd edn, K.S. Pitzer, (ed), CRC Press, Boca Raton, FL, 1991, 435–490.
7. K.S. Pitzer, *J. Phys. Chem.*, 1973, **77**, 268.
8. K.S. Pitzer, Ion interaction approach theory and data correlation, In *Activity Coefficients in Electrolyte Solutions*, 2nd edn, K.S. Pitzer (ed.), CRC Press, Boca Raton, FL, 1991, 75–153.
9. E. Königsberger, G. Eriksson, P.M. May and G. Hefter, *Ind. Eng. Chem. Res.*, 2005, **44**, 5805.
10. J.R. Loehe and M.D. Donohue, *AIChE J.*, 1997, **43**, 180.
11. D.J. Bradley and K.S. Pitzer, *J. Phys. Chem.*, 1979, **83**, 1599.
12. C.E. Harvie, N. Møller and J.H. Weare, *Geochim. Cosmochim. Acta*, 1984, **48**, 723.
13. K.S. Pitzer, *J. Solution Chem.*, 1975, **4**, 249.
14. E. Königsberger, *Monatsh. Chem.*, 2001, **132**, 1363.
15. J.M. Simonson, R.E. Mesmer and P.S.Z. Rogers, *J. Chem. Thermodyn.*, 1989, **21**, 561.
16. K.S. Pitzer, J.C. Peiper and R.H. Busey, *J. Phys. Chem. Ref. Data*, 1984, **13**, 1.
17. H.F. Holmes and R.E. Mesmer, *J. Solution Chem.*, 1986, **15**, 495.
18. C. Monnin, *Geochim. Cosmochim. Acta*, 1990, **54**, 3265.
19. E. Königsberger, *Pure Appl. Chem.*, 2002, **74**, 1831.
20. D.J. Wesolowski, *Geochim. Cosmochim. Acta*, 1992, **56**, 1065.
21. E. Königsberger, S. Bevis, G. Hefter and P.M. May, *J. Chem. Eng. Data*, 2005, **50**, 1270.
22. R. Beyer and M. Steiger, *J. Chem. Thermodyn.*, 2002, **34**, 1057.
23. E. Königsberger and G. Eriksson, *CALPHAD*, 1995, **19**, 207.
24. The International Association for the Properties of Water and Steam. *Release on the IAPWS Industrial Formulation 1997 for the Thermodynamic Properties of Water and Steam*, IAPWS Meeting, Erlangen, Germany, 1997.
25. E. Königsberger, P.M. May and G. Hefter, *Monatsh. Chem.*, 2006, **137**, in press.
26. E. Königsberger, K. Murray, C. Magalhães, A. Tromans, S. Bevis, M. Lukas, G. Hefter and P. May. *Prediction and Measurement of the Physiochemical Properties of Bayer Liquors – P507A,* Annual Progress Report to AMIRA, November 2001, 303pp, A-J. Tromans, *Ph.D. Thesis*, Murdoch University, 2003.
27. H.W. Foote and J.F. Schairer, *J. Am. Chem. Soc.*, 1930, **52**, 4202.
28. R.K. Toghiani, V.A. Phillips and J.S. Lindner, *J. Chem. Eng. Data*, 2005, **50**, 1616.

29. A.S. Russell, J.D. Edwards and C.S. Taylor, *J. Metals*, 1955, **7**, 1123.
30. A.N. Lyapunov, A.G. Khodakova and Z.G. Galkina, *Sov. J. Non-Ferrous Metals*, 1964, **3**, 48.
31. G. Bouzat and G. Philipponneau, *Light Metals*, 1991, 97.
32. S.P. Rosenberg and S.J. Healy, *Proceedings of the 4th International Alumina Quality Workshop*, Darwin, Australia, 1996, 301.

CHAPTER 16

Solubility of Gases in Polymers

JEAN-PIERRE E. GROLIER AND SEVERINE A.E. BOYER

Laboratoire de Thermodynamique des Solutions et des Polymères, Université Blaise Pascal, Clermont-Ferrand, 24 Avenue des Landais, 63177, Aubière, France

16.1 Introduction

Nowadays, polymer-based materials are at the centre of applications where they are frequently subjected to temperature variations and also to gas pressures ranging from a few MPa to 100 MPa. In particular, in the petroleum industry, taken here as an example, the transport of petroleum fluids is done using flexible hosepipes. These hosepipes are made of extruded thermoplastic or rubber sheaths and reinforcing metallic armour layers. The type of transported fluids (which may contain important amounts of dissolved gases) and the operating temperature and pressure, dictates the composition of the hosepipe sheath (*e.g.* polyethylene (PE) or polyamide (PA) and/or poly(vinylidene fluoride) (PVDF or PVF_2)). However, these thermoplastic polymers, like elastomers, are not entirely impermeable and undergo sorption/diffusion phenomena. A rupture of the thermodynamic equilibrium after a sharp pressure drop may eventually damage the polymer components. Gas concentration in the polymer, together with temperature gradients, can cause irreversible "explosive" deterioration of the polymeric structures. This blistering phenomenon, usually termed as "explosive decompression failure" (XDF) process, is actually very dramatic for the material. The resistance to physical changes is related to the influence of the gas–polymer interactions on the thermophysical properties of the polymer. The estimation of the gas sorption and of the concomitant polymer swelling as well as the measurement of the thermal effect associated with the gas–polymer interactions provide valuable and basic information for a better understanding and control of polymers' behaviour in different applications, where temperature and pressure, in combination with supercritical fluid stress, may deeply affect polymers' stability and properties. In the food industry more and more consumption products, obtained by extrusion, must respect rigorous specificities, like smell, aspect and taste, all features which are well controlled by temperature T, and pressure P, conditions of the process and by the use of supercritical fluids. Other numerous industrial activities deal with polymer modification and transformation, through different processes like extrusion, injection, and moulding. Polymer foaming, among others, is currently achieved in various ways, but typically involves elevated temperatures and pressures as well as the addition of chemicals, mostly gases that are used as blowing agents. Thermal, barometric and/or chemical stress may shift, even permanently, the polymer glass transition temperature T_g, which consequently modifies the physical properties of the material. Sorption of fluids such as gases in the supercritical state induces significant plasticization, resulting in a substantial decrease of the glass transition temperature. If such an effect is rather weak when using helium or nitrogen due to their low solubility in polymers, sufficiently high pressure should induce higher gas sorption by polymers. In this respect, gases such as carbon dioxide or hydrofluorocarbons (HFCs) are known to be good fluids for plasticization of a polymer, like polystyrene (PS). As a result of the international regulation, the blowing gases used up to now in the foaming industry, have to be replaced by less harmful to the ozone layer blowing agents. The knowledge of the influence of gas sorption and concomitant swelling on the glass transition temperature T_g of a {gas–polymer} system is of real importance in generating different types of foams. In the context of the above applications, thermophysical properties of gas saturated thermoplastic semicrystalline polymers are key elements for the development of several engineering applications.

The focus of the present chapter is on the behaviour of {gas–polymer} systems from the point of view, of gas solubility, and associated thermal effects. Depending on the temperature and pressure ranges, polymers are either in the solid or molten state, that is to say at temperatures between glass transition temperature T_g and temperature of fusion T_m; in most cases, gases are supercritical fluids. The present contribution, essentially based on current activities of the authors, is split into three parts: experimental measurements of gas solubility, evaluation of gas–polymer interactions through experimental measurements of thermal effects and thermophysical properties of polymers, and the importance of such data for engineering applications.

16.2 Experimental Measurements of Gas Solubility

Gas solubility in polymers can be measured using different techniques namely, gravimetric techniques, vibrating or oscillating techniques, pressure–volume–temperature (PVT) techniques involving the pressure decay and also gas-flow techniques. Recently, the coupling of a new gravimetric technique, using a vibrating-wire (VW) sensor, with a PVT-pressure decay technique has produced a new type of instrument to evaluate gas solubilities in polymers.

16.2.1 Gravimetric Techniques

These techniques consist in precisely weighing a polymer sample during gas sorption. They are very sensitive at low to moderate gas pressures[1,2] and a magnetic coupling used to transmit the weight to a balance,[3,4] has extended the pressure range to 35 MPa.

16.2.2 Vibrating or Oscillating Techniques

With these techniques the change of mass of a polymer sample is calculated from the resonance characteristics of a vibrating support, either a piezoelectric crystal[5,6] or a metal reed,[7] to which the polymer sample is fixed (very often this support is a spherical quartz resonator on which a thin polymer film is wrapped). Depending on the type of oscillator, the maximum pressure for these techniques is between 15 and 30 MPa.

16.2.3 PVT-Techniques and the Pressure Decay Method

In the techniques based on the pressure decay method,[8,9] a polymer sample is seating in a container of known volume acting as an equilibrium cell; the quantity of gas initially introduced in this cell is evaluated by PVT measurements in a calibrated cell from which the gas is transferred into the equilibrium cell in a series of isothermal expansions. The pressure decay in the equilibrium cell during sorption, permits the evaluation of the amount of gas penetrating in the polymer. The pressure decay principle allows a sensitivity of few hundredths of milligram of absorbed gas per gram of polymer.[10]

16.2.4 Gas–Flow Techniques

In these techniques, the solubility of gases in polymers is obtained from gas flow measurements by inverse gas chromatography[11]; in this procedure, the polymer sample (glassy or molten) acts as the chromatographic stationary phase for the measurement of retention times.

16.2.5 The Coupled VW-PVT Technique

In all above techniques where the polymer sample is immersed in the penetrating gas, the associated swelling of the polymer due to the gas sorption is an important phenomenon which needs to be accurately evaluated since it may affect the buoyancy force exerted by the gas on the polymer sample (in the case of gravimetric measurements) as well as the internal volume (in the case of PVT measurements). Usually, swelling is determined separately by various techniques using direct visual observation. The volume change is in the order of 0.3% of the volume of the initially degassed polymer.[12] Alternatively, swelling has been estimated using a theoretical model such as the Sanchez–Lacombe equation of state.[13]

Recently, Hilic *et al.*[14,15] have designed a technique permitting the simultaneous determination *in situ* of the amount of gas penetrating the polymer and the concomitant change of volume of the polymer due to gas sorption. This technique involves a vibrating-wire force sensor, acting as gravimetric device, and a pressure decay installation to evaluate the amount of gas penetrating into the polymer.

16.2.5.1 VW Sensor

The VW sensor (Figure 1) is employed as a force sensor to weigh the polymer sample during the sorption process: the buoyancy force exerted by the pressurized fluid on the polymer depends on the swollen volume of the polymer due to the gas sorption. This VW sensor is essentially a high-pressure cell in which the polymer sample is placed in a container suspended by a thin tungsten wire (diameter 25 μm, length 30 mm) in such a way that the wire is positioned in the middle of a high magnetic field generated by a square magnet placed across the high-pressure cell. Through appropriate electric circuitry and electronic control, the tungsten wire is activated to vibrate. The period of vibration which can be accurately measured is directly related to the mass of the suspended sample. The working Equation (1) for the VW sensor relates the mass m_{sol} of gas absorbed in the polymer to the change in volume ΔV_{pol} of the polymer. The natural angular frequency of the wire, through which the polymer sample holder is suspended, depends on the amount of gas absorbed. The physical characteristics of the wire are accounted for in Equation (1) as

$$m_{sol} = \rho_g \,\Delta V_{pol} + \left[\left(\omega_B^2 - \omega_0^2\right) \frac{4\,L^2\,R^2\,\rho_S}{\pi\,g} + \rho_g\left(V_C + V_{pol}\right)\right] \tag{1}$$

The volume of the degassed polymer is represented by V_{pol} and ρ_g is the density of the fluid. The terms ω_0 with ω_B represent the natural (angular) frequencies of the wire in vacuum and under pressure, respectively and V_C the volume of the container. The symbols L, R and ρ_s are, respectively, the length, the radius and the density of the wire.

16.2.5.2 PVT-Method and Pressure Decay Measurements

The three-cell principle for PVT measurements of Sato *et al.*[10] has been used (Figure 2) to determine the amount of gas solubilized in a polymer. The experimental method consists of a series of successive transfers of the gas by connecting the calibrated transfer cell V_3 to the equilibrium cell V_2 which contains the polymer. Initial P_i and final P_f pressures are recorded between each transfer. The initial methodology was based on the iterative calculation described by Hilic *et al.*[14,15] The (rigorous) working Equation (2) for the PVT-technique gives the amount of gas entering the polymer sample during the first transfer once equilibration is attained

$$m_{sol} = \frac{M_g}{R}\frac{P_f}{T_f\,Z_f}\Delta V_{pol} + \frac{M_g}{R}\left[\frac{P_i}{Z_i\,T_i}V_3 - \frac{P_f}{Z_f\,T_f}\left(V_2 + V_3 - V_{pol}\right)\right] \tag{2}$$

Equation (2) allows one to calculate the mass m_{sol} of gas dissolved in the polymer. M_g is the molar mass of the dissolved gas, Z_i with Z_f are the compression factors of the gas entering the polymer, respectively at the initial (index$_i$) and final (equilibrium sorption, index$_f$) stages. The volume of the

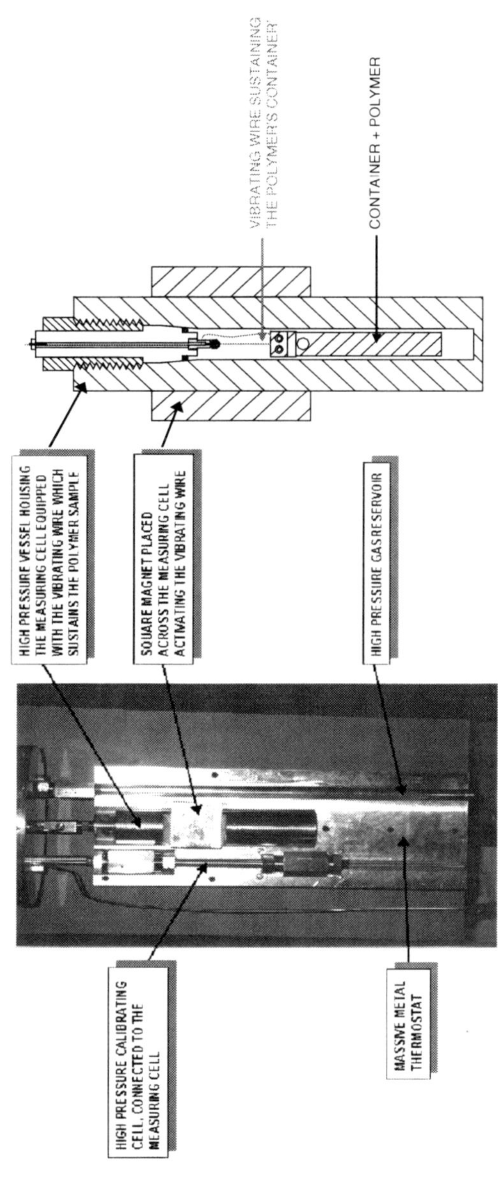

Figure 1 *The coupled VW-PVT technique. On the left-hand side is a picture of the experimental set-up showing the three high-pressure cells. On the right-hand side is a schematic view of the equilibrium cell housing the VW sensor and the container of the polymer sample.*

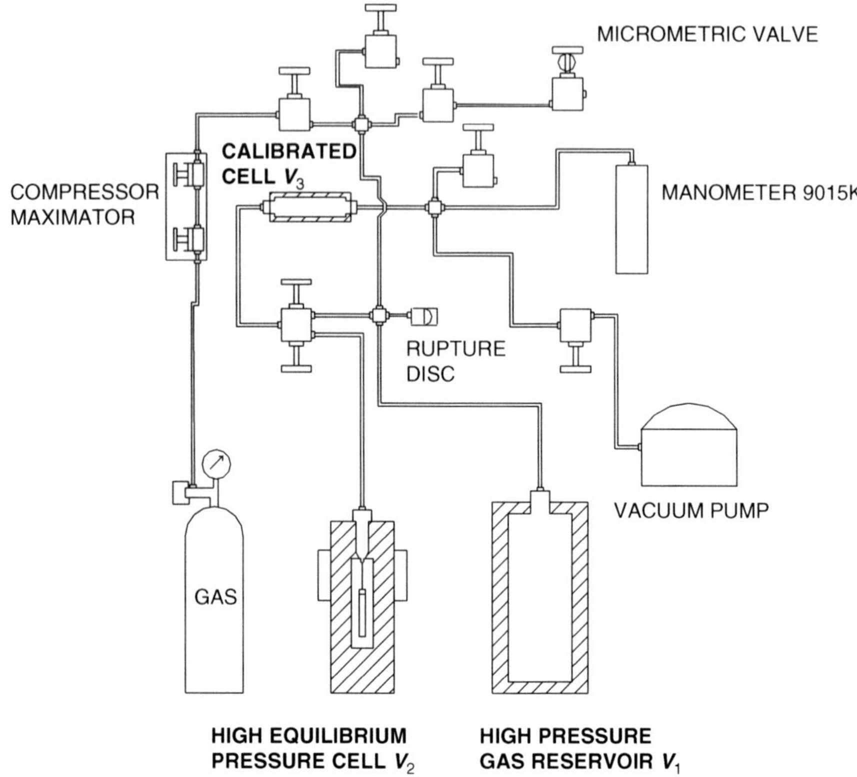

Figure 2 *Schematic view of the three-cell principle for PVT measurements after Sato et al.*[10] *The high-pressure line connects the three cells V_1, V_2 and V_3, respectively, the high-pressure reservoir cell, the high-pressure equilibrium cell housing the VW sensor, and the high-pressure calibrated transfer cell.*

degassed polymer and the volume change due to sorption are represented by V_{pol} and ΔV_{pol}, respectively. The total amount of gas absorbed by the polymer after completion of the successive transfers is given by Equation (3)

$$\begin{aligned}\Delta m_{\text{sol}}^{(k)} &= \frac{M_{\text{g}}}{R}\frac{P_{\text{f}}^{(k)}}{Z_{\text{f}}^{(k)}\,T_{\text{f}}^{(k)}}\Delta V_{\text{pol}}^{(k)} \\ &+ \frac{M_{\text{g}}}{R}\left[\frac{P_{\text{i}}^{(k)}\,V_3}{Z_{\text{i}}^{(k)}\,T_{\text{i}}^{(k)}} + \frac{P_{\text{f}}^{(k-1)}\left(V_2 - V_{\text{pol}} - \Delta V_{\text{pol}}^{(k-1)}\right)}{Z_{\text{f}}^{(k-1)}\,T_{\text{f}}^{(k-1)}}\right. \\ &\left. - \frac{P_{\text{f}}^{(k)}\left(V_2 + V_3 - V_{\text{pol}}\right)}{Z_{\text{f}}^{(k)}\,T_{\text{f}}^{(k)}}\right]\end{aligned} \quad (3)$$

where $\Delta m_{\text{sol}}^{(k)}$ is the increment in dissolved gas mass resulting from the transfer k and $\Delta V_{\text{pol}}^{(k)}$ is the change in volume after transfer k.

16.2.5.3 Evaluation of Gas Solubility and Associated Swelling

The above VW-PVT technique procedure allows one, in principle, to obtain simultaneously from two rigorous Equations (1) and (2), two unknowns, the gas solubility and the change in volume of the polymer due to sorption, at pressures up to 100 MPa, from room temperature to 473 K. However, despite its obvious advantages, the coupled technique needs further improvement.[16] As a matter of fact, the change in volume associated with high-pressure gas sorption is a complex phenomenon. On the one hand, the chemical structures of both the polymer and the gas play a major role in terms of thermal energy of gas–polymer interactions during sorption; on the other, pressure plays also an important role, depending again on the polymer's structure. For example, with the two polymers, medium density polyethylene (MDPE) and poly (vinylidene fluoride) (PVDF), it has been demonstrated[17] that supercritical carbon dioxide ($scCO_2$) affects substantially the cubic expansion coefficient of the polymers, especially at pressures ranging from 10 to 30 MPa, where the polymer–gas interactions are more marked. It seems that at lower pressures the main interactions correspond to the exothermic sorption of CO_2 by the surface and amorphous phase and possibly by some interstitial sites of the crystalline part of the polymer. At higher pressures, CO_2 is forced to enter deeply inside the interstitial or other voids in the polymer and cause their mechanical distortion, which is associated with an endothermic effect. At high pressures (above 30 MPa), the polymers saturated with CO_2 behave as pseudo-homogeneous phases and their cubic thermal expansion coefficient is only slightly higher than for pure polymers. Heats of interaction of CO_2 with PVDF are higher than with MDPE and demonstrate that CO_2 preferentially penetrates into PVDF than into MDPE. In addition to the above observed complex gas in polymer solubility phenomenon, it has been shown that if the two working Equations (1) and (2), characteristic of the VW-PVT technique, do not converge,[16] then solubility and swelling cannot be obtained simultaneously by direct experimental determination. Effectively a common term appears in both Equations (1) and (2), the density ρ_g of the gas

$$\rho_g = \frac{M_g}{R} \frac{P_f}{T_f Z_f} \tag{4}$$

As a result Equations (1)–(3) can be written as

$$\Delta m_{sol}^{(k)} = \rho_g \, \Delta V_{pol} + d \tag{5}$$

The term d represents the apparent concentration of gas in the polymer, *i.e.* when the change in volume ΔV_{pol} is zero. However, despite the different terms appearing in the two working Equations (1) and (2), these two equations can both be expressed by the same reduced Equation (5) with the slope given by Equation (4).

At this stage, it appears that the VW sensor technique is more precise than the PVT-technique since there are no cumulative errors like in the case of the

PVT-method, when the successive transfers are performed during an isothermal sorption. The technique does not require extensive calibrations. Essentially, uncertainties come from the experimentally measured resonance frequencies. Errors are reduced with the new data acquisition permitting the simultaneous recording of the phase with the frequency: effectively, the phase angle is better suited than the amplitude to detect the natural resonant frequency (and also the half-width).[16] Nevertheless, the main source of uncertainty in evaluating the gas concentration comes from the first term of Equation (5), which contains the density of the gas and the change in volume of the polymer; it was thus necessary to elaborate a procedure to estimate the apparent solubility of the gas and the associated change in volume. To this end, *"faute de mieux"*, the Sanchez–Lacombe equation of state was selected to calculate the change in volume of the polymer at different pressures and temperatures.[16]

16.3 Experimental Evaluation of Gas Polymer Interactions and Thermophysical Properties

As indicated in the introduction, gas solubility in polymers is a twofold problem with two aspects which are quite different in their consequences: either the gas penetrating the polymer generates irreversible damages of the material, or the gas sorption can help through efficient control to process polymer modifications (like extrusion, foaming, moulding). In the first case, gas penetration should be prevented; in the second case, gas sorption is an excellent partner in polymer engineering. In this context, the strict knowledge of gas solubility is not sufficient by itself. The extent of penetration of gases into polymer structures must be documented through quantitative evaluation of gas–polymer interactions as well as thermophysical properties of gas saturated polymers. This information has to be taken as an indispensable complement of solubility. To investigate thermal effects associated with the interactions of gases and of supercritical fluids with semicrystalline polymers as a function of pressure, an original technique, scanning transitiometry,[17] was used, taking full advantage of the differential mounting of experimental vessels. The experimental instrument[18] from BGR TECH (Warsaw, Poland) has been intensively used in polymer thermodynamics.[19,20] The basic principle of the instrument consists in scanning one of the three variables, *P*, *V* or *T*, a second one being kept constant, and concomitantly recording the thermal effect related to the system under investigation (polymer under gas pressure) placed in the measuring vessel as well as the change of the dependable variable. With this methodology, thermal evaluation of gas–polymers interactions are obtained on the one hand and thermo-mechanical coefficients of the gas saturated polymer are also precisely accessible on the other. This is illustrated by two typical examples; firstly, it is possible to directly compare the thermal behaviour of two different polymers submitted to the same penetrating gas in the pressure range around the critical point of the gas[17]; secondly, the cubic expansion coefficients of different polymers submitted to various high-pressure transmitting fluids, *e.g.* {CO_2 or Hg-MDPE}, {CO_2 or N_2 or Hg-PVDF}, show (see Figure 3) how more

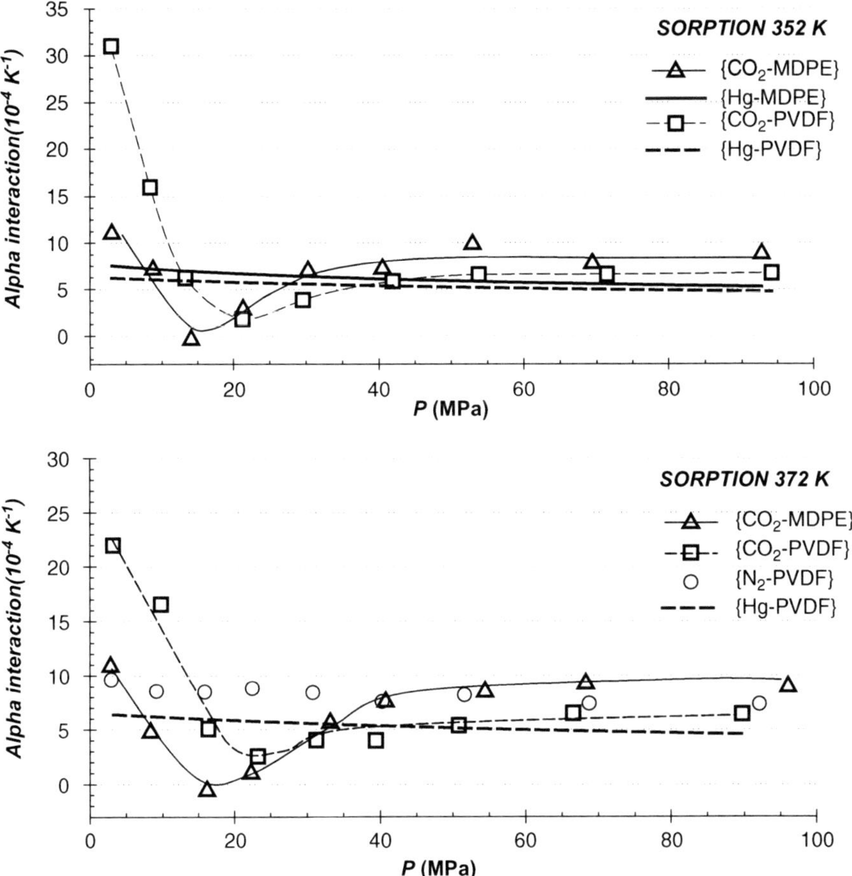

Figure 3 *Cubic expansion coefficients as functions of pressure for the two polymers MDPE and PVDF under high pressure CO_2, N_2 or Hg, at two different temperatures 352 and 372 K. The minima of the isotherms with CO_2 show the strong influence of this fluid on the polymers properties.*

chemically active carbon dioxide, CO_2 influences this property in comparison to less active nitrogen, N_2 and to "inert" mercury, Hg.

16.4 Importance of Solubility and of Associated Properties for Industrial Applications

Among industrial concerns, two major applications, such as polymer blistering and polymer foaming imply an ample knowledge of gas solubility and its consequences on polymers properties. Interestingly, the main consequence in both cases is the shift of glass transition temperatures as function of temperature and pressure. As a matter of fact, prediction of this shift is important to prevent deep penetration of gases in polymers with possible weakening of these materials as gas barriers. This prediction is also essential in developing efficient engineering processes for polymer foaming. Remarkably, in these two examples environment issues are at stake, either avoiding pollution by petroleum fluids transported by deficient hosepipes or using atmosphere-friendly blowing agents. In this context, solubility data are necessary to develop theoretical models for predicting T_g-shifts. Different models have been proposed in the literature to predict T_g. They are essentially of two types: either based on the Gibbs and Di Marzio principle (zero entropy at T_g),[21] such as the Chow's model[22] or directly, derived from lattice theory.[23] Chow's model can be used to evaluate the pressure dependence of the T_g-shift resulting from the sorption of different diluent gases. As an example (Figure 4), T_g-shifts of PS saturated with three different gases (which are potential blowing agents, *e.g.* N_2, CO_2 or still authorized HFC134a) have been computed[24] with Chow's model, at pressures higher than those usually found in literature. These calculations have been made using gas solubilities measured with the coupled VW-PVT technique described in Section 16.2.5 above. These results corroborate the observations made (Section 16.3) from the influence of pressurizing fluid on the cubic expansion coefficients.

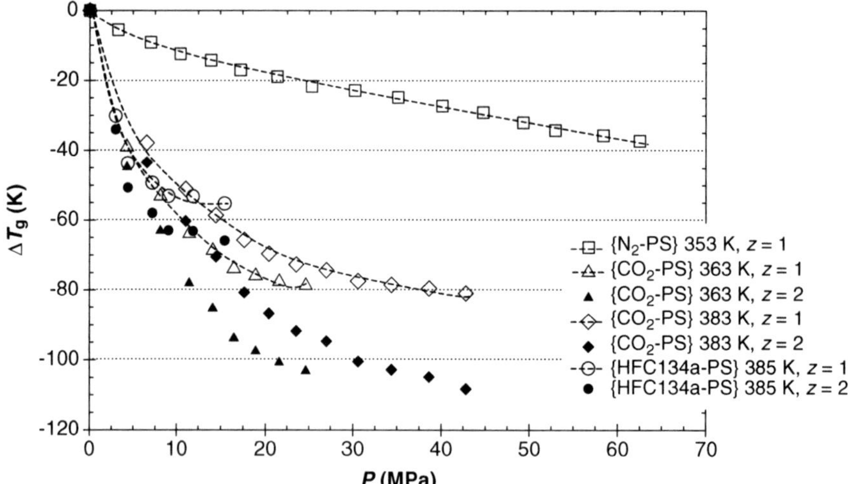

Figure 4 *Shifts of the glass transition temperature of PS as functions of pressure for the {N_2-PS}, {CO_2-PS} and {HFC134a-PS} systems at different temperatures. Points are calculated using Chow's model and solubility data obtained by the authors with the VW-PVT technique. Value of the lattice coordination number z depends on the physical state of the diluent (z = 1 for a gas or z = 2 for a liquid, respectively).*[22,24]

16.5 Conclusion

In many human and industrial activities, solubility of gases in polymers plays an important role. Comprehensive interpretation of the behaviour of a {gas–polymer} system implies the knowledge of both the amount of gas penetrating into the polymer and of the associated change in volume. Sophisticated experimental techniques allow one to determine, precisely, the gas sorption and the extent of interactions between a gas and a polymer. Direct measurements of thermophysical properties of gas saturated polymers are also possible. Polymer swelling and T_g-shifts due to gas sorption can be evaluated for engineering applications, while combining pertinent, accurately measured, experimental data and thermodynamically consistent models.

References

1. Y. Kamiya, T. Hirose, K. Mizogushi and Y. Naito, *J. Polym. Sci., Part B: Polym. Phys.*, 1986, **24**, 1525.
2. B. Wong, Z. Zhang and Y.P. Handa, *J. Polym. Sci., Part B: Polym. Phys.*, 1998, **36**, 2025.
3. B.I. Chaudary and A.I. Johns, *J. Cell. Plast.*, 1998, **34**, 312.
4. J.V. Schnitzler and R. Eggers in *Proceedings of the 5th Meeting on Supercritical Fluids, Nice France*, 23–25 March, 1998, 1, 93.
5. K.I. Miura, K. Otake, S. Kurosawa, T. Sako, T. Sugeta, T. Nakane, M. Sato, T. Tsuji, T. Hiaki and M. Hongo, *Fluid Phase Equilib.*, 1988, **144**, 181.
6. N.-H. Wang, S. Takishima and H. Matsuoka, *Int. Chem. Eng.*, 1994, **34**, 255.
7. B.J. Briscoe, O. Lorge, A. Wajs and P. Dang, *J. Polym. Sci., Part B: Polym. Phys.*, 1998, **36**, 2435.
8. W.J. Koros and D.R. Paul, *J. Polym. Sci., Polym. Phys. Ed.*, 1976, **14**, 1903.
9. S.A. Stern and A.H.D. Meringo, *J. Polym. Sci., Part B: Polym. Phys. Ed.*, 1978, **16**, 735.
10. Y. Sato, T. Iketani, S. Takishima and H. Masuoka, *Polym. Eng. Sci.*, 2000, **40**, 1369.
11. I.C. Sanchez and P.A. Rodgers, *Pure Appl. Chem.*, 1990, **62**, 2107.
12. R.G. Wissinger and M.E. Paulaitis, *J. Polym. Sci., Part B: Polym. Phys.*, 1987, **25**, 2497.
13. I.C. Sanchez and R.H. Lacombe, *Macromolecules*, 1978, **11**, 1145.
14. S. Hilic, A.A.H. Padua and J.-P.E. Grolier, *Rev. Sci. Instrum.*, 2000, **71**, 4236.
15. S. Hilic, S.A.E. Boyer, A.A.H. Padua and J.-P.E. Grolier, *J. Polym. Sci., Part B: Polym. Phys.*, 2001, **39**, 2063.
16. S.A.E. Boyer and J.-P.E. Grolier, *Polymer*, 2005, **46**, 3737.
17. S.A.E. Boyer, S.L. Randzio and J.-P.E. Grolier, *J. Polym. Sci., Part B: Polym. Phys.*, 2006, **44**, 185.

18. See for further details, *www. Transitiometry.com.*
19. J.-P.E. Grolier, F. Dan, S.A.E. Boyer, M. Orlowska and S.L. Randzio, *Int. J. Thermophys.*, 2004, **25**, 297.
20. J.-P.E. Grolier, *J. Chem. Thermodyn.*, 2005, **37**, 1226; *Pure Appl. Chem.*, 2005, **77**, 1297.
21. J.H. Gibbs and E.A. Di Marzio, *J. Chem. Phys.*, 1958, **28**, 373.
22. T.S. Chow, *Macromolecules*, 1980, **13**, 362.
23. P.D. Condo, I.C. Sanchez, C.G. Panayiotou and K.P. Johnston, *Macromolecules*, 1992, **25**, 6119.
24. S.A.E. Boyer and J.-P.E. Grolier, *Pure Appl. Chem.*, 2005, **77**, 593.

CHAPTER 17

Solubility in the Hydrometallurgical Leaching Process

TONI KASKIALA,[1] PETRI KOBYLIN[2] AND JUSTIN SALMINEN[3]

[1] *Helsinki University of Technology, Laboratory of Materials Processing, P.O.Box 6200, FIN-02015 TKK, Finland*

[2] *Helsinki University of Technology, Laboratory of Physical Chemistry and Electrochemistry, P.O.Box 6100, FIN-02015 TKK, Finland*

[3] *University of California Berkeley, Department of Chemical Engineering, Berkeley CA 94720, USA*

17.1 Mineral Processing by Aqueous Solutions

The present annual world production of iron/steel, aluminium, copper, zinc, lead, nickel, and magnesium is close to 1 billion tonnes. Most metals and their alloys are synthesized from oxide, sulfide, and halide sources. Hydrometallurgy is a specialized branch of extractive metallurgy dealing with metal recovery from ores, concentrates, and other metallurgical intermediate products by aqueous processes. These processes are usually very dependent on solubility of the reactants. In hydrometallurgical leaching the desired metal is obtained from the ore or its concentrate, and the residue, after washing, is rejected. The leach liquor, containing one or more metal ions in solution is then processed to solid metals. Hydro processes operate at lower temperature than pyro processes, usually 50–250°C.

The production of zinc is an important hydrometallurgical process and almost all the world's zinc production is derived from the treatment of sulfide concentrates in which sphalerite, (Zn,Fe)S, is the dominant zinc mineral. The concentrate, together with the slurry from the conversion process and acid from the electrolysis, is fed to the reactors where the leaching takes place by injecting oxygen into the slurry. The availability of dissolved oxygen for reactions has an important effect on the kinetics of the whole process as the oxidation of Fe^{2+} into Fe^{3+} controls the overall rate of leaching during the early stages of the process. The overall leaching reaction rate is controlled by gas diffusion into the aqueous phase, the kinetics of the oxidation reaction and leaching reaction rate. The oxidation of Fe^{2+} in acidic aqueous sulfate solutions with dissolved molecular oxygen is commonly employed in many hydrometallurgical processes such as leaching. According to several authors, the oxidation of ferrous iron is of second order with respect to $[Fe^{2+}]$ [1–4] and with low ferrous concentrations of first order with respect to $[Fe^{2+}]$.[5] The addition of ferric sulfate to the solution causes retardation in the oxidation reaction of $[Fe^{2+}]$, which may be due to the decrease in the $[SO_4^{2-}]$ ion activity through the formation of $[Fe^{3+}]$ sulfate-complexes.[2] Oxidation rates in the presence of sulfates are found far greater than without them. Cupric and zinc sulfate increases the kinetics of $[Fe^{2+}]$ oxidation with dissolved molecular oxygen.[1,6] Considering the direct leaching conditions, approximations for the reaction rate constant are difficult to estimate.

The ferrous iron-oxidation rate values range from 4 to $22 \cdot 10^{-3}$ (dm^3 mol^{-1} min^{-1}), taking into account the conditions closest to the direct sulfuric zinc concentrates leaching. The values of measured activation energies E_a vary between 51.6 and 73.7 (kJ/mol^{-1}), which indicates that the ferrous iron-oxidation rate is limited by the chemical reaction. The dependency of the rate constant upon the concentration of sulfuric acid appears to increase the concentration less and to be independent of the concentration above 1 mol dm^3.[1–3,5]

In hydrometallurgy, the knowledge of equilibrium solubilities of solids and gases in electrolyte system as well as electrochemistry, chemical thermodynamics, mass transfer, and reaction kinetics are crucial. The design of the hydrometallugical process requires the knowledge of solubilities, solubility experiments, pilot plants, and modelling. The best available techniques and continuous improvements are needed to produce environmentally acceptable, energy-efficient processes, which are capable of dealing concentrates with low precious metal grades. Rigorous multicomponent-thermodynamic models and sound measurements are required to effect the ever-smaller input of chemicals, raw materials, water, and minimal carbon dioxide emissions.

17.2 Dissolution of Sulfidic Zinc Concentrate and Gas–Liquid Mass Transfer

At atmospheric pressures, the sulfidic zinc dissolves mainly in strong oxidizers and leaching with sulfuric acid solutes has been very common. Zinc concentrate normally contains approximately 50 wt% zinc, 30 wt% sulfur and 5–15 wt% iron. In the zinc mineral sphalerite, iron occurs mainly as sulfides, replacing zinc in the crystal (Zn,Fe)S. The reactions involved in the leaching of zinc sulfide concentrates are complex and are not fully understood, although it is generally agreed that ferric sulfate attack of the sphalerite plays an important role in the overall leaching process.[7] When zinc sulfides are leached in sulfuric acid solution in oxidizing conditions in atmospheric reactors, the total reaction can be written as

$$ZnS + H_2SO_4 + 0.5O_2 \rightarrow ZnSO_4 + H_2O + S \qquad (1)$$

Iron acts as intermediary in the zinc sulfide leaching between the atmospheric oxygen and the mineral, and has an important catalytic role in the process. Ferric ions offer an efficient oxidant at moderate temperatures and suitable concentrations. The oxidation of zinc and the leaching rate is slow in direct oxygen leaching or if the concentration of acid-soluble iron in the solution is insufficient.[8–10] The reaction equations for zinc leaching can then be written as

$$2FeSO_4 + H_2SO_4 + 0.5O_2 \rightarrow Fe_2(SO_4)_3 + H_2O \quad (2)$$

$$ZnS + Fe_2(SO_4)_3 \rightarrow ZnSO_4 + 2FeSO_4 + S^o \quad (3)$$

Usually the concentrate contains small amounts of other sulfides, enabling zinc to be replaced by any of the following metals: Fe, Pb, Cu, Cd, Ca. Other sulfides either precipitate or are separated by electrochemical methods. The thermodynamic driving force of a dissolution reaction is the potential difference between the anodic and cathodic reactions. In the zinc leaching process, the potential difference, in this case mixed potential, is the equilibrium potential of a sulfide and the redox-potential of the solution.

Ferric-ions absorb on the concentrate surfaces and following oxidising-reduction reactions occur as oxygen reduces into water. The electrode and solution reactions are as follows:

$$\text{anode } S^{2-} \rightarrow 2e^- + S^o \quad (4)$$

$$\text{cathode } 2Fe^{3+} + 2e^- + 2Fe^{2+} \quad (5)$$

$$\text{solution } 0.5O_2 + 2H^+ + 2e^- \rightarrow H_2O \quad (6)$$

As the anodic-dissolution reaction proceeds, an elemental sulfur S^o product layer is formed. Other products in the solution are zinc sulfate, iron(II)sulfate, iron(III)sulfate, iron(II) and iron(III) complexes and sulfides H_2S, HS^-, and S^{2-}. Elementary sulfur S^o either moves to solution or stays as a product layer on the mineral surface according to the shrinking core model.[11–15] In this model, the particle size remains constant, but the area of the surface where the reaction takes place decreases with time. The dissolution occurs first on the surface of the mineral, and then continues by diffusion through this layer, (see Figure 1).

The core, phase I, consists of the original zinc sulfide. As the dissolution begins, the zinc ions from phase I move to the solution and in phase II a zinc deficient sulfide is formed. A higher potential is needed for the dissolution through phase II than initially in phase I. As the dissolution continues, a layer, phase III, of zinc-deficient sulfide with elemental sulfur is formed and this layer has an even higher dissolution potential. The *outermost layer* is of pure elemental sulfur, formed because all the zinc ions have moved to the solution. For dissolution to continue after each step, an increasing potential is required and the layers formed have to allow both mass and charge transfer through

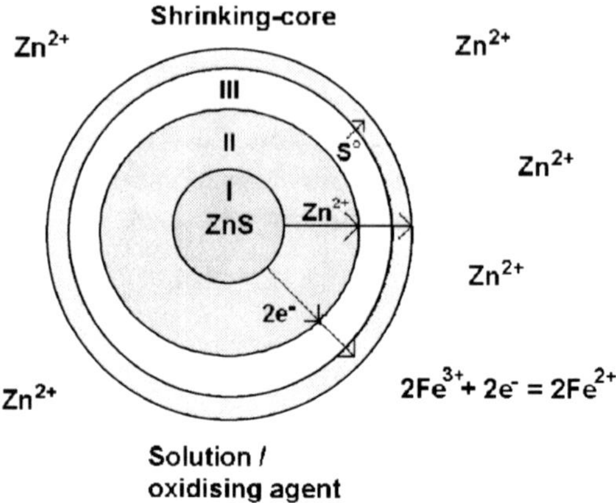

Figure 1 *A simplified shrinking-core model of the oxidative dissolution of zinc sulfide.*

them. As the sulfur layer grows, it becomes more impenetrable and eventually the dissolution ceases. Electrochemically, the elemental sulfur is almost inert and does not react and it is also a poor conductor. The dissolution rate of a sulfide mineral is affected both by charge and mass transfer through the reaction product layer on the surface of the mineral. The mass transfer of ions is affected by the porosity, thickness, and other structural factors and charge transfer, *i.e.*, negative electrons and positive holes, as well as by electrical conductivity and thickness of the layer. The dissolution of sulfidic zinc concentrate is a result of a combination of multiple kinetically constrained steps. If the mass transfer through the liquid–solid interface is ignored, the dissolution can be classified according to diffusion or reaction as the limiting factor. The possible controlling factors are the reactions on the surface or diffusion of reacting species through the porous product layer, or a combination of them. The activation energy E_a of the reaction represents the effect of temperature on the rate of reaction is given by the Arrhenius law $k_c = Ae^{\frac{-E_a}{RT}}$, where A is the pre-exponential factor, k_c the rate constant of the reaction, R the molar gas constant and T the absolute temperature. Table 1 shows some activation energy values for the zinc sulfide dissolution found in literature.

The dissolution reaction of sulfidic zinc is strongly temperature dependent. The amount of iron in sphalerite mineral has been shown to increase the rate of dissolution.[12,17,18] At lower temperatures, an elemental sulphur layer does not immediately block the surface of the particle. However, at temperatures above the melting point of sulfur, 119°C, the diffusion limiting layer appears even faster. The overall mass transfer rate is complex and influenced by a number of physical parameters, operating conditions and machine factors. Figure 2 shows simplified relations of hydrodynamic factors that effect on the total gas–liquid mass transfer rate.

Factors affecting concentration gradient ΔC include the equilibrium solubility of the gas at the given temperature, partial pressure, and metabolic activity. Higher solubilities of gases can be achieved by increasing the partial pressure of

Table 1 *Activation energies for the dissolution of zinc sulphide according to different sources*

Reference	*T (°C)*	*[Fe^{3+}] (M)*	*[H_2SO_4] (M)*	*E_a (kJ/mol^{-1})*
Halavaara[12]	50–80	0.3	0.25	29.5
Halavaara[12]	80–100	0.3	0.25	74.4
Palencia Perez[16]	50–90	0.3	0.3	41–72
Crundwell[11]	78	0.5	0.1	46
Verbaan and Crundwell[4]	25–85	0.4	0.1	56.64
Dutrizac[4]	40–100	0.3	0.3	44

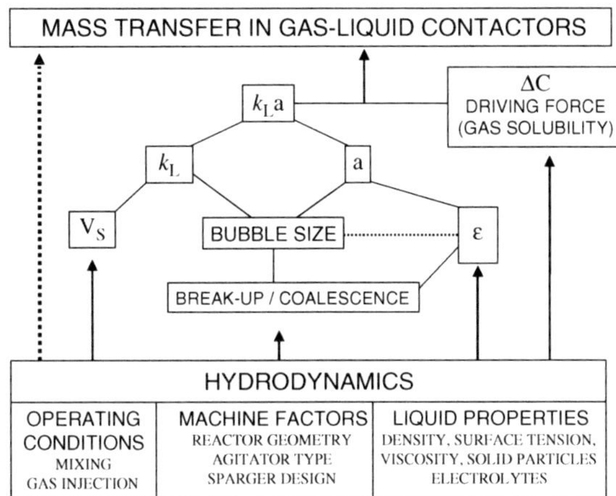

Figure 2 *Relationship between the various factors affecting the mass transfer rates in a gas–liquid reactor; k_La is the volumetric liquid-mass transfer coefficient, k_L the liquid-mass transfer coefficient, a the gas–liquid interfacial area per liquid volume, ΔC the diffusional driving force and concentration difference of oxygen between gaseous phase and liquid phase, V_S the superficial gas and liquid velocity, and ε is the gas hold-up.*

the gas. Temperature of the solution is determined or chosen by other process factors like reaction kinetics and energy consumption. Increasing the mass transfer coefficient is possible by either reducing the size of the boundary layer or increasing the rate at which molecules move through the boundary layer. Increasing the turbulence decreases the boundary layer. Increasing the temperature increases the diffusivity and reduces the boundary layer.[19,20] Temperature raise results in an increase in k_L.[21]

Numerous studies on mass transfer in the bubble column have revealed that the mass transfer coefficient k_L depends mainly on the mean bubble size, physical properties of the liquid medium, and the diffusivity of the absorbing gas component in the liquid medium.[22] The bubble size influences significantly the value of the mass transfer coefficient, k_L. One can distinguish between the effect of tiny bubbles, $d_s < 0.002$ m, and of large bubbles, $d_s > 0.002$ m.[23] For tiny bubbles, k_L values increase rapidly with bubble size from $k_L = 1 \cdot 10^{-4}$ m s^{-1}

with $d_s \leq 0.0008$ m and $k_L \cong 5 \cdot 10^{-4}$ m s^{-1} with $d_s \cong 0.002$ m. In the region of large bubbles, values of the mass transfer coefficient decrease slightly with increasing bubble diameter to the value of $k_L \cong (3\text{–}4) \times 10^{-4}$ m s^{-1}.[24] The bubble-size effect should be employed with caution, especially if bubble size is decreased with the use of a surface active agent, such as electrolytes, polymers, antifoams, oils, alcohol, and small particles. Since the addition of surface active substances reduces the rate of renewal of the surface elements at the interface, it negatively affects the mass transfer from the bubbles.[23] In general though, surface active agents increase a by increasing ε_G and decreasing d_b, by an even larger factor, so that k_La usually increases, though occasionally it has been found to decrease.[25,26] The k_L values between oxygen and water at STP typically range between $(1\text{–}6) \times 10^{-4}$ m s^{-1}.[21,27] Exact value of 1.35×10^{-4} m s^{-1} has also been reported.[21] The total gas–liquid interfacial area $a = \frac{6\varepsilon_G}{d_b}$ can be evaluated from the mean gas hold-up ε_G and the volume surface mean bubble diameter d.[28] a is determined by the size, shape, and number of the bubbles. Factors affecting the size of the bubbles include stirring speed and type of the impeller, reactor design, the way the substances are introduced, and medium composition (*e.g.*, the presence of surface active agents). The interfacial area can be increased by creating smaller bubbles or increasing the number of bubbles. For a given volume of gas, a greater interfacial area, a, is provided if the gas is dispersed into many small bubbles rather than a few large ones.

The stirrer and the mixing intensity play a major role in breaking up the bubbles. Reactor design effects the gas dispersion, hold-up, and residence time of the bubbles. Baffles are used to create turbulence and shear, which break up the bubbles. The properties of the medium also affect significantly the bubble sizes and coalescence and therefore the interfacial area. Electrolytes decrease the dissolved gas concentration, which in turn decreases the strength of the attraction between bubbles mediated by micro-bubbles; this inhibits coalescence.[29] With particles finer that 10 μm, the bubble coalescence hinders and the bubble interfacial area and hold-up increases; with particles larger than 50 μm, the effects are the opposite.[22,30] Bubble size has been shown to increase with the particle size of the ore, pulp density, and air flow rate.[31] An increase in temperature reduced bubble size, as did reduced viscosity. In literature, there are several correlations for bubble sizes, which can be divided into categories of bubbles generated at an orifice and bubbles far from the orifice. Previous studies on the mechanism of bubble formation show that, depending on the controlling mechanisms, one can distinguish between surface tension controlled by bubble detachment diameter, viscous drag controlled by bubble detachment diameter, and liquid inertia controlled by bubble diameter.[32] The surface tension and viscous forces are two major contributing forces influencing the bubble diameter and volume during its formation. The surface tension should be taken into consideration even at high gas-flow rates. The viscous force is only important at high gas-flow rates and can be ignored at low gas-flow rates. The orifice diameter d_0 influences the bubble size strongly only at very low gas-flow rates, where the bubble size is found by equating surface tension[33] and buoyancy forces.[34] Once the liquid is in turbulent motion, however, bubble break-up will also occur, and equilibrium between coalescence and break-up will determine the mean bubble size. Bubble diameter has been noticed to decrease as a function of preheated gas injected.[33] Dissolved gas concentration has an important influence on the interaction between two bubbles, but a contribution due to the Gibbs–Marangoni effect and surface elasticity cannot be ruled out. The increased pressure decreases bubble size and hold-up.[35,36] For fixed pressure and gas velocity the temperature effect on gas hold-up is complex, but an increase in temperature generally increases the gas hold-up. This general trend is due to the dominant role of the associated reduction in liquid viscosity and surface tension, which leads to smaller bubble size. The hold-up of the air–water systems increases slowly at temperature $T < 75°C$ and remarkably at $T > 75°C$ and is related to the vapour pressure of the gas.[37]

17.3 Oxygen Solubility

Dissolved gases are commonly used as reactive chemical agents in processes in atmospheric conditions as well as at elevated pressures.[38–44] Solubilities, thermodynamics, Henry's law, and experimental methods for oxygen solubility in water have been studied by many authors.[45–55]The addition of salt to water usually lower the solubilities of gases due to salting out effect.[56–58] Semi empirical correlations for estimating gas solubilities in electrolyte solutions are available.[59,60] At higher salt concentrations, the gas solubility tends to be underestimated by linear assumptions. The partial pressure of oxygen above the solution, temperature of the solution and the gas injected, organic substances and concentration of the ions (*e.g.* zinc, iron, and acid) influence oxygen solubility in solution. Figure 3 shows the solubility of oxygen in pure water, in 1.3 mol dm^{-3} $ZnSO_4$ solution and two process solutions. These process solutions 1 and 2 contained different mixtures of H_2SO_4, $ZnSO_4$, $FeSO_4$, and $Fe_2(SO_4)_3$ salts. Process solution 1 contained together 3 mol dm^{-3} sulfates and process solution 2 contained 2.7 mol dm^{-3} sulfates.

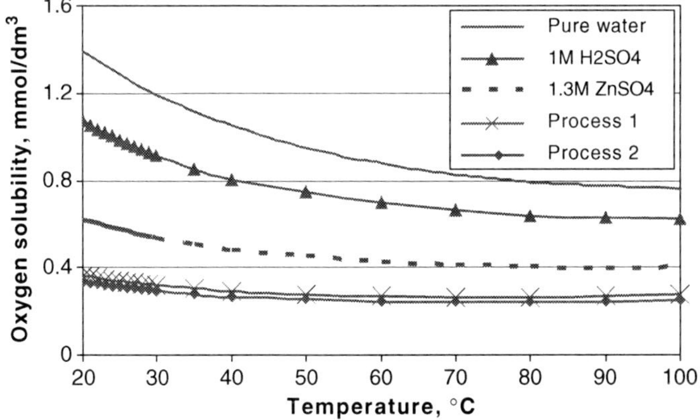

Figure 3 *Oxygen solubilities in different systems as a function of temperature.*

17.4 Solubilities of Solids in Process Solutions

The iron in the solution has the ability and tendency to form complex compounds jarosites. The most favourable conditions for the precipitation of jarosites lie in the range 90–100°C and pH in the range 1–2.5.[12] In the later stages of leaching, the iron is removed from the process when dissolved ferric iron precipitates as jarosite through the addition of ammonia (NH_4^+).[61] Overall reactions for the formation of jarosites can be written as

$$3Fe_2(SO_4)_3 + Me_2SO_4 + 12H_2O \rightarrow Me_2Fe_6(SO_4)_4(OH)_{12} + 6H_2SO_4 \quad (7)$$

where Me: NH_4^+, Na^+, K^+, Rb^+, Ag^+, Ti^+, and H_3O^+.[62]

Losses of divalent metals to alkali jarosites increase with increasing divalent metal concentrations, increasing pH or decreasing the Fe^{3+} concentration. Losses vary according to the metal in question, but rarely exceed 3 %.[62,63] The descending order for losses is: $Fe^{3+} > Cu^{2+} > Zn^{2+} > Co^{2+} > Ni^{2+} > Mn^{2+} > Cd^{2+}$. Precipitation of ferric iron can occur also as a Plumbojarosite as follows:

$$3Fe_2(SO_4)_3 + MeSO_4 + 12H_2O \rightarrow MeFe_6(SO_4)_4(OH)_{12} + 6H_2SO_4 \quad (8)$$

where Me: Pb^{2+} or Hg^{2+}.

Other options available for the treatment of the ferrite residue are hematite or goethite or smelting processes. The advantages of removing iron as a jarosite are the high recovery from various concentrates and the ability to utilize heavy metal containing acids. It is complicated to determine the kinetics of jarosite precipitation for process solutions because of the various metal ions available. The effects of temperature and seeding indicated that the precipitation reaction rate is controlled by a chemical reaction step occurring on the surface of the jarosite crystals.[64] Iron sulfate solubilities in aqueous sulfuric acid solutions have been studied for long time.[64–70] In moderate sulfuric acid concentrations $Fe_2(SO_4)_3 \cdot 7H_2O$ and $Fe_2(SO_4)_3 \cdot H_2SO_4 \cdot 8H_2O$ are the solid phases at room temperature.[66–69] Figure 4 shows the solubility of $Fe_2(SO_4)_3$ in sulfuric acid at 25°C. One can notice the large variation of the solubility data. For the aqueous phase, because of the lack of data for possible iron sulfate complexes, a very simple system was selected with the following species: Fe^{3+}-SO_4^{2-}-HSO_4^--OH^--H^+-H_2O. Interaction parameters for the interactions between Fe^{3+}-SO_4^{2-} and Fe^{3+}-HSO_4^- was optimized by Pitzer equation[74] with Baskerville's[67] data. The curve in Figure 4 was calculated by using Pitzer's model in ChemSage®, ChemSheet® programmes and with standard data obtained from HSC® database.[71–73]

The water activity data together with solubility data were used in this work to optimize the Pitzer activity coefficient parameters for this system.[65–69]

Calculations above have been made assuming no complex formation in the ferric sulfate sulfuric acid solutions. However, ferric sulphate solution may include complexes, such as $FeSO_4^+$, $Fe(SO_4)_2^-$, and $FeHSO_4^{2+}$.[75–78] Modelling of ferric ion systems is problematic because of the lack of thermodynamic data available in literature. As the hydrometallurgical process are mainly acidic and

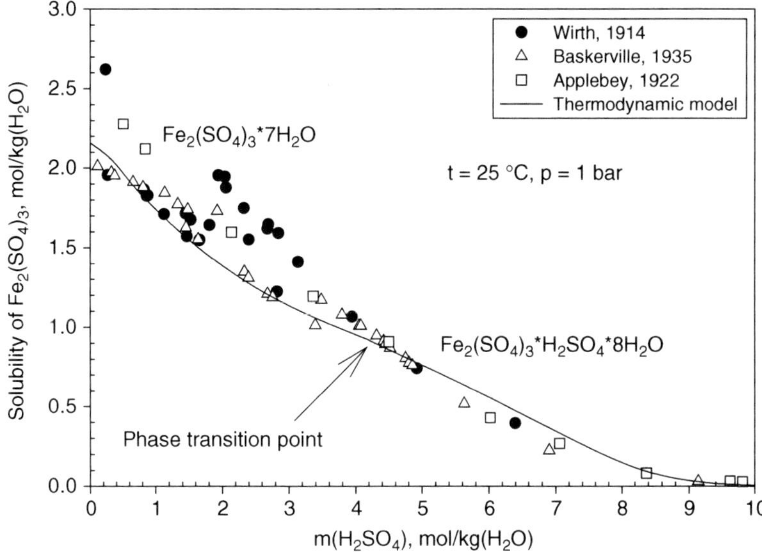

Figure 4 *Solubility of $Fe_2(SO_4)_3$ in dilute sulfuric acid solution at 25°C.*

the main interest is in the acidic area of $Fe_2(SO_4)_3$-H_2SO_4-H_2O system. $Fe_2(SO_4)_3$ is moderately soluble in water and the solubility decreases by addition of sulfuric acid. The calculations and experiments[65] also show that increase of sulfuric acid decrease the activity of water at constant molality of $Fe_2(SO_4)_3$.

17.5 Concluding Remarks

The operating conditions of a process are first defined by the method and equipment chosen. The effects of operating parameters are usually well known through practice and the main work is to hold the optimum conditions. The attempts to improve the running process are not always easy or even possible to follow consequences. For example, increasing the operating temperature usually increases the solubilities of solids and increases reaction rates. In addition, changes in the properties of both gas and liquid phases occur, which then affect the flow pattern, thickness and total contact area of the gas–liquid interface, the gas hold-up and the diffusivity, and solubility of gases. Laboratory tests and thermodynamic calculations of solubilities are useful tools and sometimes the only possible alternative to models, in order to understand the fundamentals of the processes used. This paper demonstrates an approach to the role of gas and solid solubility and related phenomena in the hydrometallurgical industry and especially in the leaching process.

References

1. Y. Awakura, in *Iron Control in Hydrometallurgy*, J.E. Dutrizac and A.J. Monhemius (eds), Ellis Horwood, Chichester England, 1986, 203.

2. M. Iwai, H. Majima and Y. Awakura, *Metall. Trans. B*, 1982, **13B**, 311.
3. K. Pohjola, *The Redox Potential of Iron in High Pressure Aqueous Solutions*. Master's Thesis, (in Finnish), Helsinki University of Technology, Finland, 1997.
4. B. Verbaan and F.K. Crundwell, *Hydrometallurgy*, 1986, **16**, 345.
5. T. Chmielewski and W.A. Charewicz, *Hydrometallurgy*, 1984, **12**, 21.
6. C.T. Mathews and R.G. Robins, The oxidation of aqueous ferrous sulphate solutions by molecular oxygen, *Proc. Aust. Inst. Min. Met.*, 1972, **47**, 242.
7. J.E. Dutrizac, The Kinetics of Sphalerite Dissolution in Ferric Sulphate-Sulphuric Acid Media, *Proceedings of Lead and Zinc '05*, Vol 2, Kioto, Japan, 2005, 833.
8. D.B. Dreisinger and E. Peters, *The mathematical modelling of the zinc pressure leach*, The Metallurgical Society Inc, Mathematical Modelling of Materials Processing Operations, Palm Springs, CA, USA, 1987.
9. S. Au-Yeung and G.L. Bolton, in *Iron Control in Hydrometallurgy*, J.E. Dutrizac and A.J. Monhemius (eds), Ellis Horwood, Chichester England, 1986, 131.
10. T. Kaskiala, Studies on gas-liquid mass transfer in atmospheric leaching of sulphidic zinc concentrates, Doctoral Thesis, Helsinki University of Technology, Finland, 2005.
11. F.K. Crundwell, *Hydrometallurgy*, 1987, **19**, 227.
12. P. Halavaara, Factors affecting on the dissolution rate of zinc sulphide, Master's Thesis, Lappeenranta University of Technology, Department of Chemical Technology, Finland, 1996.
13. P.R. Holmes, *Geochim. Cosmochim. Acta.*, 2000, **64**, 263.
14. J. Lochmann and M. Pedlík, *Hydrometallurgy*, 1995, **37**, 89.
15. D. B. Dreisinger, *Hydrometallurgy*, 1989, **22**, 101.
16. I.P. Perez, *Hydrometallurgy*, 1991, **23**, 191.
17. G. Bobeck, H. Su, 1985, *Met. Trans.*, 1985, 16 B, 413.
18. B. Verbaan and B. Mullinger, *The leaching of sphalerite in acidic ferric sulphate media in the absence of elemental sulphur*, NIM Report No. 2038, Randburg, South Africa, 1980.
19. J.C. Merchuk, *AIChEMI Modular Instruction*, 1983, Series B, 33.
20. Z. Tekie, J. Li and B.I. Morsi, *Ind. Eng. Chem. Res.*, 1997, **36**, 3879.
21. P.H. Calderbank and M.B. Moo-Young, *Chem. Eng. Sci.*, 1961, **16**, 39.
22. E. Sada, H. Kumazawa, L.H. Lee and T. Iguchi, *Ing. Eng Chem. Process Des. Dev.*, 1986, **25**, 472.
23. F. Kastanek, J. Zahradnik, J. Kratochvil and J. Vermak, *Chemical reactors for gas-liquid systems*, Ellis Horwood series in Chem. Eng., 1993.
24. T. Pedersen, *Water Res.*, 2001, **34**, 2569.
25. N. Harnby, M. Edwards and A.W. Nienow, *Mixing in the Process Industries*, Butterworth-Heinemann, Oxford, UK, 1997.
26. F. Yoshida, *Chem. Eng. Tech.*, 1988, **11**, 205.
27. P.W. Danckwerts, *Ind. Eng. Chem.*, 1951, **43**, 1460.

28. M. Fukuma, K. Muroyama and A. Yasunishi, *J. Chem. Eng. Jpn.*, 1987, **20**, 321.
29. P. K. Weissenborn and R. J. Pugh, *J. Colloid Interface Sci.*, 1996, **184**, 550.
30. E. Sada, H. Kumazawa, L.H. Lee and H. Narukawa, *Ind. Eng. Chem. Res.*, 1987, **26**, 112.
31. C.T. O'Connor, E.W. Randall and C.M. Goodall, *Int. J. Min. Process*, 1990, **28**, 139.
32. M. Jamialahmadi, M.R. Zehtaban, H. Muller-Steinhagen, A. Sarrafi and J.M. Smith, *Trans. Ichem. E.*, 2001, **79**(Part A), 523.
33. S.V. Komarov and M. Sano, *ISIJ Int.*, 1998, **38**, 1047.
34. M. Moo-Young and H.W. Blanch, *Adv. Biochem. Eng.*, 1981, **19**, 1.
35. P.K. Weissenborn and R.J. Pugh, *J. Colloid Interface Sci.*, 1996, **184**, 550.
36. T.-J. Lin, K. Tsuchiya and L.-S. Fan, *AIChE J.*, 1998, **44**, 545.
37. R. Zou, X. Jiang, B. Li, Y. Zu and L. Zhang, *Ind. Eng. Chem. Res.*, 1988, **27**, 1910.
38. E. Narita, *Hydrometallurgy*, 1983, **10**, 21.
39. D. Troman's, *Min. Eng. J.*, 2000, **13**, 487.
40. J. Salminen, P. Koukkari, J. Jäkärä and A. Paren, *J. Pulp Pap. Sci.*, 2000, **26**, 441.
41. T. Kaskiala, *Min. Eng. J.*, 2002, **15**, 853.
42. D. Tromans, *Ind. Eng. Chem. Res.*, 2000, **39**, 805.
43. W. Hayduk, *Final Report Concerning the Solubility of Oxygen in Sulphuric Acid-Zinc Pressure Leaching Solutions*, University of Ottawa, Ontario, 1991.
44. T. Kaskiala and J. Salminen, *Ind. Eng. Chem. Res.*, 2003, **42**, 1827.
45. E. Wilhelm, R. Battino and J. Wilcock, *Chem. Rew.*, 1977, **77**, 219.
46. R. Perry, *Perry's Chemical Engineers' Handbook*, 6th edn, McGraw-Hill, New York, 1984.
47. J. Prausnitz, R. Lichtenthaler and E. Azevedo, *Molecular Thermodynamics of Fluid- Phase Equilibria*, 3rd edn, Prentice-Hall, NJ, 1999.
48. D. Himmelblau, *J. Chem. Eng. Data.*, 1960, **5**, 10.
49. Å. Broden and R. Simonson, *Svensk Papperstidning*, 1978, **17**, 541.
50. H. Clever and C. Han, in *ACS Symposium Series*, S. Newman, (ed), 1980, **133**, 513.
51. R. Battino, *Oxygen and Ozone*, IUPAC Solubility Data Series, Vol 7, Pergamon Press, Oxford, 1981.
52. H. Pray, C. Schweickert and B. Minnich, *Ind. Eng. Chem.*, 1952, **44**, 1146.
53. T. Rettich, R. Battino and E. Wilhelm, *J. Chem. Thermodyn.*, 2000, **32**, 1145.
54. D. Troman's, *Hydrometallurgy.*, 1998a, **48**, 327.
55. A. Harvey, *AIChE J.*, 1997, **42**, 1491.
56. A. Setschenow, *Z. Phys. Chem.*, 1889, **IV**, 117.
57. G. Geffcken, *Z. Phys. Chem.*, 1904, **49**, 257.
58. F. A. Long and W. F. McDevit, *Chem. Rev.*, 1952, **51**, 119.
59. S. Weisenberger and A. Schumpe, *AIChE J.*, 1996, **42**, 298.

60. A. Schumpe, *Chem. Eng. Sci.*, 1993, **48**, 153.
61. H. Takala, *Erzmetall*, 1999, **52**, 37.
62. H. Arauco and F.M. Doyle, in *Hydrometallurgical Reactor Design and Kinetics*, R.G. Bautista (ed), SME-AIME, Pensylvania, 1987, 187.
63. T.J. Harvey and W.T. Yen, *Min. Eng.*, 1998, **11**, 1.
64. L.A. Teixeira, in *Iron Control in Hydrometallurgy*, J.E. Dutrizac and A.J. Monhemius (eds). Ellis Horwood, Chichester, England, 1986, 431.
65. H. Majima and Y. Awakura, Y., *Met. Transact.*, 1985, **16B**, 433.
66. M.P. Applebey and S.H. Wilkes, *J. Chem. Soc.*, 1922, **121**, 337.
67. W.H. Baskerville and F.K. Cameron, *J. Phys. Chem.*, 1935, **39**, 769.
68. F. Wirth and B. Bakke, *Z. Anorg. Chem.*, 1914, **87**, 13.
69. F. Wirth and B. Bakke, *Z. Anorg. Chem.*, 1914, **87**, 47.
70. W. Bullough, T.A. Canning and M.I. Strawbridge, *J. Appl. Chem.*, 1952, **2**, 703.
71. A. Roine, *HSC-Software version 5.11, 02103-ORC-T*, Outokumpu Research Centre, Pori, Finland, 2002.
72. G. Erikson and K. Hack, *Met. Trans. B*, 1990, **21B**, 1013.
73. P. Koukkari, K. Panttilä, K. Hack and S. Petersen, in *Microstructures Mechanical Properties and Process-Computer Simulation and Modelling*, Y. Brechet, (ed), Euromat99, Vol. 3, Wiley-VCH, New York, 2000, 323.
74. K.S. Pitzer, *J. Phys. Chem.*, 1973, **77**, 268.
75. D. Filippou, G. Demopoulos and V. Papangelakis, *AIChE J.*, 1995, **41**, 171.
76. G. Senanayake and D. Muir, *Met. Transact. B*, 1988, **19B**, 37–45.
77. V. Papangelakis and G. Demopoulos, *Can. Met. Quarterly*, 1990, **29**, 1–12.
78. J. Dousma, D. Ottelander and P. Bruyn, *J. Inorg. Nucl. Chem.*, 1979, **41**, 1565–1568.

CHAPTER 18

Solubility Related to Reaction and Process Design

RALF DOHRN,[1] RICARDA LEIBERICH[2] AND LJUDMILA FELE ŽILNIK[3]

[1] *Bayer Technology Services GmbH, Kinetics, Properties & Modelling, PT-RPT-KPM, Building B310, D-51368, Leverkusen, Germany*

[2] *Bayer Technology Services GmbH, Kinetics, Properties & Modelling, PT-RPT-KPM, Building E41, D-51368, Leverkusen, Germany*

[3] *National Institute of Chemistry, Slovenia, Laboratory for Chemical Process Engineering, Hajdrihova Ulica 19, SI-1001, Ljubljana, Slovenia*

18.1 Introduction

The mathematical simulation of entire chemical production processes, or at least of crucial unit operations, allows an optimization of the process that can lead to very significant process improvements concerning costs, product quality and environmental aspects. To develop such a simulation model, a sound chemical and physical understanding of the crucial process steps, like reactions and separations, is needed. Physicochemical property data can be regarded as raw material of chemical process design,[1] where the quality of the raw material might strongly affect the quality of the product. In addition to process design and process simulation, physicochemical property data play an increasing role in advanced process control models.

Figure 1 shows a simplified scheme of a chemical production process. Before the educts enter the reaction step, they are often purified or an additive is introduced. In both cases solubilities can be of importance. For the optimization of the chemical reaction, the knowledge of the solubilities of the components taking part in the chemical reaction is essential, including their dependence on temperature and pressure. After the reaction a separation of the products and by-products takes place and eventually a stream is recycled to the reactor. For difficult separations, small uncertainties in phase equilibrium data might have a huge impact on the design of the separation process.[2]

The knowledge and understanding of solubilities and phase equilibria is not only needed for process simulations or for the design of physicochemically

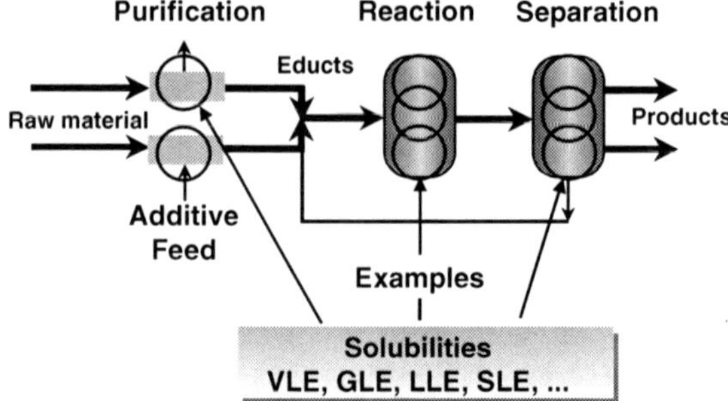

Figure 1 *Simplified scheme of a typical chemical production process.*

based process control models of reaction and separation steps, but it can play an important role, which is not always obvious at first sight, throughout the entire process chain, influencing costs and safety of the process, as well as the product quality. In this chapter, some examples from industrial practice are given.

18.2 Educt Purification and Additive Introduction

Obviously, the quality of a product of a chemical reaction is influenced by the purity of the educts. In the best case, impurities of the educts are not taking part in the reaction. If they are not wanted in the final product they have to be removed before or after the reaction. But impurities, *e.g.* a few ppm of water, might also influence the reaction scheme and lead to unwanted by-products. For polymers, a further purification after the reaction is more difficult than for chemicals of lower molecular weight, or even worse, the impurities may influence the polymerization and may change the structure of the polymer. Pure educts are of special importance for highly pure polymers, *e.g.* for optical data storage applications, where any disturbance in the polymer matrix might lead to a data loss. Therefore, for many chemical reactions, additional purification steps of the educts are implemented, with a special need for reliable phase equilibrium data in the high-purity region.

Before discussing problems of the high-purity region of a phase diagram, some basics of phase equilibrium thermodynamics should be given. In many cases, especially at pressures below 0.5 MPa, the condition for the calculation of phase equilibria in a mixture of N components can be expressed in a simplified way by using an extended form of Raoult's law

$$x_i \gamma_i P_i^{\text{sat}} = y_i P \qquad \text{for } i = 1, \ldots, N \tag{1}$$

where y_i, mole fraction of component i in the vapour phase; x_i, mole fraction of component i in the liquid phase; γ_i, activity coefficient of component i; P_i^{Sat}, vapour pressure of pure component i; and P, total pressure.

When the activity coefficients γ_i in a solution are equal to one, we call this behaviour an ideal solution (Raoult's law): the boiling pressure of an ideal solution is equal to the arithmetic mean (weighted with the mole fractions) of the pure component vapour pressures. When γ_i are greater than 1, we talk about positive deviations from Raoult's law. Usually, separations of components get "easier", the stronger the positive deviations from Raoult's law are, as long as no azeotrope (vapour and liquid phase have the same composition) occurs.

Even for "easy" separations, difficulties may be encountered in the high-purity region. This is due to the fact that at high purities the phase behaviour goes asymptotically to Raoult's law,[3,4] leading to a more difficult separation, and in some cases even coming close to an azeotrope. An example for such a system is water+acetone,[2] for which one literature source[5] even falsely reports an azeotrope in the high-purity region. Therefore, reliable phase equilibrium data are needed in the low concentration region of the impurities, *e.g.* activity coefficients at infinite dilution, measured by ebulliometry or inverse-gas chromatography.[6–8]

18.2.1 Example 1: Polyurethane Foam Quality

The following example focuses on the addition of a blowing agent to an educt stream. Polyurethane rigid (PUR) foams of the type commonly used for most refrigerator insulation are usually produced from two-component systems: component A contains a polyol, including catalysts, stabilizers and blowing agents; component B is a polyisocyanate. Polyol and polyisocyanate react to form a polyurethane.

$$\mathrm{H{-}[O{-}C_3H_6]_x{-}OH + OCN{-}R{-}NCO \rightarrow H{-}[O{-}C_3H_6]_x{-}O{-}CO{-}NH{-}R{-}NCO} \quad (2)$$

Usually low-boiling liquids and/or water are used as blowing agents. The water reacts with the polyisocyanate to form CO_2 which serves as an additional blowing agent. The pressure generated by the gaseous blowing agent and CO_2 in the closed cells has a strong influence on the stability of the foam. To produce low-density foam an additional blowing agent, with a high-vapour pressure, is needed. For this purpose, Bayer and Hennecke GmbH developed NovaFlex® technology using CO_2 as a blowing agent that allows manufacturers to make low-density PUR foam (slabstock) with densities lower than 20 kg m^{-3} without using typical blowing agents, such as CFCs or methylene chloride. CO_2 has 3.2 times the blowing effect of methylene chloride and costs only one-fifth. Since CO_2 has a high solubility in many liquids, it is dissolved in the polyol before the polyol is mixed with the polyisocyanate to react to a PUR foam. Figure 2 shows a simplified scheme of the process.

In some cases, the quality of the foam, concerning size distribution of the cells, was not stable and it was suspected that this might be influenced by the content of CO_2 that is dissolved in the polyol. To solve this problem, the solubility of CO_2 was measured in different polyols and polyisocyanates at

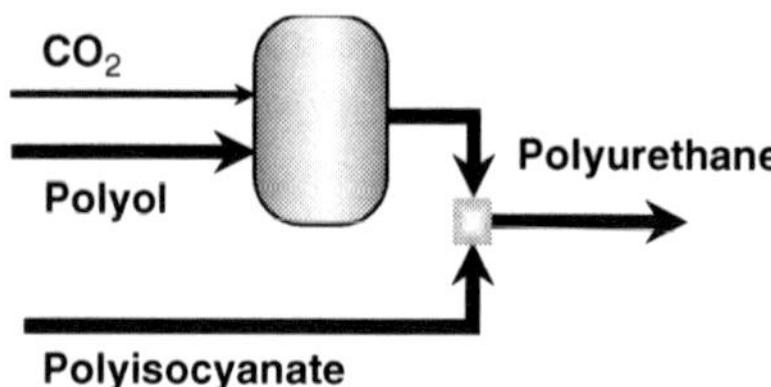

Figure 2 *Scheme of process to manufacture PUR foam using CO_2 as blowing agent.*

different pressures and temperatures. Experiments were performed isothermally in a high-pressure view cell[9] (250 cm^3 volume) using a static P-T-x method. In this method, a cell of rigorously known volume is maintained at constant temperature and pressure readings are made for different mixtures of known composition. After introduction of rigorously known amounts of substance in the cell the attainment of the equilibrium begins. The stirring device promotes the bubbling of the gas phase inside the liquid phase, assuring an efficient mixing of both phases. After the equilibrium has been reached, both pressure and temperature are registered. For the calculation of the concentration of CO_2 in the liquid phase, an iterative calculation is performed, where the density of the liquid phase and a correction for the non-ideal behaviour of the vapour phase is used. The results show, that for pressures up to 1 MPa, the solubility of CO_2 in all liquids investigated could be well presented using Henry's law. For example, at 40°C and at a CO_2 partial pressure of 0.88 MPa, 2.2 wt% CO_2 can be dissolved in a typical polyol. The solubility in a typical polyisocyanate is 15% lower. Using the information on the solubilities, temperature and pressure can be controlled in the PUR foam production process in order to have a specified CO_2 content dissolved in the polyol stream. The problem of the quality variation of the foam has been solved using basic solubility information.

18.3 Reaction Design

18.3.1 Example 2: Polyether Reaction Design

The first example to illustrate the importance of solubilities and phase equilibria in reaction design is about the production of polyethers. A starter alcohol, that can be a sugar as well, is reacted with propylene oxide (PO) to form a polyether.

$$\text{HO—R—OH} + n\,\text{C}_2\text{H—}\underset{\text{H}}{\text{C}}\text{—CH}_3\ (\text{epoxide O bridge}) \longrightarrow \text{HO—R}\left[\text{O—}\underset{\text{CH}_3}{\overset{\text{H}}{\text{C}}}\text{—}\underset{\text{H}_2}{\text{C}}\right]_n\text{OH} \quad (3)$$

The reaction is exothermic and safety precautions have to be taken. In case of a failure of the cooling system, the amount of PO in the reactor poses a potential danger, especially since PO is highly soluble in the polyether. The batch strategy (dosage of PO and temperature program) has to be adjusted accordingly, so that no damage occurs at a failure of the cooling system. To optimize the batch strategy the amount of PO that is dissolved in the liquid phase needs to be known as a function of temperature and partial pressure of PO. The problem is rather complicated, since the liquid phase is changing its thermophysical properties during the reaction significantly: at the beginning of the reaction it consists of the starter alcohol, then more and more PO groups are added to the starter molecule and at the end of the reaction, the liquid phase consists of a polyether of a molar mass of approximately 3000 kg $kmol^{-1}$. Phase equilibria of PO in different starter alcohols and in polyethers of different chain length were measured for different temperatures and partial pressures of PO. A thermodynamic model was developed, that is based on the UNIQUAC equation. But, since the molecular structure is changing during the reaction, a segment-based UNIQUAC model was developed. Binary parameters to account for the interaction between different segments of the molecules were determined by fitting them to experimental phase equilibrium data. The representation of the experimental data with the model was very good.

An interesting phenomenon was encountered: for some starter alcohols the solubility of PO in the liquid phase goes through a maximum during the reaction. This is illustrated in Figure 3.

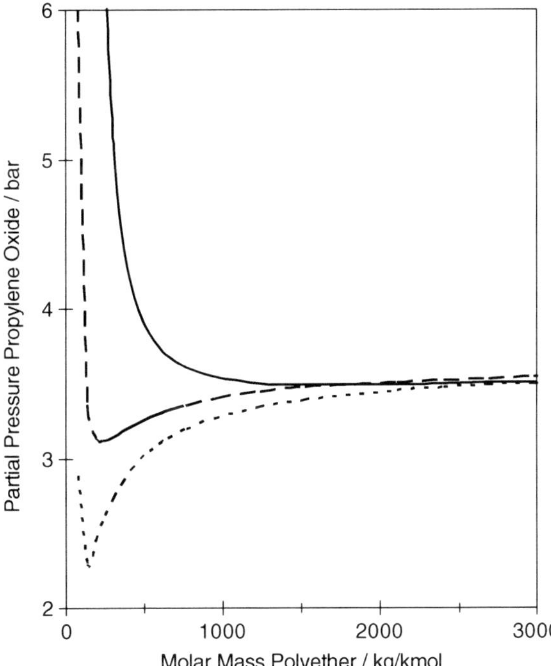

Figure 3 *Partial pressure of PO that is needed to dissolve 8 wt% PO in the polyol for a given temperature. The lines refer to different starter alcohols.*

It depicts the partial pressure of PO that is needed to have 8 wt% of PO dissolved in the liquid phase at a given temperature. The lines have been calculated with the segment-based UNIQUAC model for three different starter alcohols. The highest partial pressure is needed for the starter alcohols. Then, when more and more PO groups are added to the starter molecule, the solubility increases, meaning, a lower partial pressure of PO is needed to achieve 8 wt% PO in the liquid phase. For two of the starter molecules, the curves go through a minimum, meaning that solubility reaches a maximum. For very polar starter molecules, this effect cannot be seen. With increasing chain length the influence of the starter alcohol becomes less and less and the curves converge. It can also be seen that for high molar masses the solubility is almost constant. Using the thermodynamic model, simulations of the batch polymerization were performed. A new batch strategy was developed that lead to a very significant reduction of the batch time. The capacity of the plant could be more than doubled. Some years later, a very similar segment-based model to take into account varying chain lengths has been developed by Tritopoulou *et al.*[10] to describe the phase behaviour in polyethylene glycol+water systems.

18.3.2 Example 3: Chloroformate Reaction Design

Example 3 illustrates the importance of gas solubilities and chemical equilibria for the process design of the production of aliphatic chloroformates. To produce an aliphatic chloroformate, an alcohol is reacted with phosgene and HCl is formed as a by-product.

$$\text{R-OH} + \text{Cl}-\overset{\overset{\text{O}}{\|}}{\text{C}}-\text{Cl} \xrightarrow{k_1} \text{R}-\text{O}-\overset{\overset{\text{O}}{\|}}{\text{C}}-\text{Cl} + \text{HCl} \tag{4}$$

The exothermic reaction is fast in the beginning but after a certain conversion a strong retardation is observed and conversion rates drop. Measurements of chemical kinetics and phase equilibria were performed to get a quantitative understanding of the observed effects. The reaction rates of the heterogeneous system (liquid/gas) are mainly determinated by the solubility of phosgene and HCl in the alcohol/chloroformate mixture. The HCl concentration has a major influence on the reaction velocity.

Figure 4 shows literature data[11,12] comparing the HCl-solubility in methanol and ethanol as a function of temperature. The HCl solubility is extremely high and it decreases only slightly with increasing temperature. In contrast, the solubility in the chloroformates is significantly lower.

The solubility of phosgene in the chloroformates is high, *e.g.* at 20°C solutions containing 50 wt% phosgene can be obtained. Therefore, high phosgene concentrations are available for the reaction without using higher pressures in the process. The phosgene solubility in the alcohols cannot be measured due to the fast reaction of both components.

Reaction kinetics were measured at temperatures from 5 to 30°C by using on-line ATR (attenuated total reflection) IR spectroscopy.[13] Experiments were

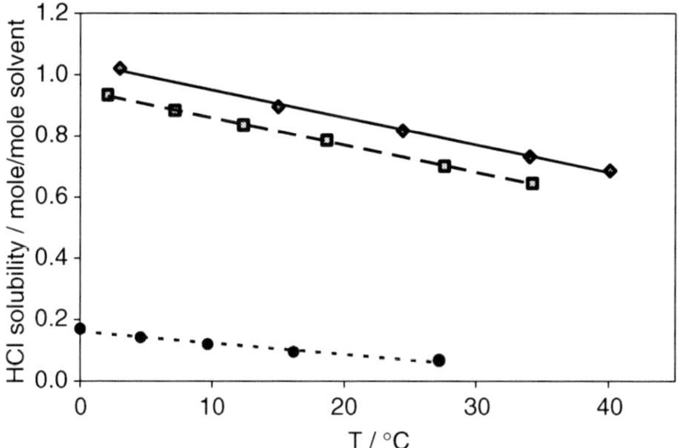

Figure 4 *HCl-solubility*[11,12] *(mole/mole solvent) in methanol, ethanol and ethyl chloroformate as a function of temperature. (—— ethanol,- - - - methanol, · · · · ethyl chloroformate).*

performed in a well-thermostated stirred glass reactor, with a horizontal ATR crystal (Si) mounted in the bottom plate. The IR-spectra were recorded with a time interval of 5 s and additionally, samples were taken and analysed off-line by HPLC. The experiments were started by injecting a defined amount of alcohol to a thermostated mixture of phosgene and chloroformate in the reactor. The chemical reactions are followed by monitoring the IR-absorption of $COCl_2$ at 1808 cm^{-1} and the IR-absorption of the chloroformates at 1778 cm^{-1} as a function of time.

The alcohol concentration is detectable by the broad O–H-absorption at 3328 cm^{-1}. This absorption is shifted and extremely broadened with increasing concentration of HCl. Additional measurements in pure alcohol/HCl-mixtures showed the same effect. This observation can be attributed to the formation of a R-OH_2Cl–complex with strong H–O–H-bonds. The data are used to estimate the constant for the R-OH_2Cl /ROH-equilibrium.

$$\text{R-OH} + \text{HCl} \underset{k_{22}}{\overset{k_2}{\rightleftharpoons}} \text{ROH}_2^+\text{Cl}^- \qquad (5)$$

Kinetic model discrimination and quantitative analysis of the experimental data indicate the following reaction mechanism. The fast and exothermic reaction of alcohol with phosgene starts immediately producing the by-product HCl. This primary reaction is a second-order reaction (nucleophile addition/elimation-type). HCl reacts with the alcohol in an equilibrium reaction, forming the R-$OH_2^+Cl^-$ complex. The protonated alcohol is less nucleophilic and not able to attack phosgene directly. Therefore the reaction rates drop and the further conversion is controlled by the R-OH/ R-OH_2Cl-equilibrium.

For an economical process, a high alcohol conversion, minimized phosgene excess and a high-space time yield are required. This is only achievable by shifting the protonation equilibrium to the free alcohol by removing HCl as the

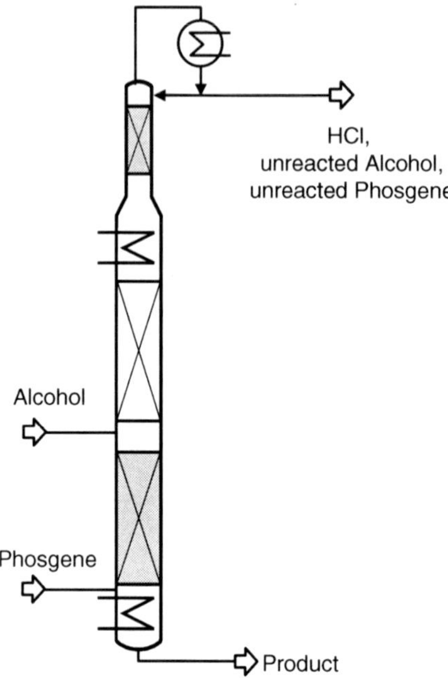

Figure 5 *Scheme of a reactive distillation column for the production of chloroformates.*

most volatile component. Therefore, reactive distillation is the most advantageous process design for this type of reaction. A simplified scheme is depicted in Figure 5.

In the bottom part of the reaction column fast conversion of the alcohol is supported by the low solubility of HCl in the chloroformates. Simulations of the column based on physical properties, solubility data and the kinetic model show that alcohol contents of <0.1 wt% in the bottom product can be reached. The best position for the phosgene addition is directly above the bottom. In this position the high phosgene concentration increases the velocity of the reaction of the still remaining alcohol. This approach is supported by the high physical phosgene solubility in the chloroformates. Separation of the product from low boiling by-products, HCl and excess phosgene, as well as further product purification can be done by additional distillation steps.

18.3.3 Example 4: Formaldehyde Production

The following example also illustrates the importance of solubilities and phase equilibria as well as reaction kinetics. The production of formaldehyde by using the silver process, as depicted in Figure 6, consists of five main steps: preparation of a methanol-air vapour mixture, partial oxidation and reduction, absorption of formaldehyde, vacuum distillation and stripping.

Phase equilibria besides kinetics play an important role in modelling of every separation unit, for instance, of the vacuum distillation column, where

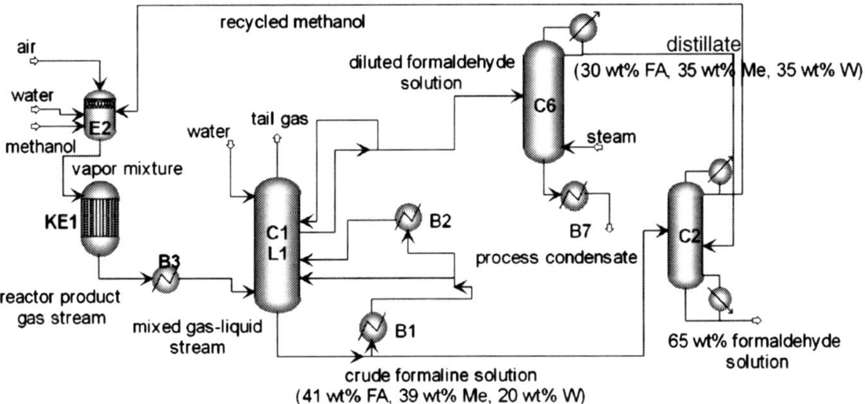

Figure 6 *Process scheme of formaldehyde production by the silver process. E2-vaporizer, KE1-reactor, C1L1-absorber, C2-vacuum distillation column, C6-stripping column, B1, B2, B3, B7-heat exchanger, FA-formaldehyde, Me-methanol and W-water.*

unconverted methanol is removed from the crude formalin solution and returned to the reaction stage. Formaldehyde mixtures are very complex, since formaldehyde reacts with water and methanol and forms different educts: methylene glycol, poly(oxymethylene) glycols, hemiformal and poly(oxymethylene) hemiformals. A successfully applied physico-chemical model was given by Maurer[14] with further extensions and revisions.[15–18] The model is based on the following reactions:

$$CH_2O + H_2O \leftrightarrow HO(CH_2O)H \quad (6)$$

$$HO(CH_2O)_{n-1}\,H + HO(CH_2O)H \leftrightarrow HO(CH_2O)_n\,H + H_2O\ (n \geq 2) \quad (7)$$

$$CH_2O + CH_3OH \leftrightarrow CH_3O(CH_2O)H \quad (8)$$

$$CH_3O(CH_2O)_{n-1}\,H + CH_3O(CH_2O)H \leftrightarrow CH_3O(CH_2O)_n\,H + CH_3OH\ (n \geq 2) \quad (9)$$

A simulation of the vacuum distillation column was carried out[19] by using the physico-chemical phase equilibrium model for multicomponent formaldehyde-containing mixtures, accounted for the presence of eight components in the liquid phase. The number of species in the simulation might be too low to give a good agreement with vapour–liquid equilbrium data in the concentration range of interest. The obtained discrepancies in overall column efficiencies using this model are most likely due to departure from equilibrium state caused by chemical reaction kinetics and mass transfer.

18.3.4 Example 5: Polyester Reaction Design

The next example is about the production of polyesters, which are formed by the direct esterification of a polyol and a polyacid:

$$\mathrm{HO-R-OH+HOOC-R'-COOH} \\ \xrightleftharpoons{230\,^{\circ}\mathrm{C,\,acid}} \mathrm{HO-(-R-OCO-R'-)-COOH+HOH} \quad (10)$$

The direct esterification reaction is reversible, with equilibrium constants from 1 to 10. If the water from the reaction is not removed from the reactor, the final molecular weight of the polyester cannot be achieved. A possible way to lower the temperature necessary to evaporate a liquid component, is to add an additional component that is immiscible with the first component. Steam distillation is taking advantage of this effect and it has been employed in the manufacture of essential oil for many centuries, though not being aware of the thermodynamic background. In our case, an aromatic solvent like toluene or xylene is used as water-removing aid. Water and toluene form a heterogeneous azeotrope at 84.1°C and 55 mol% of water at atmospheric pressure. Vapour with a composition of the heterogeneous azeotrope is then directly removed from the stirred vessel using a distillation column situated on the top of the reactor. Since the aromatic solvent also helps to volatilize the glycol, an optimization of the separation step is necessary. Therefore, thermodynamic information of the system is necessary since the system is highly non-ideal, with equilibrium constants K_x exhibiting a strong dependence on the liquid composition and on the temperature. The distillation path was assumed to move from a homogeneous 1,2-propandiol-rich region at the bottom of the column towards higher concentrations of water and toluene at the top, where two liquid phases are observed.

A number of techniques were used[20] to measure the binary and ternary equilibria in the temperature and concentration range of interest. Since the ternary LLE exhibit a phase diagram of type II at 298.15 K, with a narrow homogeneous region on the 1,2-propanediol side, ternary VLE measurements were performed starting from different initial 1,2-propanediol+water mixtures by adding small increments of toluene to reach the onset of the miscibility gap (Figure 7).

Simulation of the three-phase distillation column was performed by using the UNIQUAC[36] model to optimize the use of the aromatic solvent in the polycondensation process.

18.4 Separation Processes

Though process simulators are very valuable tools for process design and process optimization, their results can be fraught with problems if insufficient property data are used. Process simulators are usually connected to a property parameter data bank or use property estimation methods. With a few clicks the

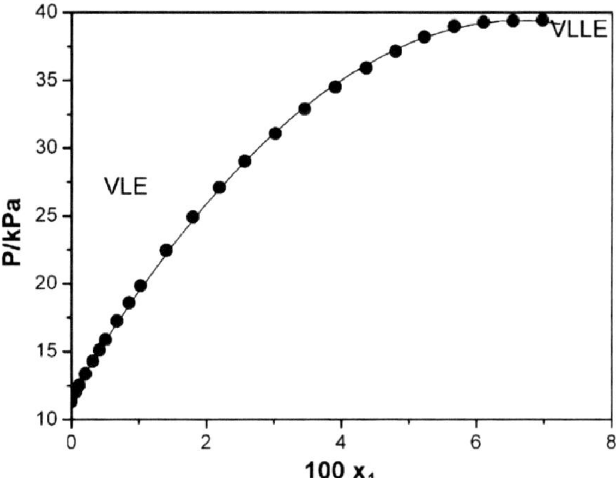

Figure 7 *Total vapour pressure for the system toluene (1) + 1,2-propanediol (2)+water (3); addition of toluene (1) to the binary mixture 1,2-propanediol (2)+water (3) of initial composition (mole fraction) $x_2 = 0.6951$, $x_3 = 0.3049$ at 343.75 K.*

process simulator is ready for a calculation. Apparently without having to worry about thermophysical property data the simulator will deliver results. The belief that the results delivered by a computer might be very wrong has been shown in the example of the classic styrene+ethyl benzene separation[21,22] calculated with three popular process simulators. Though in all three cases the standard SRK equation of state[23] was used for the calculation, the results of the three process simulators differ by a factor of 3 for the calculated mass stream of ethyl benzene in the bottom product. The main reason for the discrepancies is that different thermophysical property data have been used by the simulators.[21]

The general problem of property parameter data banks is that significant error might arise since parameters have been determined from experimental data in a temperature range different from the temperature conditions of the actual simulation problem, or by using imprecise data. Sometimes parameters have been calculated using a predictive method that is inadequate for the actual problem. When no experimental data are available process simulator might set parameters automatically to zero, assuming ideal behaviour. Even if the assumption of ideal behaviour is valid for a certain case, not all models with binary interaction parameters set to zero lead to ideal behaviour. For example, equations of state with $k_{ij} = 0$ lead to an arbitrary phase behaviour[24] that might be far away from Raoult's law and even further away from the real phase behaviour of the system. As a reaction to these problems with physical properties, a large process simulation company has written the following, very useful warning into the product documentation:[25] "Before starting a simulation study, it is important to understand the physical property and phase equilibrium behaviour of the fluids in your process, and to confirm that the behaviour predicted by the property models and data you are using is reasonable".

How small uncertainties in phase equilibrium data might have a huge impact on the design of a distillation column is shown in the following example, where the minimum number of theoretical stages of a distillation column has been calculated using the Fenske–Underwood equation[26] (with purities of the products of 99.9 mol%). The closer separation factor α lies to 1,

$$\alpha = \frac{(y_1/x_1)}{(y_2/x_2)} \tag{11}$$

the larger is the possible relative error in the number of minimum stages due to an given uncertainty of the phase equilibrium description (that is of separation factor α). This is due to the fact that the minimum number of stages is proportional to $1/(\alpha\text{-}1)$. When α approaches a value of 1, $1/(\alpha\text{-}1)$ goes to infinity.[2] An underestimation of the correct value of $\alpha = 1.1$ by 5% leads to a calculated column height, that is, more than 100% high. This is very disadvantageous, because difficult separations (α close to 1) need many theoretical stages, *e.g.* between 100 and 200 stages, and high investment costs. For easy separations (with $\alpha > 1.5$), the following approach is often used to account for the relative small influence of the separation factor on design: make the column 20% higher, *e.g.* 6 stages instead of 5. For difficult separations, 20% would might mean 20 more stages, and even that may not be enough, as can be seen from the example above, where 100 more stages would be needed, that is a second distillation column. So, for difficult separations, a 5% error of α can be very expensive.[2] Similar examples have been given in the literature.[27–32]

For calculations in multicomponent systems, it has become common practice to make simplifying assumptions about the phase behaviour of some binary subsystems for which no experimental data are known. But sometimes assuming ideal behaviour or using simple predictive methods might lead to a completely wrong description of the phase behaviour. In a previous publication,[33] the influence of increasing deviations from ideal behaviour has been illustrated in the form of several phase diagrams. The following four special cases need special attention concerning the description of the phase behaviour:

(i) *Almost ideal behaviour* (small deviations from Raoult's law) *with small differences of the boiling points* (closely boiling systems), *e.g.* mixtures of isomers. The vapour-pressure curves of the components that shall be separated are lying close to each other. Many equilibrium stages are needed for a separation. Small non-idealities (even for activity coefficients γ_i close to 1) may lead to a homogeneous azeotrope.

(ii) *Deviations from Raoult's law with a homogeneous azeotrope, e.g.* ethanol+water. When the composition of the system approaches the azeotrope composition separation factor α tends to 1 and a further separation is not possible. When an azeotrope is encountered in a distillation column one of the products will not be the desired pure component, but a mixture of components with the composition of the azeotrope. An additional problem arises in multicomponent systems. Widely used activity coefficient models, like Wilson,[34] NRTL[35] or

UNIQUAC,[36] when based on parameters obtained through binary phase equilibrium information may result in multiple sets that are not able to predict ternary and higher order azeotropes, or computed values are significantly different from experimental data. The choice of binary activity coefficient model parameters can have a strong influence on calculated ternary and higher order azeotropic points.[37]

(iii) *Strong positive deviations from Raoult's law with a liquid–liquid miscibility gap, e.g.* water+hydrocarbons, a heteroazeotrope is formed.[3] The lowest boiling "component", appearing at the top of the distillation column is – surprise – not the lowest-boiling pure component, but a mixture with the composition of the heteroazeotrope. In the above described Example 5 (Polyester Reaction Design), this effect is used intentionally; an aromatic compound, which forms a liquid–liquid miscibility gap with water is added, so that a heteroazeotrope is formed and the removal of water is possible at a lower temperature.

(iv) *Strong negative deviations from Raoult's law*, *e.g.* when large differences in the molecular weight of the substances occur, like in polymer-solvent systems. The simplifying assumption to assume ideal behaviour when the boiling points of the substances differ considerably from each other is valid for many systems with substances of comparable molecular weight, but it is misleading when the molecules in the system differ considerably in their size and entropic effects have to be taken into account. For many applications, *e.g.* for the devolatilization of oligomers or polymers, this simplification can result in the wrong design of separation apparatuses, like thin-film evaporators. With increasing difference in molar mass between the components in a binary system, increasingly strong negative deviations from Raoult's law occur. This effect has been illustrated[2] for mixtures of *n*-hexane with polydimethylsiloxane (PDMS) of different molar mass.[38] The boiling-point curve of the system decreases with increasing molar mass of the PDMS, *e.g.* for $M_n = 3000$ kg kmol^{-1} the boiling pressure for a mixture with 50 mol% *n*-hexane is only 30% of the corresponding value for an ideal solution; with $M_n = 15{,}000$ kg kmol^{-1} this value is even down to 8%. The separation of the components is far more difficult than what would be calculated assuming ideal behaviour.

18.4.1 Example 6: Furfural Production

An example for a difficult multicomponent separation occurs in the production of furfural. In the conventional furfural manufacturing process, furfural is produced by steam digestion under pressure of plant material, which contains pentosan polysaccharides. Treatment of the wood chips or other agricultural waste materials with steam produces both furfural and small amounts of acetic acid, 5-methylfurfural and methanol. Therefore, furfural has to be recovered from the condensed reactor vapours by means of a sequence of distillation steps

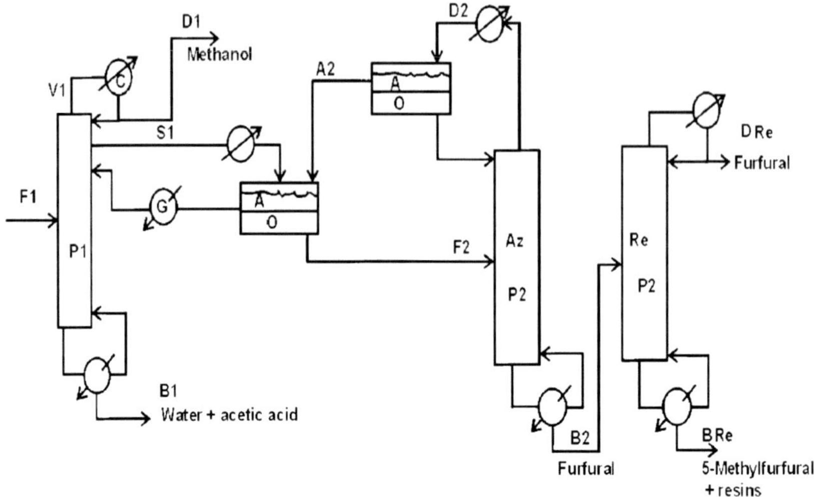

Figure 8 *Furfural recovery scheme.*

(Figure 8) or by using an effective extraction process by means of chlorinated hydrocarbons.

In the past, a lot of technologies were based on experiences, but nowadays in order to be competitive and to satisfy the quality requirements of each customer, even older production plants have to be optimised. The above-mentioned system is highly non-ideal, and without a knowlegde of the thermodynamic phase diagrams of key subsystems, an optimisation of the columns cannot be performed. Furfural as well as 5-methylfurfural forms a heterogeneous pressure maximum azeotrope with water, and both binary systems are partially miscible. The phase diagram of the first ternary system furfural+5-methylfurfural+water is of type II;[39] that of the other system, namely furfural+acetic acid+water, is of type I. For design purposes of the separation process, residue curve maps, a very effective tool, were calculated (Figure 9), showing that furfural should be recovered from the azeotropic side stream of the first column by azeotropic distillation under vacuum, followed by vacuum redistillation, where the rest of the resin and 5-methylfurfural are removed.

By knowing thermodynamic data, the proper operating conditions and explanations can be given to the operator or process engineer of the columns in order to achieve the desired purity of furfural as final product. This knowledge is especially valuable, when the producer switches from one raw material to another, thus changing the composition of the inlet flow.

18.5 Surprising Effects of Solubilities

The following examples from industrial practice show that solubilities and phase equilibria might have effects that on first sight can be quite surprising.

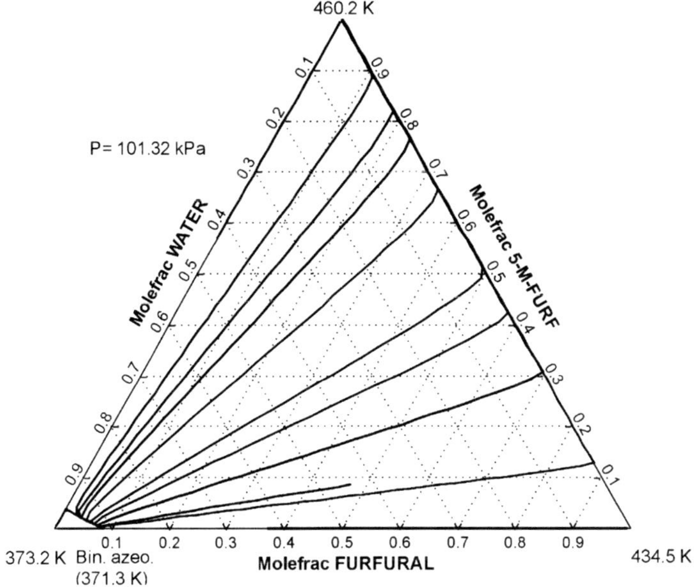

Figure 9 *Residue curves for furfural+5-methylfurfural+water system (NRTL model[35]).*

The solubility of gases in liquids can have unwanted effects for plant operation, especially when gases dissolve in large recycle streams. Owing to changing temperature and pressure conditions, gases are dissolved in the liquid recycle stream in one place of the process (*e.g.* high pressure and low temperature) and they are partly released to the vapour phase in another place of the recycle stream (*e.g.* at a lower pressure or a higher temperature). Even when this effect is very small, due to large recycle streams and many months of operation, larger quantities of gas might accumulate in a certain part of the process. An example is the accumulation of nitrogen in a pump of a closed loop with a heat transfer fluid. To reduce pressure peaks in the system and to account for the thermal expansion of the fluid, often a storage vessel under nitrogen atmosphere is used. Nitrogen dissolves in the heat transfer fluid in the storage vessel (low temperature) and might be released at a different place of the process (high temperature or low pressure). The release of gases in pumps due to a pressure reduction is a well-known problem that might lead to serious mechanical problems. Also in water cycles that are used in mining, similar problems might occur, especially since, alone from large different hydrostatic levels of the water in mining, larger pressure differences occur. In addition, the dissolution and enrichment of gases like H_2S might lead to corrosion problems. When several gases are dissolved in a recycle stream more problems might arise. In almost all fluids the main components of air, nitrogen and oxygen, have different gas solubilities. This can lead to the enrichment of oxygen in some part of the plant that in the worst case will end in an explosive mixture.

18.5.1 Example 7: Traces of Volatile Components

This example shows that the quality of a product may depend on traces of volatile substances in a liquid. After a new, very effective devolatilization process was implemented to remove traces of unwanted volatile monomers from a polyol, it turned out that the quality of the PUR foam that was made from the new, very clean polyol was of lower quality concerning mechanical stability than the old foam. A clue about the possible reason could be deducted from the hint of a production engineer: Similar problems have been seen in the past when the polyol was stored only for a short time in the storage tank, which is under nitrogen atmosphere. For the foaming process, a sufficiently high amount of traces of volatiles, *e.g.* nitrogen, is needed to help induce-cell formation. In one case, the short residence in the storage tank did not allow equilibration with nitrogen, and in the other case the very efficient devolatilization process has removed "too many" volatile components, including nitrogen. A simple solution was found, by saturating the polyol with nitrogen after the devolatilization step.

18.5.2 Example 8: Flame Ionization Detector (FID) Alarm at Fermentation Reactor

An automated FID alarm indicated repeatedly a very high alcohol content in the vent line of an aqueous fermentation reactor and lead to plant shut downs. The liquid phase in the fermentation reactor has a water content of 95 wt.%, the other main component is a C_5-alcohol. Since water is more volatile than the alcohol it was assumed by the plant manager that the vapour leaving the reactor would be even richer in water than 95 wt.%. The unexpected behaviour of the FID alarm led to the consultation of a thermodynamicist. Figure 10 shows a phase diagram of the system: temperature *versus* water content (weight fraction) at $P = 110$ kPa.

The dashed line corresponds to an ideal solution. Indeed, if ideal behaviour what be assumed, the saturated vapour of a liquid with a composition of the fermentation reactor, corresponding to Point A in the diagram, would have a higher water content (Point B). But alcohol–water systems do not behave like an ideal solution. The real phase behaviour has been calculated with the UNIQUAC model,[36] using experimental data for the liquid–liquid equilibrium. Two liquid phases occur in this system at temperatures lower than 96°C. Since the water+C_5-alcohol system shows heteroazeotropic behaviour, the vapour is much richer in alcohol (Point C) than the fermentation solution. The FID detector was positioned in the vent line, a couple of meters away from the reactor, where temperatures were lower. In Figure 10, the effect of a temperature decrease can be seen. Coming from Point C in the diagram, the vapour condenses to a liquid that, at temperatures below 96°C, splits into two liquid phases. The aqueous phase (Point E) has a similar composition as the fermentation solution, but the organic phase (Point E) has a very high alcohol content (*e.g.* more than 90 wt.% at temperatures below 70°C) that caused, after

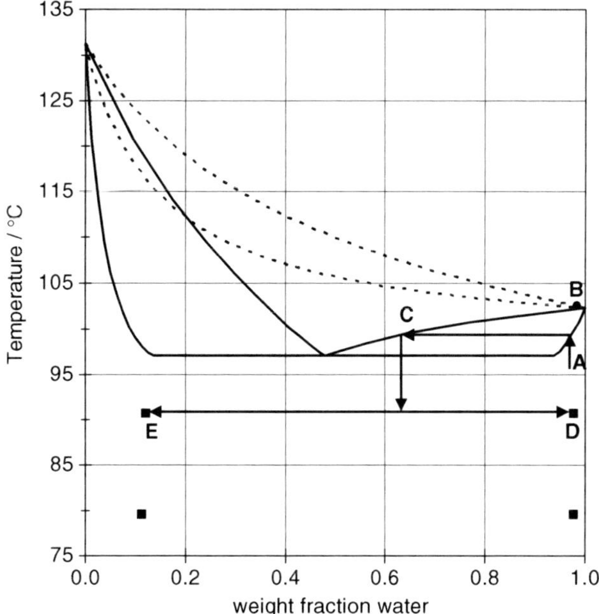

Figure 10 *Temperature versus water content (weight fraction) in a C_5-alcohol+water system at $P = 110$ kPa. Dashed line: calculated for an ideal solution, solid line: calculated using UNIQUAC,[36] ■ experimental data for the liquid–liquid equilibrium.*

evaporation due to temperature changes, the FID alarm. In this example, a simple thermodynamic analysis and a modification of the vent line lead to a higher process safety and to the avoidance of costly plant shut-downs due to FID alarms.

18.6 Conclusion

From the given examples, we can see that solubilities and phase equilibria are important for the entire process chain from the starting material that was bought from a supplier, all the way to the final product, that is, sold to a customer. Therefore, solubilities and phase equilibria can affect the daily work of almost everybody who is involved in the production of chemicals or polymers: research chemists, process developers, process simulation experts, unit operation experts, *e.g.* for distillation, reaction design, plant engineers and plant managers, people who are in charge of plant safety, quality assurance or environmental protection, as well as technical support staff who solve technical problems of customers. Since not all of these people have a deep knowledge in applied thermodynamics, it is very helpful to consult experts from a thermodynamics group where solubilities and phase equilibria can be calculated, measured and evaluated. Often precise measurements of solubilities and phase equilibria followed by process simulations are needed to optimize reaction and process design, but sometimes consulting with a sound thermodynamic background is sufficient to solve technical problems.

References

1. K.R. Cox, *Fluid Phase Equilib.*, 1993, **82**, 15.
2. R. Dohrn and O. Pfohl, *Fluid Phase Equilib.*, 2002, **194–197**, 15.
3. J.M. Prausnitz, R.N. Lichtenthaler and E. Gomes de Azevedo, *Molecular Thermodynamics of Fluid-Phase Equilibria*, Prentice-Hall, Upper Saddle River, 1999.
4. S.I. Sandler, *Chemical and Engineering Thermodynamics*, John Wiley and Sons, New York, 1999.
5. J.D. Raal and A.L. Mühlbauer, *Phase Equilibria: Measurement and Computation*, Taylor & Francis, Bristol, 1998.
6. T.A. Al-Sahhaf and N.J. Jabbar, *J. Chem. Eng. Data*, 1993, **38**, 522.
7. R. Dohrn and O. Pfohl, *Fluid Phase Equilib.*, 2004, **217**, 189.
8. K.-M. Krüger, O. Pfohl, R. Dohrn and G. Sadowski, *Fluid Phase Equilib.*, 2006, **241**, 138.
9. E. Bertakis, E. Voutsas, D. Tassios, O. Behrend and R. Dohrn, Melting point depression by using supercritical CO_2 for a novel melt-dispersion micronization process, *J. Mol. Liquids*, 2006, (accepted for publication).
10. E.A. Tritopoulou, G.D. Pappa, E.C. Voutsas, I.G. Economou and D.P. Tassios, *Ind. Eng. Chem. Res.*, 2003, **42**, 5399.
11. W. Gerrard and E. Macklen, *J. Appl. Chem.*, 1956, **6**, 241.
12. W. Gerrard, A.M.A. Mincer and P.L. Wyvill, *J. Appl. Chem.*, 1959, **9**, 89.
13. U. Wolf, R. Leiberich and J. Seeba, *Catalysis Today*, 1999, **49**, 411.
14. G. Maurer, *AIChE J.*, 1986, **32**, 932.
15. H. Hasse, I. Hahnenstein and G. Maurer, *AIChE J.*, 1990, **36**, 1807.
16. H. Hasse and G. Maurer, *Fluid Phase Equilib.*, 1991, **64**, 185.
17. M. Albert, I. Hahnenstein, H. Hasse and G. Maurer, *AIChE J.*, 1996, **42**, 1741.
18. H. Hasse and G. Maurer, *Ber. Bunsenges. Phys. Chem.*, 1992, **96**, 83.
19. L. Fele Žilnik and J. Golob, *Acta Chim. Slov.*, 2003, **50**, 451.
20. L. Fele, M. Fermeglia, P. Alessi, J.R. Rarey and J. Golob, *J. Chem. Eng. Data*, 1994, **39**, 735.
21. J. Sadeq, H.A. Duarte and R.W. Serth, *AIChE Annual Meeting*, Miami Beach, 1995, Paper 30d.
22. W.B. Whiting, *J. Chem. Eng. Data*, 1996, **41**, 935.
23. G. Soave, *Chem. Eng. Sci.*, 1972, **27**, 1197.
24. R. Dohrn, *Berechnung von Phasengleichgewichten*, Vieweg-Verlag, Wiesbaden, 1994.
25. Aspentech, *aspenONE Documentation*, Aspen Plus, Analyzing Properties, 2004, Chapter 7.
26. H.Z. Kister, *Distillation Design*, McGraw-Hill, New York, 1992.

27. W.A. Wakeham, G.S. Cholakov and R.P. Stateva, *Fluid Phase Equilib.*, 2001, **185**, 1.
28. S. Peridis, K. Magoulas and D. Tassios, *Sep. Sci. Technol.*, 1993, **28**, 1753.
29. E.C. Carlson, *Chem. Eng. Prog.*, 1996, **92**, 35.
30. R. Dohrn, R. Treckmann, and G. Olf, A centralized thermophysical-property service in the chemical industry, in *Distillation & Absorption*, R. Darton (Ed.), 1997, 1, 113.
31. W.B. Whiting, *J. Chem. Eng. Data.*, 1996, **41**, 935.
32. R.A. Nelson, H.J. Olson and S.I. Sandler, *Ind. Eng. Chem., Process Des. Dev.*, 1983, **22**, 547.
33. R. Dohrn, *Physikalische Stoffdaten und Thermodynamik*, in *Miniplant-Technik in der Prozessindustrie*, L. Deibele and R. Dohrn, (eds), Wiley, VCH Weinheim, 2006.
34. G.M. Wilson, *J.Am. Chem. Soc.*, 1964, **86**, 127.
35. H. Renon and J.M. Prausnitz, *AIChE J.*, 1968, **14**, 135.
36. D. Abrams and J.M. Prausnitz, *AIChE J.*, 1975, **21**, 116.
37. N. Aslam and A.K. Sunol, *Fluid Phase Equilib.*, 2006, **240**, 1.
38. A.J. Ashworth and G.J. Price, *J. Chem. Soc., Faraday Trans.*, 1985, **81**, 473.
39. L. Fele Žilnik and V. Grilc, *J. Chem. Eng. Data.*, 2003, **48**, 564.

CHAPTER 19

Measurements and Modelling Solid Solubilities in Supercritical Phases: Application to a Pharmaceutical Molecule, Eflucimibe

M. SAUCEAU AND J. FAGES

RAPSODEE Research Centre, UMR CNRS 2392, École des Mines d'Albi-Carmaux, 81013, Albi, France

19.1 Introduction

Supercritical fluids (SCF) are widely used for a broad field of industrial applications. The interest in using this technology is due to the special properties that are inherent to this class of fluids: the viscosity and the diffusivity of a SCF which are found in between that of a gas and liquid and the ability to vary the solvent density and the solvent properties easily and over a large extent by changing either the pressure or the temperature.

The applications often involve solutes that are in solid state at conditions where the solvent is in a supercritical condition. Among these applications, the SCF-assisted particle generation is a new and promising route to produce fine powders in mild operating conditions. It has attracted a lot of interest particularly in the pharmaceutical industry.[1] By using pressure as an operating parameter, these processes lead to the production of fine and mono-disperse powders. There exist three families of processes (RESS, SAS, and PGSS) according to the way in which the FSC – generally CO_2 – is used.

To develop them, the knowledge of the solid compound solubility in the corresponding involved medium is essential for evaluating the feasibility. Indeed, solubility is a good measurement of interactions between species. Moreover, the accurate determination of the influence of pressure and temperature on the solubility level provides insight into optimum operating conditions.

The most common SCF, carbon dioxide (CO_2), is easy to handle, inert, non-toxic, non-flammable, and has convenient critical coordinates. A limitation of CO_2 results from its lack of polarity and associated lack of capacity for specific solvent–solute interactions. For most high molecular weight compounds (non-volatile organic compounds), the solubility in supercritical CO_2 is quite low requiring high temperatures and pressures for substantial loadings. Thus, there is a great incentive to improve solvent efficiency. For these purposes, small amounts of a highly polar co-solvent can be added to CO_2 in order to increase its solvating power. The choice of a co-solvent depends not only on its ability to enhance solubility but also on its availability in high purity and its physico–chemical characteristics. In particular, for pharmaceutical purposes, the co-solvent must be also non-toxic.

Progress has been made towards the understanding of the interactions involved in dilute supercritical mixtures. It has been shown that near the critical point of a SCF solution, the solvent molecules form "clusters" around the large solute molecules to form a local density that is higher than the bulk density.[2] When a co-solvent is added, the situation is further complicated by the differences in local and bulk compositions.[3]

Several experimental techniques have been developed to investigate high-pressure equilibria.[4] There exist two types of methods according to the way the composition is measured, the synthetic methods and the analytical methods. Synthetic methods involve indirect determination of equilibrium composition without sampling. They require preparing systems of given total composition according to each point in the (T,x) or (P,x) diagrams, and therefore, are time consuming. In analytical methods, the composition of the phases in equilibrium is obtained by analyses after sampling. These methods are most widely used to determine solid-fluid equilibrium.

Because of the limited amount of experimental data dealing with solid-SCF systems, there has been considerable interest in mathematical models that can accurately predict the phase behaviour of such systems. Some of the commonly used models that have been used with some success to correlate solid solubility data include equations of state (EoS). However, such models often require properties (such as critical temperature, critical pressure, and acentric factor) that are not available for most of solid solutes. Also, the models require one or more temperature-dependent parameters, which must be obtained from solid solubility data in pure fluids. For these reasons, EoS-based models cannot be easily used to predict solubilities. Several authors have noticed that the logarithm of solid compound solubilities is approximately a linear function of the SCF density. This observation allows the representation of the solubility by using semi-empirical models based on density instead of pressure. These relations are very useful because the knowledge of the above-mentioned physical properties is not necessary.

In this chapter we will focus on the solubility of a pharmaceutical compound, called eflucimibe (Figure 1). Hypocholesterolemic properties of eflucimibe have been demonstrated on rabbits and this molecule is therefore a good candidate for becoming an effective drug for hypercholesterolemia therapy.[5] The

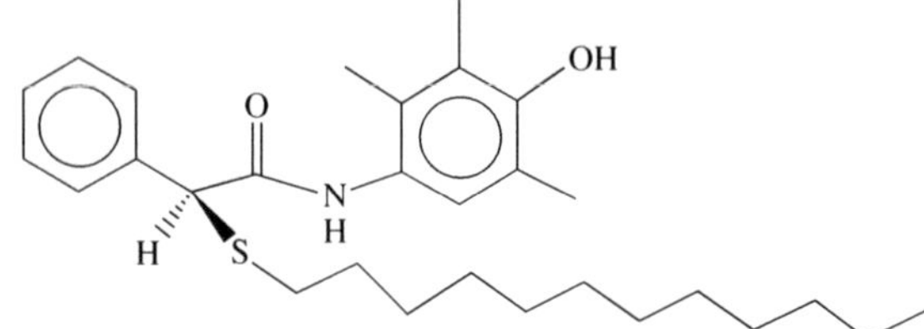

Figure 1 *Formula of eflucimibe.*

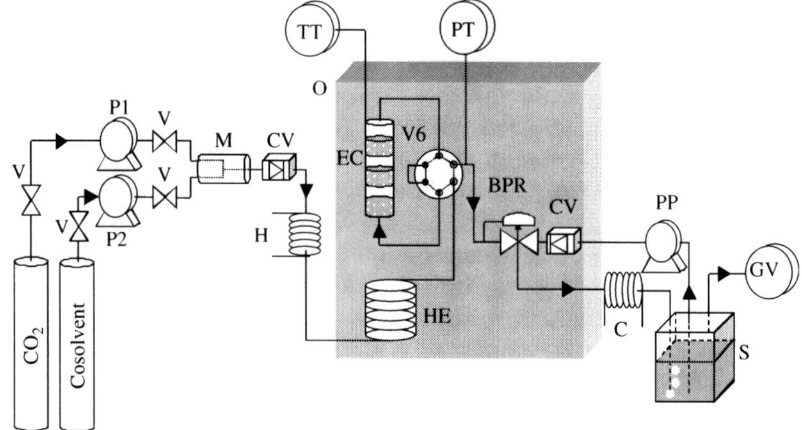

Figure 2 *Flow diagram of the apparatus: V: 2 way valves; P1, P2: high pressure pumps; M: mixer; CV: checking valve; H: heater; O: thermostated oven; HE: heat exchanger; V6:6 way-2 position valve; EC: equilibrium cell; TT: temperature transducer; PT: pressure transducer; BPR: back pressure regulator; C: cooler; S: separator; PP: peristaltic pump; GV: gas volumeter.*

empirical formula is $C_{29}H_{43}NO_2S$, with a corresponding molecular weight of 469.73 g mol^{-1}.

Its solubility was investigated by means of an original apparatus based on an open circuit analytical method[6] in pure supercritical CO_2 and in supercritical CO_2 mixtures with two co-solvents: ethanol and dimethylsulfoxide (DMSO).[7] Then, these experimental equilibrium solubilities have been correlated using two different density-based models, developed and extended to be applicable to solvent-co-solvent mixtures.[8]

19.2 Experimental: Equipment and Procedures

The flow diagram of the apparatus is shown in Figure 2. The main parts of this apparatus are: high-pressure pumps, a mixer, a heater, a heat exchanger, an equilibrium cell, a back pressure regulator and a separator.

Liquid carbon dioxide is compressed at ambient temperature by means of a syringe pump P1 (Isco, model 260D) at the desired pressure. The eventual co-solvent is introduced by means of another syringe pump P2, in a parallel flow, at a flow rate depending on the desired composition. To achieve a homogeneous mixing of the two liquid solvents, they circulate through a mixer, M. The high-pressure fluid then passes through a heater, H, which is used to heat rapidly the solvent to temperatures over its critical temperature. The SCF then enters into an oven (Spame), where the solubility cell is thermo-regulated. Owing to thermal inertia of the equilibrium cell, its internal temperature is found to be stable within 0.05 K. A heat exchanger, HE, contained in the oven, is used to set the temperature of the solvent at the desired temperature (temperature of the required solubility measurement) before it enters the solubility cell. Downstream the heat exchanger, a 6 way-2 position high pressure Valco valve is placed in the circuit to either direct the SCF to the cell or bypass it. This provides a means for removing eventual solid deposits from the line. Cylindrical in shape, the cell EC contains three compartments placed one above the other and fitted at their bottom with stainless steel fritted disks and O'rings. The solid powder, for which solubility measurements are required, is put inside the three compartments, which have a total volume of about 5 cm^3. The pressure of the supercritical phase is monitored upstream and released downstream through the BPR (Tescom, model 26-1722), which allows a pressure constancy to within 0.5% in the line. At the outlet of the BPR, the mixture pressure is reduced at the atmospheric pressure, then a recovering liquid solvent (a sufficiently good solvent at atmospheric pressure to recover all the solute) stream is used to get the solute in liquid state for collection. Then, a separator S is used to vent off the gas and collect the solvent phase. At the end of each experimental run, the liquid solvent line is washed with fresh solvent to recover all the solute. The total volume of used gaseous solvent (extraction solvent), V_1, is measured by means of a volumeter GV and the concentration of solid in the solute recovering liquid phase C_2^L by a gas or liquid chromatography. From these two data and knowing the total volume of the solute recovering liquid solvent, V^L, the solubility, y_2, of the solid in supercritical fluid can be calculated through:

$$y_2 = \frac{n_2}{\sum_i n_i} \quad \text{with } n_1 = \frac{V_1 \rho_1}{MW_1} \quad \text{and } n_2 = \frac{C_2^L V^L}{MW_2} \tag{1}$$

n_i, ρ_i, and MW_i are respectively the number of moles, the density, and the molecular weight of the compound i. The solubility uncertainty depends strongly on the experiment time length: the longer the time, the more accurate are the quantities used in the calculation of the solubility. In consequence, this time length should be optimized to ensure a high degree of accuracy of the result, without too long an experiment time.

Temperature in the cell is measured directly in its body through a 4-wires 100 Ω platinum probe, within 0.02 K as a result of a careful calibration, performed against a 25 Ω reference platinum probe. Pressure is measured in the downstream line of the cell. The pressure transducer (Druck, model PTX611) can measure pressures up to 35 MPa at temperatures up to 400 K, with an accuracy of 4.10^{-3} MPa as a result of a calibration performed using a dead weight balance (Desgranges et Huot, model 5202S CP).

An important requirement in the design of the apparatus was to obtain a saturated stream flowing outside the cell. To confirm the efficiency of the equipment over a large range of operating conditions, measurements have been performed at various flow rates. When in a given range of flow rates, no sensitive effect is observed on the measured solubility values, the equilibrium is confirmed. Separate experiments are required to confirm operating flow rates for each solute of interest.

The validity of the technique was achieved by measuring naphthalene solubility in supercritical CO_2. Naphthalene solubility data are very abundant in literature and provide a good base for quantifying tests. The solubility measurements have been found in excellent agreement with previous works.[6]

Eflucimibe was provided by IRPF (Institut de Recherche Pierre Fabre) as a white crystalline powder with a purity greater than 99%. Carbon dioxide, ethanol, and DMSO were of commercial grade.

In addition, we have checked that solvents are really in supercritical state before entering the equilibrium cell. However, little (P,T,y) data is available for the CO_2-DMSO binary mixture. Only the data proposed by Kordikowski *et al.*[9] is sufficiently complete. The authors have fitted their data using the Peng–Robinson[10] equation of state (PR EoS) with two quadratic mixing rules that include two temperature independent binary interaction parameters, k_{ij} and l_{ij}. As these authors provide also data for CO_2-ethanol binary mixture, we have chosen to use their results in order to have parameters from the same origin for the two co-solvents. The PR EoS has also allowed us to calculate the density of the supercritical fluid, for pure supercritical CO_2 or mixtures with a co-solvent.

19.3 Solubility in Pure CO_2

The eflucimibe solubility, y_2, was measured at 308.15 and 318.15 K (Figure 3). It is noticeable that the values recorded are remarkably low, giving confirmation of the accuracy of the apparatus. The effect of pressure on the solute solubility follows the expected trends, the solubility increasing with pressure for the two temperatures studied. The density of CO_2 increases with pressure, the mean inter-molecular distance between CO_2 molecules decreases, thereby increasing interaction between the solute and solvent molecules.

The existence of the crossover pressure is well known and illustrated in a number of experimental studies.[11] The pressure value where the solubility isotherms at various temperatures intersect each other is the result of the competing effects of solute vapour pressure and solvent density. From Figure 3, the crossover pressure can be estimated at about 10 MPa. Thus, over the pressure range investigated here, we can consider that solubility is an increasing function of temperature.

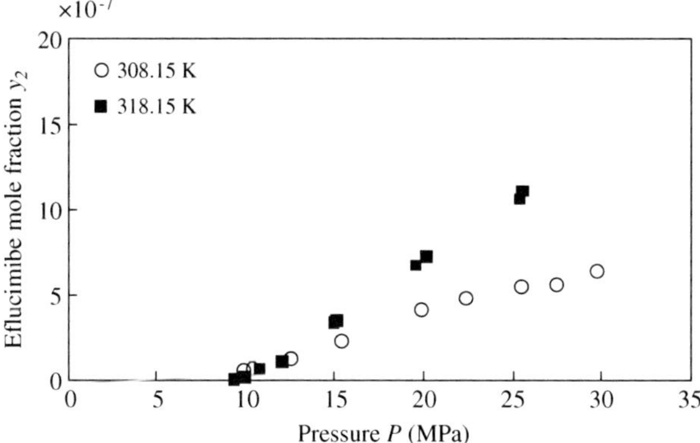

Figure 3 *Solubility of eflucimibe in pure supercritical CO_2 vs. pressure at 308.15 and 318.15 K.*

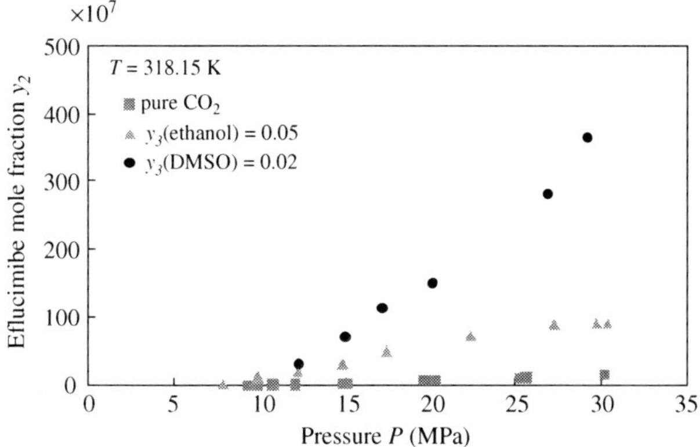

Figure 4 *Solubility of eflucimibe in supercritical co-solvent-CO_2 mixtures vs. pressure at 318.15 K and at constant co-solvent mole fraction.*

19.4 Ethanol and DMSO Co-Solvent Effects

Two series of measurements have been performed. The first concerns the influence of P on y_2, at constant T and constant co-solvent mole fraction y_3. The solubility has been measured at 318.15 K for different pressures with $y_3 = 0.05$ for ethanol and $y_3 = 0.02$ for DMSO (Figure 4).

The second series of measurements involved the variation of y_2, as a function of y_3 at constant P and T. The solubility at 318.15 K and 20 MPa has been measured for different mole fractions of the two co-solvents.

To better illustrate the solubility enhancement, a co-solvent effect A_C is defined as the ratio of the solubility obtained with co-solvent, $y_2(P,T,y_3)$, to

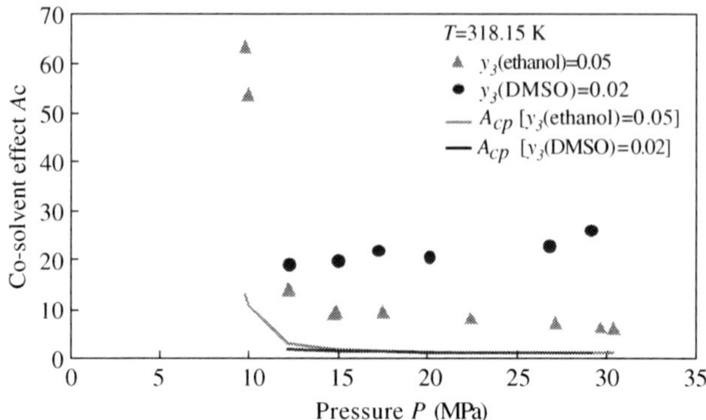

Figure 5 *The co-solvent effect in supercritical co-solvent-CO_2 mixtures vs. pressure at 318.15 K and constant co-solvent mole fraction.*

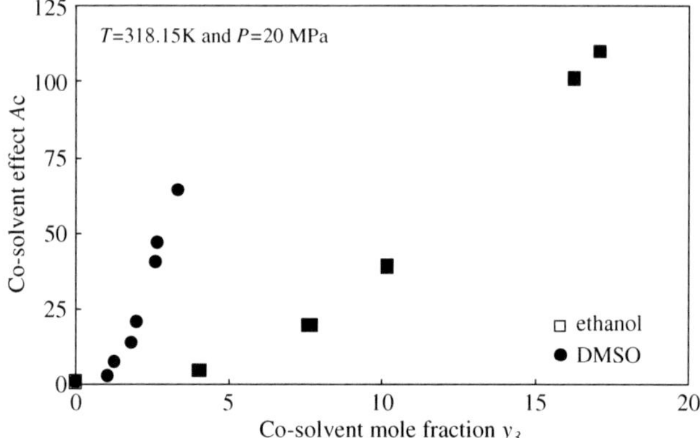

Figure 6 *The co-solvent effect in supercritical co-solvent-CO_2 mixtures vs. co-solvent mole fraction at 318.15 K and 20 MPa.*

that obtained without co-solvent at the same temperature and pressure, $y_2(P,T,y_3 = 0)$:

$$A_c(P,T,y_3) = \frac{y_2(P,T,y_3)}{y_2(P,T,y_3 = 0)} \tag{2}$$

At 318.15 K, A_C has been plotted *vs.* P (Figure 5) and *vs.* y_3 (Figure 6). From these figures, it is clear that the solubility is increased by both co-solvents, with however a higher increase with DMSO. For instance, at 20 MPa and 318.15 K, the solubility is 10 times higher with either 6% of ethanol or 2% of DMSO, and is multiplied by a factor 50 for about 11% of ethanol or 3% of DMSO in the solvent. This solubility enhancement can be attributed to three possible effects: increased density of the fluid mixture, modifications in phase equilibria and specific interactions between the solute and co-solvent.

The density contribution to the co-solvent effect is estimated by calculating a co-solvent density effect, A_{C_ρ} defined as follows:

$$A_{C_\rho}(P, T, y_3) = \frac{y_2(P, T, \rho_f, y_3 = 0)}{y_2(P, T, \rho_{CO_2}, y_3 = 0)} \tag{3}$$

$A_{C_\rho}(P, T, y_3)$ is the co-solvent density effect at P, T and y_3. It compares the solubility of the solid in pure CO_2 at T and P, $y_2(P, T, \rho_{CO_2}, y_3 = 0)$, to that also in pure CO_2 at the same temperature and pressure but at the density of the mixtures ρ_f with a co-solvent mole fraction y_3, $y_2(P, T, \rho_f, y_3 = 0)$. The co-solvent density effect is represented in Figure 5. It is clear that co-solvent effects cannot be attributed to a density effect alone, neither for ethanol nor for DMSO.

The presence of a co-solvent in a SCF can enhance the melting point depression of a solid solute in a SCF. This effect is usually accompanied by a reduction of the upper critical end point (UCEP) pressure of the system.[12] This effect is important due to the drastic enhancement of the solute solubility in the vicinity of the UCEP. Unfortunately, the contribution of this effect to the observed co-solvent effect is difficult to estimate because the relevant three phase solid–liquid–gas coexistence curves are not available. However, a qualitative indication of the importance of this effect can be obtained from the examination of the solubility isotherms.[13] When conditions are close to the UCEP, the slope of the solubility isotherms, $\partial y_2/\partial P$, becomes relatively large. An inspection of our solubility isotherms in Figure 4 reveals that as pressure increases, no dramatic variation of $\partial y_2/\partial P$ occurs. It may be concluded that co-solvent effects are the result of factors more significant than the proximity of the operating conditions to the UCEP.

The minor contribution of density to the co-solvent effect suggests that chemical forces, rather than physical forces, are responsible for the obtained solubility enhancement. These chemical forces are represented by specific interactions between the solute and the co-solvent. Several authors have observed a linear dependence of A_c with the co-solvent concentration.[14,15] However, in our experiments, A_c is not a linear function of y_3, the co-solvent effect increasing more rapidly for higher mole fractions (Figure 6). This different behaviour may be indicative of higher order interactions between the solute and the co-solvent. The type of interaction can be discussed qualitatively on the basis of pure component properties as, for instance, solubility parameters.[16] These parameters describe dispersion, common to all molecules, orientation and induction dipoles forces for polar molecules, and the ability of a species to act as a proton donor or acceptor respectively, for hydrogen bonding.

These parameters show that ethanol is polar and has a high capacity to form hydrogen bonds, being both a proton donor and acceptor (amphiprotic). DMSO is very polar, aprotic but shows a basicity close to that of ethanol.

To identify the potential interactions with the various groups of eflucimibe, co-solvent effects from literature have been examined for compounds characteristic of these groups. However, little data being available for DMSO, we have focused on the co-solvent effect of acetone, which presents some similar properties: it is polar, aprotic and basic. We have chosen results from Ekart *et al.*,[14] who have studied effects of several co-solvents in ethane. Finally, it seems that the amine and especially the phenol groups play a major role in the solubility increase of eflucimibe, by means of hydrogen bonds. However, despite its higher basicity, ethanol displays a lower co-solvent effect. Clearly, co-solvent basicity is not sufficient to explain the results, thus the dispersion parameter, which is higher for DMSO should also be considered.[16] The lower co-solvent effect of ethanol might also be explained by the self-association between amphiprotic ethanol molecules, which are no longer available to interact with the solute molecules.

19.5 Modelling

The first model was proposed by Chrastil.[17] This may be considered as a macroscopic description of the surroundings of the molecules in the fluid phase. It is based on the hypothesis that one molecule of a solute A associates with k molecules of a solvent B to form one molecule of a solvato–complex AB_k in equilibrium with the system. The definition of the equilibrium constant through thermodynamic considerations leads to the following expression for the solubility:

$$\ln(C_2) = k\ \ln(\rho_f) + \frac{\alpha}{T} + \beta \tag{4}$$

where C_2 is the concentration of the solute in the supercritical phase and ρ_f the density of the fluid phase. k is the association number, α depends on the heat of solvation and the heat of vaporization of the solute and β depends on the molecular weight of the species. k, α, and β are adjusted to solubility experimental data.

The second model is based on the theory of dilute solutions, which leads to simple expressions for many thermodynamic properties of dilute near-critical binary mixtures. In particular, Harvey[18] has obtained a simple linear relationship for the solubility of a solid in a supercritical solvent. Mendez-Santiago and Teja[19] have approximated this relationship:

$$T\ln(E) = A_1 + B_1\rho_f \tag{5}$$

E is the enhancement factor defined as the ratio between the observed equilibrium solubility and that predicted by the ideal gas law at the same temperature and pressure. A_1 and B_1 are adjustable parameters. Finally, in another paper, these authors[20] have improved the Equation (5) by taking into account the co-solvent mole fraction, y_3:

$$T \ln(E) = A_2 + B_2\rho_f + D_2 y_3 \tag{6}$$

A_2, B_2, and D_2 are three new adjustable parameters.

The quality of all data correlations is quantified by the average absolute deviation (AAD), defined as follows:

$$AAD = (1/m)\sum_{i=1}^{m} \left|(y_{2,\mathrm{cal}} - y_{2,\mathrm{exp}})/y_{2,\mathrm{exp}}\right|_i \times 100 \tag{7}$$

m is the number of data, $y_{2,\mathrm{cal}}$ the calculated solubility value and $y_{2,\mathrm{exp}}$ the experimental one.

19.6 Extension of the Chrastil Model

The Equation (4) is first applied to the solubility data of eflucimibe in pure CO_2. The two isotherms are well fitted the AAD being less than 8%. The k value obtained shows small temperature dependence. If the data of the two isotherms are gathered before parameter adjustment, the AAD remains practically constant.

The Chrastil model is applicable to pure fluids. Thus, we could apply it to mixtures at constant co-solvent mole fractions, with the hypothesis that these mixtures at constant concentration behave like pure fluids. The new values of k obtained are thus the number of molecules of solvent k_1 and co-solvent k_3 associated with one molecule of solute. These numbers are higher than that in pure CO_2: 7.2 with 5% of ethanol and 10.1 with 2% of DMSO instead of 6.5 in pure CO_2. This confirms the importance of specific interactions in the solubility enhancement phenomenon.

19.7 Generalizing the Mendez-Santiago and Teja Model

As already done by Mendez–Santiago and Teja,[19] a Clausius–Clapeyron-type equation is incorporated for the sublimation pressure in Equation (6). However, the Clausius–Clapeyron equation could be advantageously written with the dimensionless logarithm:

$$\ln \frac{P_2^{\mathrm{sat}}}{P^{\mathrm{std}}} = i - \frac{j}{T} \tag{8}$$

where P^{std} is the standard pressure (atmospheric pressure equal to 0.101325 MPa). This provides the following correlation with four adjustable parameters A_4, B_4, C_4 and D_4:

$$T \ln\left(\frac{y_2 P}{P^{\mathrm{std}}}\right) = A_4 + B_4\rho_f + C_4 T + D_4 y_3 \tag{9}$$

Equation (9) has directly been applied to all data in pure CO_2 as it takes into account the temperature. It provides a good correlation, with an *AAD* about 6%.

In a first attempt, solubility data are treated independently for each co-solvent, by gathering data at different pressures and co-solvent mole fractions. Data are well fitted, with an *AAD* about 6% for ethanol and about 19% for DMSO. However, data are available at only one temperature, which is not enough to determine correctly the value of the parameter, C_4, related to temperature. In order to have data at two different temperatures, a second correlation is carried out by gathering data for each co-solvent with that in pure CO_2. Finally, the *AAD* remains constant at about 8% for ethanol and decreased from 19 to 15% for DMSO, with coefficients attributed to density, B_4, and to temperature, C_4, close to those obtained in pure CO_2. It shows that these two coefficients can be considered to be independent of the presence of a co-solvent. It has also to be noted that the value obtained for the coefficient A_4 remains practically constant in CO_2 alone and with a co-solvent. The part of co-solvent effect due to specific interactions between solute and co-solvent is thus independent of density and temperature effects, and is quantified by the value of co-solvent mole fraction coefficient, D_4. On the basis of these observations, a correlation of all the solubility data of eflucimibe can be carried out by using the following equation with 5 adjustable parameters:

$$T \ln\left(\frac{y_2 P}{P^{\mathrm{std}}}\right) = A_5 + B_5 \rho_f + C_5 T + D_5 y_3^{\mathrm{ethanol}} + E_5 y_3^{\mathrm{DMSO}} \tag{10}$$

The data in pure CO_2 are treated with: $y_3^{\mathrm{ethanol}} = y_3^{\mathrm{DMSO}} = 0$, and the ones with a co-solvent with: $y_3^{\mathrm{DMSO}} = 0$ for ethanol as co-solvent and $y_3^{\mathrm{ethanol}} = 0$ for DMSO as co-solvent. All the data are finally correlated with a value of the *AAD* less than 13%. This correlation characterizes the solubility of the solid studied in supercritical CO_2 by using only one equation: effects of density, of temperature and of each co-solvent are quantified by means of constant values. As previously noted, the effect due to DMSO (E_5 at about 38600) is higher than that of ethanol (D_5 at about 9200). By plotting $T\ln(y_2P/P^{\mathrm{std}}) - C_5T - D_5y_3^{\mathrm{ethanol}} - E_5y_3^{\mathrm{DMSO}}$ *vs.* ρ_f all solubility data are gathered on a single line.

19.8 Conclusion

The solubility behaviour of eflucimibe was studied in pure supercritical carbon dioxide at 308 and 318 K between 8 and 30 MPa. The solubility appeared to be an increasing function of both pressure and temperature but remained at very low levels.

The effect of two co-solvents, ethanol and DMSO, was then investigated. The solubility was found to be enhanced by both co-solvents, with however a higher increase with DMSO. The co-solvent effect was found to vary nonlinearly with the co-solvent concentration, showing the importance of specific interactions between the co-solvents and the solute in comparison with density effect. These interactions could qualitatively be explained by means of solubility parameters of co-solvents and of solid functional groups. Finally, hydrogen bonds seem to play the most important role in solubility enhancement.

To extend these results, modelling appeared to be necessary to provide a tool for prediction of solid solubilities in supercritical mixtures. A study on density-based models was therefore developed.

Solubility data for pharmaceutical solid have been correlated by means of two density–based semi–empirical models: the Chrastil model and the Mendez-Santiago and Teja model.

The application of the two correlations to the data in pure CO_2 leads to expressions, which can be used for prediction purposes in a large range of pressure-temperature conditions.

In addition the Chrastil model has been extended to be applicable to solvent-co-solvent mixtures considered as pure SCF compounds. It has confirmed the importance of specific interactions in the co-solvent effect. The representation of all the data with two different co-solvents has been carried out with only one relationship by using a generalized Mendez-Santiago and Teja model, in which effects of density, temperature and co-solvent mole fraction are quantified.

Acknowledgments

The authors would like to acknowledge the financial and technical support of the Pierre Fabre Research Institute (IRPF).

References

1. J. Fages, H. Lochard, J.-J. Letourneau, M. Sauceau and E. Rodier, *Powder Technol.*, 2004, **141**, 219.
2. C. Eckert, D. Ziger, K.P. Johnston and S. Kim, *J. Phys. Chem.*, 1986, **90**, 2738.
3. S. Kim and K.P. Johnston, *AIChE J.*, 1987, **33**, 1603.
4. R. Fornari, P. Alessi and I. Kikic, *Fluid Phase Equilib.*, 1990, **57**, 1.
5. D. Junquero, F. Bruniquel, X. N'Guyen, J.-M. Autin, J.-F. Patoiseau, A.-D. Degryse, F.C. Colpaert and A. Delhon, *Atherosclerosis*, 2001, **155**, 131.
6. M. Sauceau, J. Fages, J.-J. Letourneau and D. Richon, *Ind. Eng. Chem. Res.*, 2000, **39**, 4609.
7. M. Sauceau, J.-J. Letourneau, B. Freiss, D. Richon and J. Fages, *J. Supercrit. Fluids*, 2004, **31**, 133.
8. M. Sauceau, J.-J. Letourneau, D. Richon and J. Fages, *Fluid Phase Equilib.*, 2003, **208**, 99.
9. A. Kordikowski, A.P. Schenk, R.M. Van Nielen and C.J. Peters, *J. Supercrit. Fluids*, 1995, **8**, 205.
10. D.-Y. Peng and D. Robinson, *Ind. Eng. Chem. Fundam.*, 1976, **15**, 59.
11. N. Foster, G. Gurdial, J. Yun, K. Liong, K. Tilly, K. Ting, H. Singh and J. Lee, *Ind. Eng. Chem. Res.*, 1991, **30**, 1955.

12. M. McHugh and V. Krukonis, *Supercritical Fluid Extraction: Principles and Practice*, Butterworths, Boston, 2nd edn, 1994.
13. C. Saquing, F. Lucien and N. Foster, *Ind. Eng. Chem. Res.*, 1998, **37**, 4190.
14. M. Ekart, K. Bennett, S. Ekart, G. Gurdial, C. Liotta and C. Eckert, *AIChE J.*, 1993, **39**, 235.
15. N. Foster, H. Singh, J. Yun, D. Tomasko and S. Macnaughton, *Ind. Eng. Chem. Res.*, 1993, **32**, 2849.
16. J. Dobbs, J. Wong, R. Lahiere and K.P. Johnston, *Ind. Eng. Chem. Res.*, 1987, **26**, 56.
17. J. Chrastil, *J. Phys. Chem.*, 1982, **86**, 3016.
18. A. Harvey, *J. Phys. Chem.*, 1990, **94**, 8403.
19. J. Mendez-Santiago and A. Teja, *Fluid Phase Equilib.*, 1999, **158–160**, 501.
20. J. Mendez-Santiago and A. Teja, *Ind. Eng. Chem. Res.*, 2000, **39**, 4767.

CHAPTER 20

Solubility in Food, Pharmaceutical, and Cosmetic Industries

SIMÃO PEDRO PINHO[1] AND EUGÉNIA ALMEIDA MACEDO[2]

[1] *Laboratory of Separation and Reaction Engineering, Escola Superior de Tecnologia e de Gestão, Instituto Politécnico de Bragança, Campus de Santa Apolónia, 5301-857 Bragança, Portugal*

[2] *Laboratory of Separation and Reaction Engineering, Departamento de Engenharia Química, Faculdade de Engenharia, Rua do Dr. Roberto Frias, 4200-465 Porto, Portugal*

20.1 Introduction

Solubility is well recognized as a fundamental physical property for the design of processes to separate, concentrate, and purify a targeted species. As will be discussed in the next section food, pharmaceuticals, and cosmetic industries frequently involve separation processes like precipitation, crystallization, liquid–liquid or supercritical fluid extraction (SFE). In each of these processes the choice of solvent plays an important role; for instance, it is estimated that 30% of the work of a thermodynamic group in a pharmaceutical company is directly related to the solvent selection.[1] As a result, we have decided to present this chapter in terms of the solubility of pharmaceuticals, amino acids (AA), proteins, or sugars in water, organic, and mixed solvents, liquid–liquid solubility; mostly related to aqueous two-phase systems (ATPS), or water–octanol partition coefficients, and solubility in supercritical fluids. The focus is, essentially, on the most recent developments concerning solubility correlation and prediction for substances of interest in those industries and processes. Experimental aspects, although of extreme relevance, are only highlighted for some specific cases where it is important to be aware of some particularities. In fact, several issues about the measurement of solubility were recently subject to an important edition.[2] Finally, a global overview is presented, some suggestions emphasized, and also some challenges for the near future are pointed out.

20.2 Industrial Importance

Recently, Agrawal and Noble[3] addressed some problems concerning separation needs for the 21st century. Many of them are related to the pharmaceutical, biomedical, and other biotech industries. In this context, researchers from Dow Chemical Company pointed out crystallization, ATPS, and other similar liquid–liquid extractions as processes of highest relevance.[4]

Excluding ethanol, antibiotics and AA are the major fermentation products with a market value around US$ 8 billion in 2004.[5] Their application in pharmaceutical or food industries are numerous, and AA are also used in the cosmetic industry. For instance, serine is employed for skin care cream or lotion, and some histidine derivatives act as free anti-radical agents in cosmetics.[6] After fermentation, several purification and separation techniques are applied to those highly complex broths. Crystallization is often used, for example, in glutamic acid or threonine production, for which solubility data is fundamental. Apart from key separation issues like extraction and crystallization, in pharmaceutical industries, solubility is also an essential property for the design of new drugs. Aqueous solubility gives valuable indications about the biological activity of a drug, and therefore, is most important in pre-formulation studies.[7] Water solubility, co-solvency and partition coefficients are topics under attention in many research and development groups at companies like Mitsubishi Chemical Corporation,[1] Merck and Zeneca Pharmaceuticals,[8] Hoffmann-La Roche,[9] GlaxoWellcome,[10] and Pfizer,[11] to name a few.

In the area of food processing, Agrawal and Noble[3] focused on solving such problems as the requirements of extremely high purity, and flavor and aroma capture. One of the technologies most studied in this area is the SFE with several patents and applications; some examples are the removal of cholesterol from food products,[12] de-alcoholization of beverages,[13] and concentration of flavor compounds.[14] However, even if the final sensory appreciation of flavor and aromas in food are much dependent on how the components are distributed over the different phases,[15] phase equilibria in food product design is still creating its basic foundations. Bruin,[15] and researchers at Unilever Research, applied a simple 2- or 3-suffix Margules equation[16] for the representation of the solid–liquid equilibrium of three polymorphic forms of fat crystals, sharing its success with the other well-known case, the solidification of chocolate.[17] A few final examples about research carried out for industrial needs are listed on Table 1.

20.3 Water Solubility

Water is omnipresent in many reaction and separation processes in biotechnology, and as discussed previously, solubility of biomolecules is a key equilibrium property in their production. Additionally, drug solubility in water gives general trends for rates of dissolution; poor solubility is usually synonymous with a very low dissolution velocity.[25] As a result, an administered drug will mostly be excreted without the possibility of absorption from

Table 1 *Some projects, involving solubility issues, carried out at different companies*

Problem Addressed	*Company*	*Ref.*
Effect of α-tocopherol on the solubilization of poor soluble drugs in simulated intestinal fluids	Dumex–Alpharma	Nielsen *et al.*[18]
Enrichment of Amaranth seed oil on high value lipids by SFE	Unilever	Westerman *et al.*[19]
Find an efficient excipient for rapamycin (immunosuppressor)	Schering–Plough HealthCare	Simamora *et al.*[20]
Increase the average crystal size of pharmaceuticals or agrochemicals by batch crystallization	Rhone–Poulenc	Lewiner *et al.*[21]
Influence of water content on triglycerides and their ability to be used as pharmaceutical excipients of steroids	Proctor and Gamble Novartis Pharmaceuticals	Land et al.[22]
Study of the solubility and partition coefficients of surfactants in several solvent systems to design initial extraction processes	Merck	Pollard *et al.*[23]
Study of ethanol as co-solvent in the crystallization of 1,3-dihydroxyacetone for application in the cosmetic industry	Ard–Soliance	Zhu *et al.*[24]

gastrointestinal tract into the cardiovascular system.[26] Besides the inherent complexities with experimental measurements, for this type of molecules, accuracy and reliability are specially difficult to achieve, and measurements are particularly time consuming. So, methods to predict water solubilities are an important research subject, with an extraordinary value for drug design. This task is, however, challenging because biomolecules are often very complex; they possess high molecular weight, with two or more functional groups, leading to a variety of complex molecular interactions, and are often present in different structures or isomers.[27]

In order to satisfy conditions for satisfactory water solubility and membrane permeability, drugs need to have the right balance between polarity and hydrophobicity. Empirically, if log S(S is the drug aqueous solubility in mol dm^{-3}) is in the range between -1 and -5, its adequacy is accepted.[26] Several methods to calculate log S for drugs have been proposed, but the correlations based on physicochemical properties like the octanol-water partition coefficient (P_{ow}) and the melting point are currently of little use. In fact, as will be briefly discussed in Section 5, to calculate P_{ow} several reliable methods are known, but for the melting point the opposite is true, and several reasons may hamper its experimental measurement. Another approach is based on the group-contribution concept, for which probably the most familiar is the AQUAFAC method. However, it also has the disadvantage of needing the melting point,[28] and even if several other methods that avoid this problem are available, generally, the number of groups is not enough to represent the wider variety of drugs under development nowadays. Multiple linear regression (MLR) and neural networks (NN) are two other techniques applied for solubility predictions. They are both based on a set of different descriptors like molecular weight, solvent-accessible surface area, and many other topological and electronic indices. While NN allows the introduction of non-linearity for the descriptors terms in the solubility equation, which is an advantage to MLR, it is a black box type method that cannot provide insights for drug lead optimization except by trial and error.[26] In addition, over-training is a major issue for NN techniques, and its predictive capabilities are, most of the times, no better than that of MLR. Several different equations have been proposed and reviewed,[26,28,29] but many of them do not consider AA or sugars, and some molecular descriptors are not easy to understand physically. The linear-solvation energy relationship developed by Abraham and collaborators[30] is one of the most useful and comprehensive; equations have been derived for about 50 solvents and molecular descriptors for more than 3000 common organic and pharmaceuticals compounds have been calculated for solubility predictions. Recently, Sun[9] presented a more universal method, proposing atom types molecular descriptors to build predictive models for different properties, including log *S*.

These methods are all difficult to compare since they are based on different sets of experimental data. The usual strategy is to evaluate the predictive ability of the different methods to a test set, constituted by 21 different drugs and pesticides. Very good results on deviations for log *S* were obtained using the NN[28] and Sun[9] methods. As far as accuracy is concerned one cannot ask for much better results since it strongly depends on the uncertainty of the experimental measurements of log *S* which, for complex molecules, is around 0.6 log unit.[26]

Experimental uncertainty may be attributed to substance purity, different aspects related to the solid phase, pH and temperature control, and the method used. Rousseau and collaborators[31] developed studies on the influence of isomorphic impurities in the crystal purity of AA which is intimately linked to the relative solubility ratio and type of solvent. Other studies emphasize the importance of the solid-phases analysis[32,33] and the method chosen for the measurements. In fact, as shown in Figure 1, the equilibrium solubilities obtained by the cooling and the isothermal experimental methods present quite different results for the L-isoleucine+L-valine+water system at 298 K. Analyzing the solid phase, and applying mass balances, the authors concluded that the cooling method gives more consistent results.

Several different models have been proposed to represent thermodynamic properties of aqueous solutions of AA, namely, solubility and activity coefficients, with or without a presence of an electrolyte.[34] Although some progress has been achieved, the complexities that arise from the zwitterionic nature of AA in aqueous solution, make it a difficult task, and generally, it is not possible to calculate accurately the solubilities using activity coefficient data only. The fact that AA are the building blocks of more complex molecules such as antibiotics, peptides, or proteins, makes the understanding of the effect of electrolytes on the properties of aqueous AA solutions very important and attractive, as it may give insights into processes such as salt-induced

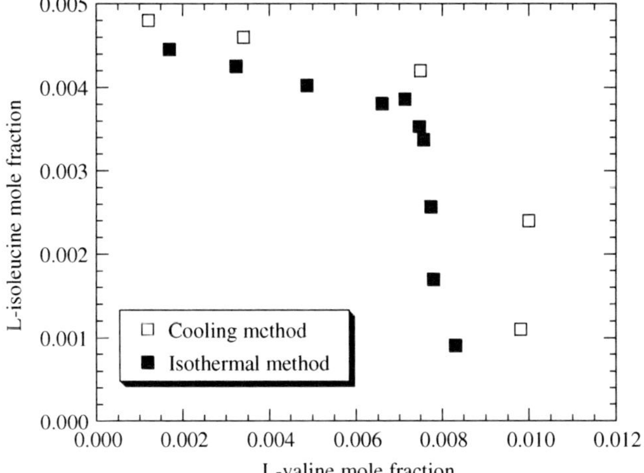

Figure 1 *Solubility in the L-isoleucine+L-valine+water system at 298 K: comparison of the cooling and the isothermal experimental methods.*[32]

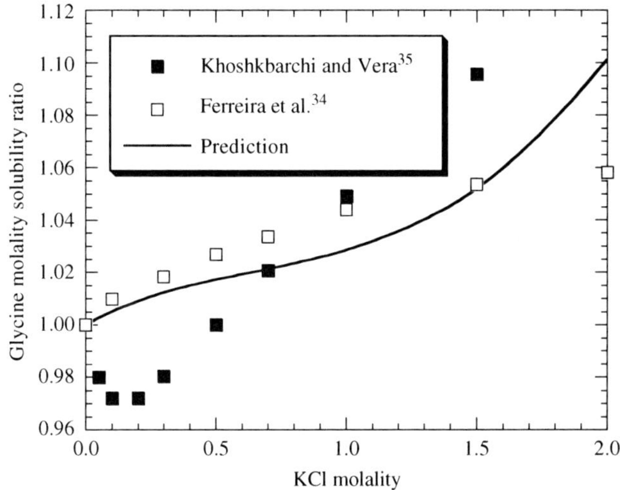

Figure 2 *Comparison of glycine solubility in water/KCl solutions at 298.15 K. The line represents prediction*[34] *using activity coefficient data only.*

precipitation of proteins. Figure 2 shows the effect of KCl on the glycine solubility in aqueous solution at 298.15 K.

The experimental results presented by Khoshkbarchi and Vera[35] and Ferreira *et al.*[34] are considerably different, which, once more, stresses the need of careful experimental planning. For that particular system, Khoshkbarchi and Vera[35] applied an equation based on the perturbation theory to correlate their activity coefficient data, but concluded they had to use an empirical correction to explain the solubility behavior, based on the observed effect of KCl on the crystallographic form of the AA. Alternatively, Ferreira *et al.*[34] correlated the same activity coefficient data, but with a modified form of the Pitzer–Simonson–Clegg equations,[36] and predicted the solubility assuming unchanged solid phase. The prediction curve is also included in Figure 2, suggesting a higher adequacy of their measured solubility data. Hamelink *et al.*,[37] in their studies about the effect of NaCl on the activity coefficients of antibiotics could not find a difference in the crystallographic structure of the solid phase formed by precipitation from electrolyte antibiotic solutions to explain the solubility behavior.

These studies are all important for a proper understanding of complex systems involving biomolecules, and might be useful for the investigation on protein solubility and crystallization. These questions are correlated, and rather complex, since protein crystallization/solubility depends on many factors such as pH, ionic strength, salt or protein type, temperature, surface hydrophobicity, and charge distribution, *etc.*, but extremely useful to identify, rationally, the optimal conditions for protein crystallization, reducing considerably the cost of a trial and error process. One interesting new effective predictive tool for protein crystallization is the introduction of the "crystallization slot" concept, which associates protein crystallization with the osmotic second virial coefficient (SVC–B_{22}). It can be briefly summarized in the following conservative way; while protein crystallization is very difficult for positive SVC values, it is favorable for negative values up to -10^{-3} mol ml g^{-2}, but do not guarantee successful crystal growth.[38] Although SVC is a thermodynamic property of dilute protein solutions, Guo *et al.*[39] have shown experimentally that it is also correlated with protein solubility. Figure 3(a) shows the surprising results when plotting these two variables for aqueous solutions of lysozyme obtained at different pH, temperature, salt type and concentration.

The link between those experimental observations and theory has been carried out by Haas *et al.*,[43] who used two different protein interaction potentials, and Rupert *et al.*,[44] who derived a two-parameter correlation based on classical thermodynamics, to represent the relation between solubility and SVC changing composition, temperature, or pH. Experimental determination of SVC by different methods like static or dynamic light scattering,[45,46] self-interaction,[38,41] or size-exclusion,[38] chromatography, can give, however, different values for the same protein under the same conditions. Figure 3(b) gives the SVC for lysozyme at different NaCl concentrations presenting considerable differences.[40–42] The subject is delicate, since anisotropy effects are much relevant,[47] for instance, the substitution of a single AA in a protein may introduce big changes in the SVC values measured. Therefore, in this active research area it will be fundamental to have the development of more reliable methods, and the extension of the conclusions for different proteins. Two different modeling approaches worth mention are the use of the UNIQUAC[48] equation to model protein solubility, and the neural network technology for protein crystallization, recently reviewed by DeLucas *et al.*[49] The lecture given by Prausnitz[50] on molecular thermodynamics for proteins in aqueous solution is highly recommended.

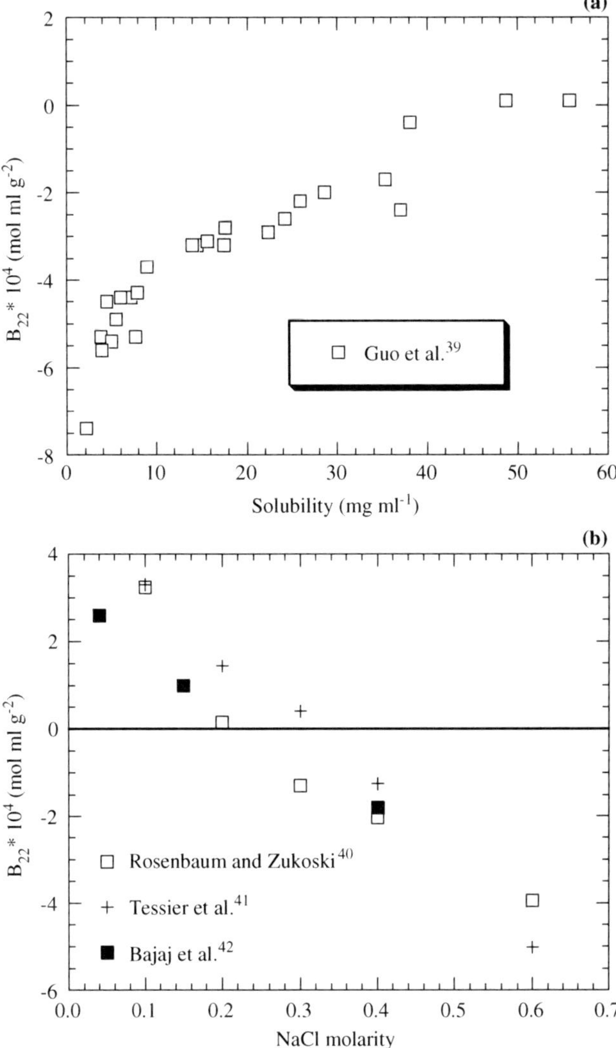

Figure 3 *(a) Experimental correlation between SVC and solubility for aqueous solutions of lysozyme; (b) Comparison of experimental results for SVC in aqueous lysozyme solutions at different NaCl molalities.*

20.4 Organic and Mixed Solvent Solubility

Either for extraction, crystallization, or drug formulation purposes the study of co-solvency is common in pharmaceutical companies. Similarly, for water solubility, several methods have been proposed to calculate, and predict, the solubility of biomolecules in organic or mixed solvent systems. One attractive approach is the so-called log-linear model;[11] it presents two specific co-solvent parameters, and as far as the substance P_{ow} and water solubility are known, the solubility of a drug can, in principle, be estimated for an aqueous mixed solvent system. It has, however, a major deficiency as it cannot predict, or even correlate, solubility in systems like water/ethanol with caffeine, that present maxima[51] over the whole solvent composition range. The application of group-contribution methods is an alternative, but many group interactions relevant for pharmaceutical compounds are missing. Thus, the MOSCED (Modified Separation of Cohesive Energy Density) developed by Lazzaroni *et al.*[52] is a good alternative since group interaction parameters are not needed. Using a large number of data for activity coefficients at infinite dilution (γ^{∞}), 5 parameters were correlated for each of the 133 solvents studied. The MOSCED parameters for a given drug can easily be obtained if a few binary solubility data (the authors suggest 5–8 data points in chemically diverse solvent set) are available. After those are used to calculate the γ^{∞}'s, and from their values the Wilson or UNIQUAC[16] interaction parameters can be obtained, making possible the calculation of the solubility in all the mixed solvent composition range. A major drawback, as explained before, is that the melting properties must be known, and most probably, for many solutes, the data used to obtain the MOSCED parameters are too far from infinite dilution conditions. Nevertheless, for 26 solutes studied, an average absolute deviation (AAD) of 24.9% was found in the correlation of 700 solubility data points. Another method, perhaps one of the most used in the pharmaceutical industry, is the regular solution theory,[16] where the solubility is a function of the solvent solubility parameter. Often, a maximum in solubility is found, which corresponds both to the ideal solubility, and to the equality between solvent and solute solubility parameters. Again, solute-melting properties must be available, and even if for some solutes, like morphine in different solvents, the prediction is of high quality, for an hetero-atomic compound, the inadequacy of the method can be extremely pronounced.[1] In fact, in some very good solvents, the solute solubility can exceed significantly the ideal solubility, which is totally impossible to predict with the model.

Avoiding some of the disadvantages pointed out in the previous paragraph, Abildskov and O'Connell[27] developed an ingenious reference solvent methodology. It involves the selection of a solvent, the "optimal solvent", which allows the calculation of the solute solubility in another solvent so long as the solubility in the optimal solvent and a predictive activity coefficient model, are available. In practice, the optimal solvent is found by a trial and error procedure, minimizing the difference (for a set of solvents) between the experimental solubility in a given solvent, and that calculated for the same solvent using the reference solvent approach. The UNIFAC[16] method was selected to calculate the activity coefficients, and for cases where the interaction parameters are unknown, a sensitivity analysis in terms of the more relevant parameters is suggested, reducing considerably the experimental measurements needed. The results are really promising except, perhaps, when the solubility is very high. Extensions for mixed solvent systems,[53] and the inclusion of the temperature influence on the solubility temperature dependency[54] were recently proposed.

For the special case of amino acids, Orella and Kirwan[55] first suggested the use of the excess solubility approach to correlate the solubility of several amino acids in water/propanol and water/isopropanol mixtures with the Wilson model, obtaining an average relative deviation (ARD) of about 15.3%. Following on, Gude *et al.*,[56,57] used the same approach, but combining the Flory–Huggins (FH) theory with a Margules residual expression. Their method is very simple and attractive since the authors claim the use of a unique specific Margules parameter for each amino acid in all aqueous alkanol solutions, which allows a straightforward prediction of amino acid solubilities in alkanol/water solvents systems. However, applying their method to the description of the solubility of amino acids in water/methanol solvents, which are usually the easiest to correlate, the ARD found was 27.7%. To the best of our knowledge, the work by Ferreira *et al.*,[58] is the more comprehensive in this subject. Within the framework of the excess solubility approach, the NRTL model was applied for the correlation of the solubility of a large number of amino acids in several alkanol/water solvents. The temperature effect was included for some specific amino acids, and some predictions were made. The ARDs were 8.4% for correlation and 15% for predictions. Figure 4(a) compares the results achieved by Gude *et al.*[56] using the FH+Margules approach, with the NRTL results obtained by Ferreira *et al.*,[58] for the ratio between the solubility of the AA in the mixed solvent to that in pure water (relative solubility). A better agreement was found with the NRTL model for the solubility of the AA in aqueous 1-butanol solutions. Figure 4(b) shows the very good results for the prediction of glycine solubility in aqueous ethanol solutions at two different temperatures outside the temperature range used in the correlation.

Regarding carbohydrates, the increasing interest for food technology applications caused a great demand for predictive methods for both aqueous and mixed solvent solutions. In the last decade two kinds of approaches were proposed in the literature: molecular models and group-contribution methods.[59] Two modified UNIQUAC equations are available: the model presented by Peres and Macedo,[60] that uses fewer parameters for each sugar–water pair and adopts the symmetric convention, and allows a straightforward extension to mixed solvent systems. This is not possible with the other model suggested by Catté *et al.*[61] These authors chose the unsymmetric convention for the activity coefficients calculations. The major trend in recent modeling research is, however, based on the group-contribution methodology.

Different UNIFAC-based models are available for the prediction of solubilities in sugar solutions.[59] Some of the UNIFAC parameters have even been predicted theoretically with methods of molecular mechanics.[62] The drawback of these models is the lack of accuracy at very high sugar concentrations (> 90%wt), as has recently been pointed out.[63] The reason for this lies in the fact that the majority of the data available does not cover this range of composition. To improve predictions of solubility in sugar solutions at these ranges of composition, new data were measured and a four-suffix Margules equation with temperature dependent parameters was presented in the literature,[63] as well as a new physical–chemical model.[64] This model takes account

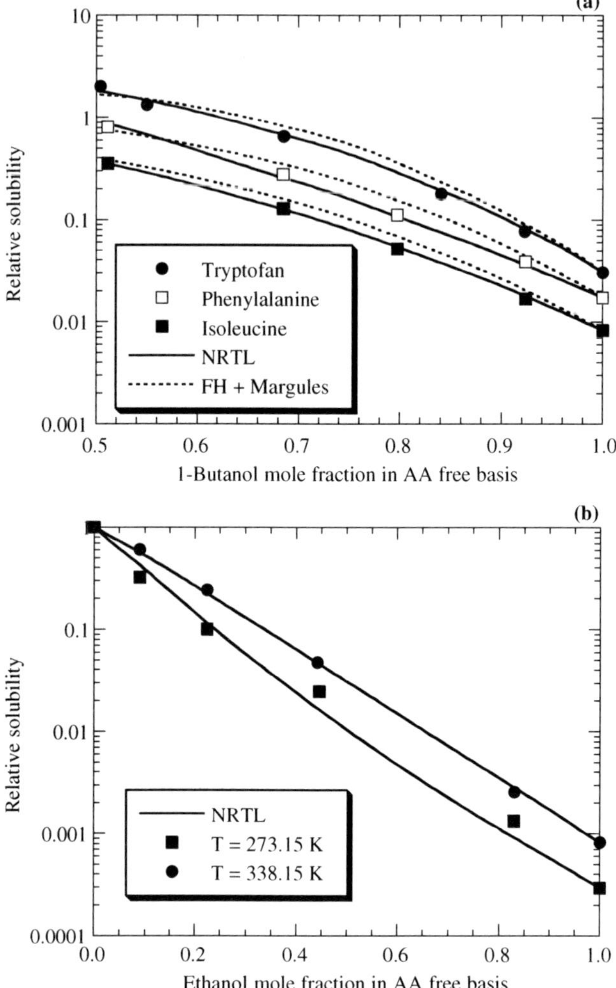

Figure 4 *(a) Relative solubilities of amino acids in water/1-butanol solutions at 298.15 K: comparison between the NRTL[58] and FH+Margules[56] models; (b) NRTL[58] predictions for the relative solubilities of glycine in water/ethanol solutions.*

for the hydration equilibrium of carbohydrates with the formation of carbohydrate *n*-water molecules and uses a UNIFAC model to describe the physical interactions. Although these very recent studies try to correct deficiencies found in other UNIFAC-based models, it is recognized[63] that the A-UNIFAC method developed by Ferreira *et al.*[65] is the tool with stronger theoretical foundations, allowing for a better capacity in predictive calculations. It incorporates a specific association term, which considers hydrogen bonding for sugar, water and other solvents molecules.

This section cannot be concluded without a brief mention of the innovative features of the NRTL-SAC model proposed by Chen and Song.[66] In this model the liquid non-idealities are described in terms of three types of conceptual segments of the molecules; hydrophobic, polar, and hydrophilic. Using reference substances for each type of segment, (hexane, water, and acetonitrile, respectively) an extensive binary VLE and LLE database, focused on the 62 solvents most used in the pharmaceutical industry, was used to estimate the number of conceptual segments required in each solvent. Following on, with a few selected solubility data values of the target solute, its number of conceptual segments can be calculated readily, and the solubility prediction in other solvents and mixed solvents is straightforward. As it requires some well-chosen data, NRTL-SAC is, like MOSCED and the reference solvent method, a hybrid-data estimation method that should be encouraged.[67] Its ability to model complex pharmaceuticals organic electrolytes has been already demonstrated,[68] and the potentialities to describe solubility of other types of solutes seems immense.

20.5 Liquid–Liquid Solubility

In the previous sections the importance of P_{ow} as a fundamental parameter for the estimation of solubilities in a variety of solvents has been stressed. Thermodynamics and extra-thermodynamics aspects of partitioning as well as its experimental and calculating methods were recently carefully reviewed by Sangster.[69] Owing to the uncertainty in the experimental P_{ow} values, Sangster also presents a list of recommended values for about 500 organic compounds. Thus, only the review by Derawi *et al.*[70] on group-contribution methods is briefly focused. Five different UNIFAC-based methods were compared, and the WATER UNIFAC,[71] and UNIFAC LLE[72] were recommended. These models, however, present a small number of interaction parameters available, and this inhibits their application for some functional groups like amines. For highly hydrophobic compounds, all the UNIFAC models underestimate P_{ow}, and generally, for AA, their derivatives, and sugars, P_{ow} is overestimated. The authors believe that for multifunctional compounds the group-contribution concept has limited capacity for further developments, and also that the atom/fragment correlation (AFC) method[73] showed superior performance in all cases studied. This method, similarly to the one proposed by Sun[9] (Section 20.3), allows the calculation of both P_{ow} and solubility by building a substance from atom descriptors. In the AFC method MLR was applied to derive fragment coefficients and correction factors using 2,473 P_{ow} in the training set, and around 10,600 for the validation of the method. The results seem really remarkable as it is possible to take into account steric interactions, hydrogen bondings, and even for zwitterionic species like ampicillin, amoxycillin, or peptides, values of P_{ow} can be estimated. A free online interactive demonstration to calculate P_{ow} is available at http://www.syrres.com/esc/kowdemo.htm.

Despite the increase and progresses achieved in the research work on ATPS, so far the studies are rather scattered, making the knowledge of the mechanisms of solute partitioning, limited. This is probably one of the main reasons for the reluctance in its commercial exploitation.[74] Traditionally, protein partitioning has been studied in polyethyleneglycol (PEG)/dextran or PEG/(phosphate or sulfate) salt, and the factors to consider, beyond those mentioned earlier for protein crystallization, must now include some characteristics of the polymer(s). The implementation of general rules to choose the best ATPS and the best operating conditions for a given separation, will make practical applications simpler. However, making those rules accessible depends much on how these different factors are understood. Recently, some interesting attempts have been made: Lin *et al.*[75] studied the influence of polymer concentration and molecular weight; Andrews *et al.*[76] focused on the protein charge and surface hydrophobicity, which was also done by Tubio et al.[77] However, no general trend was found. Even if a relationship between the hydrophobic character of the partitioned substance and its partitioning coefficient was found the general picture is, when studying polymer molecular weight effects the conclusions are limited to certain proteins, and studying the protein surface hydrophobicity effects, the results are restricted to certain values of the polymer molecular weight.

Though much more experimental work is needed, the application of molecular thermodynamics to this kind of problems must have the highest priority. In the recent past, several different approaches have been proposed concerning protein, peptides, and AA partition in ATPS. This was recently reviewed briefly by Jiang and Prausnitz,[78] who also derived a model that takes into account, successfully, the different partitioning behavior of native and denatured proteins. One of the most recent studies on protein partitioning, and perhaps the most comprehensive, is due to Madeira *et al.*[79] Their modified Wilson model, based on the lattice theory and the two-fluid theory, was successively applied to the representation of electrolyte solutions, water activity in aqueous polymer solutions, and polymer/polymer or polymer/salt ATPS. A Debye-Hückel term was included to take into consideration the long-range nature of the electrostatic forces in solution, and the authors end up with a model where only the interactions involving proteins are needed to calculate protein partitioning. To simplify, Madeira *et al.*[79] fixed those at zero, and calculated the partition of four different proteins in Na_2SO_4/PEG6000 and K_2HPO_4/PEG6000 by adjusting the protein net charge. Globally, the results may be considered very reasonable even if in some cases large discrepancies were found between the experimental and the calculated net charge. That is not the case for the partitioning behavior of lysozyme in K_2HPO_4/PEG6000 aqueous system at 298.15 K shown in Figure 5. Here the published experimental value for the net charge is two, and it produces much higher deviations on the calculated partition coefficient than that obtained using the fitted value of four for the net charge. The complexity of the problem and the lack of data remain as the major reasons for the development of more efficient predictive tools for protein partitioning on ATPS. However, some useful insights from protein crystallization must also be considered, and as it is expected that ATPS will be extend into food and cosmetic industries,[74] these problems will continue to draw attention in the near future.

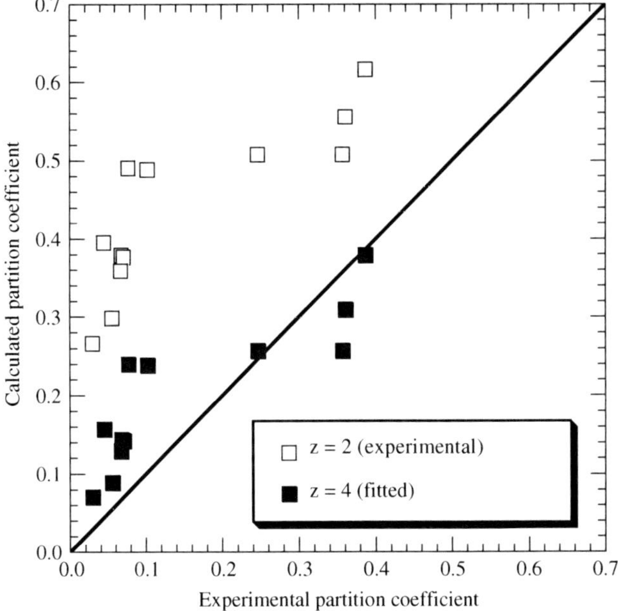

Figure 5 *Influence of the protein net charge (z) on the lysozyme partition coefficient calculation, by a modified Wilson model,*[79] *in* $K_2HPO_4/PEG6000$ *aqueous system at 298.15 K.*

20.6 Solubility in Supercritical Fluids

Contrary to the work on the ATPS, most of the studies on SFE started in the area of food technology. In fact, around 125 industrial scale SFE units are in operation,[80] and some of those applications were reviewed by Knox.[81] The studies are now spreading into the area of drug processing that is currently a very popular research area, namely, on the purification, crystallization, or micronization of pharmaceuticals. In this context, a review on techniques such as rapid expansion supercritical solution, or gas anti-solvent system is given elsewhere.[80,82]

The large majority of the studies concern the solubility of a solute in carbon dioxide. Some supercritical measurements on other systems such as the solubility of solid mixtures and the solubility in solvents other than carbon dioxide (and co-solvency) are also available.[83–85] To correlate solubilities, empirical equations have usually been applied, and the Chrastil equation[86] is one of the most often used. Several other empirical equations have been proposed through the years, and some were recently compared for solute solubility in carbon dioxide by Jouyban *et al.*[87] Avoiding the difficulties of calculating some physicochemical properties, the authors only compared equations for which the independent variables are readily accessible, *e.g.*, temperature, pressure, and pure solvent density promoting, in this way, their usefulness. A six parameter equation, developed by the authors, showed the best performance with an AAD of 21.4% for the 106 systems compiled in their database. This is comparable to the experimental uncertainty.

Several different equations of state (EoS) have also been applied, but a major difficulty arises from the need to know solute critical properties, vapor pressure, and density. Unfortunately, for many substances those are impossible to measure because the solute decomposes, and estimating methods must be applied for their evaluation. However, different methods give, naturally, different values for those properties, which may have a lot of influence in the correlation abilities of the EoS and, what is worse, can produce poor and sometimes impossible results.[88] Furthermore, relatively small variations in the properties can cause large differences in the predicted solubilities,[89] and so, a lot of caution must be taken in the choice of methods used. Another issue that must be considered carefully when modeling solubilities is the stability of the results. In a very enlighten work, Xu *et al.*[90] developed a strategy, and gave some good examples, about the need of considering the iso-fugacity condition simultaneously with a check on global thermodynamic phase stability by applying tests such as tangent plane analysis and global minimization methodologies. Cubic EoS like Soave–Redlich–Kwong and Peng–Robinson are, surely, the most used, but for rigorous calculations of solubilities in these type of systems much more work is imperative. Nevertheless, even if accurate general conclusions are not possible at this time, cubic EoS that uses free energy models in its parameters and non-quadratic mixing rules, with interaction parameters in the volume constants, give the best results.[91] Finally, taking into consideration the results already achieved with the group-contribution associating EoS,[92] the research on the potential of association fluid theories is also highly recommended.

20.7 Conclusions

A global overview about current solubility issues for food, pharmaceutical, and cosmetic industries has been given. Great progress has been achieved for solute solubility in water and organic solvents as well as for water–octanol partition coefficients, but the potentialities of some very recent models and methods, *e.g.*, reference solvent methodology or NRTL-SAC model, should be extensively explored. Nevertheless, constant evolution in those industries will stress the need for new measurements and advances for innovative experimental techniques. The development of a solubility database and a measurement strategy, perhaps, as suggested by Kolář *et al.*,[1] is highly recommended, but applications of relatively novel compounds like ionic liquids or cyclodextrins should also be taken into account for the development of new processes. A very interesting progress on protein crystallization has also been achieved with the "crystallization slot" concept that should be applied to several different systems. Understanding the behavior of simple molecules like AA and small peptides in aqueous electrolyte solutions can also be useful for further developments. As far as ATPS are concerned some interesting studies have briefly been discussed.

However, the development of alternative ATPS as well as much more informative models capable of explaining mechanisms under protein partitioning is fundamental to make the technique attractive to industries. That is also an issue for simulation SFE processes, but EoS for associating fluids might be a very useful tool. In fact, some particular difficulties pointed out for solubility modeling in supercritical fluids make it a hard task, and an analysis for methods to estimate solute properties must be taken into account. Finally, it is consensual that, generally, understanding solubility phenomena may benefit very much from molecular simulation data.

References

1. P. Kolář, J.-W. Shen, A. Tsuboi and T. Ishikawa, *Fluid Phase Equilib.*, 2002, **194–197**, 771.
2. G.T. Hefter and R.P.T Tomkins, (eds), *The Experimental Determination of Solubilities*, Wiley, Chichester, 2003.
3. R.D. Noble and R. Agrawal, *Ind. Eng. Chem. Res.*, 2005, **44**, 2887.
4. S. Gupta and J.D. Olson, *Ind. Eng. Chem. Res.*, 2003, **42**, 6359.
5. W. Leuchtenberger, K. Huthmacher and K. Drauz, *Appl. Microbiol. Biotechnol.*, 2005, **69**, 1.
6. K. Araki and T. Ozeki, Amino acids, *Kirk-Othmer Encyclopedia of Chemical Technology*, Wiley, Chichester, 2003.
7. A. Blasko, A. Leahy-Dios, W.O. Nelson, S.A. Austin, R.B. Killion, G.C. Visor and I.J. Massey, *Monatsh. Chem.*, 2001, **132**, 789.
8. B. C. Hancock, P. York and R. C. Rowe, *Int. J. Pharm.*, 1997, **148**, 1.
9. H. Sun, *J. Chem. Inf. Comput. Sci.*, 2004, **44**, 748.
10. M.H. Abraham, J.A. Platts, A. Hersey, A.J. Leo and R.W. Taft, *J. Pharm. Sci.*, 1999, **88**, 670.
11. J.W. Millard, F.A. Alvarez-Núñez and S.H. Yalkowsky, *Int. J. Pharm.*, 2002, **245**, 153.
12. R. Hartono, G.A. Mansoori and A. Suwono, *Chem. Eng. Sci.*, 2001, **56**, 6949;and references therein.
13. F. J. Señoráns, A. Ruiz-Rodríguez, E. Ibáñez, J. Tabera and G. Reglero, *J.Supercrit. Fluids*, 2003, **26**, 129.
14. Z. Shen, V. Mishra, B. Imison, M. Palmer and R. Fairclough, *J. Agric. Food Chem.*, 2002, **50**, 154.
15. S. Bruin, *Fluid Phase Equilib.*, 1999, **158–160**, 657.
16. J.M. Prausnitz, R.N. Lichtenthaler and E.G. Azevedo, *Molecular Thermodynamics of Fluid-Phase Equilibriua*, 3rd edn, Prentice-Hall, Upper Saddle River, NJ, 1999.
17. T. Morwood, *CAPE, Eureka Project*, 2001, 2311.
18. P.B. Nielsen, A. Müllertz, T. Norling and H.G. Kristensen, *Int. J. Pharm.*, 2001, **222**, 217.
19. D. Westerman, R.C.D. Santos, J.A. Bosley, J.S. Rogers and B. Al-Duri, *J. Supercrit. Fluids*, 2006, **37**, 38–52.
20. P. Simamora, J.M. Alvarez and S.H. Yalkowsky, *Int. J. Pharm.*, 2001, **213**, 25.

21. F. Lewiner, G. Févotte, J.P. Klein and F. Puel, *Ind. Eng. Chem. Res.*, 2002, **41**, 1321.
22. L. M. Land, P. Li and P.M. Bummer, *Pharm. Res.*, 2005, **22**, 784.
23. J. M. Pollard, A. J. Shi and K. E. Göklen, *J. Chem. Eng. Data*, 2006, **51**, 230.
24. Y. Zhu, D. Youssef, C. Porte, A. Rannou, M.P. Delplancke-Ogletree and B.L.M. Lung-Somarriba, *J. Cryst. Growth*, 2003, **257**, 370.
25. R.H. Muller and C.M. Keck, *J. Biotechnol.*, 2004, **113**, 151.
26. W.L. Jorgensen and E.M. Duffy, *Adv. Drug Deliv. Rev.*, 2002, **54**, 355.
27. J. Abildskov and J.P. O'Connell, *Ind. Eng. Chem. Res.*, 2003, **42**, 5622.
28. J. Huuskonen, *Comb. Chem. High Throughput Screening*, 2001, **4**, 311.
29. J.S. Delaney, *Drug Discov. Today*, 2005, **10**, 289.
30. K.R. Hoover, W.E. Acree, Jr. and M.H. Abraham, *J. Solution Chem.*, 2005, **34**, 1121.
31. J.C. Givand, A.S. Teja and R.W. Rousseau, *AIChE J.*, 2001, **47**, 2705.
32. I. Kurosawa, A.S. Teja and R.W. Rousseau, *Ind. Eng. Chem. Res.*, 2005, **44**, 3284.
33. I. Kurosawa, A.S. Teja and R.W. Rousseau, *Fluid Phase Equilib.*, 2004, **224**, 245.
34. L.A. Ferreira, E.A. Macedo, E.A. and S.P. Pinho, *Ind. Eng. Chem. Res.*, 2005, **44**, 8892; and references therein.
35. M.K. Khoshkbarchi and J.H. Vera, *Ind. Eng. Chem. Res.*, 1997, **36**, 2445.
36. S.L. Clegg and K.S. Pitzer, *J. Chem. Phys.*, 1992, **96**, 3513.
37. J.M. Hamelink, E.S.J. Rudolph, L.A.M. van der Wielen and J.H. Vera, *Biophys. Chemist.*, 2002, **95**, 97.
38. T. Ahamed, M. Ottens, G.W.K. van Dedem and L.A.M. van der Wielen, *J. Chromatogr. A*, 2005, **1089**, 111.
39. B. Guo, S. Kao, H. McDonald, A. Asanov, L.L. Combs and W.W. Wilson, *J. Cryst. Growth*, 1999, **196**, 424; and references therein.
40. D. F. Rosenbaum and C. F. Zukoski, *J. Cryst. Growth*, 1996, **169**, 752.
41. P.M. Tessier, A.M. Lenhoff and S.I. Sandler, *Biophys. J.*, 2002, **82**, 1620.
42. H. Bajaj, V.K. Sharma and D.S. Kalonia, *Biophys. J.*, 2004, **87**, 4048.
43. C. Haas, J. Drenth and W.W. Wilson, *J. Phys. Chem. B*, 1999, **103**, 2808.
44. S. Ruppert, S.I. Sandler and A.M. Lenhoff, *Biotechnol. Prog.*, 2001, **17**, 182.
45. R.A. Curtis, J.M. Prausnitz and H.W. Blanch, *Biotechnol. Bioeng.*, 1998, **57**, 11.
46. W. Liu, T. Cellmer, D. Keerl, J.M. Prausnitz and H.W. Blanch, *Biotechnol. Bioeng.*, 2005, **90**, 482.
47. R. A Curtis and L. Lue, *Chem. Eng. Sci.*, 2006, **61**, 907.
48. S. M. Agena, M. L. Pusey and I. D. L. Bogle, *Biotechnol. Bioeng.*, 1999, **64**, 144.
49. L.J. DeLucas, D. Hamrick, L. Cosenza, L. Nagy, D. McCombs, T. Bray, A. Chait, B. Stoops, A. Belgovskiy, W.W. Wilson, M. Parham and N. Chernov, *Progr. Biophys. Mol. Biol.*, 2005, **88**, 285.
50. J.M. Prausnitz, *Pure Appl. Chem.*, 2003, **75**, 859.
51. P. Bustamante, J. Navarro, S. Romero and B. Escalera, *J. Pharm. Sci.*, 2001, **91**, 874.

52. M.J. Lazzaroni, D. Bush, C.A. Eckert, T.C. Frank, S. Gupta and J.D. Olson, *Ind. Eng. Chem. Res.*, 2005, **44**, 4075.
53. J. Abildskov and J.P. O'Connell, *Mol. Simul.*, 2004, **30**, 367.
54. J. Abildskov and J.P. O'Connell, *Fluid Phase Equilib.*, 2005, **228**, 395.
55. C.J. Orella and D.J. Kirwan, *Ind. Eng. Chem. Res.*, 1991, **30**, 1040.
56. M.T. Gude, H.H.J. Meuwissen, L.A.M. van der Wielen and K. Ch. A.M. Luyben, *Ind. Eng. Chem. Res.*, 1996, **35**, 4700.
57. M.T. Gude, L.A.M. van der Wielen and K. Ch. A.M. Luyben, *Fluid Phase Equilib.*, 1996, **116**, 110.
58. L.A. Ferreira, E.A. Macedo and S.P. Pinho, *Chem. Eng. Sci.*, 2004, **59**, 3117; and references therein.
59. E.A. Macedo, *Pure Appl. Chem.*, 2005, **77**, 559; and references therein.
60. A.M. Peres and E.A. Macedo, *Fluid Phase Equilib.*, 1996, **123**, 71.
61. M. Catté, C.G. Dussap, C. Achard and J.B. Gros, *Fluid Phase Equilib.*, 1994, **96**, 33.
62. S.Ó. Jónsdóttir and P. Rasmussen, *Fluid Phase Equilib.*, 1999, **158–160**, 411.
63. M. Starzak and M. Mathlouthi, *Food Chem.*, 2006, **96**, 346.
64. L.B. Gaïda, G.G. Dussap and J.B. Gros, *Food Chem.*, 2006, **96**, 387.
65. O. Ferreira, E.A. Brignole and E.A. Macedo, *Ind. Eng. Chem. Res.*, 2003, **42**, 6212.
66. C.-C. Chen and Y. Song, *Ind. Eng. Chem. Res.*, 2004, **43**, 8354.
67. P. M. Mathias, *Fluid Phase Equilib.*, 2005, **228–229**, 49.
68. C.-C. Chen and Y. Song, *Ind. Eng. Chem. Res.*, 2005, **44**, 8909–8921.
69. J. Sangster, *Octanol-Water Partition Coefficients: Fundamentals and Physical Chemistry*, Wiley, Chichester, 1997.
70. S.O. Derawi, G.M. Kontogeorgis and E.H. Stenby, *Ind. Eng. Chem. Res.*, 2001, **40**, 434.
71. F. Chen, J. Holten-Andersen and H. Tyle, *Chemosphere*, 1993, **26**, 1325.
72. T. Magnussen, P. Rasmussen and A. Fredenslund, *Ind. Eng. Chem. Process Des. Dev.*, 1981, **20**, 331.
73. W.M. Meylan and P.H. Howard, *Perspect. Drug Discov. Des.*, 2000, **19**, 67.
74. M. Rito-Palomares, *J. Chromatogr. B*, 2004, **807**, 3.
75. D.-Q. Lin, Y.-T. Wu, L.-H. Mei, Z.-Q. Zhu and S.-J. Yao, *Chem. Eng. Sci.*, 2003, **58**, 2963.
76. B.A. Andrews, A.S. Schmidt and J.A. Asenjo, *Biotechnol. Bioeng.*, 2005, **90**, 380.
77. G. Tubio, B. Nerli and G. Picó, *J. Chromatogr. B*, 2004, **799**, 293.
78. J. Jiang and J.M. Prausnitz, *J. Phys. Chem. B*, 2000, **104**, 7197; and references therein.
79. P.P. Madeira, X.Xu, J.A. Teixeira and E.A. Macedo, *Biochem. Eng. J.*, 2005, **24**, 147; and references therein.
80. M. Sihvonen, E. Järvenpää, V. Hietaniemi and R. Huopalahti, *Trends Food Sci. Technol.*, 1999, **10**, 217.
81. D. E. Knox, *Pure Appl. Chem.*, 2005, **77**, 513.
82. F. Dehghani and N. R. Foster, *Curr. Opin. Solid State Mater. Sci.*, 2003, **7**, 363.

83. M. Christov and R. Dohrn, *Fluid Phase Equilib.*, 2002, **202**, 153.
84. F.P. Lucien and N.R. Foster, *J. Supercrit. Fluids.*, 2000, **17**, 111.
85. W.H. Hauthal, *Chemosphere*, 2001, **43**, 123.
86. J. Chrastil, *J. Phys. Chem.*, 1982, **86**, 3016.
87. A. Jouyban, H.-K. Chan and N. R. Foster, *J. Supercrit. Fluids*, 2002, **24**, 19.
88. P. Coimbra, C.M.M. Duarte and H.C. Sousa, *Fluid Phase Equilib.*, 2006, **239**, 188.
89. G.I. Burgos-Solórzano, J.F. Brennecke and M.A. Stadtherr, *Fluid Phase Equilib.*, 2004, **220**, 57.
90. G. Xu, A.M. Scurto, M. Castier, J.F. Brennecke and M.A. Stadtherr, *Ind. Eng. Chem. Res.*, 2000, **39**, 1624.
91. J.O. Valderrama, *Ind. Eng. Chem. Res.*, 2003, **42**, 1603.
92. T. Fornari, A. Chafer, R.P. Stateva and G. Reglero, *Ind. Eng. Chem. Res.*, 2005, **44**, 8147.

CHAPTER 21

Solubility of Solids in Radioactive Waste Repositories

WOLFGANG HUMMEL

Laboratory for Waste Management, Paul Scherrer Institut, 5232 Villigen PSI, Switzerland

21.1 Introduction

Worldwide a significant amount of nuclear waste exists today, and will continue to arise in the future. This waste stems from a range of sources, including electricity production in nuclear power plants and applications of radioactive substances in medicine, industry, and research. A key principle of radioactive waste management is that the waste must be disposed of in such a way that the safety of man and the environment is ensured. It is widely accepted that one possibility to meet this obligation is to emplace the waste in a carefully sited and well-designed geological repository.

Internationally, a large number of nuclear waste repositories for low-level waste (LLW), low- and intermediate-level waste (L/ILW), and intermediate-level waste (ILW) have been in operation for many years; for example in Finland, France, Germany, Japan, Spain, Sweden, the UK, and the USA. Mainly for technical reasons, the situation is different for spent fuel (SF) and vitrified high-level waste (HLW) from reprocessing of SF. These waste types are currently in interim storage facilities to allow the radiogenic heat production to decline to such levels that the waste can be disposed of in deep geological repositories so that temperatures stay below specification limits set to ensure the good performance of the engineered barriers (specifically bentonite, a natural clay-based material that is foreseen in many SF/HLW repository concepts as a buffer material between the waste canisters and the host rock). Typical required minimal interim storage times are in the order of 40 years. Even though there is no repository for SF or HLW in operation yet, detailed concepts have been developed and refined over the last years, and implementation is well under way in several countries.

A few selected examples for different repository concepts for SF and HLW are given in Table 1. It is beyond the scope of this article to discuss these in

Table 1 *Examples for different repository concepts for SF/HLW*

Country	*Organisation*	*Waste*	*Canister*	*Buffer*	*Host rock*	*Concept*	*References*
Belgium	ONDRAF	SF/HLW	Steel	Bentonite	Boom Clay	Horizontal emplacement in tubes	1
Finland	POSIVA	SF	Copper with steel insert	Bentonite	Crystalline basement	Individual vertical boreholes in the floor of emplacement tunnels	2
France	ANDRA	HLW	Steel	Bentonite	Callovo-oxfordian clay	Short horizontal tunnels starting from main gallery	3
Japan	JNC	HLW	Steel	Bentonite-sand	2 host rocks considered: "hard"/"soft"	2 concepts considered: horizontal/vertical	4
Sweden	SKB	SF	Copper with steel insert	Bentonite	Crystalline basement	Individual vertical boreholes in the floor of emplacement tunnels	5
Switzerland	Nagra	SF/HLW/ILW	Steel	SF/HLW: Bentonite ILW: Cement-based mortar	Opalinus clay	Horizontal emplacement	6
USA	DoE	SF/HLW	Steel	None	Tuff	Horizontal emplacement in host rock above the water table, use of drip shields	7

detail here; instead, the interested reader is referred to the references given in Table 1. A broader overview of national repository concepts for all waste types is given by, for example, Witherspoon and Bodvarsson.[8]

As an example, the possible layout for a Swiss deep geological repository for SF/HLW/ILW in Opalinus Clay is shown in Figure 1. The repository will consist of a tunnel system for SF and HLW and separate tunnels for ILW, with access *via* a ramp and/or vertical shaft, depending on the repository location and the host rock that is selected.

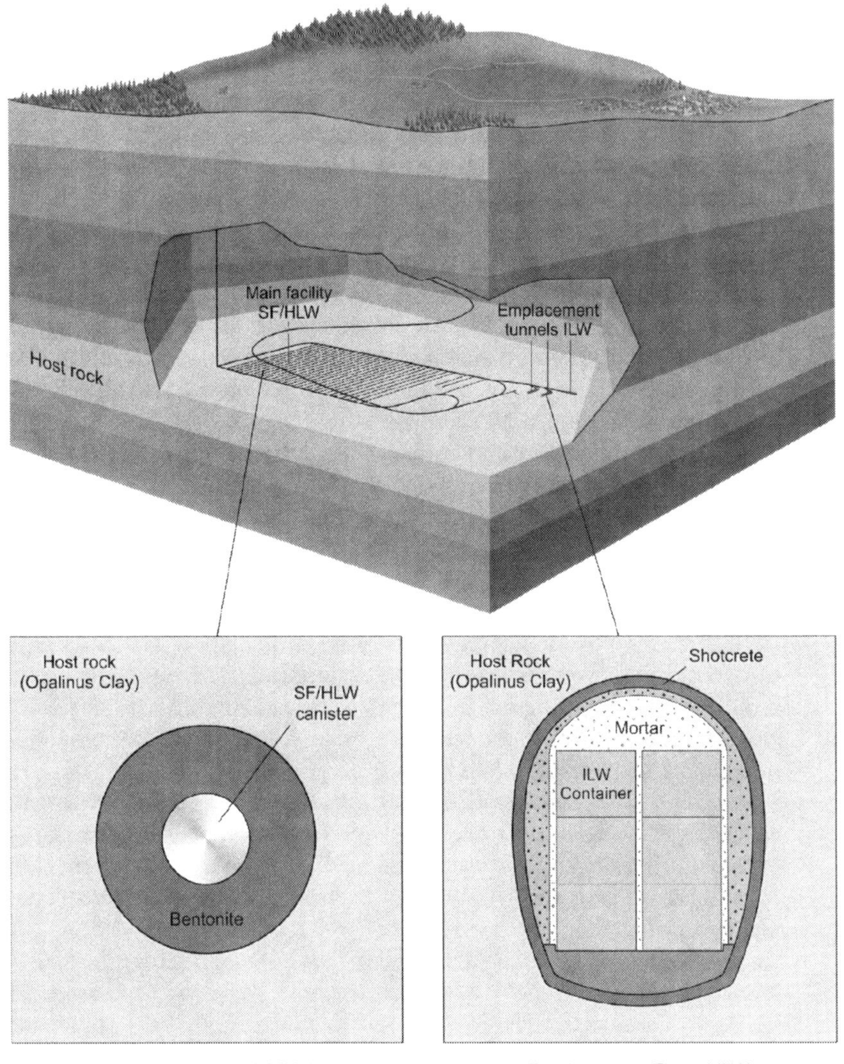

Figure 1 *Possible layout for a deep geological repository for SF/HLW/ILW in Opalinus Clay.*[6]

21.2 The Safety Concept of a Geological Repository

The safety concept explains how the repository system provides long-term safety. This usually involves a multi-barrier system where each barrier contributes to safety by providing multiple safety-relevant functions. For the case of vitrified HLW, this is illustrated in Figure 2, where the innermost barrier (the solidification matrix) is shown at the top and the outermost barrier (the geosphere with the host rock) is shown at the bottom. For each barrier, the safety-relevant functions are given, and it is indicated by which processes each function operates. Similar functional descriptions of the safety-barrier system for SF and long-lived ILW are given elsewhere.[6]

Such a repository system provides the following safety functions:

- Isolation of the waste from the human environment: the safety and security of the waste, including fissile material, is ensured by placing it in a repository located deep underground, with all access routes backfilled and sealed, thus isolating it from the human environment and reducing the likelihood of any undesirable intrusion and misapplication of the materials. Furthermore, the absence of any currently recognised and economically viable natural resources and the lack of conflict with future infrastructure projects that can be conceived at present reduce the likelihood of inadvertent human intrusion. Finally, appropriate siting ensures that the site is not prone to disruptive events and to processes unfavourable to long-term stability.
- Long-term confinement and radioactive decay within the disposal system: much of the activity initially present decays while the wastes are totally contained within the primary waste containers, particularly in the case of SF and HLW, for which the high-integrity steel canisters are expected to remain unbreached for at least 10 000 years. Even after the canisters are breached, the stability of the SF and HLW waste forms in the expected environment, the slowness of groundwater flow and a range of geochemical immobilisation and retardation processes ensure that radionuclides continue to be largely confined within the engineered barrier system and the immediately surrounding rock, so that further radioactive decay takes place.
- Attenuation of releases to the environment: although complete confinement cannot be provided over all relevant times for all radionuclides, release rates of radionuclides from the waste forms are low, particularly from the stable SF and HLW waste forms. Furthermore, a number of processes attenuate releases during transport towards the surface environment, and limit the concentrations of radionuclides in that environment. These include (a) radioactive decay during slow transport through the barrier provided by the host rock and (b) the spreading of released radionuclides in time and space by, for example, diffusion, hydrodynamic dispersion, and dilution.

Safety barrier system for vitrified HLW

Glass matrix (in steel flask)

- Confinement
  - Containment of radionuclides in glass
- Attenuation of releases
  - Low corrosion rate of glass

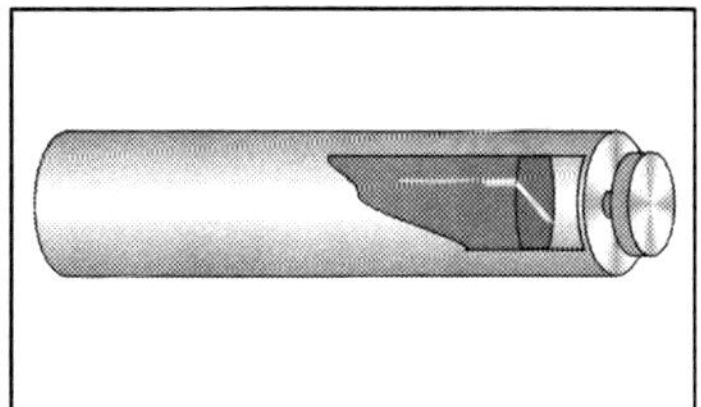

Steel canister

- Confinement
  - Prevents inflow of water and release of radionuclides from waste for several thousand years
- Attenuation of releases
  - Corrosion products act as reducing agent (giving low radionuclide solubilities)
  - Corrosion products take up radionuclides

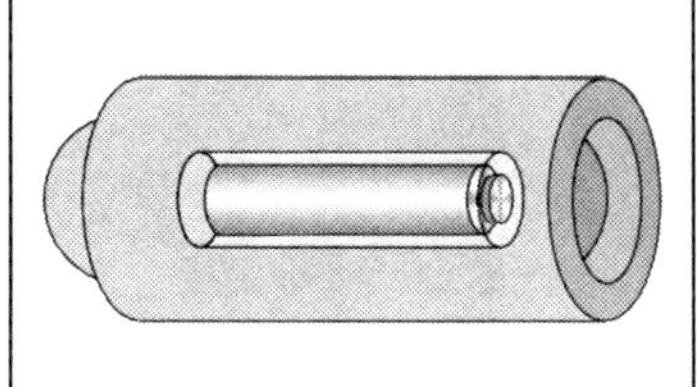

Bentonite backfill

- Confinement
  - Long resaturation time
  - Plasticity (self-sealing following physical disturbance)
- Attenuation of releases
  - Low solute transport rates (diffusion)
  - Retardation of radionuclide transport (sorption)
  - Low radionuclide solubility in pore water

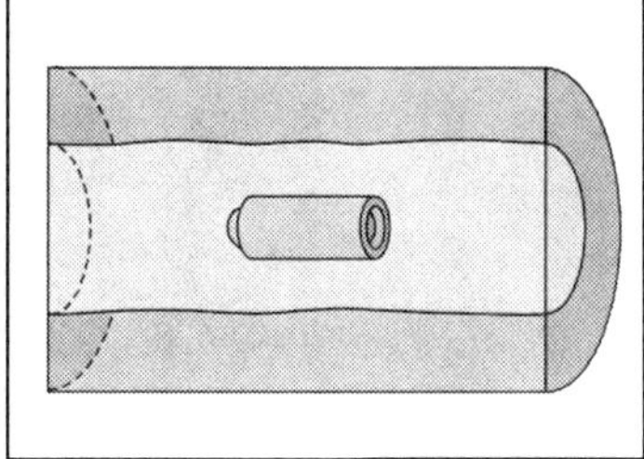

Geological barriers

Host rock

- Confinement
  - Absence of water-conducting features
  - Mechanical stability
- Attenuation of releases
  - Low groundwater flux
  - Retardation of radionuclide transport (sorption and colloid filtration)

Geosphere

- Confinement
  - Physical protection of the engineered barriers (e.g. from glacial erosion)
- Attenuation of releases
  - Retardation of radionuclide transport (sorption)
  - Dispersion

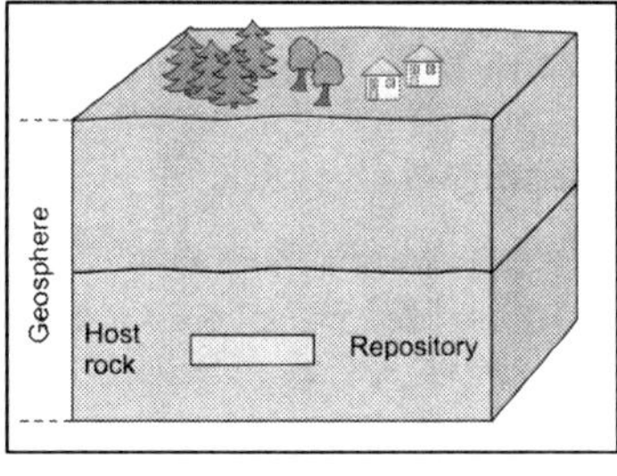

Figure 2 *The system of safety barriers and the multiple safety-relevant functions that each barrier provides in the case of vitrified HLW.*[6]

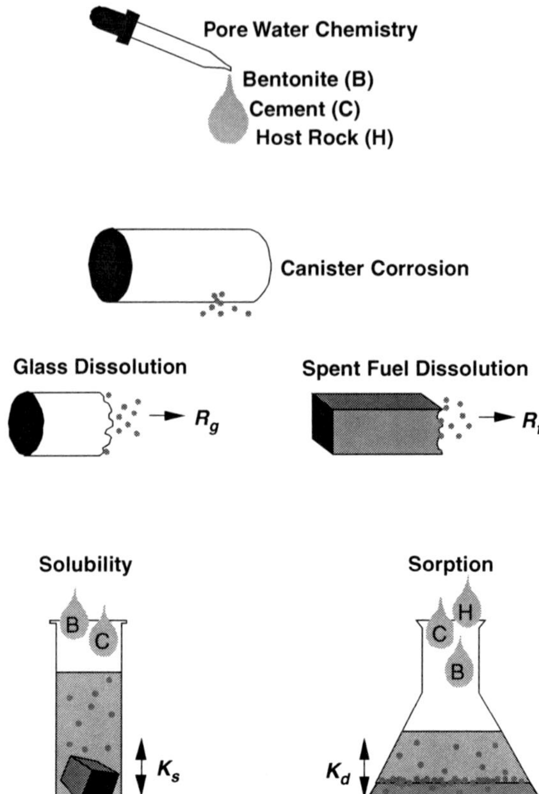

Figure 3 *Important geochemical parameters and processes for repository chemistry. The dissolution parameters R_g and R_f are case specific kinetic rate constants, whereas the elemental solubility limits K_s and the sorption coefficients K_d are element and case specific equilibrium constants.*

From the viewpoint of (geo)chemistry a number of important parameters and processes have been and still are topics of scientific studies relevant for repository safety. As shown as a graphical summary in Figure 3, these parameters and processes are:

(i) Porewater chemistry
(ii) Canister/waste package corrosion
(iii) SF/HLW/waste form dissolution with radionuclides entering the aqueous phase
(iv) Solubility limits for radionuclides in the near field and the geosphere
(v) Diffusion (+ advection) and sorption of radionuclides in the near field
(vi) Diffusion (+ advection) and sorption of radionuclides in the geosphere

In the following, some selected aspects concerning solubility limits are discussed in detail with specific examples.

21.3 Solubility of Solids in Repository Safety Assessments

An important feature of repository safety is the fact that radionuclides cannot be dissolved in unlimited quantities in the aqueous phase. Their maximum concentrations are limited by the precipitation of solid phases when the solution becomes oversaturated with respect to a certain radionuclide. These so-called elemental solubility limits, K_s, can be calculated by thermodynamic modelling assuming that in the long-term chemical equilibrium is reached between solids and the aqueous phase.

If for a given radionuclide a true thermodynamic equilibrium with a well-known solid is expected to establish, and all thermodynamic data related to this equilibrium solid phase are available, chemical equilibrium calculations straightforwardly result in an estimation of the maximum concentration of the radionuclide of interest in a specified pore fluid of a geological repository. However, all three aspects, the composition of the pore fluid, the relevant thermodynamic data, and the nature of the equilibrium solid phase, pose scientific challenges. These challenges are elucidated in the following sections.

21.4 What is the Composition of the Solution?

Potential repository sites are foreseen in geological formations with low groundwater flow. Specifically, clay-host rocks (see Table 1) are characterised by very low groundwater flow. This implies difficulties with respect to the prerequisite of all geochemical modelling for repository safety, *i.e.* the definition of the porewater chemistry. Whereas sampling and subsequent chemical analyses of surface water, water from wells or groundwater in deep aquifers are routine procedures, the sampling and analysis of porewater in material with a low hydraulic conductivity is an exceedingly difficult task.

If, for example, the water "flow" in the sampling site of a dense clay formation is a few drops per day, there is a high probability that the small quantity of water obtained after several weeks of sampling is no longer identical with the original porewater. The original porewater perhaps is not in thermodynamic equilibrium with carbon dioxide of the atmosphere and hence, either some outgassing of CO_2 or dissolution of atmospheric CO_2 in the water may occur during the sampling. In both cases not only the carbonate concentration in the sampled water changes but also the pH of the water. These and other sampling artefacts may change the water chemistry of the finally analysed sample considerably. Similar artefacts are encountered in the attempt to obtain porewater by high pressure squeezing of rock samples or bentonite backfill material. The latter is foreseen in most SF/HLW repository concepts (Table 1).

In summary, the definition of porewater chemistry involves the scrutinising of field data, laboratory experiments and geochemical modelling, as discussed in several reports.[9,10] Despite all these efforts some uncertainty in key

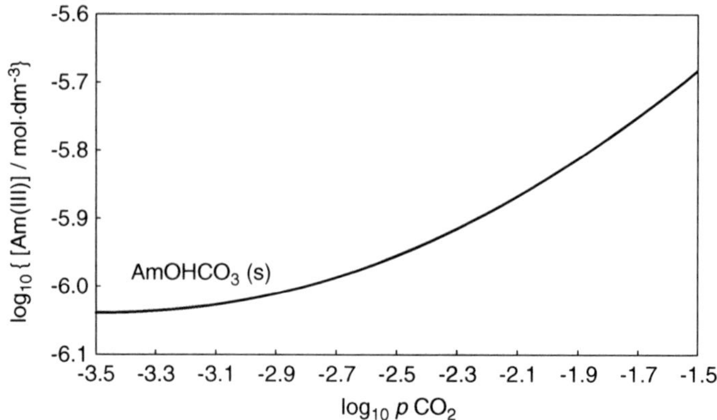

Figure 4 *Americium solubility in bentonite porewater.*[11]

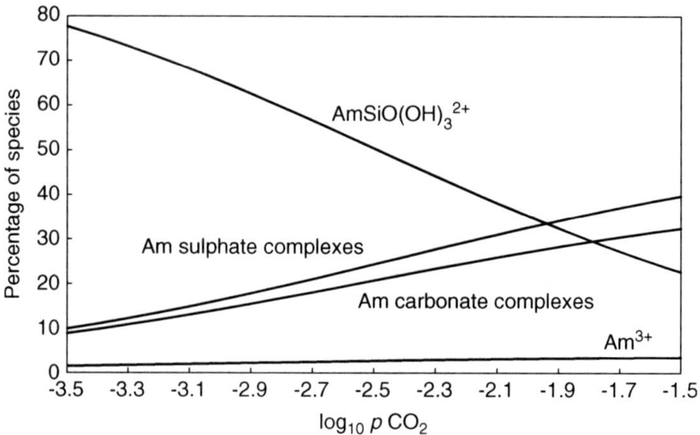

Figure 5 *Americium speciation in bentonite porewater.*[11]

parameters such as CO_2 partial pressure and pH remains and has to be considered in subsequent modelling of solubilities for repository safety purposes.

As an example, the solubility of americium in bentonite porewater[11] is shown in Figure 4. The above mentioned uncertainty in the CO_2 partial pressure of the bentonite reference porewater is reflected in the $\log_{10}pCO_2$ parameter range of Figure 4. As a consequence, the calculated solubility of the solid $AmOHCO_3(s)$ (Figure 4) and the aqueous speciation of americium (Figure 5) vary within certain bounds as shown in the figures. These uncertainties are considered in subsequent safety calculations.[6]

Recent research activities in the field of molecular modelling aim at an understanding of the properties of water in such highly compacted clay systems on a molecular level with the goal of reducing the uncertainties in the definition of porewaters in these systems.

21.5 Which are the Relevant Thermodynamic Data?

Thermodynamic modelling of solubility limits not only needs parameter ranges of the water chemistry but also a critically reviewed database of chemical-thermodynamic constants[12] and careful considerations whether this database is complete with respect to the chemical systems to be modelled.[13] For example, a new feature in the americium example (Figure 5), resulting from these critical data review efforts, is the dominance of an aqueous americium silicate complex in a certain parameter range. Experimental investigations of aqueous silicate complexes of radionuclides commenced a few years ago and are still ongoing.

The Radioactive Waste Management Committee of the OECD Nuclear Energy Agency (NEA) recognised the need for an internationally acknowledged, high-quality thermochemical database for application in the safety assessment of radioactive waste disposal, and initiated the development of the NEA Thermochemical Data Base (TDB) project in 1984.[14] The first four books in the series on the chemical thermodynamics of uranium,[15] americium,[16] technetium,[17] and neptunium and plutonium[18] originated from this initiative.

In 1998, Phase II of the NEA TDB Project was started[14] to provide for the further needs of the radioactive waste management programs by updating the existing database[19] and applying the TDB review methodology to other elements (nickel,[20] selenium,[21] zirconium[22]) and to simple organic compounds and complexes.[23]

In the ongoing Phase III of the NEA TDB Project, started in 2003, the elements iron, thorium, and tin are being reviewed.

Despite all these efforts, the chemical thermodynamic database necessary for modelling solubility limits is still far from being sufficiently complete.[24] If chemical thermodynamic constants are too scarce for modelling of a certain system, solubility limits are estimated by scrutinising experimental evidence related to solubility phenomena[13,25] or by chemical analogies.[11,26]

21.6 Which are the Relevant Solid Phases?

From a purely thermodynamic point of view, the relevant solid phase determining the solubility of a certain element is the one leading to the lowest solubility of this element under the specified conditions. However, in the context of repository safety we need elemental solubility limits, *i.e.* the maximum concentration of this element to be expected in the aqueous phase due to the precipitation of a solid phase. This solid phase not necessarily is the thermodynamically stable phase. The following example concerning the solubility of thorium illustrates this point.[25]

The solubility of ThO_2 as a function of pH has been studied extensively by several groups with the aim of providing basic data for the safety assessments of planned geological repositories.[27–31] The growing number of published

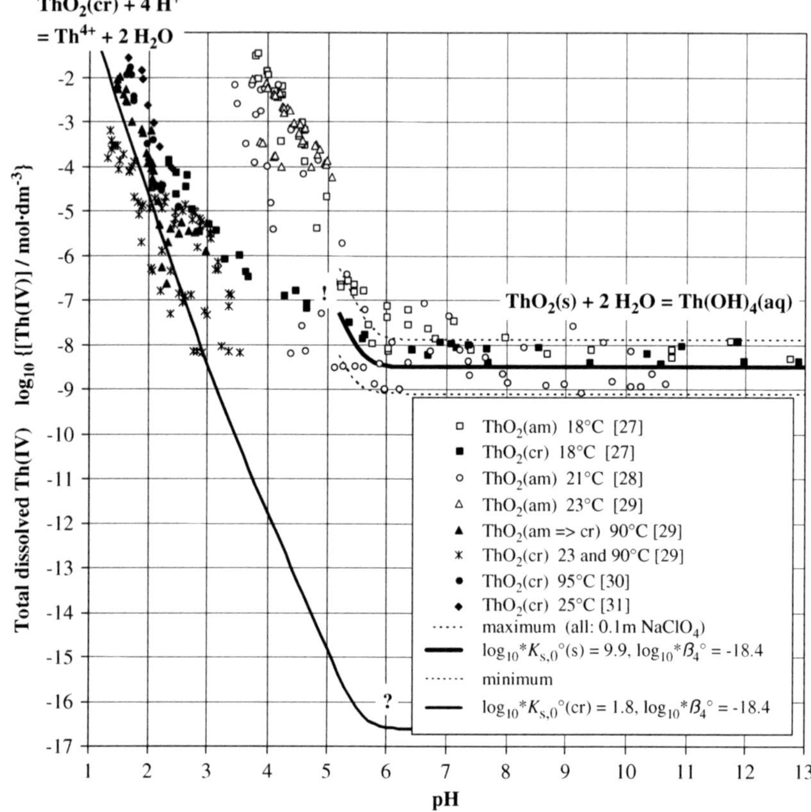

Figure 6 *The enigma of ThO_2 solubility: Data derived from the solubility of crystalline ThO_2 at low pH cannot describe $ThO_2(s)$ solubility above pH 6 when using independent $Th(OH)_4(aq)$ formation data (question mark). Vice versa, measured solubilities in alkaline solutions do not point back to the behaviour of $ThO_2(cr)$ at acidic conditions (exclamation mark).*

experimental solubility data reveals a consistent but puzzling picture of the system ThO_2–H_2O (Figure 6).

At low pH the solubility of ThO_2 strongly depends on the crystallinity of the solid. Differences in solubility of several orders of magnitude have been found between freshly precipitated amorphous and well crystalline solids (Figure 6). However, with increasing pH the measured concentrations converge, and at neutral and alkaline pH the ThO_2 solubility is found to be independent of (bulk) crystallinity (Figure 6). The solution always seems to "see" the same solid in neutral and alkaline solutions. The following problems were encountered by the attempt to describe all these experimental data by a unique set of thermodynamic constants:

Solubility data for ThO_2(cr) agree fairly well with the solubility predicted by calorimetric data in the range pH < 3 (line in Figure 6). However, combining the solubility product $\log_{10}{}^*K_{s,0}^{\circ}(cr)$ for $ThO_2(cr) + 4H^+ \rightleftharpoons Th^{4+} + 2H_2O$ with the independently determined hydrolysis constant[12] $\log_{10}{}^*\beta_4^{\circ} = -18.4$ for $Th^{4+} + 4H_2O \rightleftharpoons Th(OH)_4(aq) + 4H^+$ results in predicted Th concentrations far from any measured values in neutral and alkaline solutions. The concentration of dissolved Th(IV) should fall below any detection limit to $[Th] < 10^{-16}$ M (see question mark in Figure 6).

All $ThO_2(s)$ solubility data measured in the laboratory at pH > 6 have been found in the range 10^{-7} M > [Th] > 10^{-9} M. A mean value of $10^{-8.5\pm0.6}$ M represents ${}^*K_{s,4}$ (s) for $ThO_2(s) + 2H_2O \rightleftharpoons Th(OH)_4(aq)$. If we combine this constant with $\log_{10}{}^*\beta_4^{\circ} = -18.4$ of $Th(OH)_4(aq)$ a solubility product for $ThO_2(s)$ is calculated which produces a solubility curve somewhere in the "cloud" of solubility data for $ThO_2(am)$ and $ThO_2(cr)$ (Figure 6). This set of parameters now describes the measured solubilities at pH > 6 but cannot account for the solubility variation of several orders of magnitude at lower pH (see exclamation mark in Figure 6).

Furthermore, this behaviour is not restricted to Th(IV). Similar patterns have also been found for U(IV), Np(IV), and Pu(IV).[32,33] In all cases the constant values measured for tetravalent actinides, An(IV), posed the same enigma as illustrated here for Th(IV). In a seminal effort Neck and co-workers[32–36] tried to resolve this enigma and came up with the following resume.[36]

At very low pH, the thermodynamically stable crystalline dioxides $AnO_2(cr)$ may actually represent the solubility limiting solid phase, in particular at higher temperature.[29,35] However, experimental solubility data in neutral and alkaline solutions are 6–7 orders of magnitude higher than the low values of $\leq 10^{-15}$ M calculated from the known thermodynamic data[19,32] for $AnO_2(cr)$ and An(IV) hydroxide complexes. At pH values above the onset of hydrolysis, the sorption or precipitation of monomeric or polynuclear hydroxide complexes $An_x(OH)_y^{4x-y}(aq)$ on the surface of the crystalline $AnO_2(cr)$ will result in an amorphous solubility limiting surface layer:

$$AnO_2(cr) \rightarrow An_x(OH)_y^{4x-y}(aq) \rightleftharpoons \text{“}An(OH)_4(am)\text{”}$$

Hence, performance assessment calculations on the long-time behaviour of tetravalent actinides in natural systems must not take credit of the extremely low solubilities resulting from the thermodynamic data of the crystalline An(IV) dioxides.

Currently, thermodynamic modelling of solubility limits in safety analyses is based on solubility data for pure solid phases. However, most radionuclides are not expected to form pure solid phases but to take part in solid solutions with major host minerals in the repository surroundings. Considering these solid solutions in thermodynamic modelling would result in lower radionuclide concentrations in the aqueous phase.

For example, recent studies of europium forming solid solutions with calcite[37] indicate that the solubility of the chemical analogue americium may not be limited by the pure phase $AmOHCO_3(s)$ (Figure 4) but by an americium – calcite solid solution.

Likewise, the solubility of radium may not be determined by the precipitation of pure $RaSO_4(s)$ but by the formation of a $(Ra,Ba)SO_4(s)$ solid solution.[38]

Ongoing research activities aim at obtaining reliable data for safety-relevant solid solution systems with the goal of establishing a thermodynamic database for solid solutions.

Acknowledgments

Partial financial support by the Swiss National Co-operative for the Disposal of Radioactive Waste (Nagra) is gratefully acknowledged.

References

1. *Technical overview of the SAFIR 2 report: Safety Assessment and Feasibility Interim Report 2*, NIROND-2001-05E. ONDRAF, Brussels, Belgium, 2001.
2. T. Vieno and H. Nordman, *Safety Assessment of Spent Fuel Disposal in Hästholmen, Kivetty, Olkiluoto and Romuvaara*: TILA-99, Posiva Report 99-07. Posiva Oy, Helsinki, Finland, 1999.
3. *Dossier 2001 Argile sur l'avancement des études et recherches relatives à la faisabilité d'un stockage de déchets à haute activité et à vie longue en formation géologique profonde: Rapport de synthèse*. ANDRA, Châtenay-Malabry, France, 2001.
4. *H12: Project to Establish the Scientific and Technical Basis for HLW Disposal in Japan: Second Progress Report on Research and Development for the Geological Disposal of HLW in Japan*, JNC TN1410 2000–001. JNC, Tokyo, Japan, 2000.
5. *Deep Repository for Spent Nuclear Fuel: SR 97: Post-Closure Safety: Main Report*, Technical Report TR-99-06. SKB, Stockholm, Sweden, 1999.
6. *Project Opalinus Clay: Safety Report, Demonstration of Disposal Feasibility for Spent Fuel, Vitrified High-Level Waste and Long-Lived Intermediate-Level Waste (Entsorgungsnachweis)*, Nagra Technical Report NTB 02-05. Nagra, Wettingen, Switzerland, 2002.
7. J.A. McNeisch, *Total System Performance Assessment for the Site Recommendation: Yucca Mountain Project*, TDR-WIS-PA-000001 REV 00 ICN 01. U.S. Department of Energy, Yucca Mountain Project, USA (CD), 2000.
8. P.A. Witherspoon and G.S. Bodvarsson (eds), *Geological Challenges in Radioactive Waste Isolation: Third Worldwide Review*, LBNL-49767. Ernest Orlando Lawrence Berkeley National Laboratory, Berkeley, 2001.
9. E. Curti and P. Wersin, *Assessment of Porewater Chemistry in the Bentonite Backfill for the Swiss SF/HLW Repository*, Nagra Technical Report NTB 02-09. Nagra, Wettingen, Switzerland, 2002.
10. M.H. Bradbury and B. Baeyens, Porewater chemistry in compacted resaturated MX-80 bentonite, *J. Cont. Hydr.*, 2003, **61**, 329–338.
11. U. Berner, *Project Opalinus Clay: Radionuclide Concentration Limits in the Near Field of a Repository for Spent Fuel and Vitrified High-Level Waste*,

Nagra Technical Report NTB 02-10 and PSI Bericht Nr. 02-22. Paul Scherrer Institut, Villigen, Switzerland, 2002.

12. W. Hummel, U. Berner, E. Curti, F.J. Pearson and T. Thoenen, *Nagra/PSI Chemical Thermodynamic Database 01/01*, Nagra Technical Report NTB 02-16 and Universal Publishers/uPublish.com, Parkland, Florida, USA, 2002.
13. W. Hummel and U. Berner, *Application of the Nagra/PSI TDB 01/01: Solubility of Th, U, Np and Pu*, Nagra Technical Report NTB 02-12. Nagra, Wettingen, Switzerland, 2002.
14. F.J. Mompeán and H. Wanner, The OECD nuclear energy agency thermochemical database project, *Radiochim. Acta*, 2003, **91**, 617–621.
15. I. Grenthe, J. Fuger, R.J.M. Konings, R.J. Lemire, A.B. Muller, C. Nguyen-Trung and H. Wanner, *Chemical Thermodynamics of Uranium*, OECD/NEA, Paris, 1992, 715.
16. R.J. Silva, G. Bidoglio, M.H. Rand, P.B. Robouch, H. Wanner and I. Puigdomènech, *Chemical Thermodynamics of Americium*, OECD/NEA, Paris, 1995, 374.
17. J.A. Rard, M.H. Rand, G. Anderegg and H. Wanner, *Chemical Thermodynamics of Technetium*, Elsevier, Amsterdam, 1999, 544.
18. R.J. Lemire, J. Fuger, H. Nitsche, P. Potter, M.H. Rand, J. Rydberg, K. Spahiu, J.C. Sullivan, W.J. Ullman, P. Vitorge and H. Wanner, *Chemical Thermodynamics of Neptunium & Plutonium*, Elsevier, Amsterdam, 2001, 845.
19. R. Guillaumont, T. Fanghänel, J. Fuger, I. Grenthe, V. Neck, D.A. Palmer and M.H. Rand, *Update on the Chemical Thermodynamics of Uranium, Neptunium, Plutonium, Americium and Technetium*, Elsevier, Amsterdam, 2003, 919.
20. H. Gamsjäger, J. Bugajski, T. Gaida, R.J. Lemire and W. Preis, *Chemical Thermodynamics of Nickel*, Elsevier, Amsterdam, 2005, 617.
21. Å. Olin, B. Noläng, L.-O. Öhman, E.G. Osadchii and E. Rosén, *Chemical Thermodynamics of Selenium*, Elsevier, Amsterdam, 2005, 851.
22. P.L. Brown, E. Curti and B. Grambow, *Chemical Thermodynamics of Zirconium*, Elsevier, Amsterdam, 2005, 512.
23. W. Hummel, G. Anderegg, I. Puigdomènech, L. Rao and O. Tochiyama, *Chemical Thermodynamics of Compounds and Complexes of U, Np, Pu, Am, Tc, Zr, Ni and Se with Selected Organic Ligands*, Elsevier, Amsterdam, 2005, 1088.
24. W. Hummel and E. Curti, Nickel aqueous speciation and solubility at ambient conditions: a thermodynamic elegy, *Monatsh. Chem./Chem. Monthly*, 2003, **134**, 941–973.
25. W. Hummel, Solubility equilibria and geochemical modelling in the field of radioactive waste disposal, *Pure Appl. Chem.*, 2005, **77**, 631–641.
26. U. Berner, *Project Opalinus Clay: Radionuclide Concentration Limits in the Near Field of a Repository for Long-Lived Intermediate-Level Waste*, Nagra Technical Report NTB 02-22 and PSI Bericht Nr. 02-26. Paul Scherrer Institut, Villigen, Switzerland, 2003.
27. H.C. Moon, *Bull. Korean Chem. Soc.*, 1989, **10**, 270–272.

28. A.R. Felmy, D. Rai and M.J. Mason, *Radiochim. Acta*, 1991, **55**, 177–185.
29. D. Rai, D.A. Moore, C.S. Oakes and M. Yui, *Radiochim. Acta*, 2000, **88**, 297–306.
30. C.F. Baes, Jr., N.J. Meyer and C.E. Roberts, *Inorg. Chem.*, 1965, **4**, 518–527.
31. T. Bundschuh, R. Knopp, R. Müller, J.I. Kim, V. Neck and T. Fanghänel, *Radiochim. Acta*, 2000, **88**, 625–629.
32. V. Neck and J.I. Kim, Solubility and hydrolysis of tetravalent actinides, *Radiochim. Acta*, 2001, **89**, 1–16.
33. T. Fanghänel and V. Neck, Aquatic chemistry and solubility phenomena of actinide oxides/hydroxides, *Pure Appl. Chem.*, 2002, **74**, 1895–1907.
34. V. Neck, R. Müller, M. Bouby, M. Altmaier, J. Rothe, M.A. Denecke and J.I. Kim, Solubility of amorphous Th(IV) hydroxide – application of LIBD to determine the solubility product and EXAFS for aqueous speciation, *Radiochim. Acta*, 2002, **90**, 485–494.
35. V. Neck, M. Altmaier, R. Müller, A. Bauer, T. Fanghänel and J.I. Kim, Solubility of crystalline thorium dioxide, *Radiochim. Acta*, 2003, **91**, 253–262.
36. M. Altmaier, V. Neck and T. Fanghänel, Solubility and colloid formation of Th(IV) in concentrated NaCl and $MgCl_2$ solution, *Radiochim. Acta*, 2004, **92**, 537–543.
37. E. Curti, D. Kulik and J. Tits, Solid solutions of trace Eu(III) in calcite: thermodynamic evaluation of experimental data over a wide range of pH and pCO_2, *Geochim. Cosmochim. Acta*, 2005, **69**, 1721–1737.
38. U. Berner, E. Curti, *Radium Solubilities from SF/HLW Wastes using Solid Solution and Co-Precipitation Models*, PSI Technical Report TM-44-02-04. Paul Scherrer Institut, Villigen, Switzerland, 2002.

CHAPTER 22
Carbon Dioxide in Chemical Processes

JUSTIN SALMINEN[1,2] AND JOHN PRAUSNITZ[1,3]

[1] *Department of Chemical Engineering, University of California, Berkeley, Berkeley CA 94720, USA*

[2] *Environmental Energy Technology Division, Lawrence Berkeley National Laboratory, Berkeley CA 94720, USA*

[3] *Chemical Sciences Division, Lawrence Berkeley National Laboratory, Berkeley CA 94720, USA*

22.1 Applications of CO_2

Carbon dioxide as a useful reactive ingredient has increasingly gained interest in several process industries. These include buffered carbonate and alkaline systems that depend on the solubility and reactivity of the reactants. CO_2 gas is commonly used for pH control that facilitates dissolution and precipitation of carbonates.[1–8] Techniques for removing acid gases like CO_2 and H_2S from flue gases by absorption are commonly used in coal gasification and sweetening of natural gases.[9,10] Aqueous alkanolamine is often used for absorption. These well-studied systems include aqueous mixtures of monoethanolamine (MEA), *n*-methyldiethanolamine (MDEA), diethanolamine (DEA), and 2-amino-2-methyl-1-propanol (AMP).[5,10–52] Ionic liquids have recently been studied for CO_2 absorption and other industrial applications.[53–67] For process design, we require knowledge of molecular interactions, speciation, and solubilities as well as chemical equilibria in reactive multiphase systems. At specified temperature and liquid composition, the total pressure of the system is mainly determined by the concentration of non-dissociated dissolved CO_2 (aq) according to Henry's law. As discussed later, the non-dissociated CO_2 (aq) is in equilibrium with ionic bicarbonate, carbonate and other ionic species in the system, and with solid phases. Modeling of both phase equilibia and chemical equilibria provides a rigorous tool for describing aqueous industrial and environmental processes that contain CO_2.[8–10,14,35,44]

For process optimization, chemical equilibrium data are increasingly needed to interpret complex industrial and natural processes that contain aqueous electrolytes and solid precipitates. Thermodynamic analysis provides quantitative relations among chemical energy, chemical reactions, solubilities of gases and salts, and important online process parameters like pH, temperature, and pressure. Supported by laboratory experiments, and by pilot- and full-scale experience, models have provided applications in different industries where CO_2 is a reactant. For example, in the pulp-and-paper industry the properties of wood fibre combined with the results of model calculations can be used to reduce heating and raw-materials costs.[68–74] New industrial applications include cleaning waste, and process waters as well as recovering valuable or environmentally harmful metal ions as carbonates. The current trend towards using carbon dioxide is partly due to increased demand for making chemical processes environmental friendly. Often a process that uses CO_2 is preferred because it replaces some harmful chemicals or because it is economically superior.

22.2 CO_2 In Multiphase Aqueous Systems

Carbon dioxide and carbonate minerals play an important role in geochemistry including natural waters. In low-mineral freshwaters, atmospheric CO_2 has a significant influence on pH. The effect of dissolved atmospheric CO_2 on the pH of seawater follows from its buffer capacity but more important, from interaction with dissolved mineral carbonates. Seawater pH is *ca.* 8.2 while low-mineral lakes can exhibit a pH as low as 3, partly due to acid rain. Owing to its weak acid character, the solubility of carbon dioxide in water is larger than that of air; unlike air, in water, CO_2 partly dissociates into ions. Figure 1 shows the temperature dependence of Henry's constant for CO_2 dissolved in water.

Equation (1) shows the relation between Henry's constant H_{CO_2} and equilibrium constant K_H for the reaction shown in Equation (2). Here, x is molefraction, p partial pressure. Henry's constant has units of pressure (MPa). Henry's law is strictly valid only at infinite dilution of CO_2. Assuming that the gas phase

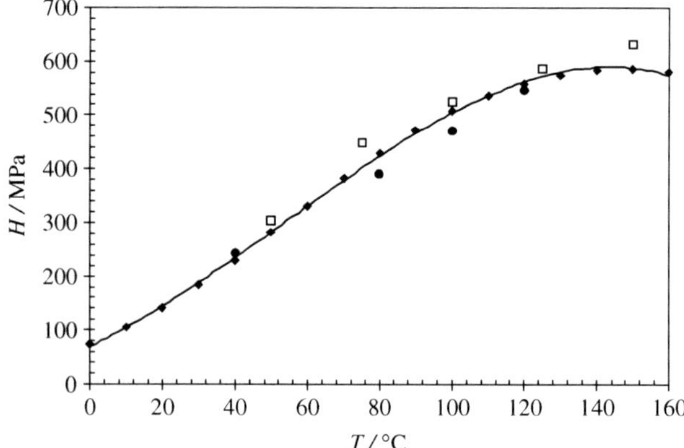

Figure 1 *(Henry's constant for CO_2 in water as a function of temperature. Experimental results (◆) and solid line,[75] (□),[76] (●).[77]*

is ideal, and for $x_{CO_2} \ll 1$,

$$K_H = \frac{x_{CO_2(aq)}}{p_{CO_2}} = \frac{1}{H_{CO_2}} \tag{1}$$

The physical equilibria for carbon dioxide dissolved in water is

$$CO_2\ (g) \leftrightarrow CO_2\ (aq) \tag{2}$$

and the essential chemical equilibria are

$$CO_2\ (aq) + H_2O \leftrightarrow H^+ + HCO_3^- \tag{3}$$

$$HCO_3^- \leftrightarrow H^+ + CO_3^{2-} \tag{4}$$

The first- and second-dissociation reactions define the acid–base chemistry of aqueous carbon dioxide. The chemical dissociation reactions (3) and (4) are related to the physical reaction Equation (2) as given by Henry's law.[75–78] Dissolved salts affect the solubility of CO_2 through chemical reactions and by the salting-out or salting-in effect.[79–84]

Carbon dioxide gas can be used as an acidifying agent in chemical- or biochemical processes. Carbon dioxide gas can be used to dissolve or precipitate metal carbonate salts through the common-ion effect. Equations (5)–(8) show that an increase of proton activity $a(H^+)$ in the solution raises the solubility of carbonate minerals and lowers the solubility of CO_2. A decrease in proton activity and subsequent increase in basicity raises the overall solubility of CO_2 causing precipitation of mineral carbonates. Sparingly soluble metal carbonates include calcium carbonate ($CaCO_3$), magnesium carbonate ($MgCO_3$), zinc carbonate ($ZnCO_3$), lead carbonate ($PbCO_3$), and cadmium carbonate ($CdCO_3$); sparingly soluble salts are commonly present in chemical processes. In aqueous metal-carbonate systems, the following main reactions occur. Let M^{2+} represent metal cations (Mg^{2+}, Ca^{2+}, Zn^{2+}, Cd^{2+}, Pb^{2+}). The overall reaction is

$$M^{2+}\ CO_3{}^{2-}\ (s) + 2\ H^+\ (aq) \leftrightarrow M^{2+}\ (aq) + H_2O + CO_2\ (g)\uparrow \tag{5}$$

The dissociation of a metal carbonate salt $M^{2+}CO_3{}^{2-}$ can be written as

$$M^{2+}\ CO_3{}^{2-}\ (s) \leftrightarrow M^{2+}\ (aq) + CO_3{}^{2-}\ (aq) \tag{6}$$

Four equilibrium relations describe a multiphase aqueous system containing CO_2 and MCO_3. Equilibria for reaction Equations (2)–(6) are written using total pressure P, gas-phase molefraction y_i, liquid-phase molality m_i, activity a_i, fugacity coefficient ϕ_i, and activity coefficient γ_i, where i stands for a species that takes part in the chemical reaction. An activity coefficient model is needed to relate the liquid-phase activities of individual species to their molalities. The solubility product K_{SP} for reaction (6) is:

$$K_{sp} = a(M^{2+})\ a\ (CO_3^{2-}) = m(M^{2+})\ \gamma(M^{2+})\ m\ (CO_3^{2-})\ \gamma(CO_3^{2-}) \tag{7}$$

The solubility of a M^{2+} carbonate salt can be increased by adding an acid. Further, the M^{2+} carbonate salt can be precipitated by adding electrolytes with anions like OH^-,

HCO_3^-, HS^-, and CO_3^{2-}. The relation among solubility product, CO_2 partial pressure, proton activity, and M^{2+} ion activity is:

$$K_{sp} = a(M^{2+})\, P(CO_2)\, a\,(H^+)^{-2} \tag{8}$$

Solubility products have been measured by several authors for $CaCO_3$ (calcite, aragonite), $ZnCO_3$ (smithsonite), $CdCO_3$ (otavite), and $PbCO_3$ as a function of temperature.[85–87] The descending order for solubility of carbonates is: $Mg^{2+} > Ca^{2+} > Zn^{2+} > Cd^{2+} > Pb^{2+}$. Figure 2 shows the solubility products for Ca^{2+}, Zn^{2+}, Cd^{2+}, and Pb^{2+} carbonates as a function of temperature.

Depending on CO_2 partial pressure, the pH, and the ionic strength of the solution, the phase diagram of aqueous Zn^{2+} is in equilibrium with zinc carbonate ($ZnCO_3$), hydrozincite ($Zn_5(OH)_6(CO_3)_2$), and zinc oxide (ZnO).[88] In dilute electrolyte solution, Zn^{2+} is in equilibrium with $ZnCO_3$ above ca. 0.001 bar CO_2 partial pressures and pH range of 5.5–7.2. As the CO_2 pressure decreases hydrozincite appears above pH 7.2. In a strong ionic-strength solution such as 4 mol(NaCl)/kg(H_2O) at 1 bar CO_2 partial pressure, Na_2CO_3 tends to co-precipitate with $ZnCO_3$.[89]

An effective way to estimate the chemical state of a reactive multiphase system is provided by thermodynamic calculations using an applicable activity-coefficient model. The total Gibbs energy of the system is related to chemical potentials μ_i for every species i. The chemical potentials of the pure species are derived from tabulated values of the enthalpy of formation, heat capacity and the entropy of formation.[90,91] The chemical potentials in a multicomponent mixture are then obtained by including the activities of all individual species in

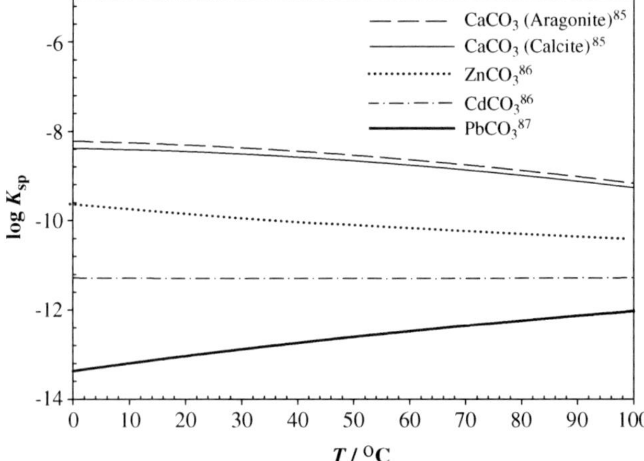

Figure 2 *Experimental solubility products for common M^{2+} carbonate salts show large differences depending on the metal cation.*[85–87]

the system.[92–101] The total Gibbs energy of the system is obtained as a sum over all species i and all stable gas, liquid and solid phases,

$$G = \sum_{\alpha} \sum_{i} n_i^{\alpha} \mu_i^{\alpha} \tag{9}$$

where n is a number of moles, i stands for species and α stands for phase.

Chemical equilibrium in a closed system at constant temperature and pressure is achieved at the minimum of the total Gibbs energy, $min(G)$ constrained by material-balance and electro-neutrality conditions. Computer routines have been published for various practical applications.[91–93] For aqueous electrolyte solutions, we require activity coefficients for all species in the mixture. Well-established models, *e.g.* Debye–Hückel, extended Debye-Hückel, Pitzer, and the Harvie–Weare modification of Pitzer's activity coefficient model, are commonly combined with a minimization routine to derive the activities of individual ions.[98–101] Modeling of phase and chemical equilibria in a multiphase system provides a practical tool for describing aqueous chemical processes containing CO_2.

Figure 3 shows calculations for the multiphase calcite system at 25 and 50°C. We see how dissolved calcite at constant 1 atm CO_2 partial pressure depends on pH up to saturation. After saturation, added calcite does not change the pH because it goes to the solid phase. Chemical equilibrium is reached at the minimum of the total Gibbs energy of the system. Calculated phase diagrams, obtained by minimization of Gibbs energy G, are useful for systems that involve many chemical reactions because all speciation, concentrations, solubilities, activities including pH, partial pressures, and osmotic coefficient are obtained simultaneously at *min* (G).

22.3 Applications in the Process Industries

The annual production of paper and paperboard is approximately 300 million tons worldwide. Paper products are made from natural fibres, mostly from wood processed with water, selected chemicals, and heat. New processes are more environmentally acceptable, more energy- efficient and require an ever-smaller input of chemicals and water but a larger amount of re-circulated matter.[68–74] Water consumption in paper production has declined from 350 to ca. 14 m^3/1000 kg (dry pulp), indicating that the aqueous solutions that circulate in the process are more concentrated in ionic species. Increased ionic strength influences the solubilities of gases and therefore, precipitates appear in the process. Precipitation and dissolution of solid phases like carbonates, oxalates, and sulfates must be controlled in the process. Higher ionic strength also influences the physical properties of the wood fibre and can have a detrimental effect on the final quality of the paper products.

Dissolution and precipitation processes of aqueous $CaCO_3$ solutions are controlled by acid–base chemistry; these processes are thermodynamically driven. Chemical thermodynamics provides quantitative knowledge concerning

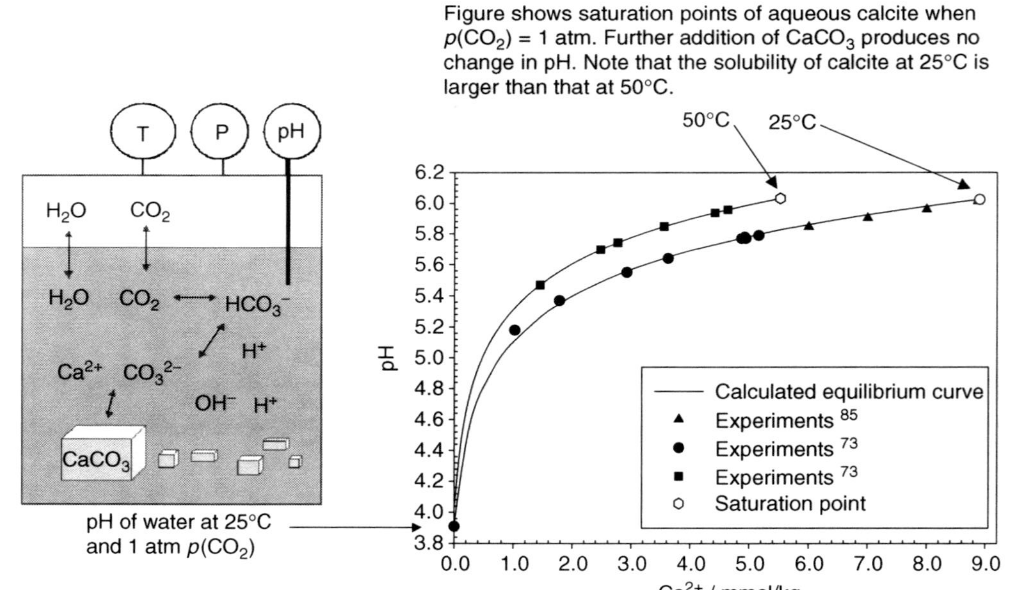

Figure 3 *Thermodynamic calculations for the multiphase calcite system. We see the effect of dissolved calcite on pH at constant 1 atm CO_2 partial pressure, at 25 and 50°C up to saturation. After saturation, added calcite does not change pH because calcite goes to the solid phase.*

the final equilibrium state of a multiphase $CaCO_3$ solution as we change temperature, CO_2 -pressure or composition. Maintenance and control of process pH have a major effect on paper-production rates. Calcium carbonate, used as a filler in making fine paper grades like office paper, exists in recycled paper.[8] Carbon dioxide has been used to partially replace sulfuric acid (H_2SO_4) as a reactive acidifying chemical in the papermaking process and in chemical pulp mills. Typical range where CO_2 is used to acidify calcite buffered or alkaline pulp suspensions are from pH 10 to 6. These typically include 1–2wt % wood fibres.

The commonly used pH agent, sulfuric acid, dissociates in water forming bisulphate HSO_4^- and sulfate SO_4^{2-} ions. The ions like SO_4^{2-} tend to build up, causing precipitation of unwanted salts as the water is circulated to reduce the amount of wastewater. Traditionally, dissolved gases have not been used in the papermaking process due to undesired foaming or de-aeration. Because carbon dioxide contributes H^+ ions to the system, $CaCO_3$ is dissolved according to the overall reaction (5). Solubility models may be utilized for evaluating process circulation; it is advantageous to dissolve inorganic material in the re-circulating parts to enhance precipitation of calcium in the input. To decrease the solubility of $CaCO_3$, it is useful to introduce a common anion as found in sodium or potassium carbonate or bicarbonate when carbon dioxide is chosen to control pH.

Absorption and reaction of CO_2 is governed by two fundamental processes: first, the equilibrium constraints at given pressure, temperature, and composition, and second, chemical kinetics. Dissolution or precipitation of solids do not occur at instantaneously; some characteristic time is needed for the process to achieve a new equilibrium state. This characteristic time is often longer than the timeconstant for the overall process.

Figure 4 shows results from laboratory experiments for two systems at 25 and 50°C. These systems approach the equilibrium state from different initial states: (a) calcite-saturated fresh distilled water mixed with CO_2; (b) CO_2-saturated distilled water mixed with calcite. The time-dependent pH curves show the influence of dissolved calcite ($CaCO_3$) and subsequent introduction of Ca^{2+} and ${CO_3}^{2-}$ ions into the carbon dioxide-bicarbonate mixture, as indicated by Equations (2)–(6). The total pressure was 101.3 kPa. The partial pressures of CO_2 were calculated by subtracting the water vapour pressure from the total pressure.

In Figure 4, the introduction of CO_2 to a $CaCO_3$–H_2O system [case (a)] increases the solubility of calcite as pH decreases. The pH value remains constant after the system reaches equilibrium. The reactions in solution change the pH until equilibrium is reached at constant $p(CO_2)$ and temperature. Addition of $CaCO_3$ into previously CO_2-acidified water [case (b)] increases pH as calcite dissolves and more HCO_3^- is formed. The system pH levels off towards the same equilibrium pH as that in case (a).

The rate of equilibration can vary over orders of magnitude as can the rates of dissolution. By measuring pH, we can follow online the equilibration process of a heterogeneous ionic system. Addition of soluble sodium bicarbonate for

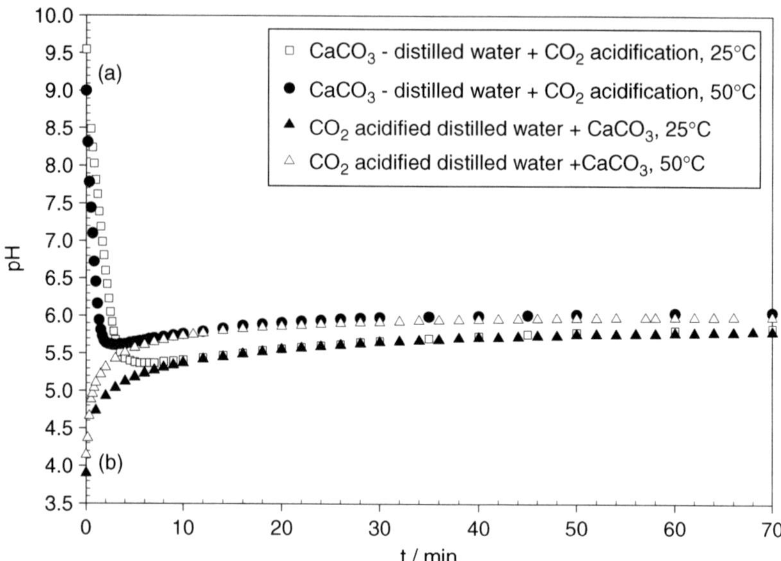

Figure 4 *Observed pH changes in the $CaCO_3$–CO_2–H_2O system from two different starting conditions, one at basic- and the other at acidic pH. (a) Freshly prepared calcite –distilled water solution, i.e. a basic solution that is acidified with CO_2 at total pressure 101.3 kPa (CO_2 + H_2O). (b) Distilled water acidified by CO_2 at total pressure 101.3 kPa (CO_2 + H_2O), i.e. an acidic solution where calcite was added. The experiments were carried out at 25 and 50°C. The rate of pH change decreases as the solution approaches equilibrium at constant temperature and constant CO_2 pressure.*

example in [case(b)], raises pH. In the solution, mixtures of divalent metal M^{2+} carbonates or hydroxide carbonates then precipitate due to the common-anion effect.

22.4 Dynamic Systems

Aqueous inorganic carbonate systems are found in industrial processes and in nature.[102–105] Thermodynamic-calculation methods can be combined with known overall kinetics or known time-dependent compositions (*e.g.* Ca^{2+} dissolution rate) to calculate concentrations of other species and pH.[8,97,106–108] Calculations are subject to material-balance constraints and to the electron-eutrality condition. Some natural sources of CO_2 exist at conditions favourable for biological activity. Natural CO_2 sources, like those from the decomposition of organic matter in water, exhibit proton activity and bicarbonate levels different from those achieved by dissolution of atmospheric CO_2 only.[109]

Many aqueous processes operate at 20–50°C and close to a neutral pH favourable for unwanted biochemical activity. For chemical pulping, many current processes use low alkaline levels; the pH is close to neutral, for example,

7.5–9. Carbon dioxide can be used to acidify process waters, e.g. below the pH levels that are favourable for typical biological activity.

Aerobic decomposition of organic matter forms CO_2 gas while anaerobic decomposition produces ammonia (NH_3) that reacts to $NH_4^+OH^-$ and further to NH_4HCO_3, NH_4^+ and HCO_3^- species in the presence of CO_2.[110,111] While CO_2 lowers pH, ammonia increases pH. Because pH often has a large influence on a bio-reaction, CO_2 can be used to prevent or facilitate biochemical reactions.

22.5 Concluding Remarks

Aqueous electrolyte systems and carbon-dioxide electrolyte solutions cover a wide spectrum of applications in industrial and in environmental systems. Carbon dioxide can be used as an acidifying agent in processes, often replacing more harmful chemicals. Applications include cleaning of waste- and process waters as well as recovering valuable or environmentally harmful metal ions as carbonates.

References

1. W. Preis and H. Gamsjäger, in *Highlights in Solute–Solvent Interactions*, W. Linert, (ed), Springer-Verlag, New York, 2002, 91.
2. G.T. Hefter and R.P.T. Tomkins, *The Experimental Determination of Solubilities*, Vol 3, Wiley Series in Solution Chemistry, London, 2003.
3. S.A. Newman, (ed), *Thermodynamics of Aqueous Systems with Industrial Applications*, ASC Symposium Series 133, ACS, Washington, 1980.
4. J. Salminen, in *McGraw-Hill Yearbook of Science and Technology*, McGraw-Hill, New York, 2004, 245.
5. Q. Chen, *Chemical Thermodynamics of Carbon Dioxide with Mixed Aqueous Alkanolamine Solutions and Metalcarbonates*, Master's Thesis, Helsinki University of Technology, Finland, 2005.
6. S.-B. Park, H. Lee and K.-L. Lee, *Int. J. Thermophys.*, 1998, **19**, 1421.
7. I.L. Leites, J.G. Karpova and V.M. Berchenko, *Gas. Sep. Purif.*, 1996, **10**, 35.
8. P. Koukkari, R. Pajarre, H. Pakarinen and J. Salminen, *Ind. Eng. Chem. Res.*, 2001, **40**, 5014.
9. T. Edwards, G. Mauer, J. Newman and J. Prausnitz, *AIChE J.*, 1978, **24**, 966.
10. J. Prausnitz, R. Lichtenthaler and E. Azevedo, *Molecular Thermodynamics of Fluid-Phase Equilibria*, 3rd edn, Prentice-Hall, New Jersey, 1999, 563.
11. J. Mason and B.F. Dodge, *Am. Inst. Chem. Eng.*, 1936, **32**, 27.
12. J.I. Lee, F.D. Otto and A.E. Mather, *Can. J. Chem. Eng.*, 1974, **52**, 803.
13. J.I. Lee, F.D. Otto and A.E. Mather, *J. Appl. Chem. Biotechnol.*, 1976, **26**, 541.

14. R.D. Deshmukh and A.E. Mather, *J. Chem. Eng. Sci.*, 1981, **36**, 355.
15. D. Barth, C. Tondre, G. Lappal and J.J. Delpuech, *J. Phys. Chem.*, 1981, **85**, 3660.
16. F.-Y. Jou, A.E. Mather and F.D. Otto, *Ind. Eng. Chem. Proc. Des. Dev.*, 1982, **21**, 539.
17. G. Sartori, W.S. Ho and D.W. Savage, *Ind. Eng. Chem. Fundam.*, 1983, **22**, 239.
18. A. Bhairi, *Experimental Equilibrium Between Acid Gases and Alkanoamine Solutions*, Doctoral Thesis, University of Oklahoma, USA, 1984.
19. M.H. Oyevaar, H.J. Fontein and K.R. Westerterp, *J. Chem. Eng. Data*, 1989, **34**, 405.
20. T.T. Teng and A.E. Mather, *J. Chem. Eng. Data*, 1990, **35**, 410.
21. K.-P. Shen and M.-H. Li, *J. Chem. Eng. Data*, 1992, **37**, 96.
22. S. Xu, Y.-W. Wang, F.D. Otto and A.E. Mather, *Chem. Eng. Proc.*, 1992, **31**, 7.
23. M.D. Anderson, M.J. Hegarty and J.E. Johnson, *Proc. Annu. Conv. Gas Proc. Assoc.*, 1992, **71**, 292.
24. F.-Y. Jou, J.J. Carroll, A.E. Mather and F.D. Otto, *J. Chem. Eng. Data*, 1993, **38**, 75.
25. O.F. Dawodu and A. Meisen, *J. Chem. Eng. Data*, 1994, **39**, 548.
26. E.J. Stewart and R.A. Lanning, *Hydrocarb. Processing*, 1994, **5**, 67.
27. M.H. Li and B.C. Chang, *J. Chem. Eng. Data*, 1994, **39**, 361.
28. A. Henni and A.E. Mather, *J. Chem. Eng. Data*, 1995, **40**, 493.
29. M.L. Posey, K.G. Tapperson and G.T. Rochelle, *Gas Sep. Purif.*, 1996, **10**, 181.
30. M.Z. Haji-Sulaiman and M.K. Aroua, *Chem. Eng. Commun.*, 1996, **140**, 157.
31. S.-W. Rho, K.-P. Yoo, J.S. Lee, S.C. Nam, J.E. Son and B.-M. Min, *J. Chem. Eng. Data*, 1997, **42**, 1161.
32. I.S. Jane and M.-H. Li, *J. Chem. Eng. Data*, 1997, **42**, 98.
33. C. Mathonat, V. Majer, A.E. Mather and J.-P.E. Grolier, *Fluid Phase Equilibr.*, 1997, **140**, 171.
34. S.M. Yih and K.P. Shen, *Ind. Eng. Chem. Res.*, 1998, **27**, 2237.
35. D. Silkenbäumer, B. Rumpf and R. Lichtenthaler, *Ind. Eng. Chem. Res.*, 1998, **37**, 3133.
36. K.A.-A. Mousa, M.A.-J. Asem, E.-E. Mohammed and A. Tamimi, *J. Chem. Eng. Data*, 2001, **46**, 516.
37. M.E. Rebolledo-Libreros and A. Trejo, *Fluid Phase Equilibr.*, 2002, **200**, 203.
38. M.K. Aroua and R. Salleh, *Chem. Eng. Tech.*, 2004, **27**, 65.
39. N. Daneshvar, M.T. Zaafarani Moattar, A.M. Abedinzadegan and S. Aber, *Sep. Sci. Tech.*, 2004, **37**, 135.
40. M.E. Rebolledo-Libreros and A. Trejo, *Fluid Phase Equilib.*, 2004, **218**, 261.
41. R. Sidi-Boumedine, S. Horstmann, K. Fischer, E. Provost, W. Furst and J. Gmehling, *Fluid Phase Equilibr.*, 2004, **218**, 85.

42. H.-M. Wang and M.-H. Li, *J. Chem. Eng. Jpn.*, 2004, **37**, 267.
43. B.S. Ali and M.K. Aroua, *Int. J. Thermophys.*, 2004, **25**, 1863.
44. D. Silkenbaeumer, B. Rumpf, R. Lichtenthaler and N. Ruediger, in *Thermodynamic Properties of Complex Fluid Mixtures*, G. Maurer, (ed), Wiley-VCH Verlag, Weinheim Germany, 2004, 48.
45. K. Tounsi, B.A. Habchi, E. Le Corre, P. Mougin and E. Neau, *Ind. Eng. Chem. Res.*, 2005, **44**, 9239.
46. J. Gabrielsen, M. Michelsen, E. Stenby and G. Kontogeorgis, *Ind. Eng. Chem. Res.*, 2005, **44**, 3348.
47. A. Benamor and M.K. Aroua, *Fluid Phase Equilibr.*, 2005, **231**, 150.
48. P.W.J. Derks, H.B.S. Dijkstra, J.A. Hogendoorn and G.F. Versteeg, *AIChE J.*, 2005, **51**, 2311.
49. N. Ai, J. Chen and W. Fei, *J. Chem. Eng. Data*, 2005, **50**, 492.
50. F.-Y. Jou and A.E. Mather, *Fluid Phase Equilibr.*, 2005, **228–229**, 465.
51. K.N. Tounsi, B.A. Habchi, E. Le Corre, P. Mougin and E. Neau, *Ind. Eng. Chem. Res.*, 2005, **44**, 9239.
52. M. Hosseini, A.M. Abedinzadegan, S.H. Najibi, M. Vahidi and N.S. Matin, *J. Chem. Eng. Data*, 2005, **50**, 583.
53. J.L. Anthony, E. Maginn and J.F. Brennecke, *J. Phys. Chem. B*, 2002, **106**, 7315.
54. E.D. Bates, R.D. Mayton, I. Ntai and J.H. Davis, Jr., *J. Am. Chem. Soc.*, 2002, **124**, 926.
55. H.M. Zerth, N.M. Leonard and R.S. Mohan, *Org. Lett.*, 2003, **5**, 55.
56. J. Zhang et al., *New J. Chem.*, 2003, **27**, 333.
57. K.A. Perez-Salado, D. Tuma, J. Xia and G. Maurer, *J. Chem. Eng. Data*, 2003, **48**, 746.
58. M. Alvaro, C. Baleizao, D. Das, E. Carbonell and H. Garcia, *J. Catal.*, 2004, **228**, 254.
59. F. Li, L. Xiao, C. Xia and B. Hu, *Tetrahedron Lett.*, 2004, **45**, 8307.
60. R.E. Baltus, B.H. Culbertson, S. Dai, H. Luo and D.W. DePaoli, *J. Phys. Chem. B*, 2004, **108**, 721.
61. C. Cadena, J.L. Anthony, J.K. Shah, T.I. Morrow, J.F. Brennecke and E.J. Maginn, *J. Am. Chem. Soc*, 2004, **126**, 5300.
62. D. Camper, P. Scovazzo, K. Koval and R. Noble, *Ind. Eng. Chem. Res.*, 2004, **43**, 3049.
63. Y.S. Kim, W. Choi, J. Jang, K.-P. Yoo and C. Lee, *Fluid Phase Equilbr.*, 2005, **228–229**, 439.
64. J. Anthony, J. Anderson, E. Maginn and J. Brennecke, 2005, **109**, 6366.
65. H. Kawanami, H. Matsumoto and Y. Ikushima, *Chem. Lett.*, 2005, **34**, 60.
66. Y.S. Kim, W.Y. Choi, J.H. Jang, K.-P. Yoo and C.S. Lee, *Fluid Phase Equilibr.*, 2005, **228–229**, 439.
67. J. Tang, W. Sun, H. Tang, M. Radosz and Y. Shen, *Macromolecules*, 2005, **38**, 2037.
68. G.A. Smook, *Handbook for Pulp & Paper Technologists*, 3rd edn, Angus Wilde Publications Inc., Vancouver, 2002.

69. R. Kirk and M.T. Othmer, *Encyclopedia of Chemical Technology*, 4th edn, Vol 18, Wiley, New York, 1996, 1–60.
70. J. Gullichsen and H. Paulapuro (eds), *Papermaking Science and Technology*, Vol 1–19, TAPPI Press, ISBN 952-5216-00-4, Atlanta, Georgia, 1999–2001.
71. C.V. Dence and D.W. Reeve, *Pulp Bleaching Principles and Practice*, TAPPI Press, Atlanta, Georgia, 1996.
72. P. Koukkari, R. Pajarre and E. Räsänen, in *Thermodynamics and Industry*, T. Letcher (ed), RSC, Cambridge, 2004, 23.
73. J. Salminen, *Chemical Thermodynamics of Aqueous Electrolyte Systems for Industrial and Environmental Applications*, Doctoral Thesis Helsinki University of Technology, Finland, 2004.
74. J. Salminen and O. Antson, *Ind. Eng.Chem. Res.*, 2002, **41**, 3312.
75. J.J. Carroll, J.D. Slupsky and A.E. Mather, *J. Phys. Chem. Ref. Data*, 1991, **20**, 1201.
76. A. Zawisza and B. Maleslnska, *J. Chem. Eng. Data*, 1981, **26**, 388.
77. J. Kiepe, S. Horstmann, K. Fischer and J. Gmehling, *Ind. Eng. Chem. Res.*, 2002, **41**, 4393.
78. A. Harvey, *AIChE J.*, 1997, **42**, 1491.
79. A. Setschenow, *Z. Phys. Chem.*, 1889, **IV**, 117.
80. F.A. Long and W.F. McDevit, *Chem. Rev.*, 1952, **51**, 119.
81. H. Corti, J. Pablo and J. Prausnitz, *J. Phys. Chem.*, 1990, **94**, 7876.
82. H. Corti, M. Krenzer, J. Pablo and J. Prausnitz, *Ind. Eng. Chem. Res.*, 1990, **29**, 1043.
83. A. Schumpe and S. Weisenberger, *AIChE J.*, 1996, **42**, 298.
84. P. Scharlin, *Carbon Dioxide in Water and Aqueous electrolyte Solutions*, IUPAC Solubility data Series, Vol 62, Oxford University Press, Oxford, 1996.
85. L.N. Plummer and E. Busenberg, *Geochim. Cosmochim. Acta*, 1982, **46**, 1011.
86. H.C. Lawrence, M.D. Elizabeth and A.J. Susan, *J. Phys. Chem. Ref. Data*, 1992, **21**, 941.
87. H.C. Lawrence and F.J. Johnston, *J. Phys. Chem. Ref. Data*, 1980, **9**, 751.
88. W. Preis and H. Gamsjäger, *J. Chem. Thermodyn.*, 2001, **33**, 803.
89. N. Kivekäs, *Modelling of Concentrated Electrolyte Solutions with Metal Carbonates and Carbon Dioxide*, Master's Thesis, (in Finnish) Helsinki University of Technology, Finland, 2005.
90. D.D. Wagman, *J. Phys. Chem. Ref. Data*, 1982, **11**, 1.
91. A. Roine, *HSC-Software ver. 5.11, 02103-ORC-T*, Outokumpu Research Centre, Pori, Finland, 2002.
92. G. Erikson and K. Hack, *Met. Trans. B*, 1990, **21B**, 1013.
93. P. Koukkari, K. Penttilä, K. Hack and S. Petersen, in *Microstructures Mechanical Properties and Process-Computer Simulation and Modelling*, Brechet Y. (ed), Euromat99 Vol 3, Wiley-VCH, 2000.
94. H. Gamsjäger, E. Königsberger and W. Preis, *Pure Appl. Chem.*, 1998, **70**, 1913.

95. E. Königsberger and G. Eriksson, *J. Solution Chem.*, 1999, **28**, 721.
96. R.J. Silbey and R.A. Alberty, *Physical Chemistry*, 3rd edn, Wiley, New York, 2001.
97. J. Salminen, P. Kobylin, O. Chiavone-Filho and S. Liukkonen, *AIChE J.*, 2004, **50**, 1942.
98. P. Debye and E. Huckel, *Physik. Z.*, 1923, 24, 185. Ref. from, H. Harned and B. Owen, *The Physical Chemistry of Electrolyte Solutions*, 2nd edn, Reinhold Publishing Corp., New York, 1950, 18.
99. K.S. Pitzer, Thermodynamics, 3rd edn, McGraw-Hill, New York, 1995.
100. C. Harvie and J. Weare, *Geochim. Cosmochim. Acta*, 1980, **44**, 997.
101. C. Harvie, N. Möller and J. Weare, *Geochim. Cosmochim. Acta*, 1984, **48**, 723.
102. E. J. Beckman, *Ind. Eng. Chem. Res.*, 2003, **42**, 1598.
103. M.E.Q. Pilson, *Introduction to the Chemistry of the Sea*, 1st edn, Pearson Education POD, 1998.
104. J.N. Butler, *Ionic Equilibrium. Solubility and pH Calculations*, Wiley, New York, 1998.
105. D. Langmuir, *Aqueous Environmental Geochemistry*, Prentice-Hall, New Jersey, 1997.
106. R. Haase, *Z. Phys. Chem.*, 1982, **132**, 1.
107. D. Kondepudi and I. Prigogine, *Modern Thermodynamics – From Heat Engines to Dissipative Structures*, Wiley, 1998.
108. P. Koukkari and R. Pajarre, *CALPHAD.*, 2006, **30**, 18.
109. P. Kobylin, J. Salminen, A. Ojala and S. Liukkonen, *Sixth Finnish Conference of Environmental Sciences*, Joensuu, Finland, 2003, 145.
110. K. Thomsen in *Thermodynamics and Industry*, T. Letcher, (ed), RSC, Cambridge, 2004.
111. T.J. Edwards, *Thermodynamics of Aqueous Solutions Containing One or More Volatile Weak Electrolytes*, Master's Thesis, University of California, Berkeley, USA, 1974.

CHAPTER 23

Solubility and the Oil Industry

ANTHONY R.H. GOODWIN,[1] KENNETH N. MARSH[2] AND COR J. PETERS[3]

[1] *Schlumberger Technology Corporation, 125 Industrial Blvd., Sugar Land TX 77478, USA*

[2] *Department of Chemical and Process Engineering, University of Canterbury, Christchurch, New Zealand*

[3] *DelftChemTech, Physical Chemistry & Molecular Thermodynamics, Julianalaan 136, 2628 BL Delft, The Netherlands*

23.1 Introduction

This chapter focuses on the energy sector and more specifically the oil and gas industries. These industries use the disciplines of both chemistry and physics and also require specialists with the following training: petroleum engineers, geologists, geophysicists, environmental scientists, geochemists, and chemical engineers. Not surprisingly, the literature on solubility for this chapter is as diverse in source as it is in the range of precision of the data. The topics of interest to the industry and relevant to this chapter are numerous and to name but only a few include coal gasification, enhanced oil recovery (EOR), CO_2 sequestration, flow assurance, hydrate inhibition, and natural and synthetic gas de-sulfurization. The oil and gas industry is in the business of extracting hydrocarbon but as a by-product obtains an aqueous phase from the formation. Globally, the volume of aqueous phase produced is greater than the volume of hydrocarbon and so in this chapter we cannot ignore the equilibria between partially immiscible phases of hydrocarbon and water. Ultimately, these industries are interested in (vapor+liquid+solid) equilibria (VLSE) between partially immiscible aqueous electrolyte solutions, hydrocarbon phases, and substances that are normally gases.

Figure 1 shows a schematic of a constant composition (p, T) section of a phase diagram for a reservoir fluid and illustrates the location of the (liquid+gas) phase border and (liquid+solid) phase transitions relative to each other. The solid phases found in the petroleum industry include hydrates, wax, and asphaltene. Wax and hydrates are predominantly formed by a decrease in temperature, whereas asphaltenes are formed by a pressure decrease at reservoir temperature. The (solid+liquid) phase behavior of petroleum fluids depends on

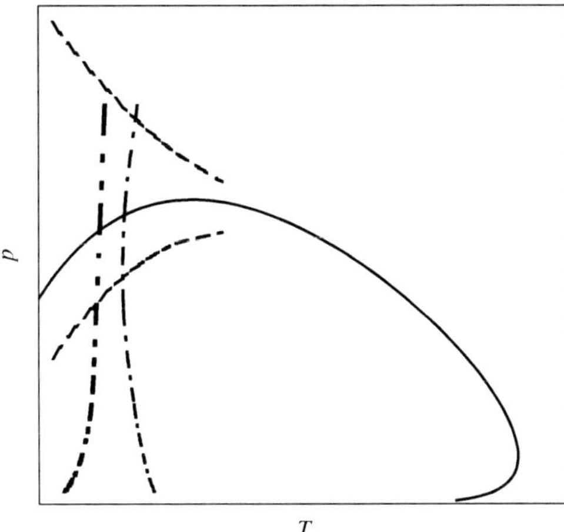

Figure 1 *$(p, T)_x$ section for a typical reservoir fluid showing the location of the following phase transition: ———, gas to liquid phase boundary;- - - - -, upper and lower asphaltene formation lines; — -— -, wax formation loci; and — - - — - -, hydrate formation loci.*

the distribution of the higher $[M(C_{25}H_{52}) \approx 0.350\ \mathrm{kg \cdot mol^{-1}}]$ molar mass hydrocarbons, such as asphaltenes, paraffins, aromatics, and resins, in the fluids. The formation of hydrates depends on the mole fraction of gaseous components such as N_2, CO_2, CH_4 through C_4H_{10} and the presence of an aqueous phase.

The issues of solubility are not confined to the reservoir and production tubulars but are also present at the separator where liquid can be entrained in the gas stream, with foamy oils that do not completely separate into gas and liquid. Further, when the liquid to gas volume ratio is low, gas can be found within the liquid stream. Thus knowledge of the ratio of the volume of the gas to that of the liquid at a temperature of 288 K and pressure of 0.1 MPa, known as the gas–oil ratio (GOR), is required. This ratio has many ramifications.

The naturally occurring fluids are often also exposed to other fluids during the production to either enhance the recovery of hydrocarbon or minimize formation of undesired phases, for example, gas hydrates. This VLSE can include supercritical components that are of profound theoretical interest, covered elsewhere in this and other volumes. From an applied perspective, the ability of supercritical fluids to attract low-volatility materials from mixtures has made supercritical fluid extraction an effective tool for EOR processes.[1–4] At the center of this technology lies the enhanced solubility of the solute in a supercritical solvent. Thus solubility is one of the most important thermophysical properties that must be determined and modeled for any supercritical fluid extraction process design.

EOR processes could include the injection of CO_2, water, including steam, or gases stripped from the produced reservoir fluid. In addition to the presence of three, or even more, phases the fluid can contain significant (mole fractions greater than 0.1) of hydrogen sulfide and thus be corrosive, particularly in the presence of an aqueous phase.

Hydrocarbon reserves can be at temperatures exceeding 473 K and pressures above 250 MPa. The task of describing solubility in the oil industry requires reservoir characteristics of temperature, pressure, VLSE, pH, and chemical composition. Therefore, the problem is, to say the least, complex. Consequently, there are a plethora of either empirical or semiempirical models that are used to describe solubility in systems relevant to the oil industry. Hydrocarbon mixtures can also experience temperatures below 273.15 K during processing, transportation, and storage.

The need for knowledge of hydrocarbon solubility in water arises from regulatory requirements that govern, among many aspects, the disposal of the produced aqueous phase. The removal of carbon dioxide from hydrocarbon fluids has received renewed attention with the implementation of the Kyoto Protocol and the desire to sequester greenhouse gases. The solubility of metallic mercury[5–9] as well as its complexes[7] is also significant to the production and processing of petroleum as are the solubility of other metals but these will not be discussed further in this chapter.

The purpose of this chapter is to focus on the types of solubility that are relevant to the oil industry and in this context we have, arbitrarily, chosen to consider only the following: (1) organic and inorganic gases in a liquid organic phase; (2) organic liquids in a gaseous phase; (3) organic and inorganic gases in an aqueous phase; (4) organic and inorganic gases in a solid organic phase; (5) organic solids in an organic liquid; (6) organic solids in an organic gas; (7) organic solids in an aqueous phase; (8) organic and inorganic gases in a solid aqueous phase; (9) liquid organic in liquid organic phase; and (10) liquid organic in liquid aqueous phase. In this chapter the requirement and application of measurements and correlations for each of these solubility categories are considered from a petroleum processing perspective along with some of the relevant data sources from the archival literature. We also present results that were obtained by one of us (Cor Peters) for ternary mixtures that have not been published in the archival literature. These results are significant because they are a comprehensive set of measurements and provide a means of validating either correlations or equations of state that might be used in the petroleum industry.

Other chapters of this book present additional material that is also relevant to the petroleum industry and includes the theory of solubility presented in Chapter 1 and the methods of measurement described in Chapters 4 and 8. The oil industry has also developed their own measurement schemes that include the determination of minimum miscibility pressures and so-called slim-tube measurements, which were designed to provide data for EOR, and these are described in this chapter. Solubility of substances in supercritical fluids is described in Chapter 9 while environmental issues are discussed in Chapter 21. Other chapters in this book that are of particular relevance to the oil industry, particularly hydrocarbon processing at pressures, relative to reservoir values, close to ambient, are Chapter 6, which describes the solubility of gases in water and seawater, and Chapter 8, which presents solubility of organic solids and is also concerned with (liquid+liquid) phase equilibrium and separation. For CO_2 sequestration this chapter presents recent developments.

The calculations performed in reservoirs are generally described by simulators in which the fluid containing formation is segmented into blocks. A reservoir simulation will require on the order of 10^6 calls to a package that calculates the thermophysical properties of the fluid and so the methods chosen to present these properties must not contribute significantly to the time required to perform the simulation. This requirement precludes, at least for routine work, the use of intensive calculation methods that are based on molecular models. Because of the requirement for simple correlations, for a particular process, often over a limited temperature and pressure range, the industry makes frequent utilization of both empirical and semiempirical methods.

23.2 Solubility Theories used in the Oil Industry

For a heterogeneous system of $\mathscr{P}$ phases containing ζ chemical components in complete thermal, hydrostatic, and diffusive equilibrium there are $(\mathscr{P} - 1)$ restrictions imposed on it so that the number of degrees of freedom of the system $\mathscr{I}$ of independent intensive variables is given by

$$\mathscr{I} = (\zeta + 1) - (\mathscr{P} - 1) = \zeta - \mathscr{P} + 2 \tag{1}$$

Equation (1) is the phase rule. In oil reservoirs, there could be a gas, liquid hydrocarbon, and aqueous electrolyte phase so that $\mathscr{P} = 3$. Limiting our discussion solely to the liquid hydrocarbon phase, ζ might exceed 100 so that $\mathscr{I} > 99$ and the use of the Gibbs–Duhem equation to estimate the phase behavior does not provide a viable solution. The reliability of any method chosen to describe the system is also limited by knowledge of the chemical composition of the fluid. Indeed, it is highly unlikely the chemical composition of the hydrocarbon phase will be known so that a unique solution to the equations cannot be found. Petroleum fluids were formed by anaerobic decay of prehistoric organic material, at elevated temperature and pressure. Hence if the geology of the formation is known then there is often sufficient information to predict, based on information from similar formations, the properties of the fluid within a sufficient degree of certainty.

The methods of determining the fluid composition have been discussed by Danesh.[10] Broadly speaking, the chemical composition of molecules with up to six carbon atoms are measured and hydrocarbons containing a greater number of carbon atoms, and higher molar mass, are determined in sub-sets. A subset might be defined as all hydrocarbons eluted from a gas chromatograph between the retention times for C_7H_{16} to that for C_8H_{18}.[11] From a thermodynamic perspective, these fractions are represented by so-called pseudo-components with associated critical properties. As the molar mass of the molecules increase so does the complexity of the analysis and it is more common in the oil industry to determine the ratio of saturates to aromatics to resins to asphaltenes, known by the acronym SARA analysis.

To ameliorate the hydrocarbon composition lumped pseudocomponents, methods of continuous thermodynamics or the thermodynamics of polydisperse fluids (as described by Halpin and Quirke[12] and Danesh[10]) have been applied to generate the required compositional profile as a function of species molar mass. The reader interested in these topics should refer to Danesh[10] and references cited therein.

It is often the case that substances are injected into the reservoir either to increase hydrocarbon recovery, for example, injection of CO_2, or to inhibit the formation of undesirable phases, for example, injecting methanol to prevent the formation of hydrates. These additional compounds, not necessarily those listed above, may also give rise to a chemical reaction. The $\mathscr{R}$ independent chemical reactions reduce $\mathscr{I}$ of Equation (1). During production, neither the fluid composition nor the state of the phases remain constant and these additional variables place even further demand on the predictive schemes used to estimate the thermophysical properties.

A solution is a special description of a mixture for which it is convenient to distinguish between the solvent A and the solutes B, C, *etc*. The solvent is often present in great excess compared to the solute and in that case the solution is then called a dilute solution; materials injected into reservoirs are not in excess. The colligative properties of a dilute solution follow from the theorem: in any sufficiently dilute solution the solvent behaves ideally. A binary mixture of solvent A and solute B is described by the absolute ideal activity of A, λ_A^{id}; the absolute activity of pure A, λ_A^*; the absolute ideal activity of B, λ_B^{id}; and the absolute activity of pure B, λ_B^*, provided the mole fraction $x_B \ll 1$.

The oil business, as do many others, resorts to approximations to describe the mixture of interest and invoke Raoult's and Henry's laws. For a binary liquid mixture $\{(1-x)A+xB\}$ Raoult's law states

$$(1-y)p^{1+g} = (1-x)\,p_A^{1+g}, \quad yp^{1+g} = xp_B^{1+g} \tag{2}$$

where the superscripts g and l denote the gas and liquid phases, and the subscripts A and B refer to the solvent and the solute, respectively. In equation (2), the x refers to the mole fraction of the condensed phase while y to the mole fraction of the gaseous phase.[13] A description of the use of Raoult's and Henry's laws in the petroleum industry is given by Danesh.[14]

The osmotic coefficient $\phi_{m,A}$ of the solvent A on the molality scale in a solution is defined by

$$\phi_{m,A} = \ln\left(\frac{\lambda_A}{\lambda_A^*}\right) \Big/ \left(M_A \sum_B m_B\right) \tag{3}$$

where λ_A is the absolute activity of A, λ_A^* the absolute activity of pure A, m_B the molality of solute B, and M_A the molar mass of A. A solution is called ideal-dilute if the osmotic coefficient $\phi_{m,A} = 1$ and if the activity coefficient of each solute $\gamma_B = 1$. γ_B is defined by

$$\lambda_B = m_B\,\gamma_B(\lambda_B/m_B)^\infty, \tag{4}$$

and ∞ denotes infinite dilution or $\sum_B m_B \to 0$. For an ideal-dilute solution then

$$\lambda_A = \lambda_A^* \exp\left(-M_A \sum_B m_B\right) \tag{5}$$

and

$$\lambda_B = (\lambda_B/m_B)^{\infty}\ m_B, \tag{6}$$

for all B. Equation (6) is Henry's law.

For solutions of electrolytes the Gibbs–Duhem equation is restricted to electrical neutral systems for which $\sum_i m_i z_i = 0$, where m denotes the molality and z the charge number of the ion i. The Gibbs–Duhem equation applies to a solution of electrolytes when the ions are treated as independent components. However, both the absolute activities λ and the activity coefficients of the individual ions are experimentally inaccessible.

For a dilute solution of nonelectrolytes, that is for $\sum_B m_B < 1\ \text{mol} \cdot \text{kg}^{-1}$, $\phi_{m,A} = 1$ and $\gamma_B = 1$ are good approximations because the pair-interaction energy decreases on dilution approximately to the square of the concentration.[15] In a solution of electrolytes, the pair-interaction energy decreases on dilution roughly to the cube root of concentration,[15] and thus even in extremely dilute solutions $\sum_B m_B < 10^{-4}\ \text{mol} \cdot \text{kg}^{-1}$ we cannot assume $\phi_{m,A} = 1$ and $\gamma_B = 1$. Fortunately, the exact forms of ϕ and the activity coefficients $\gamma_\pm$'s of the ions in the limit $\sum_i m_i \to 0$ are known as are approximations at finite molalities. The Debye–Hückel limiting law applies for dilute solution of electrolytes while the systems with finite molalities are described by the extended Debye–Hückel approximation. The latter includes an additional term $(1 + \beta d I^{1/2})$ in the denominator of the limiting law, where d, the mean diameter of the ions, is treated as an adjustable parameter and β is given by

$$\beta = 2\left(2\pi L \rho_A^*\right)^{1/2}\left(\frac{e^2}{4\pi\varepsilon k T}\right)^{3/2} \tag{7}$$

In equation (7), ρ_A^* is the density of pure solvent A, e the charge on a proton, and ε the electric permittivity of the solvent; $\varepsilon = \varepsilon_0\varepsilon_r$ where ε_0 is the permittivity of vaccum and ε_r the relative permittivity or dielectric constant. The product βd is about unity and is often simplified by setting $\beta d = 1$. Equation (11) has been extended, empirically, to higher molalities by the addition of terms proportional to I, $I^{3/2}$, I^2, and so on.

The activity coefficients of mixtures of solutions with more than one electrolyte are usually expressed in terms of the activity coefficient of corresponding solutions, each of a single electrolyte. These theorems assume that only the interactions of oppositely charged ions need to be taken into account. The reader interested in the thermodynamics of electrolyte solutions should consult, for example, the work of Robinson and Stokes for further details.[16] For concentrated multicomponent electrolyte solutions, the extended Pitzer ion interaction model, although complicated to use, is often the preferred approach;[17] ref. 17 provides an account of the thermodynamics of electrolytes solutions.

There are numerous models and theories for solubility that have been documented in the literature and the reader is referred to ref. 17 for a complete account; Chapters 5 to 11 of ref. 17 are of particular interest. These include models applicable to solubility of gases in liquids,[18–25] which are based on hard spheres,[26] perturbed hard chain,[27] and even neural networks.[28] Reference 20 includes a discussion of the effect of kinetics on solubility. Reid *et al.*[29] provide a review of the methods used to predict fluid-phase equilibria in multicomponent mixtures and include a discussion on solubility of gases in liquids and liquids in solids. The description of solid solubility in cryogenic solvents has been provided by Preston and Prausnitz.[30] The theories of Scatchard and Hildebrand[31] and O'Connell and Prausnitz[32] provide a thermodynamic treatment for mixed liquid solvents. There are also articles that discuss methods of predicting the solubility of organic compounds in aqueous solvents.[33,34] For the solubility of solids in liquids, Hansen parameters,[35] Flory–Huggins theory,[36] and neural networks[37,38] have been discussed. More recently, a general treatment of solubility has been presented by Katritzky *et al.*[39] The methods of statistical associating fluid theory (SAFT) have also been used to describe the solubility of gases in hydrocarbons.[40]

In the petroleum industry, as in most others, solubility is predicted using models based on activity coefficients[41] that are usually obtained from the Scatchard–Hildebrand approach; Danesh[10] provides an overview of these models. For example, the solubility of methane, ethane, carbon dioxide, and hydrogen in coal liquid fractions and solids has been described by Riazi and Vera.[42] In their model, solely the ratio of paraffin to naphthalene to aromatic composition of the fluid is required. Some other methods rely upon so-called cubic equations of state,[43,44] which are all derivatives of the van der Waals equation of state for hard-spheres, even when water is the solvent.[45] Other publications report models for the solubility of solids in liquids specifically intended for asphaltene precipitation and deposition. The model of Almehaideb[46] is for fluids that are near to the well-bore conditions. Predictive schemes, such as COSMO-RS, have been used to predict the mutual solubilities of hydrocarbons in water,[47] while the UNIFAC method has been used to estimate solubility of gases,[48] hydrocarbons in water,[49] industrially important solvents,[50] and of carbon dioxide, oxygen, methane, and ethane in liquid alkanes, albeit at a pressure of 0.1 MPa that is only relevant to fluids that are stored in tanks after transmission and processing.[51] The dependent local compound (DDLC)[52] model for the Helmholtz energy of a fluid mixture has been shown to accurately represent the solubility of inert and acidic gases and of hydrocarbons from CH_4 to C_6H_{14} in water at pressures up to 30 MPa.

23.3 Experimental Methods

The oil industry, just like any other, has its own terminology for experiments that have been designed to answer specific questions posed by the environment in which they operate. In the petroleum industry, there are experiments that are known by either the term shrinkage or the term swelling. In thermodynamists language, these terms refer to the determination of the molar volume of mixing two miscible phases and, thus, shrinkage and swelling refer to negative and positive volume of mixing, respectively.

Petroleum reservoirs at a vertical depth are essentially at constant temperature, thus engineers usually discuss minimum miscibility pressure rather than either solubility or for that matter (vapor+liquid+solid) equilibrium. The minimum miscibility pressure is required to engineer miscible gas injection projects, for example, EOR with carbon dioxide. The measurement methods used to provide the data required to design this EOR process are called swelling, multicontact, slim tube, rising bubble, and core floods. Here we will confine our discussion to rising bubble and slim tube, which is also often referred to as a single contact experiment; in this context, contact refers to the number of times the solvent, in this case a gas, is exposed to the liquid.

In a rising bubble experiment, the shape and size of a gas bubble injected into a reservoir liquid are observed. Usually, the apparatus consists of a high-pressure visual cell with a glass tube mounted along the axis.[53] The tube is filled with reservoir oil and, after equilibration at the desired temperature and pressure, a small, relative to that of the oil, volume of gas is injected into the oil column, often through water, at the base of the tube. It is usual to assume that this represents the first contact of gaseous solvent with reservoir oil and that during the bubbles passage upward, relative to gravity, it will, depending on the pressure of the reservoir fluid, either reach equilibrium with the oil or be miscible. At pressures less than the minimum miscibility pressure the bubble retains an almost spherical shape, albeit with a reduced size because some of the gas dissolves into the oil. At pressures above the minimum miscibility pressure the bubble disperses rapidly and disappears. A gas bubble that achieves miscibility will also disappear into an under saturated oil but will not disperse thus providing a region of, usually, lower density. The rising bubble apparatus experiment is considered to be a rapid means of estimating the minimum miscibility pressure for (liquid+gas). The values obtained generally agree with those obtained from other methods, including the slim tube.

Gas injection processes are often simulated in batch type tests, known as multiple contact experiments, which are conducted within a (p, V, T) apparatus. In this method, finite volumes of reservoir fluid and injected gas are repeatedly contacted and shrinkage or swelling (volume of mixing) of the oil occurs and the density and composition of the equilibrated oil and gas are measured. The most common experiment is the swelling test or single contact gas injection. In this approach, a known volume of oil is added to a (p, V, T) cell and gas is then injected in a stepwise process. After each injection, the bubble pressure and volume of the liquid are determined.

A slim-tube experiment[54] is effectively a one-dimensional model of a reservoir. The tube consists of a narrow tube, of length between (5 and 40) m, the inside of which is filled (packed) with either sand or glass beads. Initially, the tube is filled with oil at reservoir temperature at a pressure above the bubble pressure. The oil is then displaced by injecting gas into the tube at a constant pressure; practically, this is achieved by controlling the outlet pressure with a backpressure regulator. The slim-tube effluent is expanded to $T \approx 293$ K and $p \approx 0.1$ MPa and the fluid phases collected. The volumes of the fluid phases recovered are determined as a function of time and the density and composition of each phase also determined. Visual observation of the fluid eluted from the column is often used to aid the phase identification process. The chemical composition of the effluent is monitored for the first appearance of the solvent, in this case a gas, to obtain the time at which the solvent first exited the column. As an alternative to performing analyses of the chemical composition, measurements of the ratio of the gas-to-liquid (oil) volumes, at $T \approx 293$ K and $p \approx 0.1$ MPa, can be used to determine elution of solvent from the column, often referred to as break-through. In the slim-tube experiment, miscibility conditions are determined by conducting the displacement at either a series of pressures or different gas volume injection rates and monitoring the volumes of oil and gas recovered.

Both the rising bubble and slim-tube experiments provide minimum miscibility pressure but often not the volumetric and compositional data required for evaluation and adjustment of the models used in the industry to predict reservoir fluid phase behavior. The maximum solubility of a gas in a liquid is the bubble curve while the maximum solubility of a liquid in a gas is the dew curve. There are many other methods of obtaining information on the phase behavior of fluids and these have been reviewed in other publications including the most recent accounts given in the *Experimental Thermodynamics Volume VII*, an International Union of Pure and Applied Chemistry (IUPAC) volume, edited by Weir and de Loos.[55]

There are numerous other methods reported in the literature for the determination of the solubility of gases in liquids,[56–58] solids in liquids and gases, and liquids in liquids,[59] some of which could be utilized by the oil industry. These methods include the use of radioactive sources for organic solvents in water,[60] gas chromatographic methods to determine gas solubility in liquids,[58,61–63] quartz microbalances,[64] measurements of refractive index,[65,66] relative electric permittivity measurements,[67] and with differential scanning calorimetry.[68] Markham and Kobe[69] have reviewed the solubility of gases in liquids and also discuss in detail the array of experimental methods that have been reported up to the year 1941 and include work from 1808. Thus solubility has been a topic of scientific interest for over 200 years. However, only recent publications[57,70–72] provide an estimate of the uncertainty of the measurements, which range from (1 to 3) %.

For the processing of oil, the industry requires measurements of solubility. The methods developed for these measurements are often required to operate at temperature exceeding 473 K and pressure up to 250 MPa.[73] Others are required to provide results with fluids that are only partially miscible.[74,75] Most of the techniques that have been reported in the archival literature use visual cells,[76] or flow[77] or volume measurements to determine solubility.[78] However, ref. 78 concludes that methods developed for phase equilibria measurements are to be preferred over those designed specifically to obtain solubility. In general, determining solubility is a "poor mans" alternative to measurements of (vapor+liquid+solid) equilibria; measurements of solubility have been used to predict (vapor+liquid) equilibria for C_3H_8 to C_5H_{12} hydrocarbons.[79] However, when the solubility of volatile components within a nonvolatile material are required, volumetric methods have the distinct advantage of providing (vapor+liquid) equilibria from (p, V, T) measurements, without recourse to sampling and chemical analysis of either phase, provided in the analysis it is assumed the vapor contains essentially none of the nonvolatile components.

23.4 Relevance of Solubility Measurements

In this section, we intend to demonstrate the need for solubility data by reference to examples of where the data has been used to design processes and to reduce the formation of undesirable phases within the production system. The most extensive compilation of data relevant to the oil and gas industry can be found in the series of publication known as the *Solubility Data Series* and more recently those published in the *Journal of Physical and Chemical Reference Data* by IUPAC and the National Institute of Standards and Technology (NIST), known under the acronym IUPAC-NIST solubility data publications. An on-line source of recommended values is available from the NIST chemistry web book, which has been summarized by Linstrom and Mallard.[80] The output of the IUPAC-NIST collaboration between the years 1973 and 2001 has been reviewed by Clever.[81] The NIST Standard Reference Database 106 contains the IUPAC-NIST solubility data series. Some of the evaluations documented in both the IUPAC-NIST and *Solubility Data Series* publications are cited in the relevant subsections below. Because of the *Solubility Data Series* it is neither essential nor desirable to even attempt to review the archival literature; our intent is to simply cite some useful articles for both the measurements and their use.

23.4.1 Gases in Liquids

The use of (methane+ethane+propane) as the gas used to push (drive) the oil out of a reservoir in the enhanced (or improved) recovery of crude oil has been investigated extensively by the petroleum recovery industry. This approach to extracting oil is attractive, particularly in remote locations, where, because of the absence of either a natural gas pipeline or a liquefaction plant, the alternative to re-injection of reservoir gas might be to burn it (referred to as flaring). For this reason, there have been extensive measurements of the solubility of hydrocarbon gases in liquid hydrocarbons at temperatures and pressures encountered in the reservoir to provide data against which equations of state may be adjusted and then used to model rich-gas injection processes. This topic is considered further in Section 23.4.2. The solubility of gases in liquids has been reviewed by Markham and Kobe[69] and Battino and Clever.[82] Infinite dilution Henry's constants of CH_4 to C_4H_{10} in hexadecane, octadecane, and 2,2,4,4,6,8,8-heptamethylnonane has been measured for the purpose of inert gas stripping.[83]

The solubility of methane has been measured at a function of temperature up to 423 K and pressure below 42 MPa in binary mixtures with benzene,[84] toluene,[85] cyclohexane,[86] hexane, decane, and dodecane,[87] bicyclo[4.4.0]decane,[88] naphthalene, phenanthrene, and pyrene,[89] hexatriacontane,[90] and eicosane, octacosane, hexatiacontane, and tetrateracontane;[91] for some of these substances that are normally solid the melting temperature determined the lower bound for the solubility measurements. The results reported in ref. 90 cover temperatures between (373 and 573) K and pressures up to 5.1 MPa. Young[92] has provided a critical review of the solubility data for methane in *m*-xylene and naphthalene and for methane in ethane,[93] while Clever[94] has reviewed the solubility of methane in benzene, methylbenzene, diphenylmethane, 1-methylnaphthalene, and decahydronaphthalene at temperatures between (283.15 and 473.15) K. The solubility of ethane in decanes at temperatures between (278 and 411) K and pressures below 8.2 MPa have been measured by Bufkin and Robinson.[95] Young[96] has also reviewed the solubility measurements for ethane in propene, benzene, hydrogen sulfide, and methanol including temperatures in the range (270 to 540) K and pressures up to 41 MPa. Chapoy *et al.*[97] have reported measurements of, and a correlation for, the solubility of (methane+water) and (methane+ethane+butane+water) at temperatures between (275 and 313) K and pressure up to 18 MPa. The solubility data for propane, butane, and 2-methylpropane in kerosene and crude oil have been evaluated in a IUPAC project.[98] Ethene solubility in eicosane, octacosane, and hexatriacontane has been reported by Chou and Chao[99] at temperatures up to 573 K and pressures below 5.1 MPa, and was primarily acquired to model bubble columns within the Fischer–Tropsch synthesis; the Fischer–Tropsch process is used to convert coal to diesel, waxes, and alcohols.

When the reservoir produces insufficient rich-gas and the reserve is geographically located on land, in some cases the only viable alternative is to inject air. Therefore measurements of the solubility of oxygen, nitrogen, and argon in hydrocarbons are significant and there are measurements for O_2 in benzene;[100] nitrogen, or oxygen or argon in benzene;[101] hydrogen in methylbenzene;[102] hydrogen and carbon monoxide in octacosane;[103] and He, Ne, Ar, Kr, N_2, O_2, and CH_4 in C_2H_6 to $C_{16}H_{34}$ albeit at $T = 298$ K and $p = 0.1$ MPa;[104] and crude oil.[105] The measurements reported by Simnick *et al.*[102] are at temperatures of (462, 502, 542 and 575) K at pressures between (2 and 25) MPa. Reference 101 is also of interest because it provides measurements of diffusivity, which is another significant factor to be understood for the injection process, while the results presented in ref. 103 were measured, for Fischer–Tropsch synthesis, at a temperature of 528 K and pressures between (1 and 3) MPa. There are also critical evaluations of the data for argon with methane and ethane;[106] tetradecane, pentadecane, and hexadecane;[107] benzene;[108] methylbenzene and dimethylbenzene;[109] methylcyclohexane, dimethylcyclohexane, and cyclooctane;[110] cyclohexane;[111] and dodecane and tridecane.[112] A critical evaluation of the solubility measurements for nitrogen in cyclohexane at temperatures between (283.15 and 308.15) K and a pressure of 101.3 kPa has been reported by Battino.[113] Battino[114] has also provided a critical evaluation of the solubility of nitrogen in benzene, xylene, and 1-methylnaphthalene at a pressure of 0.1 MPa and temperatures between (278 and 475) K. Young[115] has reviewed the solubility of helium in methane, ethane, and propane while Cargill the solubility of carbon monoxide in cyclic hydrocarbons,[116] with aromatic hydrocarbons[117] and aliphatic hydrocarbons.[118] Data on the solubility of hydrogen are required for understanding coal liquefaction. Solubilities have been reported in butane,[119] hexane,[120] 2-phenylpropane, cyclohexane, 2-phenyl-1-propene,[121] and octane, octanol, and squalene.[122] Hydrogen solubility has also been measured at temperatures of (398, 473, 548, 523, and 673) K and pressures of (5, 10, 15, 20, and 25) MPa in 1,2,3,4-tetrahydronaphthalene, in (1,2,3,4-tetrahydronaphthalene+2-methylnapthalene+1-hydroxy-4-methylbenzene+4-methylpyridine) and two coal-based liquids[123] obtained from a distillation. Riazi and Al-Roomi[124] have described a model to predict the solubility of hydrogen in petroleum fuels. The solubility of (hydrogen+methane) and (hydrogen+methane+ethane) in toluene and (toluene+eicosane) have been measured by Peramanu *et al.*[125] at pressures below 17.3 MPa and a temperature of 295 K.

The solubility of hydrogen sulfide in eicosane has been measured by Feng and Mather[126] at temperatures up to 423 K and pressures below 7.6 MPa while Sciamanna and Lynn[127] have measured the solubility of hydrogen sulfide, sulfur dioxide, carbon dioxide, propane, and butane in diethylene glycol dimethyl ether, triethylene glycol dimethyl ether, tetraethylene glycol dimethyl ether, diethylene glycol dimethyl ether, and triethylene glycol butyl ether for the purpose of developing a sulfur recovery process for gas streams that produce elemental sulfur. Yokoyama *et al.*[128] have reported the solubility of hydrogen sulfide in 2,2,4-trimethylpentane, decane, tridecane, hexadecane, and squalane at temperatures from (323 to 523) K and pressures up to 1.6 MPa. The solubility of ethane in hexane at temperatures between (273 and 303) K and of ethane in C_6H_{14} through C_9H_{20} were determined by Waters and Mortimer,[129] while solubility in hexatriacontane is reported in ref. 90 at temperatures up to 573 K and pressures below 5.1 MPa.

Carbon dioxide is used in EOR because its solubility in oil is greater than that of either methane or ethane. Thus the solubility of CO_2 in hydrocarbons has received considerable attention in the literature and with that data expert systems have been developed to design EOR processes.[130] In addition to EOR, issues of global warming have generated interest in CO_2 sequestration and one option to achieve this is by injecting CO_2 into aquifers and oil reservoirs.[131] Aycaguer *et al.*[132] has provided a life-cycle analysis for sequestering CO_2 in a Texas, USA, oil well. CO_2 EOR is a topic for further discussion for the extraction of heavy oil that is confined in this work to Section 23.4.4 even though heavy oil can be in either a liquid or a solid phase.[133] Correlations for the solubility of gas in heavy liquid oil have been presented.[134]

Quail *et al.*[135] have measured the solubility, viscosity, and density of 59 heavy crude oil samples, taken from different producing areas of Saskatchewan, Canada, as a function of the concentration of CO_2 at temperatures between (293 and 383) K and pressures from (0.1 to 14) MPa. These measurements were used to obtain empirical correlations for the solubility, viscosity, and density as a function of pressure and CO_2 concentration; the viscosity decreased at a temperature with increasing CO_2 concentration. This is important data because, as we will discuss further in Section 23.4.4, CO_2 can be used to increase the mobility of viscous heavy oil that contain significant solid.

For EOR, the solubility of CO_2 and the volume of mixing (swelling) have been discussed by Mulliken and Sandler[136] and Moore *et al.*[137] Henson *et al.*[138] have measured the solubility of CO_2 in tetralin, 1-methylnaphthalene, and liquid hydrocarbons (defined as creosote), which was derived from coal, at temperatures up to 673 K and pressures below 20 MPa. Hong and Kobayashi[139] have shown that propane breaks the (carbon dioxide+ethane) azeotrope, which is significant for CO_2 EOR gas processing, while the same authors also provide the (vapor+liquid) equilibria data required to design extraction distillation for the resulting gas mixture.[140]

The solubility of CO_2 in heptane, dodecane, and hexadecane has been reported at a pressure of about 0.1 MPa and temperatures between (273 and 323) K by Hayduk *et al.*[141] while data for CO_2 in hexatriacontane are given in ref. 90. Huang *et al.*[142,143] have measured the solubility of CO_2 in octacosane[142] and eicosane,[143] at temperatures between (373 and 573) K and pressures up to 5 MPa. Such data are required to correlate gas solubility in wax slurries used in the Fischer–Tropsch process. Tanaka *et al.*[144] have determined, at $T = 313$ K and $p = 0.1$ MPa, the solubility of CO_2 in pentadecane, hexadecane, and (pentadecane+hexadecane) and Tsai and Yau[145] the solubility of CO_2 in tetracosane and dotriacontane also for the Fischer–Tropsch process.

The importance of the solubility of carbon dioxide in water arises from the use of carbon dioxide as a solvent in EOR processes and, more recently, because of the interest in sequestering carbon dioxide in aquifers. At a temperature of 373 K and pressure of 40 MPa, up to 33 cm^3 of CO_2 will dissolve in 1 g of water. If a formation has a porosity (the ratio of the volume of interstices of a material to the volume of the material) of 0.2 and it is permeable then 6.6 m^3 of CO_2 at ambient temperature and pressure can be dissolved in 1 m^3 of aquifer; this estimated value for solubility was arrived at assuming diffusion is instantaneous and there is a seal on the aquifer (often called caprock) that prevents CO_2 from leaking back into the atmosphere through the ground. Fortunately, there are numerous measurements of the solubility of carbon dioxide in water. The most recent results were presented by Chapoy *et al.*[146] at temperature between (274 and 351) K and pressures in the range (0.2 to 9.3) MPa. Over these ranges of temperature and pressure the maximum mole fraction of CO_2 dissolved was about 0.025 at $T = 292$ K and $p = 5.2$ MPa. Chapoy *et al.*[146] have also reviewed the literature data and models used to describe the solubility of carbon dioxide in water. Munjal and Stewart[147] have reported an equation for the solubility of the gas in seawater, which is based on their measurements at temperatures between (268 and 298) K at pressures below 4.5 MPa.[148] Wiebe and Gaddy[149] have reported the solubility of carbon dioxide in brine at temperatures up to 373 K and pressures below 70 MPa and there are also data for the solubility in the vicinity of the critical solution pressure.[150] Kiepe *et al.*[151] have reported measurements of CO_2 in water and brine and a method of predicting the solubility. The solubility of carbon dioxide and methane, or ethane, or propane or butane in water was reported by Dhima *et al.*[152] at $T = 344$ K and pressures from (10 to 100) MPa, conditions where the mutual solubility of the hydrocarbon in water becomes significant.

The solubility of methane in brine has been measured at temperatures up to 573 K and pressures of over 200 MPa.[153–156] Solubility measurements have been reported for methane, ethane, butane, and mixtures of these components in water at temperatures of 344.3 K and pressures between (2.5 and 100) MPa.[157] Henry's constants for methane and ethane in water have also been reported.[70] The solubility of nitrogen,[158,159] oxygen,[160] and (hydrogen+nitrogen)[161] in water has been reported at pressures up to 100 MPa and temperatures exceeding 473 K. The results reported in ref. 161 showed the solubility of (hydrogen+nitrogen) could be estimated to be within a few percent from values for the pure gases. The rate at which oxygen dissolves in water has also been determined.[162] Butane solubility in water and brines has been reported by Rice *et al.*[163] There are also measurements of the mutual solubility of (methane+ethane+butane) in water.[164] Models for the solubility of gases in brine, based on cubic equations of state, have been discussed in the literature.[165]

When carbon dioxide, hydrogen sulfide, and nitrous oxide dissolve in an aqueous solution they form acidic fluids. The solubility of these gases has been measured usually in aqueous solutions of alkanolamines[166–169] because they are relevant to their removal in the so-called sweetening of sour gases.

Many oil-wells are drilled with a hydrocarbon based drill-bit lubricant (often referred to as mud), to which are added emulsifiers, viscosifiers, bridging solids, and weighting agents to form the complete drilling fluid. These drilling fluids are referred to as either oil-based drilling fluids (OBDF) or water-based drilling fluids (WBDF), which uses water as the solvent. WBDF is essentially immiscible with hydrocarbons. In most cases, to satisfy environmental regulations regarding the disposal of OBDF and the rock cuttings carried with it from the drill bit, the solvent used for OBDF is $\{(1-x)C_{16}H_{34} + xC_{18}H_{38}\}$ but diesel and other mineral oils are also used. Sodium bentonite is added to the OBDF to increase the density of the drilling mud. Gases are soluble in drilling fluids, particularly OBDF, and measurements of the solubility of methane in OBDF have been performed.[170]

The solubility of oxygen and nitrogen in molten petroleum fractions and microcrystalline waxes, which may be significant for *in situ* combustion,[137] has been reported by Ridenour *et al.*,[171,172] while the solubility of hydrogen in bitumen has been documented by Lal *et al.*[173] and Cai *et al.*[174] Their results for methane extend to temperatures of 353 K and pressures up to 20.1 MPa.

23.4.2 Liquids in Gases

The solubility of crude oil, with a density of about $806\ kg \cdot m^{-3}$, in methane has been reported by Price *et al.*[175] at temperatures between (323 and 523) K and pressures in the range (5.1 to 102) MPa. Not surprisingly, the solubility increased with increase in both temperature and pressure with values of 1 g of oil in $10^{-3}\ m^3$ (at $T = 293$ K and $p = 0.1$ MPa) of methane at temperatures in the range (323 to 473) K and pressures between (35 and 100) MPa. At $T = 373$ K and $p = 104$ MPa and $T = 473$ K and $p = 52$ MPa the solubility was greater than 5 g of liquid in $10^{-3}\ m^3$ (at $T = 293$ K and $p = 0.1$ MPa). Qualitative chemical analyses of the crude-oil solute at low pressure and temperature showed that the mole fractions of C_5H_{12} to $C_{15}H_{32}$ as well as for the saturated hydrocarbons $C_nH_{(2n+2)}$ with $n > 15$ had increased relative to the original oil. As the temperature and especially pressure increased, the composition of the solute approached that of the starting crude oil. When the results of this study were compared with the solubility of alkanes in methane, it was found that (methane+water) could dissolve more crude oil than methane alone and that the presence of water also drastically lowered the temperature and pressure required for a given solubility. Price *et al.*[175] suggested that primary migration by gaseous solution could "strip" a source rock of crude-oil-like components, leaving behind a bitumen-like substance very unlike the original crude oil. The results showed that methane by itself could neither dissolve a sufficient amount of crude oil nor dissolve the higher molar mass components such as tars and asphaltenes. This information is considered significant by the petroleum industry particularly when oil fields are owned by more than one company and the equity is split with one company owning the gas cap and the other the liquid phase, and gas cap re-injection into the reservoir is made for the purpose of pressure maintenance.

Price *et al.*[175] also measured the solubilities of two high molar mass petroleum distillation fractions in methane at temperatures from (323 to 523) K and pressures from (31 to 174) MPa. The results for the petroleum distillation fractions, one of which was the highest molecular weight material of petroleum (boiling temperature greater than 539 K at a pressure of 800 Pa), were similar to those described for crude oil. These results when combined with values from the literature demonstrate the addition of carbon dioxide, ethane, propane, or butane to methane also increases the solubility of crude, as does the presence of fine-grained rocks. The solubility of C_7H_{16} to $C_{36}H_{74}$ in methane in the presence of water has been measured at temperatures in the range (293.2 to 423.2) K, and pressures between (12 and 24) MPa.[176]

23.4.3 Liquids in Liquids

Hydrocarbon-bearing formations can also contain water that is referred to as connate water. An aquifer is often, but not always, located beneath a hydrocarbon-bearing zone. Water or brine is also injected into the reservoir, in a method known as water-flooding, to increase the recovery of oil from reserves for which the produced fluid volume decreases with increasing time (these are said to be depleted). Indeed, on a global basis, it has been estimated that about 4 m^3 of water are produced for every 1 m^3 of oil. Therefore the chemistry of water, both connate and injected, and hydrocarbon is important to the petroleum industry. In particular, the propensity of water to deposit calcite, barite, or halite scale is a problem that occurs when aquifer water mixes with injected water.[177] This scale forms on the flow-line walls and consequently reduces the cross-sectional area of the tubulars and ultimately decreases productivity. There have been numerous studies of the solubility of hydrocarbons in water and water in hydrocarbons.

Because hydrocarbon-bearing formations are in contact with an aqueous phase there have been numerous studies of the solubility of liquid hydrocarbons in water.[178–183] Assessing the environmental impact of hydrocarbon spills has been another motive for these studies[184] that have also included the additives used in gasoline,[185] and the solubility of naphthalene in seawater[186] and water.[187] In these studies, particular attention has been given to aromatic hydrocarbons in water[188–195] and aqueous electrolyte solutions[196–198] including seawater.[199–201] There have also been measurements of the solubility of water in nonane and decane[202] and reservoir fluids at pressures up to 1 GPa and temperatures of 473 K.[203] Connolly[204] reported the solubility of heptane, pentane, 2-methylpentane, methylbenzene, and benzene in water at temperatures from 260 K to near the critical solution temperature of about 300 K at pressures between (10 and 82) MPa. Sharp maxima and double branches were observed in the solubility isotherms near to the critical solution temperature of these hydrocarbons in water. The solubility of alkanes[205,206] and cycloalkanes[207] in water and cycloalkanes in seawater[208] has been measured as well as the solubility of mixtures including (hexane+heptane)[209] and (aniline+cyclohexane) in water.[210] Economou *et al.*[211] report the solubility of water with 1-hexene, 1-octene, decane, butylcyclohexane, *m*-diethylbenzene, *p*-diisopropylbenzene, *cis*-decalin, tetralin, 1-methylnaphthalene, and 1-ethylnaphthalene. Yaws[212,213] has reviewed the solubility of 400 organic compounds in water and has also provided an empirical correlation of the solubility of C_5H_{12} hydrocarbons in salt water.[214] (Liquid+liquid) equilibria of hydrocarbons in water have been reviewed in the literature.[215] Recent publications from the IUPAC-NIST solubility data series report the solubility of hydrocarbons in water and seawater. These include the aqueous solubility of hydrocarbons with five carbon atoms,[216] benzene,[217] C_6H_8 to C_6H_{12},[218] C_6H_{14},[219] hydrocarbons with seven carbon atoms,[220] C_8H_8 to C_8H_{10},[221] hydrocarbons with nine carbon atoms,[222] and C_8H_{12} to C_8H_{18}.[223]

Not surprisingly, the study of the solubility of liquid hydrocarbons in each other is rather more limited.[224] The solubility of normal alkanes ($C_{20}H_{42}$ to $C_{24}H_{50}$) and some of their binary mixtures ($C_{22}H_{46}+C_{24}H_{50}$) and ($C_{23}H_{48}+C_{24}H_{50}$) has been determined in ethylbenzene.[225] (Liquid+liquid) equilibria for ($H_2O+C_2H_5OH+C_6H_{12}$) and ($C_2H_5OH+C_6H_6+C_6H_{12}+H_2O$) have been reported by Gramajo de Doz *et al.*[226] who in another article[227] also report the phase behavior of ($H_2O+CH_3OH+C_7H_8$),

($CH_3OH+C_7H_8+C_8H_{18}$), ($H_2O+C_8H_{18}+C_7H_8$), and ($H_2O+C_2H_5OH+C_7H_8$), ($CH_3OH+C_7H_8+C_8H_{18}+H_2O$) and ($C_2H_5OH+C_7H_8+C_8H_{18}+H_2O$). The measurements from refs. 226 and 227 were correlated by the UNIQUAC method.

Skzecz[228] has evaluated the solubility of methanol with butane through hexadecane, while Mabery[229] has reported the solubility of petroleum in alcohols, both of which are relevant to hydrate inhibition. The solubility of water in mixtures of (ethanol+2,2,4-trimethylpentane), (ethanol+2,2-dimethyl-butane), (ethanol+2,2,4-trimethylpentane), (ethanol+2,4,4-trimethyl-1-pentene), (ethanol+2,2,4-cyclohexane), (ethanol+methylcyclohexane), and (ethanol+cyclohexene) has also been studied at temperatures between (228 and 298) K.[230] Ethanol is added to gasoline (petrol) to reduce emission especially during the winter months in the USA and, therefore, the solubility of ethanol in gasoline as a function of temperature has also been determined.[231,232] The effect of methanol and ethanol on the solubility of hydrocarbons in water has been reported.[233]

Chemicals added to both production streams and processing equipment might have a negative impact on the marine environment and also ultimately be a potential threat to humans. Thus there has been a drive to know the solubility of these chemicals in the gas, oil, and water phases. Alcohols are added to inhibit the formation of hydrates and in gas dehydration systems. These environmental concerns have driven Derawi *et al.*[234] to measure the (liquid+liquid) equilibria of (monoethylene glycol+heptane), (methylcyclohexane+hexane), (propylene glycol+heptane), (diethylene glycol+heptane), (triethylene glycol+heptane), and (tetraethylene glycol+heptane) at a pressure of 0.1 MPa and temperatures between (303 and 353) K; monoethylene glycol is used as a hydrate inhibitor and triethylene glycol in gas dehydration systems. These results were used to test the validity of NRTL and UNIQUAC models for these systems.

Water soluble hydrophobically associating polymers are used in oilfield applications including drilling as well as water and chemical flooding for improved oil recovery. Taylor and Nasr-El-Din[235] have reviewed these compounds and their solubility in water.

23.4.4 Gases in Solids

As conventional oil reserves are depleted, the focus for future production leans toward deposits that are in deep water, or exist as heavy oil or deposits of methane adsorbed in coal. Heavy oil includes substances that are more commonly known as bitumen and tar. These materials are often located in unconsolidated sand and, when sufficiently shallow, can be produced with methods similar to those used in open-cast coal mines and are referred to as strip mining. These sources of hydrocarbons are often referred to as heavy because their molar mass is higher than the mean of a typical liquid hydrocarbon reserve. Indeed, substances are usually referred to as heavy when their density is greater than 925 $kg \cdot m^{-3}$, about a factor of 1.2 times that of octane at $T = 293$ K and $p = 0.1$ MPa; bitumen has a density of about 1050 $kg \cdot m^{-3}$.[236] Not surprisingly, these substances also have viscosities that can be several orders of magnitude greater than octane (and for that matter the viscosity of the majority of reserves in production at the time of writing) for which the viscosity is about 1 mPa · s at $T = 293$ K and $p = 0.1$ MPa. Both bitumens and heavy oils contain significant quantities of high molar mass. In the oil industry, asphaltenes are defined by their solubility class: asphaltenes are polar molecules and are soluble in aromatic but not aliphatic hydrocarbons. Asphaltenes are self-associating and form aggregates.[237] The presence of asphaltenes has been reported to significantly increase the oil viscosity. In some cases asphaltenes have been deliberately precipitated to decrease the viscosity and hence increase the productivity of a reservoir. Heron and Spady[238] as well as Oblad *et al.*[239] have described oil and tar sands, respectively, while Speight[240] and Ovalles *et al.*[241] have discussed the issues of upgrading a heavy feedstock. Mullins[242] has edited a book that describes some of the properties of asphaltenes.

The solubility of gases in solids is a subject of much interest to the oil and coal industries because these solutes can change the phase of the solvent and therefore provide a mobile phase[243] and enhance the production of both wax and asphaltenes. The solubility of wax in hydrocarbon gases is of interest for retrograde condensate fields.[244]

The high solubility of carbon dioxide in heavy petroleum fractions is significant[245] because CO_2 is often used as a solvent to enhance the rate of recovery from a hydrocarbon reserve; CO_2 when compared with nitrogen and methane provides the greatest decrease in viscosity according to Svrcek and Mehrotra.[246] The solubility of CO_2 in bitumen and models for it that are based on the Peng Robinson cubic equation of state have been reported by Deo *et al.*[247] and commented on in ref. 248. Gasem and Robinson[249] have measured the solubility of CO_2 in $C_{20}H_{42}$ through $C_{44}H_{90}$ at temperatures between (323 and 423) K at pressures up to 9.6 MPa. These systems are of interest to petroleum processing, production of coal liquids, and EOR. The solubility of (methane+ethane+carbon dioxide) in hexatriacontane was reported by Tsai *et al.*[250]

23.4.5 Solids in Gases

Solid hydrocarbons dissolved in natural gas can deposit on the walls of pipelines. For this reason, the solubility of diamondoids, adamantine, and diamantine in carbon dioxide, ethane, and methane has been measured by Smith and Teja and a correlation developed as a function of solvent density.[251] Measurements of the solubility of naphthalene in ethane have been reported,[252] while Gopal *et al.*[253] report the solubility of naphthalene, phenanthrene, phenol, and biphenol in supercritical CO_2. Others have reported the solubility of heavy aromatic hydrocarbons[254] and waxes[255] in CO_2 at pressures up to 190 MPa.

23.4.6 Solids in Liquids

Here we refer to substances that are, at $T = 293$ K and $p = 0.1$ MPa, solids and liquids. At reservoir temperatures and pressures these may not be the phases in which the substances exist. The formation of a solid phase in the pores of a formation may significantly reduce the permeability and porosity of the rock and thus the production from it, and are to be avoided if at all possible. Solids can also deposit, as they do for natural gas discussed in Section 23.4.5, in the production tubulars and reduce the volume throughput of the system and ultimately can halt production. These issues are part of a broad theme in petroleum engineering called flow assurance. Not surprisingly, in the 1920s and 1930s there were numerous measurements of the solubility of alkane waxes in oil,[256] pure hydrocarbons,[257] and petroleum fractions.[258,259] More recently measurements of solubility have been reported for long-chain alkanes in a narrow alkane hydrocarbon distillate fraction,[260] which was obtained from crude oil with a boiling temperature range between (333 and 353) K that is predominately a hexane fraction; this distillate is commonly referred to as petroleum ether. Domanska and Morawski[261] have reported measurements of the (solid+liquid) phase equilibria, and therefore solubility, for (hexadecane+3-methylpentane), (hexadecane+2,2-dimethylbutane), (hexadecane+benzene), (octadecane+3-methylpentane), (octadecane+2,2-dimethylbutane), and (octadecane+benzene) at temperatures in the range (293 to 353) K and pressures up to 1 GPa. There are measurements of the solubility of long-chain alkanes in heptane,[262,263] toluene,[264] tetradecane,[265] as well as of the (solid+liquid) and (solid+solid) equilibria.[266] From an environmental perspective, the solubility of hydrocarbons, particularly multiple ring aromatic substances, in water is important for hydrocarbon spills and has been discussed in the literature.[267] The solubility of heavy oil in water has been discussed by Wiehe.[268]

The solubility of asphaltene, which is often found in coal and heavy oil reserves, has been discussed by Speight.[269] The solubility of asphaltene in heptane has been determined with spectroscopic methods such as Fourier transform infrared and nuclear magnetic resonance.[270] The solubility of tetracosane, octacosane, and dotriacontane in ethane at a temperature of 308 K and at pressures between (5 and 20) MPa has been reported.[71] Asphaltene precipitation in alkanes has been the focus of at least two sets of measurements[271,272] and models for the solubility based on the regular solution approach have been proposed by Akbarzadeh *et al.*[271,273] A Flory–Huggins model for the prediction of asphaltenes precipitation in crude oil has been presented by Pazuki and Nikookar.[274] There are also several experimental studies reported in the literature directly related to the solubility of asphaltenes, obtained from natural fluids from specific geographic locations, in other hydrocarbons.[275–277] Studies have also reported the solubility of asphaltenes that include the formation of emulsions with water.[278]

The solubility of naphthalene, a significant component of coal liquid (often referred to as tar), in methylbenzene and 1,3-dimethylbenzene[279] and cyclohexane has been reported.[280] Several theories have been presented for the volume of mixing of organic liquids with coal.[281–283] The solubility of C_{60} fullerene in hydrocarbons has been reviewed by Marcus *et al.*[284]

Heavy oil can also be produced with processes such as steam assisted gravity drainage and thus the solubility of oil in water is significant from the perspective of oil production and the environmental challenge of water disposal.

At temperatures below 273 K the solubility of solid toluene has been determined in liquid nitrogen[285] and the results compared with estimates obtained from the Preston–Prausnitz[30] model, which has also been used for solid alkanes in liquid oxygen[286] and for carbon dioxide in low molar mass hydrocarbons.[287]

23.4.7 Gases, Liquids, and Solids

As we have mentioned in Section 23.1, a petroleum reservoir potentially contains gases, solids, and liquids, of which the latter also includes an essentially immiscible aqueous phase. Thus studies of the mutual solubility in these three phase systems more closely resemble reality. The aqueous solubility of crude oil, containing CH_4 through $C_{34}H_{70}$, at temperatures from (373 to 673) K and pressures up to 200 MPa as a function of the molality of NaCl has been reported by Price.[288] Tiffin *et al.*[289] have reported solid hydrocarbon solubility in liquid (methane+ethane+octane) and (methane+ethane+cyclohexane) along the three phase (solid+liquid+vapor) loci. These results are significant because they demonstrate that modest additions of ethane to liquid methane enhance the solubility of other hydrocarbons significantly beyond the values anticipated based on simple combinatorial rules. Cai *et al.*[290] have measured the solubility of hydrogen in hexadecane, tetralin, and a representative heavy reservoir hydrocarbon fluid in the presence of hydrotreating catalysts, such as Ni and Pd, and substances representative of the strata such as alumina and silica. These results were compared and the effect of the presence of a hydrocarbon liquid on the sorption of hydrogen on the solid phase determined and it was found that silica and used hydrotreating catalysts adsorb significant quantities of hydrogen and their presence raises the apparent solubility of hydrogen in these liquids. The impact of the solubility on reservoir production including the sediment has been discussed by Kuo.[291]

The solubility of a hydrocarbon in water in equilibrium with its hydrate has been studied for methane,[292,293] carbon dioxide,[294] and ethane.[295] Part of the study of the inhibition of hydrates with alcohols[296] requires data for the solubility of gases in alcohols.[297] The solubility of gaseous methane, liquid ethane, and liquid propane in equilibrium with hydrate at cryogenic conditions has been reported by Song *et al.*[298] Measurements reported in the literature include the solubility of gaseous alkanes in methanol and 2-hydroxypropane,[299] ethane in 2,2-dihydroxyethyl ether,[72] and heptane in ethanol.[300] The reader interested in the subject of hydrates may wish to consult Sloan,[301] or Makogon[302] or Volume 912 *Annals of the New York Academy of Sciences* entitled Gas Hydrates Challenges for the Future[303] that includes papers presented at the Third International Conference on Gas Hydrates that was held in Salt Lake City, UT, USA, July 18 to 22, 1999.

23.5 New Results: Solubility of Hydrogen in Normal Alkanes

Hydrogen is a key compound in the production of fuels for the automotive industry and much more importance will be acquired in the future, as the free carbon energetic sources tend to emerge for pollution and environmentally evident reasons. Nowadays, hydrogen is mainly obtained from the catalytic steam reforming of nafta and natural gas, but renewable sources of energy seem to be promising in the near future. In many industrial processes where molecular hydrogen plays an important role, its solubility in different hydrocarbon solutions (*e.g.*, fuels) is among the major factors required for design and optimal operation of these processes. It is also a key parameter in process models, such as the ones used in hydrogenation and hydrotreatment processes, where hydrogen solubility in selected liquid hydrocarbons is a good estimation for the hydrogen concentration in the liquid phase, a variable to which kinetics is often related to.

More than one decade ago, the thermodynamics group of Delft University of Technology was asked by Shell Global Solutions International in Amsterdam to provide them with an extensive database on hydrogen solubilities in a number of selected higher molecular weight *n*-alkanes. For that purpose hydrogen solubility data in binary and ternary mixtures of *n*-alkanes were extensively measured, resulting in a unique data collection. For proprietary reasons, until recently this data collection was not available for the open literature. However, the first part of this data collection, comprising the binary systems, was published recently by Florusse *et al.*[304] In addition to the experimental work, a modeling approach to this data collection was presented using a molecular-based equation of state based on the Statistical Associating Fluid Theory (SAFT). For the modeling work one is also referred to Florusse *et al.*[304]

A Cailletet apparatus was used to measure the solubility of hydrogen in decane ($C_{10}H_{22}$), hexadecane ($C_{16}H_{34}$), octacosane ($C_{28}H_{58}$), hexatriacontane ($C_{36}H_{74}$), and hexatetracontane ($C_{46}H_{94}$). The apparatus has been described in detail by Raeissi and Peters[305] and only the important features are described here. In this apparatus, a sample of fixed and known composition is held within a Pyrex glass tube, one end of which is sealed the other open end of the tube is placed in a stainless steel autoclave and in contact with mercury that acts to seal the sample within the tube and transmit to the sample. The sample is stirred by a stainless steel ball moved by reciprocating magnets. The autoclave and mercury are connected to a pressure generating system by hydraulic oil. The temperature of the sample is kept constant by circulating thermostated liquid through a glass jacket surrounding the tube. The solubility was determined by varying the pressure of a sample of constant composition at constant temperature until a phase change was observed visually. The solubility measurements were conducted at temperature between (280 and 450) K at pressures up to 16 MPa.

The results obtained from the experiments are shown in Figures 2 through 11 and over this range of temperature, pressure, and composition studied, the hydrogen mole fractions up to 0.25 were found soluble in the alkanes $C_{10}H_{22}$,

$C_{16}H_{34}$, $C_{28}H_{58}$, $C_{36}H_{74}$, and $C_{46}H_{94}$. For $\{xH_2+(1-x)C_{10}H_{22})\}$ with $0.016<x<0.0881$ at temperature between (275 and 450) K the (p, T) section is shown in Figure 2 while the (p, x) section is shown in Figure 3. For $\{xH_2+(1-x)C_{16}H_{34})\}$ with $0.018<x<0.113$ at temperature between (320 and 450) K the (p, T) section is shown in Figure 4 while the (p, x) section is shown in Figure 5. For $\{xH_2+(1-x)C_{28}H_{58})\}$ with $0.0295<x<0.1777$ at temperature between (340 and 450) K the (p, T) section is shown in Figure 6 while the (p, x) section is shown in Figure 7. For $\{xH_2+(1-x)C_{36}H_{74})\}$ with $0.033<x<0.210$ at temperature between (360 and 450) K the (p, T) section is shown in Figure 8 while the (p, x) section is shown in Figure 9. For $\{xH_2+(1-x)C_{46}H_{94})\}$ with $0.065<x<0.257$ at temperature between (370 and 450) K the (p, T) section is shown in Figure 10 while the (p, x) section is shown in Figure 11.

In addition to the experimental data on the binary mixtures, measurements were also performed of the solubility of hydrogen in $(C_{10}H_{22} + C_{20}H_{42})$, $(C_{10}H_{22} + C_{28}H_{58})$, $(C_{10}H_{22} + C_{36}H_{74})$, and $(C_{10}H_{22} + C_{46}H_{94})$.[306] The (p, T) sections are shown in Figure 12 for $(xH_2 + x_1C_{10}H_{22} + x_2C_{20}H_{42})$ with $0.06<x<0.096$, Figure 13 for $(xH_2 + x_1C_{10}H_{22} + x_2C_{28}H_{58})$ with $0.092<x<0.153$, Figure 14 for $(xH_2 + x_1C_{10}H_{22} + x_2C_{36}H_{74})$ with $0.0763<x<0.0798$, and Figure 15 for $(xH_2 + x_1C_{10}H_{22} + x_2C_{46}H_{94})$ with $0.065<x<0.153$. The results show that the solubility of hydrogen in the alkanes mixtures increases with increasing number of carbon atoms in the chain of the alkane and that for $C_{46}H_{94}$ the hydrogen solubility may reach values as high as 0.25 % on a molar basis. Furthermore, the hydrogen solubility in alkanes increases with increasing temperature.

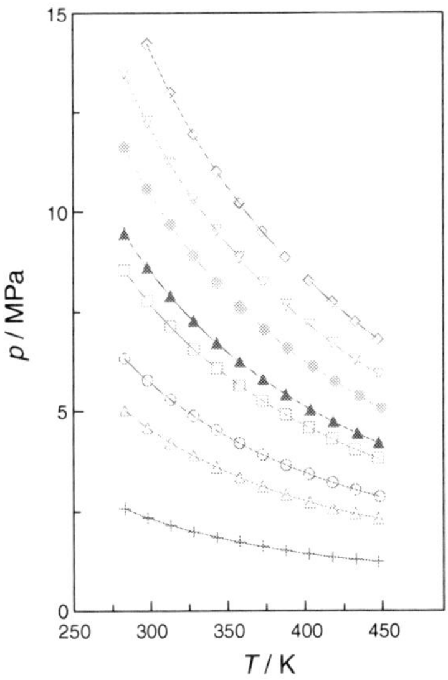

Figure 2 *(p, T) section for $\{xH_2 + (1-x)C_{10}H_{22})\}$. ◇, $x = 0.0881$; ▽, $x = 0.0780$; ●, $x = 0.0674$; ▲, $x = 0.0560$; □, $x = 0.0510$; ○, $x = 0.0380$; △, $x = 0.0312$; and +, $x = 0.0160$. For each mole fraction the curves are best fits to the experimental values.*

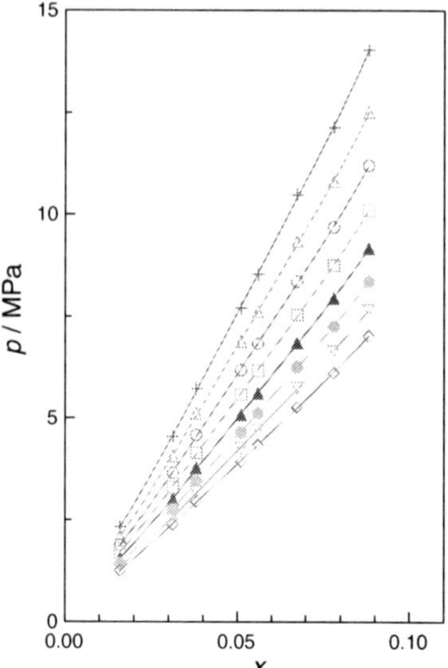

Figure 3 *(p, x) section for* $\{xH_2 + (1-x)C_{10}H_{22})\}$. ◊, $T = 440$ K; ▽, $T = 420$ K; ●, $T = 400$ K; ▲, $T = 380$ K; □, $T = 360$ K; ○, $T = 340$ K; △, $T = 320$ K; *and* +, $T = 300$ K. *For each mole fraction the curves are best fits to the experimental values.*

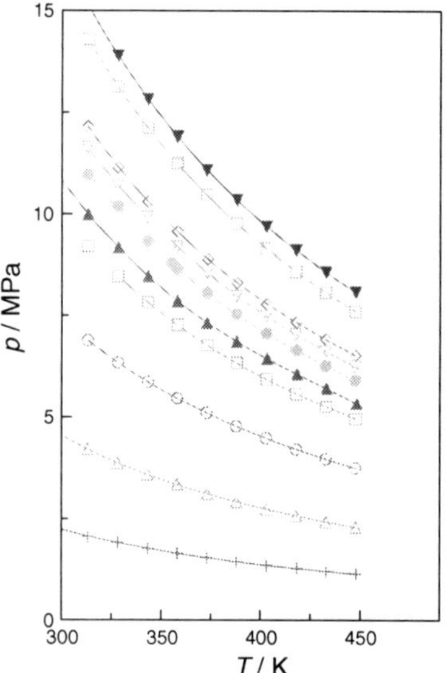

Figure 4 *(p, T) section for* $\{xH_2 + (1-x)C_{16}H_{34})\}$. ▼, $x = 0.1130$; □, $x = 0.1087$; ◊, $x = 0.0936$; ▽, $x = 0.0910$; ●, $x = 0.0856$; ▲, $x = 0.0781$; ○, $x = 0.0560$; △, $x = 0.0354$; *and* +, $x = 0.0181$. *For each mole fraction the curves are best fits to the experimental values.*

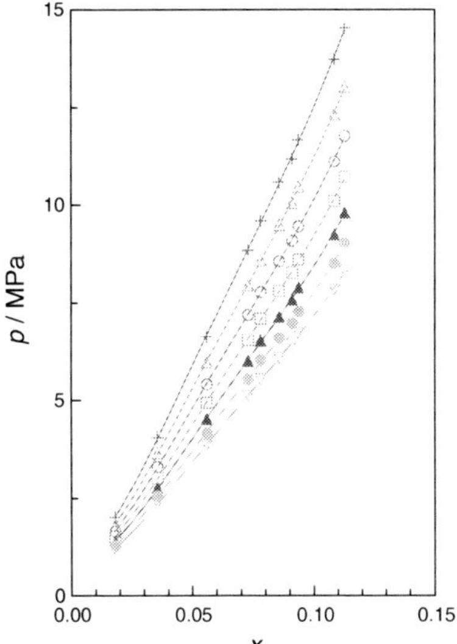

Figure 5 *(p, x) section for* $\{xH_2 + (1-x)C_{16}H_{34})\}$. ▽, $T = 440$ K; ●, $T = 420$ K; ▲, $T = 400$ K; □, $T = 380$ K; ○, $T = 360$ K; △, $T = 340$ K; *and* +, $T = 320$ K. *For each mole fraction the curves are best fits to the experimental values.*

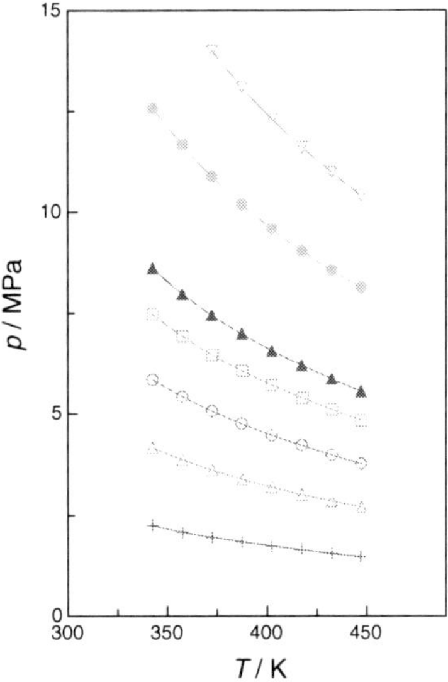

Figure 6 *(p, T) section for* $\{xH_2 + (1-x)C_{28}H_{58})\}$. ▽, $x = 0.1777$; ●, $x = 0.1432$; ▲, $x = 0.1057$; □, $x = 0.0908$; ○, $x = 0.0709$; △, $x = 0.0536$; *and* +, $x = 0.0295$. *For each mole fraction the curves are best fits to the experimental values.*

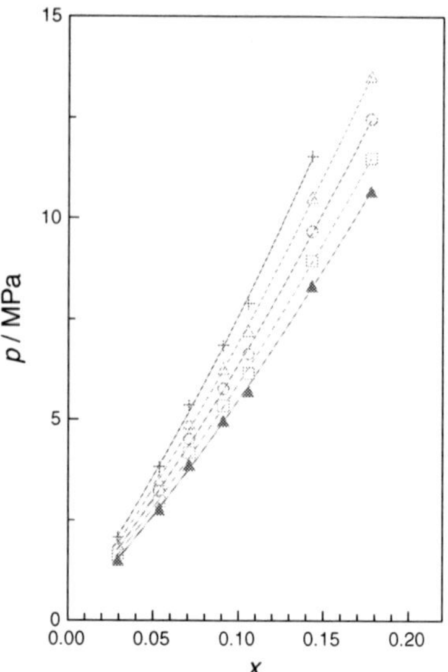

Figure 7 *(p, x) section for* $\{xH_2 + (1-x)C_{28}H_{58})\}$. ▲, $T = 440$ K; □, $T = 420$ K; ○, $T = 400$ K; △, $T = 380$ K; *and* +, $T = 360$ K. *For each mole fraction the curves are best fits to the experimental values.*

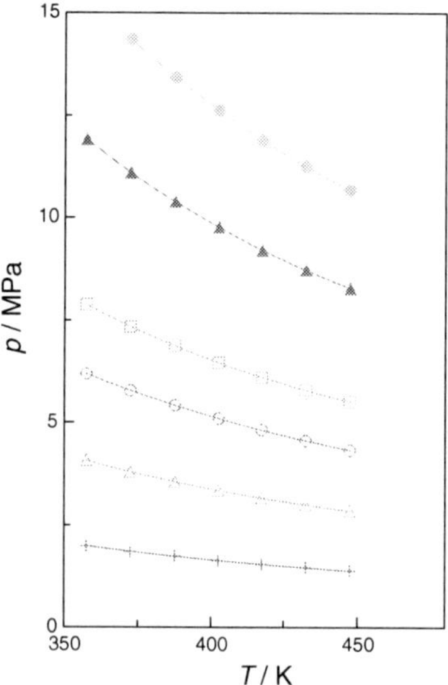

Figure 8 *(p, T) section for* $\{xH_2 + (1-x)C_{36}H_{74})\}$. ●, $x = 0.2102$; ▲, $x = 0.1690$; □, $x = 0.1180$; ○, $x = 0.0970$; △, $x = 0.0655$; *and* +, $x = 0.0330$. *For each mole fraction the curves are best fits to the experimental values.*

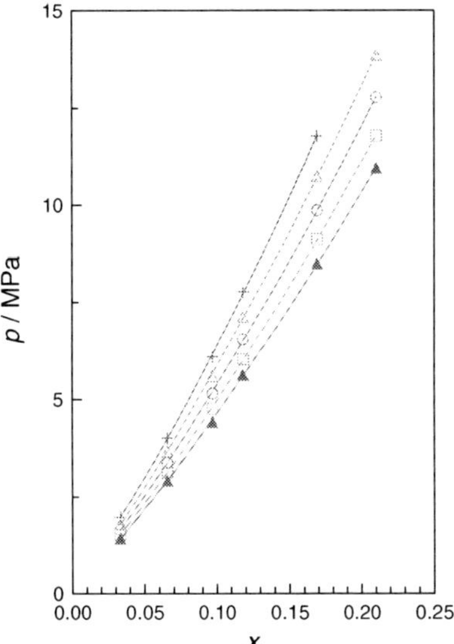

Figure 9 *(p, T) section for* $\{xH_2 + (1-x)C_{36}H_{74})\}$. ▲, $T = 440$ K; □, $T = 420$ K; ○, $T = 400$ K; △, $T = 380$ K; *and* +, $T = 360$ K. *For each mole fraction the curves are best fits to the experimental values.*

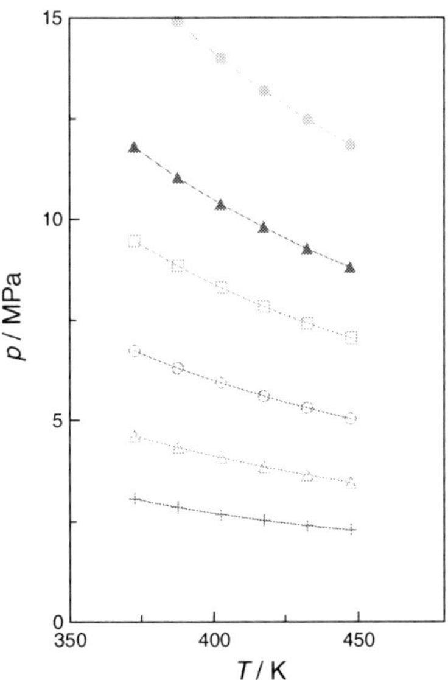

Figure 10 *(p, T) section for* $\{xH_2 + (1-x)C_{46}H_{94})\}$. ●, $x = 0.257$; ▲, $x = 0.204$; □, $x = 0.173$; ○, $x = 0.129$; △, $x = 0.095$; *and* +, $x = 0.065$. *For each mole fraction the curves are best fits to the experimental values.*

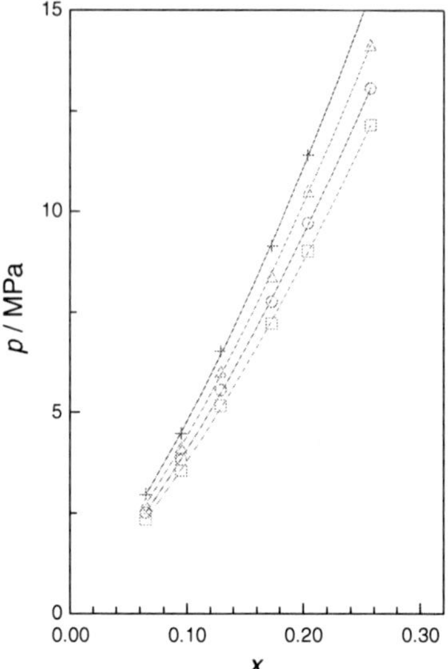

Figure 11 *(p, x) section for* $\{xH_2+(1-x)C_{46}H_{94})\}$. □, $T = 440$ K; ○, $T = 420$ K; △, $T = 400$ K; *and* +, $T = 380$ K. *For each mole fraction the curves are best fits to the experimental values.*

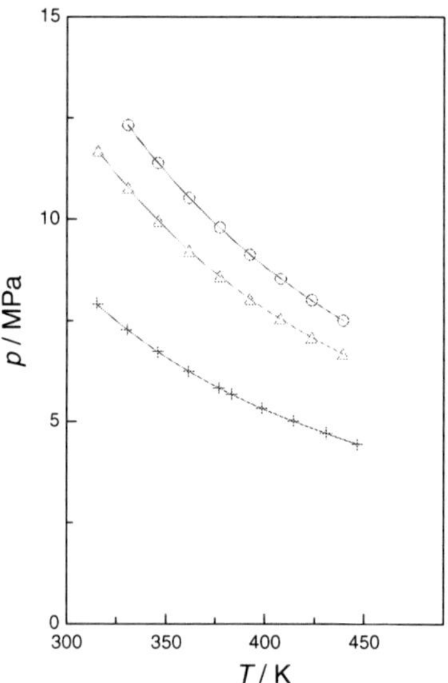

Figure 12 *(p, T) section for* $(xH_2 + x_1C_{10}H_{22} + x_2C_{20}H_{42})$ *where* $x + x_1 + x_2 = 1$ *and the mole-fraction-averaged number of carbon atoms (excluding hydrogen) in the alkane mixture* $n = x_1n_1 + x_2n_2$. ○, $x = 0.0958$ *and* $n = 12.22$; △, $x = 0.0911$ *and* $n = 15.95$; *and* +, $x = 0.0632$ *and* $n = 15.06$. *For each mole fraction the curves are best fits to the experimental values.*

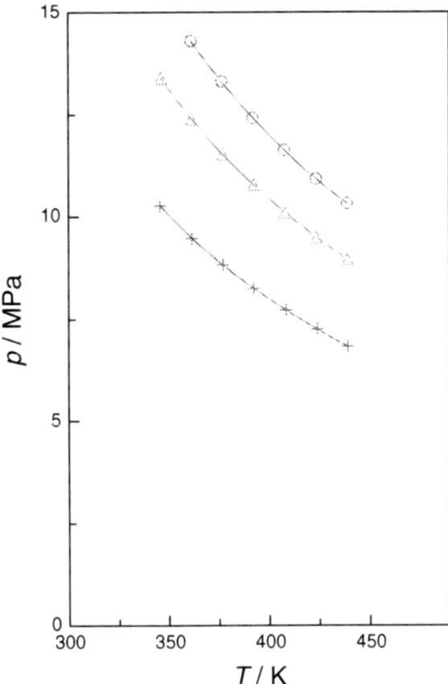

Figure 13 *(p, T) section for* $(xH_2 + x_1C_{10}H_{22} + x_2C_{28}H_{58})$ *where* $x + x_1 + x_2 = 1$ *and the mole-fraction-averaged number of carbon atoms (excluding hydrogen) in the alkane mixture* $n = x_1n_1 + x_2n_2$. ○, $x = 0.1532$ *and* $n = 22.15$; △, $x = 0.1284$ *and* $n = 20.00$; *and* +, $x = 0.0917$ *and* $n = 14.77$. *For each mole fraction the curves are best fits to the experimental values.*

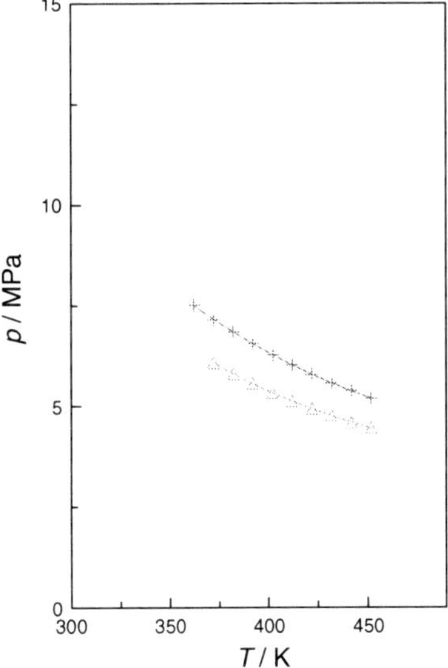

Figure 14 *(p, T) section for* $(xH_2 + x_1C_{10}H_{22} + x_2C_{36}H_{74})$ *where* $x + x_1 + x_2 = 1$ *and the mole-fraction-averaged number of carbon atoms (excluding hydrogen) in the alkane mixture* $n = x_1n_1 + x_2n_2$. +, $x = 0.0763$ *and* $n = 15.87$; *and* △, $x = 0.0798$ *and* $n = 24.53$. *For each mole fraction the curves are best fits to the experimental values.*

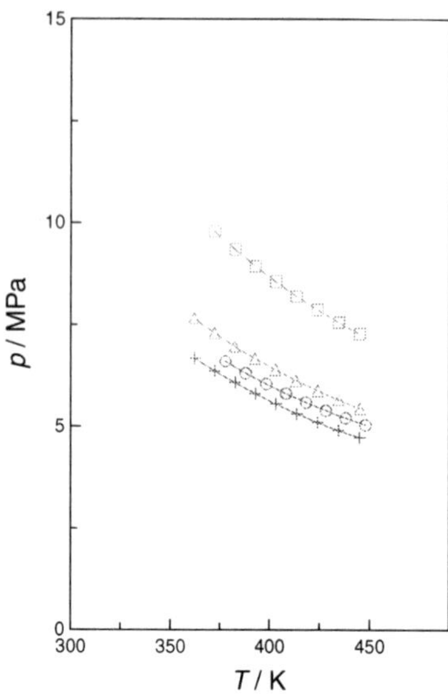

Figure 15 *(p, T) section for* $(xH_2 + x_1C_{10}H_{22} + x_2C_{46}H_{94})$ *where* $x + x_1 + x_2 = 1$ *and the mole-fraction-averaged number of carbon atoms (excluding hydrogen) in the alkane mixture* $n = x_1n_1 + x_2n_2$. □, $x = 0.1533$ *and* $n = 35.88$; △, $x = 0.0776$ *and* $n = 16.02$; ○, $x = 0.0901$ *and* $n = 25.34$; *and* +, $x = 0.0649$ *and* $n = 13.68$. *For each mole fraction the curves are best fits to the experimental values.*

Acknowledgment

One of us (CJP) is grateful to Shell Global Solutions International in Amsterdam for giving permission to include the unpublished data on the various ternary systems in this chapter.

References

1. W.O. Eisenbach, K. Niemann and P.J. Gottsch, *Supercritical fluid extraction of oil sands and residues from coal hydrogenation*, in *Chemical Engineering at Supercritical Conditions*, M.E. Paulaitis, R.D. Gray and P. Davidson (eds), Ann Arbor Science, Ann Arbor, MI, 1983, 419–433.
2. F.M. Orr, C.L. Lien and M.T. Pelletier, *Prepr. Pap. Am. Chem. Soc. Div. Fuel Chem.*, 1981, **26**, 132–145.
3. F.M. Orr, M.K. Silva and C.L. Lien, *Soc. Petrol. Eng. J.*, 1983, **23**, 281–291.
4. M.D. Deo, J. Hwang and F.V. Hanson, *Fuel*, 1992, **71**, 1519–1526.
5. R.R. Kuntz and G.J. Mains, *J. Phys. Chem.*, 1964, **68**, 408–410.
6. H.L. Clever and M. Iwanmoto, *Ind. Eng. Chem. Res.*, 1987, **26**, 336–337.
7. M.M. Miedaner, A.A. Migdisov and A.E. Williams-Jones, *Geochim. Cosmochim. Acta*, 2005, **69**, 5511–5516.

8. S.M. Wilhelm, L. Liang and D. Kirchgessner, *Energy Fuels*, 2006, **20**, 180–186.
9. H.L. Clever, *Solubility Data Series*, 1987, **29**, 149–162.
10. A. Danesh, *PVT and Phase Behaviour of Petroleum Reservoir Fluids*, vol. 47, Developments in Petroleum Science, Elsevier, Amsterdam, 1998.
11. A. Danesh, *PVT and Phase Behaviour of Petroleum Reservoir Fluids, Developments in Petroleum Science*, vol. 47, Elsevier, Amsterdam, 1998, 209–252.
12. T.P.J. Halpin and N. Quirke, *SPE Res. Eng.*, 1990, **5**, 617–622.
13. T. Cvitaš, *Metrologia*, 1996, **33**, 35–40.
14. A. Danesh, *PVT and Phase Behaviour of Petroleum Reservoir Fluids, Developments in Petroleum Science*, vol. 47, Elsevier, Amsterdam, 1998, 112–116.
15. M.L. McGlashan, *Chemical Thermodynamics*, Academic Press, London, 1979, 303.
16. R.A. Robinson and R.H. Stokes, *Electrolyte Solutions*, 2nd edn, Butterworths, London, 1959.
17. J.M. Prausnitz, R.N. Lichtenthaler and E. Gomez de Azevedo, *Molecular Thermodynamics of Fluid Phase Equilibria*, 3rd edn, Prentice Hall, New Jersey, 1999, 543–552.
18. J.E. Trevor, *J. Phys. Chem.*, 1906, **10**, 99–107.
19. D. Tyrer, *J. Phys. Chem.*, 1912, **16**, 69–85.
20. J.H. Hilberbrand, *J. Am. Chem. Soc.*, 1916, **38**, 1452–1473.
21. N.W. Taylor and J.H. Hilderbrand, *J. Am. Chem. Soc.*, 1923, **45**, 682–694.
22. J.H. Hilderbrand, *J. Am. Chem. Soc.*, 1935, **57**, 866–871.
23. S.K. Lachowicz and K.E. Weale, *Ind. Eng. Chem. Eng. Data Ser.*, 1958, **3**, 162–166.
24. J.H. Hilderbrand and R.H. Lamoreaux, *Ind. Eng. Chem. Fundam.*, 1974, **13**, 110–115.
25. K. Fischer and J. Gmehling, *Fluid Phase Equilib.*, 1996, **121**, 185–206.
26. R.A. Pierotti, *J. Phys. Chem.*, 1963, **67**, 1840–1845.
27. P. Alessi, A. Cortesi, M. Fermeglia, M. Fontana and I. Kikic, *Fluid Phase Equilib.*, 1989, **53**, 397–406.
28. T. Khayamian and M. Esteki, *J. Supercrit. Fluid*, 2004, **32**, 73–78.
29. R.C. Reid, J.M. Prausnitz and B.E. Poling, *The Properties of Gases and Liquids*, 4th edn, McGraw Hill, New York, 1987, 241–387.
30. G.T. Preston and J.M. Prausnitz, *Ind. Eng. Chem. Process Des. Dev.*, 1970, **9**, 264–271.
31. L.L. Lee, *Fluid Phase Equilib.*, 1987, **35**, 77–92.
32. J.P. O'Connell and J.M. Prausnitz, *Ind. Eng. Chem. Fundam.*, 1964, **3**, 347–351.
33. Y. Ran, N. Jain and S.H. Yalkowsky, *J. Chem. Inf. Comput. Sci.*, 2001, **41**, 1208–1217.
34. Y. Iwai, Y. Mori and Y. Arai, *Fluid Phase Equilib.*, 2000, **167**, 33–40.
35. P. Redelius, *Energy Fuels*, 2004, **18**, 1087–1092.
36. G.R. Pazuki and M. Nikookar, *Fuel*, 2006, **85**, 1083–1086.

37. I.Z. Kiss, G. Mándi and M.T. Beck, *J. Phys. Chem. A*, 2000, **104**, 8081–8088.
38. I.Z. Kiss, G. Mándi and M.T. Beck, *J. Phys. Chem. A*, 2000, **104**, 10994.
39. A.R. Katritzky, A.A. Oliferenko, P.V. Oliferenko, R. Petrukhin, D.B. Tatham, U. Maran, A. Lomaka and W.E. Acree, Jr., *J. Chem. Inf. Comput. Sci.*, 2003, **43**, 1794–1805.
40. A. Ghosh, W.G. Chapman and R.N. French, *Fluid Phase Equilib.*, 2003, **209**, 229–243.
41. S.H. Ali, F.S. Al-Mutairi and M.A. Fahim, *Fluid Phase Equilib.*, 2005, **230**, 176–183.
42. M.R. Riazi and J.H. Vera, *Ind. Eng. Chem. Res.*, 2005, **44**, 186–192.
43. D.M. Kassim and S.F. Al-Doury, *J. Pet. Res.*, 1985, **4**, 1–19.
44. J.M. Moysan, M.J. Huron, H. Paradowski and J. Vidal, *Chem. Eng. Sci.*, 1983, **38**, 1085–1092.
45. H. Sorensen, K.S. Pedersen and P.L. Christensen, *Org. Geochem.*, 2002, **33**, 635–642.
46. R.A. Almehaideb, *J. Pet. Sci. Eng.*, 2004, **42**, 157–170.
47. A. Klamt, *Fluid Phase Equilib.*, 2003, **206**, 223–235.
48. G. Nocon, U. Weidlich, J. Gmehling, J. Menke and U. Onken, *Fluid Phase Equilib.*, 1983, **13**, 381–392.
49. T.A. Al-Sahhaf, *J. Environ. Sci. Health A*, 1989, **A24**, 49–56.
50. R.S. Basu, H. Pham and D.P. Wilson, *Int. J. Thermophys.*, 1986, **7**, 319–330.
51. G. Nocon, U. Weidlich, J. Gmehling and U. Onken, *Berich. Bunsen. Gesell.*, 1983, **87**, 17–23.
52. J.M. Mollerup and W.M. Clark, *Fluid Phase Equilib.*, 1989, **51**, 257–268.
53. N.D. Sanders, *Ind. Eng. Chem. Fundam.*, 1986, **25**, 169–171.
54. A. Danesh, *PVT and Phase Behaviour of Petroluem Reservoir Fluids*, Elsevier, Amsterdam, 1998, 253–266.
55. R.D. Weir and T.H. W de Loos (eds), Measurement of the Thermodynamic Properties of Multiple Phases, Experimental Thermodynamics Vol VII, IUPAC Elsevier, Amsterdam, 2005.
56. J. Livingston, R. Morgan and H.R. Pyne, *J. Phys. Chem.*, 1930, **34**, 1578–1582.
57. J. Dymond and J.H. Hilderbrand, *Ind. Eng. Chem. Fundam.*, 1967, **6**, 130–131.
58. K.E. Gubbins, S.N. Carden and R.D. Walker, Jr., *J. Gas Chromatogr.*, 1965, **3**, 330.
59. E.J. Farkas, *Anal. Chem.*, 1965, **37**, 1173–1175.
60. J.M. Lo, C.L. Tseng and J.Y. Yang, *Anal. Chem.*, 1986, **58**, 1596–1597.
61. M.J. Horvath, H.M. Sebastian and K.-C. Chao, *Ind. Eng. Chem. Fundam.*, 1981, **20**, 394–396.
62. A.G. Vitenberg, *Russ. J. Appl. Chem.*, 2001, **74**, 241–246.
63. K.D. Bartle, A.A. Clifford and S.A. Jafar, *J. Chem. Eng. Data*, 1990, **35**, 355–360.
64. R.E. Baltus, B.H. Culbertson, S. Dai, H. Luo and D.W. DePaoli, *J. Phys. Chem. B.*, 2004, **108**, 721–727.

65. D.D. Lawson and J.D. Ingham, *Nature*, 1969, **223**, 614–615.
66. S. Wingefors, *J. Chem. Technol. Biol.*, 1981, **31**, 530–534.
67. A.N. Fedotov, A.P. Simonov, V.K. Popov and V.N. Bagratashvili, *J Phys. Chem. B*, 1997, **101**, 2929–2932.
68. P.H. Young and C.A. Schall, *Thermochim. Acta*, 2001, **367–368**, 387–392.
69. A.E. Markham and K.E. Kobe, *Chem. Rev.*, 1941, **28**, 519–588.
70. T.R. Rettlch, Y.P. Handa, R. Battino and E. Wilhelm, *J. Phys. Chem.*, 1981, **85**, 3230–3237.
71. A. Kalaga and M. Trebble, *J. Chem. Eng. Data*, 1997, **42**, 368–370.
72. F.-Y. Jou, K.A.G. Schmidt and A.E. Mather, *J. Chem. Eng. Data*, 2005, **50**, 1983–1985.
73. V.R. Choudhary, M.G. Parande and P.H. Brahme, *Ind. Eng. Chem. Fundam.*, 1982, **21**, 472–474.
74. K.T. Koonce and R. Kobayashi, *J. Chem. Eng. Data*, 1964, **9**, 490–494.
75. W. Chey and G.V. Calder, *J. Chem. Eng. Data*, 1972, **17**, 199–200.
76. S. Rahman and M.A. Barrufet, *J. Pet. Sci. Eng.*, 1995, **14**, 23–34.
77. H. Chen and J. Wagner, *J. Chem. Eng. Data*, 1994, **39**, 470–474.
78. T.H. Vaughn and E.G. Nutting, Jr., *Ind. Eng. Chem. Anal. Ed.*, 1942, **14**, 454–456.
79. P. Barton, R.E. Holland and R.H. McCormick, *Ind. Eng. Chem. Process Des. Dev.*, 1974, **13**, 378–383.
80. P.J. Linstrom and W.G. Mallard, *J. Chem. Eng. Data*, 2001, **46**, 1059–1063.
81. H.L. Clever, *J. Chem. Eng. Data*, 2004, **49**, 1521–1529.
82. R. Battino and H.L. Clever, *Chem. Rev.*, 1966, **66**, 395–463.
83. D. Richon and H. Renon, *J. Chem. Eng. Data*, 1980, **25**, 59–60.
84. E.P. Schooh, A.E. Hoffmann, A.S. Kasperik, J.H. Lightfoot and F.D. Mayfield, *Ind. Eng. Chem.*, 1940, **32**, 788–791.
85. S. Srivatsan, W. Gao, K.A.M. Gasem and R.L. Robinson, Jr., *J. Chem. Eng. Data*, 1998, **43**, 623–625.
86. E.P. Schooh, A.E. Hoffmann and F.D. Mayfield, *Ind. Eng. Chem.*, 1940, **32**, 1351–1353.
87. S. Srivastan, N.A. Darwish, K.A.M. Gasem and R.L. Robinson, Jr., *J. Chem. Eng. Data*, 1992, **37**, 516–520.
88. N.A. Darwish, K.A.M. Gasem and R.L. Robinson, Jr., *J. Chem. Eng. Data*, 1998, **43**, 238–240.
89. N.A. Darwish, K.A.M. Gasem and R.L. Robinson, Jr., *J. Chem. Eng. Data*, 1994, **39**, 781–784.
90. F.-N. Tsai, S.H. Huang, H.-M. Lin and K.-C. Chao, *J. Chem. Eng. Data*, 1987, **32**, 467–469.
91. N.A. Darwish, J. Fathikalajahi, K.A.M. Gasem and R.L. Robinson, Jr., *J. Chem. Eng. Data*, 1993, **38**, 44–48.
92. C.L. Young, *Solubility Data Series*, 1987, **27–28**, 533–581.
93. C.L. Young, *Solubility Data Series*, 1987, **27–28**, 477–490.
94. H.L. Clever, *Solubility Data Series*, 1987, **27–28**, 493–510.
95. B.A. Bufkin and R.L. Robinson, Jr., *J. Chem. Eng. Data*, 1986, **31**, 421–423.
96. C.L. Young, *Solubility Data Series*, 1982, **9**, 232–251.

97. A. Chapoy, A.H. Mohammadi, D. Richon and B. Tohidi, *Fluid Phase Equilib.*, 2004, **220**, 113–121.
98. IUPAC Commission on Solubility Data, *Solubility Data Series*, 1986, 24, 422–429.
99. J.S. Chou and K.-C. Chao, *J. Chem. Eng. Data*, 1989, **34**, 68–70.
100. J. Livingston, R. Morgan and H.R. Pyne, *J. Phys. Chem.*, 1930, **34**, 2045–2048.
101. R.J. Combs and P.E. Field, *J. Phys. Chem.*, 1987, **91**, 1663–1668.
102. J.J. Simnick, H.M. Sebastian, H.M. Lin and K.-C. Chao, *J. Chem. Eng. Data*, 1978, **23**, 339–340.
103. A.A. Miller, A. Ekstrom and N.R. Foster, *J. Chem. Eng. Data*, 1990, **35**, 125–127.
104. P.J. Hesse, R. Battino, P. Scharlin and E. Wilhelm, *J. Chem. Eng. Data*, 1996, **41**, 195–201.
105. J.L. Vogel and L. Yarborough, The effect of nitrogen on the phase behavior and physical properties of reservoir fluid, SPE Paper number 815-MS at *SPE/DOE Enhanced Oil recovery Symposium*, 20 to 23 April, Tulsa, Oklahoma, 1980.
106. C.L. Young, *Solubility Data Series*, 1980, **4**, 261–270.
107. H.L. Clever, *Solubility Data Series*, 1980, **4**, 137–144.
108. H.L. Clever, *Solubility Data Series*, 1980, **4**, 158–161.
109. H.L. Clever, *Solubility Data Series*, 1980, **4**, 162–168.
110. H.L. Clever, *Solubility Data Series*, 1980, **4**, 149–157.
111. H.L. Clever, *Solubility Data Series*, 1980, **4**, 145–148.
112. H.L. Clever, *Solubility Data Series*, 1980, **4**, 133–136.
113. R. Battino, *Solubility Data Series*, 1982, **11**, 148–161.
114. R. Battino, *Solubility Data Series*, 1982, **11**, 162–173.
115. C.L. Young, *Solubility Data Series*, 1979, **1**, 263–281.
116. R.W. Cargill, *Solubility Data Series*, 1990, **43**, 100–109.
117. R.W. Cargill, *Solubility Data Series*, 1990, **43**, 110–151.
118. R.W. Cargill, *Solubility Data Series*, 1990, **43**, 67–99.
119. E.E. Nelson and W.S. Bonnell, *Ind. Eng. Chem.*, 1943, **35**, 204–206.
120. W. Gao, R.L. Robinson, Jr. and K.A.M. Gasem, *J. Chem. Eng. Data*, 2001, **46**, 609–612.
121. M. Herskowitz, J. Wisniak and L. Skladman, *J. Chem. Eng. Data*, 1983, **28**, 164–166.
122. K.J. Kim, T.R. Way and K.T. Feldman, Jr., *J. Chem. Eng. Data*, 1997, **42**, 214–215.
123. R.H. Harrison, S.E. Scheppele, G.P. Strum, Jr. and P.L. Grizzle, *J. Chem. Eng. Data*, 1985, **30**, 183–189.
124. M.R. Riazi and Y.A. AlRoomi, *Am. Chem. Soc. Div. Pet. Chem.*, 2005, **50**, 389–390.
125. S. Peramanu, B.B. Pruden and P.F. Clarke, *J. Chem. Eng. Data*, 1998, **43**, 865–867.
126. G.-X. Feng and A.E. Mather, *J. Chem. Eng. Data*, 1992, **37**, 412–413.
127. S.F. Sciamanna and S. Lynn, *Ind. Eng. Chem. Res.*, 1988, **27**, 492–499.

128. C. Yokoyama, A. Usui and S. Takahashi, *Fluid Phase Equilib.*, 1993, **85**, 257–269.
129. J.A. Waters and G.A. Mortimer, *J. Chem. Eng. Data*, 1972, **17**, 156–157.
130. R.N.C. Gharbi, *J. Pet. Sci. Eng.*, 2000, **27**, 33–47.
131. J. Suebsiri, M. Wilson and P. Tontiwachwuthikul, *Ind. Eng. Chem. Res.*, 2005, **45**, 2483–2488.
132. A.-C. Aycaguer, M. Lev-On and A.W. Winer, *Energy Fuels*, 2001, **15**, 303–308.
133. T. Babadagli, *J. Pet. Sci. Eng.*, 2003, **37**, 25–37.
134. B.A. Mandagaran and E.A. Campanella, *Chem. Eng. Tech.*, 1993, **16**, 399–404.
135. B. Quail, G.A. Hill and K.N. Jha, *Ind. Eng. Chem. Res.*, 1988, **27**, 519–523.
136. C.A. Mulliken and S.I. Sandler, *Ind. Eng. Chem. Process Des. Dev.*, 1980, **19**, 709–711.
137. R.G. Moore, S.A. Mehta and M.G. Ursenbach, A guide to high pressure air injection (HPAI) based oil recovery, SPE paper number 75207 at *SPE/DOE Improved Oil Recovery Symposium*, 13 to 17 April, Tulsa, Oklahoma, 2002.
138. B.J. Henson, A.R. Tarrer, C.W. Curtis and J.A. Guin, *Ind. Eng. Chem. Process Des. Dev.*, 1982, **21**, 575–579.
139. J.H. Hong and R. Kobayashi, *Ind. Eng. Chem. Process Des. Dev.*, 1986, **25**, 736–741.
140. J.H. Hong and R. Kobayashi, *Ind. Eng. Chem. Process Des. Dev.*, 1987, **26**, 296–299.
141. W. Hayduk, E.B. Walter and P. Simpson, *J. Chem. Eng. Data*, 1972, **17**, 59–61.
142. S.H. Huang, H.M. Lin and K.C. Chao, *J. Chem. Eng. Data*, 1988, **33**, 143–145.
143. S.H. Huang, H.M. Lin and K.C. Chao, *J. Chem. Eng. Data*, 1988, **33**, 145–147.
144. H. Tanaka, Y. Yamaki and M. Kato, *J. Chem. Eng. Data*, 1993, **38**, 386–388.
145. F.-N. Tsai and J.-S. Yau, *J. Chem. Eng. Data*, 1990, **35**, 43–45.
146. A. Chapoy, A.H. Mohammadi, A. Chareton, B. Tohidi and D. Richon, *Ind. Eng. Chem. Res.*, 2004, **43**, 1794–1802.
147. P. Munjal and P.B. Stewart, *J. Chem. Eng. Data*, 1971, **16**, 170–172.
148. P.B. Stewart and P. Munjal, *J. Chem. Eng. Data*, 1970, **15**, 67–71.
149. R. Wiebe and V.L. Gaddy, *J. Am. Chem. Soc.*, 1939, **61**, 315–318.
150. R. Wiebe and V.L. Gaddy, *J. Am. Chem. Soc.*, 1940, **62**, 815–817.
151. J. Kiepe, S. Horstmann, K. Fischer and J. Gmehling, *Ind. Eng. Chem. Res.*, 2003, **42**, 3851–3856.
152. A. Dhima, J.-C. de Hemptinne and J. Jose, *Ind. Eng. Chem. Res.*, 1999, **38**, 3144–3161.
153. S. Yamamoto, J.B. Alcauskas and T.E. Crozier, *J. Chem. Eng. Data*, 1976, **21**, 78–80.
154. S.D. Cramer, *Ind. Eng. Chem. Process Des. Dev.*, 1984, **23**, 533–538.

155. J. Kiepe, S. Horstmann, K. Fischer and J. Gmehling, *Ind. Eng. Chem. Res.*, 2003, **42**, 5392–5398.
156. J. Kiepe, S. Horstmann, K. Fischer and J. Gmehling, *Ind. Eng. Chem. Res.*, 2004, **43**, 3216.
157. A. Dhima, J.-C. de Hemptinne and G. Moracchini, *Fluid Phase Equilib.*, 1998, **145**, 129–150.
158. R. Wiebe, V.L. Gaddy and C. Heins, Jr., *J. Am. Chem. Soc.*, 1933, **55**, 947–953.
159. M. Rigby and J.M. Prausnitz, *J. Phys. Chem.*, 1968, **72**, 330–334.
160. J. Livingston, R. Morgan and A.H. Richardson, *J. Phys. Chem.*, 1930, **34**, 2356–2366.
161. E. Wiebe and V.L. Gaddy, *J. Am. Chem. Soc.*, 1935, **57**, 1487–1488.
162. J. Livingston, R. Morgan and H.R. Pyne, *J. Phys. Chem.*, 1930, **34**, 1818–1821.
163. P.A. Rice, R.P. Gale and A.J. Barduhn, *J. Chem. Eng. Data*, 1976, **21**, 204–206.
164. A. Chapoy, A.H. Mohammadi, A. Chareton, B. Tohidi and D. Richon, *J. Chem. Eng. Data*, 2005, **50**, 1157–1161.
165. H. Sørensen, K.S. Pedersen and P.L. Christiansen, *Org. Geochem.*, 2002, **33**, 635–642.
166. J.D. Lawson and A.W. Garst, *J. Chem. Eng. Data*, 1976, **21**, 30–32.
167. A.E. Mather and K.N. Marsh, *J. Chem. Eng. Data*, 1996, **41**, 1210.
168. G.F. Versteeg and W.P.M. van Swaalj, *J. Chem. Eng. Data*, 1988, **33**, 29–34.
169. A. Henni, P. Tontiwachwuthikul and A. Chakma, *J. Chem. Eng. Data*, 2006, **51**, 64–67.
170. N. Berthezene, J.-C. de Hemptinne, A. Sudibert and J.-F. Argiller, *J. Pet. Sci. Eng.*, 1999, **23**, 71–81.
171. W.P. Ridenour, W.D. Weatherford, Jr. and R.G. Capell, *Ind. Eng. Chem.*, 1954, **46**, 2376–2381.
172. J. Tong, W. Gao, R.L. Robinson, Jr. and K.A.M. Gasem, *J. Chem. Eng. Data*, 1999, **44**, 784–787.
173. D. Lal, F.D. Otto and A.E. Mather, *Fuels*, 1999, **78**, 1437–1441.
174. H.Y. Cai, J.M. Shaw and K.H. Chung, *Fuels*, 2001, **80**, 1055–1063.
175. L.C. Price, L.M. Wenger, T. Ging and C.W. Blount, *Org. Geochem.*, 1983, **4**, 201–221.
176. D. Suleiman and C.A. Eckert, *J. Chem. Eng. Data*, 1995, **40**, 2–11.
177. K.S. Sorbie and E.J. Mackay, *J. Pet. Sci. Eng.*, 2000, **27**, 85–106.
178. A. Braibanti, E. Fisicaro, A. Ghiozzi, C. Compari and M. Panelli, *J. Sol. Chem.*, 1995, **24**, 703–718.
179. C.A. Page, J.S. Bonner, P.L. Sumner and R.L. Autenrieth, *Mar. Chem.*, 2000, **70**, 79–87.
180. C.L. Yaws, *Chem. Eng.*, 1993, **100**, 108–111.
181. J.-C. De Hemptinne, H. Delepine, C. Jose and J. Jose, *Rev. Instit. Francais Petrole*, 1998, **53**, 409–419.
182. C.E.B. Conybeare, *J. Geo. Soc. Australia*, 1970, **16**, 667–681.

183. A. Maaczynski, M. Goral, B. Wisniewska-Goclowska, A. Skrzecz and D. Shaw, *Monatshefte fur Chemie*, 2003, **134**, 633–653.
184. W.Y. Shiu, M. Bobra, A.M. Bobra, A. Maijanen, L. Suntio and D. Mackay, *Oil Chem Pollut.*, 1990, **7**, 57–84.
185. J.R. Mihelcic, *Ground Water Monit. Rev.*, 1990, **10**, 132–137.
186. C.L. Yaws and X. Lin, *Pollut. Eng.*, 1994, **26**, 70–72.
187. C.L. Yaws and X. Lin, *Chem. Eng.*, 1993, **100**, 122–123.
188. E.F.G. Herrington, *J. Am. Chem. Soc.*, 1948, **70**, 1995–1996.
189. R. Karlsson, *J. Chem. Eng. Data*, 1973, **18**, 290–292.
190. R.L. Bohen and W.F. Claussen, *J. Am. Chem. Soc.*, 1951, **73**, 1571–1578.
191. W.F. Claussen, *J. Am. Chem. Soc.*, 1952, **74**, 3937–3738.
192. C.L. Yaws, L. Bu and S. Nijhawan, *Pollut. Eng.*, 1995, **27**, 78–80.
193. Y. Hashimoto, K. Tokura, H. Kishi and W.M.J. Strachan, *Chemosphere*, 1984, **13**, 881–888.
194. R.S. Pearlman, S.H. Yalkowsky and S. Banerjee, *J. Phys. Chem. Ref. Data*, 1984, **13**, 555–562.
195. C. Tsonopoulos and G.M. Wilson, *AIChE J.*, 1983, **29**, 990–999.
196. J.F. Connoly, *J. Chem. Eng. Data*, 1966, **11**, 13–16.
197. W.L. Masterton, T.P. Lee and R.L. Boyington, *J. Phys. Chem.*, 1969, **73**, 2761–2763.
198. D.F. Keeley, M.A. Hoffpauir and J.R. Meriwether, *J. Chem. Eng. Data*, 1988, **33**, 87–89.
199. C. Sutton and J.A. Calder, *J. Chem. Eng. Data*, 1975, **20**, 320–322.
200. D.F. Keeley, M.A. Hoffpauir and J.R. Meriwether, *J. Chem. Eng. Data*, 1991, **36**, 456–459.
201. W.-H. Xie, W.-Y. Shiu and D. Mackay, *Mar. Environ. Res.*, 1997, **44**, 429–444.
202. C. McAuliffe, *Science*, 1969, **163**, 478–479.
203. K.S. Pedersen, J. Milter, C.P. Rasmussen and P. Claus, *Fluid Phase Equilib.*, 2001, **189**, 85–97.
204. J.F. Connolly, *J. Chem. Eng. Data*, 1966, **11**, 13–16.
205. C. Marche, C. Ferronato and J. Jose, *J. Chem. Eng. Data*, 2003, **48**, 967–971.
206. S. Brandani, V. Brandani and D. Flammini, *J. Chem. Eng. Data*, 1994, **39**, 201–202.
207. C. McAuliffe, *J. Phys. Chem.*, 1966, **70**, 1267–1275.
208. F.R. Groves, Jr., *J. Chem. Eng. Data*, 1988, **33**, 136–138.
209. L.H. Milligan, *J. Phys. Chem.*, 1924, **28**, 494–497.
210. A.P. Sazonov, V.V. Filippov and N.V. Sazonov, *J. Chem. Eng. Data*, 2001, **46**, 959–961.
211. I.G. Economou, J.L. Heidman, C. Tsonopoulos and G.M. Wilson, *AIChE J.*, 1997, **43**, 535–546.
212. C.L. Yaws, *Pollut. Eng.*, 1990, **22**, 70–75.
213. C.L. Yaws, H.C. Yang, J.R. Hopper and K.C. Hansen, *Chem. Eng.*, 1990, **97**, 177–178.

214. C.L. Yaws, *Pollut. Eng.*, 1992, **24**, 46–50.
215. M. Goral, A. Maczynski and B. Wisniewska-Goclowska, *J. Phys. Chem. Ref. Data*, 2004, **33**, 579–591.
216. A. Maczynski, D.G. Shaw, M. Goral, B. Wisniewska-Goclowska, A. Skrzecz, Z. Maczynska, I. Owczarek, K. Blazej, M.-C. Haulait-Pirson, F. Kapuku, G.T. Hefter and A. Szafranski, *J. Phys. Chem. Ref. Data*, 2005, **34**, 441–476.
217. A. Maczynski, D.G. Shaw, M. Goral, B. Wisniewska-Goclowska, A. Skrzecz, I. Owczarek, K. Blazej, M.-C. Haulait-Pirson, G.T. Hefter, Z. Maczynska, A. Szafranski, C. Tsonopoulos and C.L. Young, *J. Phys. Chem. Ref. Data*, 2005, **34**, 477–552.
218. A. Maczynski, D.G. Shaw, M. Goral, B. Wisniewska-Goclowska, A. Skrzecz, I. Owczarek, K. Blazej, M.-C. Haulait-Pirson, G.T. Hefter, Z. Maczynska, A. Szafranski and C.L. Young, *J. Phys. Chem. Ref. Data*, 2005, **34**, 657–708.
219. A. Maczynski, D.G. Shaw, M. Goral, B. Wisniewska-Goclowska, A. Skrzecz, I. Owczarek, K. Blazej, M.-C. Haulait-Pirson, G.T. Hefter, F. Kapuku, Z. Maczynska and C.L. Young, *J. Phys. Chem. Ref. Data*, 2005, **34**, 709–753.
220. A. Maczynski, D.G. Shaw, M. Goral, B. Wisniewska-Goclowska, A. Skrzecz, I. Owczarek, K. Blazej, M.-C. Haulait-Pirson, G.T. Hefter, F. Kapuku, Z. Maczynska, A. Szafranski and C.L. Young, *J. Phys. Chem. Ref. Data*, 2005, **34**, 1399–1487.
221. A. Maczynski, D.G. Shaw, M. Goral, B. Wisniewska-Goclowska, A. Skrzecz, I. Owczarek, K. Blazej, M.-C. Haulait-Pirson, G.T. Hefter, Z. Maczynska and A. Szafranski, *J. Phys. Chem. Ref. Data*, 2005, **34**, 1489–1553.
222. D.G. Shaw, A. Maczynski, M. Goral, B. Wisniewska-Goclowska, A. Skrzecz, I. Owczarek, K. Blazej, M.-C. Haulait-Pirson, G.T. Hefter, F. Kapuku, Z. Maczynska and A. Szafranski, *J. Phys. Chem. Ref. Data*, 2005, **34**, 2299–2345.
223. D.G. Shaw, A. Maczynski, M. Goral, B. Wisniewska-Goclowska, A. Skrzecz, I. Owczarek, K. Blazej, M.-C. Haulait-Pirson, G.T. Hefter, F. Kapuku, Z. Maczynska and A. Szafranski, *J. Phys. Chem. Ref. Data*, 2005, **34**, 2261–2298.
224. K. Kniaż, *J. Chem. Eng. Data*, 1991, **36**, 471–472.
225. P.M. Ghogomu, J. Dellacherie and D. Balesdent, *J. Chem. Thermodyn.*, 1989, **21**, 925–934.
226. M.B. Gramajo de Doz, C.M. Bonatti and H.N. Solimo, *J. Chem. Thermodyn.*, 2003, **35**, 2055–2065.
227. M.B. Gramajo de Doz, C.M. Bonatti and H.N. Solimo, *Fluid Phase Equilib.*, 2003, **205**, 53–67.
228. A. Skrzecz, *Thermochim. Acta*, 1991, **182**, 123–131.
229. C.F. Mabery, *Ind. Eng. Chem.*, 1924, **16**, 937.
230. C.B. Kretschmer and R. Wiebe, *Ind. Eng. Chem.*, 1945, **37**, 1130–1132.

231. O.C. Bridgeman and D. Querfeld, *Ind. Eng. Chem.*, 1933, **25**, 523–525.
232. Y. Yang, D.J. Miller and S.B. Hawthorne, *J. Chem. Eng. Data*, 1997, **42**, 908–913.
233. F. Groves, Jr., *Environ. Sci. Technol.*, 1988, **22**, 282–286.
234. S.O. Derawi, G.M. Kontogeorgis, E.H. Stenby, T. Haugum and A.O. Fredheim, *J. Chem. Eng. Data*, 2002, **47**, 169–173.
235. K.C. Taylor and H.A. Nasr-El-Din, *J. Pet. Sci. Eng.*, 1998, **19**, 265–280.
236. R.F. Meyer and E. Attanasi, *Natural Bitumen and Extra Heavy Oil*, in *Survey of Energy Resources*, J. Trinnaman and A. Clarke (eds), for the World Energy Council, Elsevier, Amsterdam, 2004.
237. E. Rogel, *Langmuir*, 2004, **20**, 1003–1012.
238. J.J. Heron and E.K. Spady, *Ann. Rev. Energy*, 1983, **8**, 137–163.
239. A.G. Oblad, J.W. Bunger, F.V. Hanson, J.D. Miller, H.R. Ritzma and J.D. Seader, *Ann. Rev. Energy*, 1987, **12**, 283–356.
240. J.G. Speight, *Ann. Rev. Energy*, 1986, **11**, 253–274.
241. C. Ovalles, A. Hamana, I. Rojas and R.A. Bolivar, *Fuel*, 1995, **74**, 1162–1168.
242. O.C. Mullins and E.Y. Sheu (eds), *Structure and Dynamics of Asphaltenes*, Springer, New York, 1999.
243. R.A. DeRuiter, L.J. Nash and M.S. Singletary, *SPE Res. Eng.*, 1994, **9**, 101–106.
244. D.V. Nichita, L. Goual and A. Firoozabadi, *SPE Prod. Fac.*, 2001, **16**, 250–259.
245. H. Tanaka and M. Kato, *Netsu Bussei*, 1994, **8**, 74–78.
246. Y.W. Svrcek and K.A. Mehrotra, *J. Can. Pet. Tech.*, 1982, **21**, 31–38.
247. M.D. Deo, C.J. Wang and F.V. Hanson, *Ind. Eng. Chem. Res.*, 1991, **30**, 532–536.
248. M.D. Deo, P.E. Rose and F.V. Hanson, *Ind. Eng. Chem. Res.*, 1992, **31**, 1424.
249. K.A.M. Gasem and R.L. Robinson, Jr., *J. Chem. Eng. Data*, 1985, **30**, 53–56.
250. F.-N. Tsai, S.H. Huang, H.-M. Lin and K.-C. Chao, *J. Chem. Eng. Data*, 1987, **32**, 467–469.
251. V.S. Smith and A.S. Teja, *J. Chem. Eng. Data*, 1996, **41**, 923–925.
252. G.S.A. van Welie and G.A.M. Diepen, *J. Phys. Chem.*, 1963, **67**, 755–757.
253. J.S. Gopal, G.D. Holder and E. Kosal, *Ind. Eng. Chem. Process Des. Dev.*, 1985, **24**, 697–701.
254. S. Mitra, J.W. Chen and D.S. Viswanath, *J. Chem. Eng. Data*, 1988, **33**, 35–37.
255. J.J. Czubryt, M.N. Myers and J.C. Giddings, *J. Phys. Chem.*, 1970, **74**, 4260–4266.
256. F.W. Sullivan, Jr., W.J. McGill and A. French, *Ind. Eng. Chem.*, 1927, **19**, 1042–1045.
257. P. Weber and H.L. Dunlap, *Ind. Eng. Chem.*, 1928, **20**, 383–384.
258. A. Bernie-Allen Jr. and L.T. Work, *Ind. Eng. Chem.*, 1930, **22**, 874–878.
259. D.S. Davis, *Ind. Eng. Chem.*, 1940, **32**, 1293.

260. R. Boistelle and H.E. Lundager Madsen, *J. Chem. Eng. Data*, 1978, **23**, 28–29.
261. U. Domanska and P. Morawski, *J. Chem. Thermodyn.*, 2005, **37**, 1276–1287.
262. K.L. Roberts, R.W. Rousseau and A.S. Teja, *J. Chem. Eng. Data*, 1994, **39**, 793–795.
263. K.C.M. Alves, R. Condotta and M. Giulietti, *J. Chem. Eng. Data*, 2001, **46**, 1516–1519.
264. E. Provost, V. Chevallier, M. Bouroukba, D. Petitjean and M. Dirand, *J. Chem. Eng. Data*, 1998, **43**, 745–749.
265. R. Rakotosaona, M. Bouroukba, D. Petitjean, N. Hubert, J.C. Moise and M. Dirand, *Fuel*, 2004, **83**, 851–857.
266. E. Flöter, B. Hollanders, T.W. de Loos and J. de Swaan Arons, *J. Chem. Eng. Data*, 1997, **42**, 562–565.
267. L.E. Sverdrup, T. Nielsen and P.H. Krogh, *Environ. Sci. Tech.*, 2002, **36**, 2429–2435.
268. I.A. Wiehe, *Fuel Sci. Techn. Int.*, 1996, **14**, 289–312.
269. J.G. Speight, *The Chemistry and Technology of Petroleum*, 3rd edn, Marcel Dekker, New York, 1998, 452–453.
270. H.H. Ibrahim and R.O. Idem, *Energy Fuels*, 2004, **18**, 1354–1369.
271. K. Akbarzadeh, A. Dhillon, W.Y. Svrcek and H.W. Yarranton, *Energy Fuels*, 2004, **18**, 1434–1441.
272. I.A. Wiehe, H.W. Yarranton, K. Akbarzadeh, P.M. Rahimi and A. Teclemariam, *Energy Fuels*, 2005, **19**, 1261–1267.
273. K. Akbarzadeh, H. Alboudwarej, W.Y. Svrcek and H.W. Yarranton, *Fluid Phase Equilib.*, 2005, **232**, 159–170.
274. G.R. Pazuki and M. Nikookar, *Fuel*, 2006, **85**, 1083–1086.
275. F. Mutelet, G. Ekulu, R. Solimando and M. Rogalski, *Energy Fuels*, 2004, **18**, 667–673.
276. E. Hong and P. Watkinson, *Fuel*, 2004, **83**, 1881–1887.
277. C.W. Angle, Y. Long, H. Hamza and L. Lue, *Fuel*, 2006, **85**, 492–506.
278. P.M. Spiecker, K.L. Gawrys and P.K. Kilpatrick, *J. Colloid Interface Sci.*, 2003, **267**, 178–193.
279. F.H. Rhodes and F.S. Eisenhauer, *Ind. Eng. Chem.*, 1927, **19**, 414–416.
280. E.L. Heric and K.-N. Yeh, *J. Chem. Eng. Data*, 1970, **15**, 13–17.
281. P.C. Painter, J. Graf and M.M. Coleman, *Energy Fuels*, 1990, **4**, 379–384.
282. P.C. Painter, J. Graf and M.M. Coleman, *Energy Fuels*, 1990, **4**, 393–397.
283. P.C. Painter, Y. Park, M. Sobkowiak and M.M. Coleman, *Energy Fuels*, 1990, **4**, 384–393.
284. Y. Marcus, A.L. Smith, M.V. Korobov, A.L. Mirakyan, N.V. Avramenko and E.B. Stukalin, *J. Phys. Chem. B*, 2001, **105**, 2499–2506.
285. E. Szczepaniec-Cieciak and K. Nagraba, *Cryogenics*, 1978, **18**, 601–603.
286. E. Szczepaniec-Cieciak and B. Ciejek, *Cryogenics*, 1979, **19**, 639–648.
287. R.J.J. Chen, V.W. Liaw and D.G. Elliot, *Adv. Cryo. Eng.*, 1980, **25**, 620–628.
288. L.C. Price, *J. Petrol. Geol.*, 1981, **4**, 195–223.

289. D.L. Tiffin, J.P. Kohn and K.D. Luks, *J. Chem. Eng. Data*, 1979, **24**, 306–310.
290. H.-Y. Cai, J.M. Shaw and K.H. Chung, *Fuel*, 2001, **80**, 1065–1077.
291. K.-C. Kuo, *Mar. Pet. Geo.*, 1997, **14**, 221–229.
292. D.N. Glew, *J. Phys. Chem.*, 1962, **66**, 605–609.
293. P. Servio and P. Englezos, *J. Chem. Eng. Data*, 2002, **47**, 87–90.
294. J. Xia, M. Jödecke, Á.P.-S. Kamps and G. Maurer, *J. Chem. Eng. Data*, 2004, **49**, 1756–1759.
295. A. Chapoy, C. Coquelet and D. Richon, *J. Chem. Eng. Data*, 2003, **48**, 957–966.
296. H.A. Al-Anazi, J.G. Walker, G.A. Pope, M.M. Sharma and D.F. Hackney, *SPE Prod. Facil.*, 2005, **20**, 60–69.
297. M.H. Abraham, G.S. Whiting, W.J. Shuely and R.M. Doherty, *Can J. Chem.*, 1998, **76**, 703–709.
298. K.Y. Song, M. Yarrison and W. Chapman, *Fluid Phase Equilib.*, 2004, **224**, 271–277.
299. C.B. Kretschmer and R. Wiebe, *J. Am. Chem. Soc.*, 1952, **74**, 1276–1277.
300. X. Shuqian, M. Peisheng, G. Yugao and H. Chao, *J. Chem. Eng. Data*, 2004, **49**, 479–482.
301. E.D. Sloan, *Clathrate Hydrates of Natural Gas*, 2nd edn, Marcel Dekker, New York, 1998.
302. Y.F. Makogon, *Hydrates of Hydrocarbons*, PennWell Books, Tulsa, 1997.
303. G.D. Holder and P.R. Bishnoi (Eds), *Gas Hydrates Challenges for the Future*, Annals of the New York Academy of Sciences, vol. 912, New York, NY, 2000.
304. L.J. Florusse, C.J. Peters, J.C. Pàmies, L.F. Vega and H. Meijer, *AIChE J.*, 2003, **49**, 3260–3269.
305. S. Raeissi and C.J. Peters, *J. Supercrit. Fluid*, 2001, **20**, 221–228.
306. L.J. Florusse, C.J. Peters and H. Meijer, in preparation.

CHAPTER 24

Solubility of Inorganic Salts and their Industrial Importance

WOLFGANG VOIGT

Institut für Anorganische Chemie, TU Bergakademie Freiberg, 09596, Freiberg, Germany

24.1 Introduction

Combinations of metal cations with simple anions like halides (fluoride, chloride, bromide, iodide), sulfates, nitrates, carbonates or phosphates yield an overwhelming number of salts, double salts and salt hydrates. The industrial interest in a large number of these salts results from their specific properties as final products (anti-freezing agents, fertilizer, oxidation reagents, *etc.*) or as a suited storage and transportation medium for later chemical or electrochemical conversions. In general, salts are produced by crystallization after steps of extraction or leaching from ores as is done in hydrometallurgy (Chapter 17). The largest amounts of salts are produced and used in the form of chlorides, nitrates and phosphates of sodium, potassium, magnesium and calcium. These compounds are needed in enormous quantities as fertilizers or basic chemicals in chemical, metallurgical and building material industries. Lithium salts and borates are becoming very important in industry for the production of batteries and high-tech materials. The oceans and their ancient evaporitic salt deposits as well as several salt lakes represent the main sources for their recovery. Many of the simple salts like NaCl, Na_2SO_4, KCl, $MgCl_2$ or $CaCl_2$ are very soluble. An understanding of the formation of evaporitic salt deposits (rock salt, potash, carnallite, gypsum, *etc.*), of crystallization sequences in evaporation of sea water in salt ponds and of the development of technologies for recovering of salts from deposits, is based on the complex solubility equilibria in the multicomponent system Li^+, Na^+, K^+, Mg^{2+}, Ca^{2+} // Cl^-, SO_4^{2-}, CO_3^{2-}, OH^-, $(B_4O_7^{2-})$–H_2O and its subsystems. Thus the prediction of occurrence and quality of salt deposits is based on the same solubility diagrams as the technology for their exploitation, although the technological schemes often need the consideration of extended temperature ranges. In addition, when considering economic production schemes, energy balances have to be included, which requires complementary information about the enthalpy changes during crystallization/evaporation processes.

24.2 Oceanic Salts

The major components of oceanic salts can be derived from sea water composition, which has changed only slightly over millions of years (Table 1). Neglecting the minor elements Li^+, Br^-, I^-, CO_3^{2-}, the crystallization sequence can be described by solubility equilibria in the hexary system Na^+, K^+, Mg^{2+}, Ca^{2+} // Cl^-, SO_4^{2-}–H_2O. Typically a temperature range between 0 and 110°C is of interest. The stoichiometry and mineral names of stable salt phases occurring in this system are listed in Table 2. In certain cases, temperatures up to 250°C are of interest, where also new double salts appear.[1–3]

Despite more than a 100 years of experimental work on phase equilibria in the hexary system, beginning with van't Hoff [4,5] and including recent experimental work[1,4–10] not all temperature – composition regions of the equilibria are known in respect to phase stability and solution composition. For most technological purposes (potash and magnesium chloride recovery) considerations within the quinary system Na^+, K^+, Mg^{2+}, // Cl^-, SO_4^{2-}–H_2O, are sufficient. This excludes the calcium salts. For the presentation of phase relations and crystallization processes projection diagrams involving additional constraints are used. Jänecke diagrams are often used in oceanic salt chemistry. In Figure 1 such diagrams are given for the quinary system between 25 and 55°C. Both diagrams assume saturation with NaCl and therefore Na^+ and Cl^- are not part of the diagrams coordinates. From the evaporation line drawn in Figure 1(a) the sequence of crystallization from sea water can be seen: epsomite, kainite, carnallite, bischofite. From the diagram it cannot be recognized that gypsum (anhydrite) and halite are deposited before and halite also simultaneously by an evaporation process of sea water. All these minerals are found in large quantities in evaporitic deposits and can be recovered from them by mining.

Table 1 *Chemical composition of seawater (SW)*

Ion/Element	*g/1000 g SW*	*mol/1000 mol H_2O*
Na^+	10.76	8.737
Mg^{2+}	1.29	0.991
Ca^{2+}	0.413	0.192
K^+	0.399	0.19
Sr^{2+}	0.0079	0.0017
Cl^-	19.357	10.19
SO_4^{2-}	2.709	0.526
HCO_3^-	0.14	0.043
Br^-	0.0671	0.0157
F^-	0.0013	0.0013
Rb^+	0.00012	0.000025
B	0.0045	0.0078

Table 2 *Salt minerals crystallizing in the hexary system Na^+, K^+, Mg^{2+}, Ca^{2+}// Cl^-, SO_4^{2-}–H_2O*

Mineral name	*Composition*	*Abbreviation (Figure 1)*
Anhydrite	$CaSO_4$	
Arcanite	K_2SO_4	
Bloedite	$Na_2SO_4 \cdot MgSO_4 \cdot 4H_2O$	bl
Chlorocalcite	$KCl \cdot CaCl_2$	
Bischofite	$MgCl_2 \cdot 6H_2O$	bi
Epsomite	$MgSO_4 \cdot 7H_2O$	ep
Carnallite	$KCl \cdot MgCl_2 \cdot 6H_2O$	ca
D'Ansite	$3NaCl \cdot 9Na_2SO_4 \cdot MgSO_4$	da
Goergeyite	$K_2SO_4 \cdot 5CaSO_4 \cdot H_2O$	
Glaserite	$3K_2SO_4 \cdot Na_2SO_4$	gs
Glauberite	$Na_2SO_4 \cdot CaSO_4$	
Gypsum	$CaSO_4 \cdot 2H_2O$	
Halite	$NaCl$	
Hexahydrite	$MgSO_4 \cdot 6H_2O$	hx
Labile salt	$2Na_2SO_4 \cdot CaSO_4 \cdot 2H_2O$	
Kainite	$4KCl \cdot 4MgSO_4 \cdot 11H_2O$	ka
Kieserite	$MgSO_4 \cdot H_2O$	ks
Langbeinite	$K_2SO_4 \cdot 2MgSO_4$	lg
Leonite	$K_2SO_4 \cdot MgSO_4 \cdot 4H_2O$	le
Loeweite	$6Na_2SO_4 \cdot 7MgSO_4 \cdot 15H_2O$	lw
Mirabilite	$Na_2SO_4 \cdot 10H_2O$	
Polyhalite	$K_2SO_4 \cdot MgSO_4\ 2CaSO_4 \cdot 2H_2O$	
Picromerite, schönite	$K_2SO_4 \cdot MgSO_4 \cdot 6H_2O$	pc
Sylvite	KCl	sy
Syngenite	$K_2SO_4 \cdot CaSO_4 \cdot H_2O$	
Tachhydrite	$2MgCl_2 \cdot CaCl_2 \cdot 12H_2O$	
Thenardite	Na_2SO_4	th
Vanthoffite	$3Na_2SO_4 \cdot MgSO_4$	vh

The coordinates SO_4^{2-}–Mg^{2+}–$(K^+)_2$ do not allow one to draw conclusions in respect to absolute salt concentrations or the content of Na^+ or Cl^- in the solutions. However, plotting the solution compositions observed during the process of evaporation into this type of diagram gives a quick overview of the process. For instance which salts should crystallize or which has already crystallized. Besides the stable fields of crystallization it is also important to know the so-called meta-stable fields of crystallization. Owing to kinetic reasons the stable phases often does not crystallize, but the meta-stable ones do. These are the underlying phases in the diagram. Typical examples in the quinary system are kainite, bloedite and kieserite, which need seeding or a long time in order to crystallize.

Complete process information requires knowledge about the degrees of saturation of the respective salts, the absolute ion concentrations in relation to the concentration solubility product, mass balances and all its changes when temperature is varied. To achieve this aim the complete multidimensional phase diagram has to be stored in a computerized form as a numerical model. Using this model and exploiting generalized geometrical relationships of phase

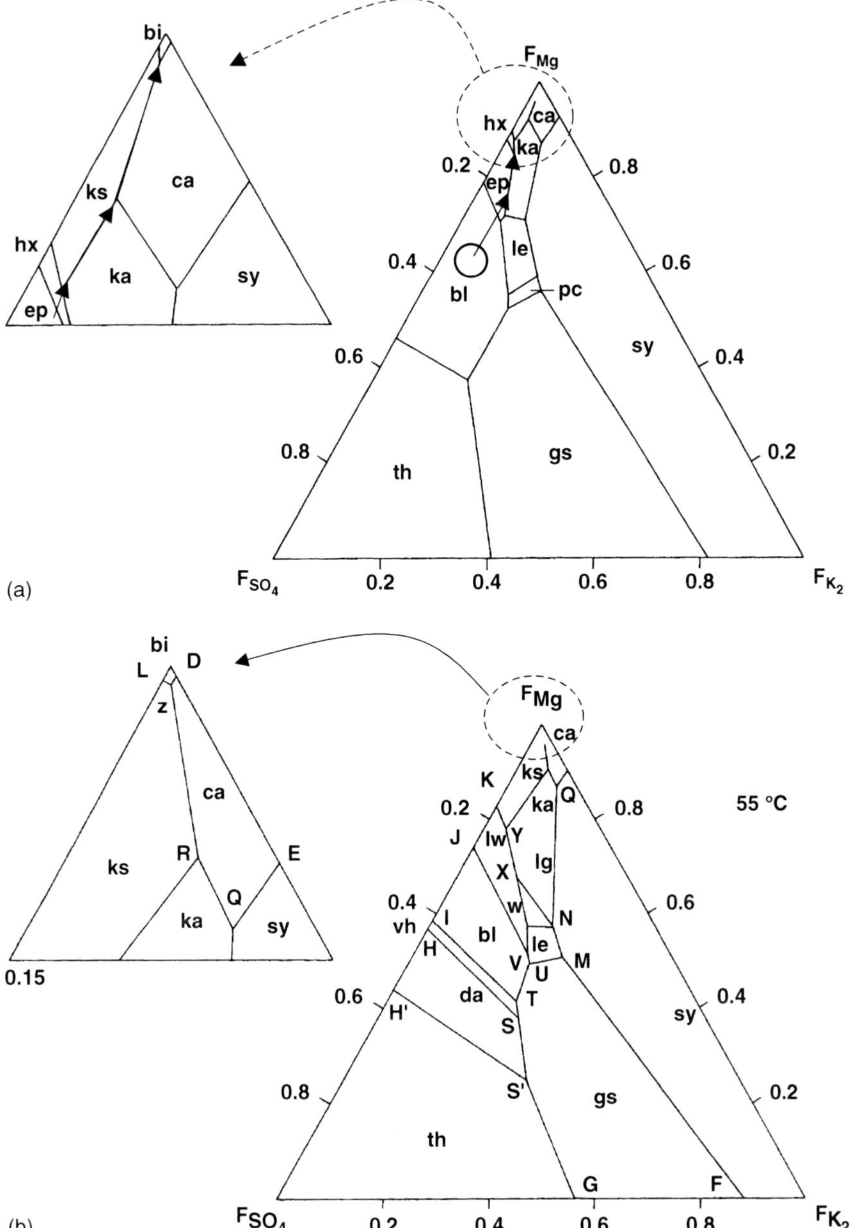

Figure 1 *(a) Jänecke diagram of the quinary ocenic salt system at 25°C*[46]*, arrows: path of crystallization when sea water is evaporated under equilibrium conditions at 25°C (b) Jänecke diagram of the quinary ocenic salt system at 55°C.*[46]

diagrams, all necessary concentrations, phase relations and mass balances are at hand. Such an approach was suggested and developed by Cohen–Adad.[11,12] The basis of the numerical model is represented by a series of the form

$$A_i\,(X_i, Y_j) = \sum\sum X_i Y_j\,\{a^{ij} + \sum\sum U_k{}^n\,\{(a^{ijk})_n + \ldots\}\} \tag{1}$$

where A_i denotes a coefficient in Equation (2) describing the temperature dependence of the concentration solubility product K_S of a salt consisting of the cation i and anion j. X_i, Y_j and U_k represent concentration variables, a^{ij} and a^{ijk} are adjustable parameters.

$$\ln(K_S) = A_0 + A_1{*}(1/T) + A_2{*}\ln(T) + A_3{*}T + A_4{*}T^2 \tag{2}$$

More common in use are thermodynamic models based on the Pitzer equations of ion interactions (see Chapter 2), which are often integrated in process simulators like CHEMSAGE-FactSage or ASPEN-Plus. The number of adjustable parameters is comparable to the Cohen-Adad and Pitzer approach if only solid-liquid equilibria are considered.[33] However, the Pitzer type models describe also the thermodynamic properties of unsaturated solutions including enthalpies and heat capacities. Thus, with this type of thermodynamic models complete technological schemes can be evaluated in respect to material and energy balances, if all model coefficients are known and the model parameters had been determined in an appropriate manner supported by a sufficient number of reliable experimental data points. At present, neither for the hexary nor for the quinary system of oceanic salts, does a parameter set exist which describes all temperature – composition regions with an accuracy corresponding to practical demands. For a certain case of application specific data fits of the Pitzer model have to be selected. Table 3 contains a collection of several parameter sets.

24.2.1 Production of K_2SO_4

Besides potash, K_2SO_4 (arcanite) represents the second most important potassium fertilizer. It is applied for growing a number of chloride-sensitive plants (potatoes, vegetables) – world production is of the order of 4 million tonnes per annum. K_2SO_4 does not primarily crystallize from seawater and therefore it can only be produced by conversion from other salts. The preferred process in Germany[13] is based on the following gross reaction

$$2\ KCl + MgSO_4 \rightarrow K_2SO_4 + MgCl_2 \tag{I}$$

From the solubility diagram different technological schemes can be derived, where intermediately the double salts leonite or picromerite (schönite) are separated.[13] The principle of the two-stage "schönite process" is illustrated in Figure 2.

Epsomite and sylvite are mixed in a ratio, which corresponds to point S1 in the diagram. In the presence of a defined amount of water this mixture is converted to solid picromerite and a solution with a composition given at

Table 3 *Pitzer parameter sets for multi-component oceanic salt systems*

System *T range, p range* — *Remarks*	*Source*
Na^+, K^+, Mg^{2+}, Ca^{2+}, H^+ // Cl^-, SO_4^{2-}, OH^-, CO_3^{2-}, CO_2–H_2O 25°C; 1 bar; still considered as the standards model for seawater and brines at ambient conditions	14
Na^+, K^+, Mg^{2+} // Cl^-, SO_4^{2-}–H_2O 25°C–250°C; 1 bar or sat.; only ternary systems tested with the model	15
Na^+, Ca^{2+} // Cl^-, SO_4^{2-}–H_2O 25°C–250°C; 1 bar or sat.; model at the reciprocal system careful tested	16
Na^+, K^+, Ca^{2+} // Cl^-, SO_4^{2-}–H_2O 25°C–250°C; 1 bar or sat.; model at higher order systems tested	17
Na^+, K^+, Mg^{2+}, Ca^{2+}, Sr^{2+}, Ba^{2+} // Cl^-, SO_4^{2-}, CO_3^{2-}–H_2O 25°C–110°C; 1 bat or sat.; focus on T and p dependence of solubility of sulfates	18
Na^+, K^+, Mg^{2+} // Cl^-, SO_4^{2-}–H_2O 90°C; 1 bar complete system study	19
Na^+, K^+, Mg^{2+} // Cl^-, SO_4^{2-}–H_2O 50°C–140°C; 1 bar or sat.; extension of earlier work[19]	20
Na^+, K^+, Ca^{2+} // Cl^-, SO_4^{2-}, CO_3^{2-}, CO_2–H_2O 0°C–90°C; 1 bar; solubility of CO_2 and $CaCO_3$ in the system	21
Na^+, K^+, Mg^{2+} // Cl^-, SO_4^{2-}–H_2O 0°C–110°C; 1 bar; model tested against all available higher order system equilibria	22
Na^+, K^+, Mg^{2+}, Ca^{2+} // Cl^-, SO_4^{2-}–H_2O 25°C–200°C; 1 bar or sat.; new high-T solubility data in $MgSO_4$ containing systems included	10
Na^+, K^+, Mg^{2+}, Ca^{2+} // Cl^-, SO_4^{2-}–H_2O −60°C–25°C; 1 bar; focus on re-evaluation of the sulfate equilibria	23
Na^+, K^+, Mg^{2+}, Ca^{2+}, Sr^{2+}, Ba^{2+} // Cl^-, SO_4^{2-}–H_2O 0°C–200°C; 1 bar – 1 kbar; pressure dependence of solubility of alkaline earth sulfates	24
Na^+, K^+, Mg^{2+}, Ca^{2+}, H^+ // Cl^-, SO_4^{2-}, OH^-, CO_3^{2-}, CO_2–H_2O −30°C–25°C; 1 bar; focus on solubility equilibria of carbonates	25
Na^+, K^+, H^+ // Cl^-, SO_4^{2-}, OH^-–H_2O 0°C–200°C; 1bar or sat.; ternary and reciprocal systems tested with model	26
Na^+, K^+, Ca^{2+}, H^+ // Cl^-, SO_4^{2-}, OH^-–H_2O 0°C–250°C; 1 bar or sat.; ternary and reciprocal systems tested with model	27
Na^+, K^+, Mg^{2+}, Ca^{2+} // Cl^-, SO_4^{2-}–H_2O 25°C–125°C; 1 bar or sat.; focus on equilibria with $MgCl_2$ and $CaCl_2$ containing solids	28

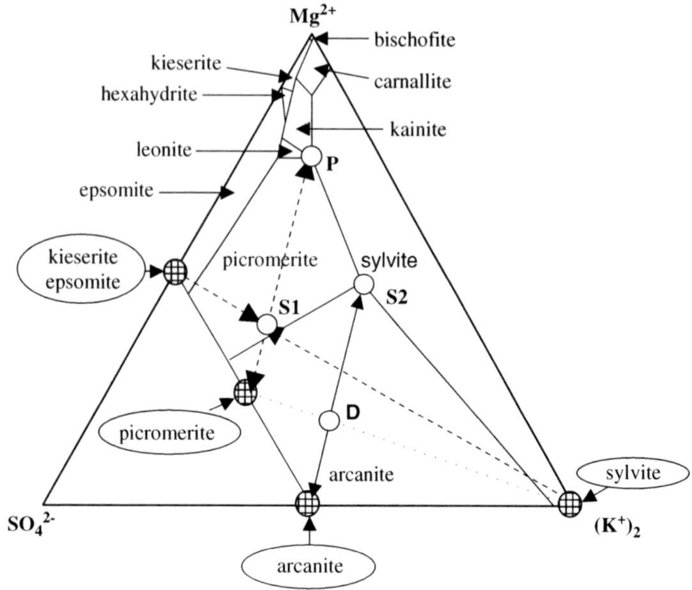

Figure 2 *Representation of the two-stage schönite process for production of K_2SO_4 in a Jänecke diagram (25°C). ⊞ composition of solids; ○ process points.*

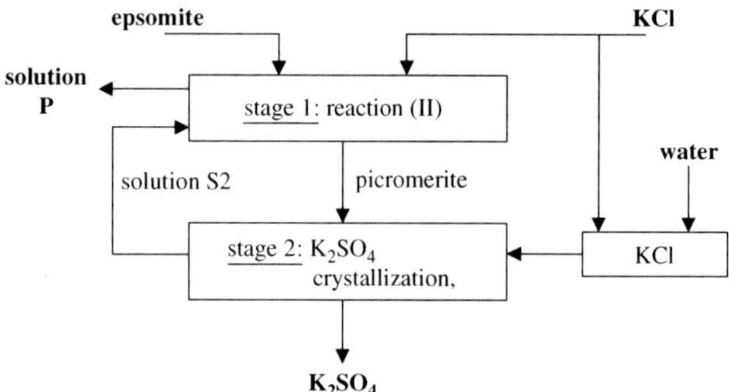

Figure 3 *Technological scheme of the two-stage schönite process for K_2SO_4 production.*

point P. This conversion can be formulated as a stoichiometric reaction:

$$2\,KCl_{(s)} + 2\,MgSO_4 \cdot 7\,H_2O_{(s)} + n\,H_2O \rightarrow K_2SO_4 \cdot MgSO_4 \cdot 6H_2O(s) + \{\ MgCl_2 + (n+8)H_2O\}_{(aq)} \qquad (II)$$

The picromerite is filtered off and the solution P containing 180–200 g/L $MgCl_2$ leaves the process. In a second stage picromerite is agitated with warm (55°C) KCl solution (point D) whereby K_2SO_4 (arcanite) is formed according to reaction (III) and the solution temperature decreased to about 35°C. The solution S2 is returned into process stage 1 for the generation of mixture S1 as shown in the technological scheme in Figure 3. In the two-stage process a theoretical yield between 68% and 84% is reached for potassium and sulfate, respectively.

$$2\ KCl_{(s)} + K_2SO_4 \cdot MgSO_4 \cdot 6H_2O(s) + n\ H_2O \rightarrow 2\ K_2SO4 + \{MgCl_2 + (n + 8)\ H_2O\}_{(aq)} \quad \text{(III)}$$

24.2.2 Solution Mining of Carnallitite

If salts have to be processed through dissolution/crystallization steps, solution mining represents a cost-effective method saving the costs for mining. The method is used for soda production from NaCl solutions in the classical Solvay process or if NaCl solutions are prepared for electrolysis to produce chlorine and caustic soda. Another more recent example is the solution mining of carnallitite. Carnallitite denotes a mineral assemblage consisting of carnallite, halite, kieserite and often containing some sylvite and anhydrite. Typical composition ranges are: 35–70% carnallite, 20–40% halite, 0–20% kieserite, 1–10% anhydrite, 0–1% sylvite.

Because of the disadvantageous geo-mechanical and hygroscopic properties, the mining of carnallitite results in high costs. In Bleicherode (former Kombinat "Kali" presently DEUSA International, Germany) a solution mining process was developed. At present approximately 200,000 m^3/a solution is processed to produce KCl and $MgCl_2 \cdot 6H_2O$. For other regions in the world with extended underground deposits of carnallite as in the Congo or Thailand, similar recovery technologies are planned. This technology requires a careful control of the solution composition transported from the mine to the factory, ensuring appropriate actions in the factory in accordance with the composition of the brine. Owing to irregularities in the underground dissolution process composition changes can occur. With the help of Figure 4 the main characteristics of the process shall be explained. Carnallite has the particular property to dissolve incongruently, that is before reaching saturation in respect to carnallite, solution compositions are located in the crystallization field of KCl (Figure 4). As long as carnallite remains in contact with such solutions $MgCl_2$ is dissolved preferentially and KCl is crystallized. Thereby the $MgCl_2$ concentration of the solution increases and the composition approaches line E (Figure 4). Only at $MgCl_2$ concentrations above the line E, can carnallite be dissolved congruently. In practice the underground dissolution is performed at a temperature of about 55°C and solution compositions near point S in Figure 4 are reached. One litre of solution at 55°C contains on average: 319 g $MgCl_2$, 34 g $MgSO_4$, 60 g KCl, 29 g NaCl.[29] The solutions are all the time saturated with NaCl and increasing the $MgCl_2$ concentration during evaporation will cause simultaneous crystallization of NaCl, which in turn affects the purity of KCl. The relative small $MgSO_4$ concentration also has a remarkable influence on the solubility of KCl and carnallite. The shift of the solubility isotherms with increasing $MgSO_4$ content is depicted by means of thin lines in

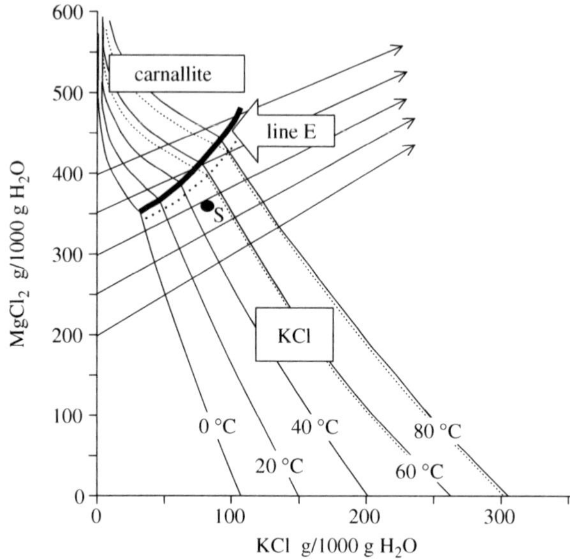

Figure 4 *Solubility diagram of the system NaCl–KCl–$MgCl_2$–($MgSO_4$)–H_2O; continuous lines: 10 g $MgSO_4$ kg^{-1} H_2O thin dotted lines: 40 g $MgSO_4$ kg^{-1} H_2O; S: composition of mining solution; arrows: direction of the figurative point of carnallite.*

Figure 4. Therefore the technological scheme consists of a multi-step crystallization, carnallite decomposition and vacuum evaporation/crystallization process as given in Figure 5. The final products are potash and $MgCl_2 \cdot 6H_2O$ (bischofite).

24.3 Salts from Non-Oceanic Salt Lakes

Salt lakes, which are located in deserts or high mountain plateaux in many parts of the world, represent isolated reservoirs, where dissolution (weathering) of rocks and intensive evaporation takes place. This gives rise to solution compositions very different from seawater, often containing considerable amounts of carbonate, borate and lithium salts (Table 4). Especially because of the borate and lithium contents these lakes are of increasing interest for industrial use. As a result Chile has became the third largest producer of lithium compounds due to exploitation of the Salar de Atacama and China has announced large investments to recover lithium, borate and potassium from the salt lakes in the Qinghai – Tibet area. For the economic production of these salts the formation of the salt lakes brines and their dependence on climatic conditions must be well understood and suited technological extraction schemes have to be developed. As can be seen from Table 4 in one and the same lake, the salt composition varies strongly, depending on the location of the samples. Both require the determination of the solid-liquid phase diagrams

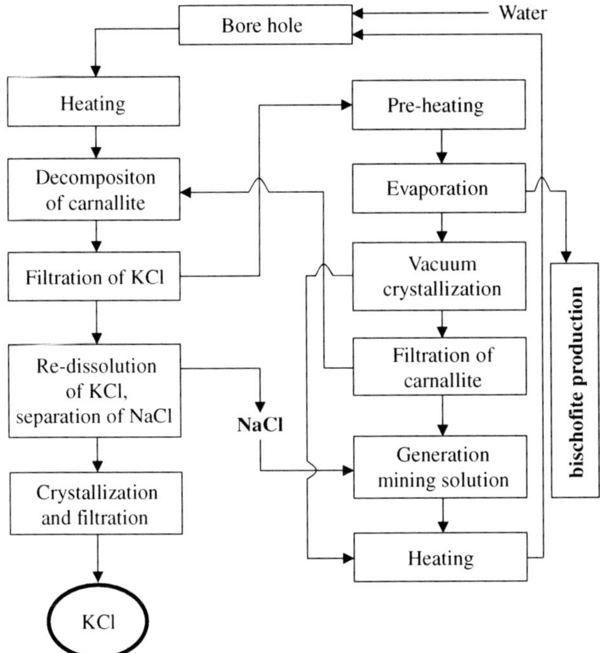

Figure 5 *Technological scheme of processing solutions from carnallitite solution mining.*

of the corresponding multi-component system involving Li^+, $B_4O_7^{2-}$ and CO_3^{2-} besides the major ions Na^+, K^+, Mg^{2+}, Cl^- and SO_4^{2-}. In certain circumstances the silicate equilibria has also to be considered. As can be seen from Table 4, the lithium content seldom exceed 500 ppm. Also the small lithium cation possesses a tendency to be absorbed in or on the crystals of other salts. Therefore, the development of economic extraction schemes for lithium presumes a thorough knowledge of the corresponding phase equilibria. In recent years a number of papers have been published dealing with experimental determination of solubilities and thermodynamic modelling of lithium and borate containing systems.[47–59] Although 45 pure borate minerals with Na^+, K^+, Mg^{2+} and Ca^{2+} as cations are known[30] mostly the equilibria with tetraborates seem to be important for the crystallization processes from aqueous solutions at ambient conditions. Often lithium is separated as a carbonate salt possessing a relatively low solubility of 1.3 mass-% in water at room temperature. However, borates and boric acid also have solubilities within the same order of magnitude (see Table 5) and all solubilities are very dependent on the actual solution composition and temperature. In borate solutions the pH dependent condensation/hydrolysis equilibria have to be taken into account. Considering only sodium as the cation, 27 borate phases with 9 different Na_2O/B_2O_3 ratios and various contents of hydrate water can crystallize from aqueous solutions[31] between 0 and 100°C.

Table 4 *Composition of brings in several salt lakes*[31]

Lake	*Na (wt%)*	*K (wt%)*	*Mg (wt%)*	*Ca (ppm)*	*Li (ppm)*	*Cl (wt%)*	*SO_4 (wt%)*	*HCO_3 (ppm)*	*B (ppm)*
Salar de Uyuni (Bolivia)	8.2	0.66	0.64	–	321	14.8	1.1	–	187
	8.7	0.72	0.65	463	349	15.7	0.85	333	204
	–	1.10	–	–	540	–	–	–	525
Salar de Coipasa (Bolivia)	7.5	1.10	1.36	156	350	15.1	2.5	747	786
Salar de Atacama (Chile)	91.1	23.6	9.65	0.45	1.57	189.5	15.9	0.23	0.440
(data in g/L)	85.5	13.0	6.30	1.10	0.94	163.9	8.5	0.28	0.360
	45.1	9.0	5.3	0.90	0.52	83.8	18.2	0.24	0.360
	14.8	2.9	1.9	1.10	0.19	27.5	7.9	0.10	0.088
Da Qaidam(China)									
Intercryst.	5.63	0.44	2.02	200	310	13.4	3.4	600	621
Surface	7.77	0.36	1.17	300	182	14.2	2.0	2100	867
Quinghai Lake (China)	3.93	0.16	0.79	100	0.84	5.8	2.3	6800	5
Mahai(China)									
Intercryst.	8.08	0.16	0.96	700	51	10.8	2.3	–	145

Table 5 *Solubility of some compounds relevant to salt lake composition*

Compound	*T(°C)*	*Solubility (wt%)*
$Li_2B_4O_7 \cdot 3H_2O$	15	2.55
$Na_2B_2O_7 \cdot 10H_2O$	15	2.13
$K_2B_2O_7 \cdot 4H_2O15$	15	11.5
Li_2CO_3	25	1.3
H_3BO_3	25	5.5

24.4 Salt Phase Formation in Building Materials

Solubility driven dissolution – precipitation reactions occur during setting and hardening of gypsum and cement-based binders and building materials. Phase formations in the primary, relatively fast proceeding, setting reactions are kinetically controlled. However, for the long-term durability of concrete buildings, subjected to an aqueous or humid environment, the stability of the primarily formed silicates, aluminates and sulfates depends on solubility relationships. Deterioration of Portland cement-based products by intrusion of sulfate containing solution is a well-known phenomenon. Until recently, the formation of gypsum and ettringite (composition: $3CaO*Al_2O_3*3CaSO_4*32H_2O$) were assumed to be the only sources of disintegration of concrete by sulfate attack. Situation changed when the UK government's Thaumasite Expert Group stated the formation of thaumasite (composition: $CaSiO_3*CaSO_4*CaCO_3*15H_2O$) was the main factor for destructive sulfate attack in a number of UK motorway bridge foundations in 1998.[32] Later on it was shown that damages due to thaumasite formation happen much more frequently, especially in tunnel buildings at temperatures below 20°C.[34] In recent years much effort has been made to understand the chemical conditions of thaumasite formation. Thaumasite is structurally similar to ettringite as can be recognized by comparing the structure-based formulas: thaumasite[35] $\{Ca_3[Si(OH)_6](SO_4)(CO_3)(H_2O)_{12}\}$ ettringite[36] $\{Ca_3[Al(OH)_6](SO_4)_{1.5}(H_2O)_{13}\}$ Thus, solid solution formation was observed[37] and investigated in equilibrium with aqueous solutions.[38] Experimental determinations of solubility equilibria of thaumasite at 8°C[39] and at 5, 15 and 30°C[38] established the basis for thermodynamic analyses of formation and stability relations of thaumasite in the system $CaO–Al_2O_3–SiO_2–CaSO_4–CaCO_3–H_2O$.[40,41] Further experimental work was stimulated,[42,43] which can be discussed now in the framework of thermodynamic models derived from solubility equilibria. The first conclusions on measures to prevent damages caused by thaumasite were drawn up and new standards for testing the sulfate resistance of concrete have to be introduced.[34]

Another category of inorganic binders includes sorel cement, also called magnesia cement. It is often used in potash mines to erect dams or to make them tight against intruding salt solutions. Salt cement is prepared from mixtures of MgO, concentrated $MgCl_2$ solution and filler materials (calcite, dolomite, halite, sand). The durability of such a dam is governed by the stability of the basic magnesium chloride phase, $Mg_2(OH)_3Cl \cdot 4H_2O$, which is formed in the sorel cement. Solutions with very low or no dissolved $MgCl_2$ will extract the $MgCl_2$ from the this binder phase and decompose it to $Mg(OH)_2$, brucite. The question arises, what is the lowest $MgCl_2$ concentration still compatible with the sorel phase. New systematic solubility experiments in the system $MgCl_2$–$Mg(OH)_2$–NaCl–NaOH–H_2O at 25°C and subsequent thermodynamic analysis applying Pitzer' equations up to ionic strengths of 15 mol $(kg\ H_2O)^{-1}$ yielded the solubility constants[44]

$$\log K_S^o\,(Mg_2(OH)_3Cl \cdot 4H_2O) = 26.0 \pm 0.2$$
$$\log K_S^o\,(Mg(OH)_2) = 17.1 \pm 0.2$$

These new constants remove doubts from old experimental data and confirm earlier estimations of the solubility constants.[14] At the same time, conclusions drawn from the data of Harvie, Möller and Weare[14] are supported by the new results that the stability of sorel cement in saturated NaCl solutions can be guaranteed down to $MgCl_2$ concentrations of 0.5 $mol(kg\ H_2O)^{-1}$.[45]

24.5 Salt Hydrates for Heat Storage

In considerations concerning solubility, attention is directed to the amount of solid dissolved in a certain amount of solvent. If the solubility increases with temperature, less and less solvent is necessary to obtain a homogeneous solution. For many salt hydrates at a certain temperature no additional water is necessary to get a homogenous liquid, that is, the hydrate congruently melts in its own hydrate water. Focussing on this melting temperature, solubility diagrams are inverted to fusibility diagrams by exchange of the axes: temperature becomes the dependent variable and composition is used as an independent variable. Figure 6 gives an example for the system $Mg(NO_3)_2$–H_2O with the congruent melting salt hydrates $Mg(NO_3)_2 \cdot 6H_2O$ and $Mg(NO_3)_2 \cdot 2H_2O$. Most of the salt hydrates show incongruent (peritectic) melting that is simultaneous with the formation of a liquid phase, a new solid phase is librated. The latter is an anhydrous salt or a lower hydrate. Glauber salt ($Na_2SO_4 \cdot 10H_2O$) represents a practical important example for an incongruent melting hydrate (see Figure 7). In this case melting occurs at 32°C (305 K) and most of the sodium sulfate settles down in form of the anhydrous salt. Regardless of whether the melting proceeds congruently or incongruently, it is generally accompanied by the absorption of a relative large amount of heat (enthalpy of fusion). The structural reason can be seen in a partial breakdown of the hydrogen network and the partial dehydration of ions on melting. Typical enthalpies of melting[60] are in the range 150–320 kJ kg^{-1}. Since salt hydrates have densities much larger than one (often between 1.2 and 2.5 g cm^3) the volume specific heat is much higher than for other compounds melting at temperatures below 150°C.[60] Crystallizing the salt hydrate just below the melting temperature would liberate the same amount of heat in the form of heat of crystallization. Thus, salt hydrates represent an interesting class of heat storage materials, also named

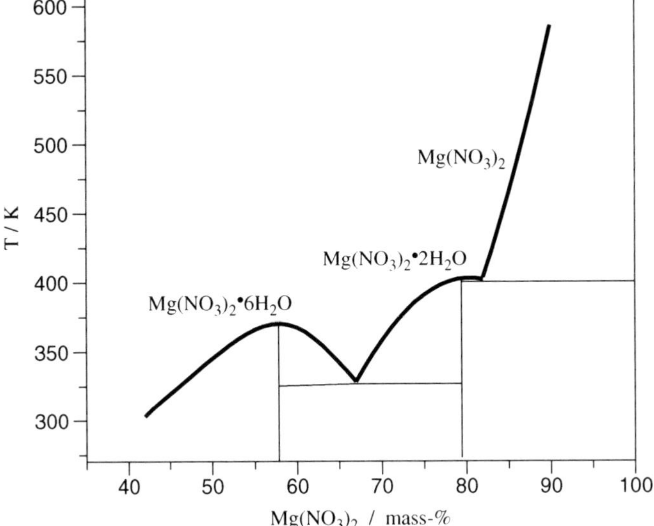

Figure 6 *Part of the melting diagram of the system $Mg(NO_3)_2$–H_2O.*

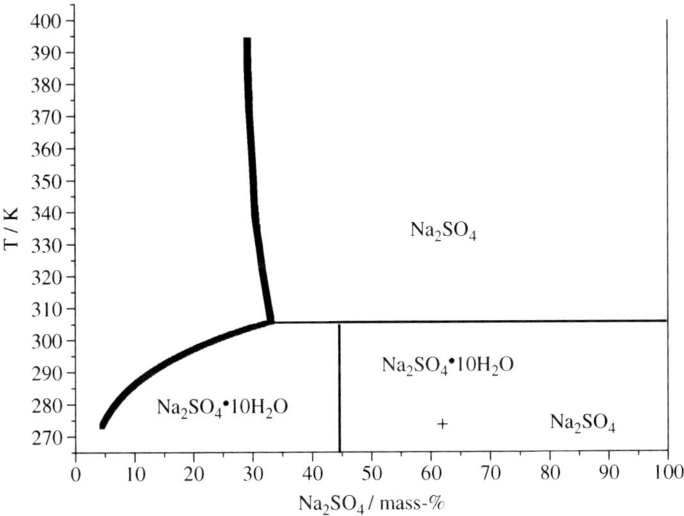

Figure 7 *Part of the melting diagram of the system Na_2SO_4–H_2O.*

"latent heat storage materials" or more often simply "phase change materials" (PCM). PCM's can only be applied in the range of its melting-crystallization temperature. However, for practical applications of a PCM further properties have to meet practical demands, for example chemical stability, minimal super-cooling of the melt, vapour pressure, price, toxicity, *etc.*[61] The number of simple fusible salt hydrates is quite limited. In order to increase the number of possible PCM candidates in a given temperature range, a systematic search for eutectic mixtures of salt hydrates have been started on the basis of the modified BET model.[62] The idea of transferring the formalism of the BET adsorption model for gases on solid surfaces to water of hydration in very concentrated salt solutions was first formulated by Stokes and Robinson[63] and resulted in the simple Equation (3) describing the water activity of binary salt solutions at low water activities a_{H} (mostly at or below 0.5).

$$\frac{(1-x_{\mathrm{H}})\cdot a_{\mathrm{H}}}{x_{\mathrm{H}}(1-a_{\mathrm{H}})}=\frac{1}{c_{\mathrm{A}}\cdot r_{\mathrm{A}}}+\frac{c_{\mathrm{A}}-1}{c_{\mathrm{A}}\cdot r_{\mathrm{A}}}a_{\mathrm{H}} \tag{3}$$

with $c_A = \exp[(E_A - E_L)/RT] = \exp[(-\varepsilon_A)/(RT)]$, x_{H} the mole fraction of water.

The model needs only two adjustable parameters, r and ΔE, which should be determined from activity data of binary salt–water systems at high concentrations. Later derivations yielded expressions for the salt activity in binary systems[64,65] and finally for all activities in ternary and higher component systems.[66] The model has been applied to predict eutectic compositions in a number of ternary systems.[62,67] Several of them could be used in heat storage applications.

References

1. D. Freyer and W. Voigt, *Geochim. Cosmochim. Acta*, 2004, **68**, 307.
2. Th. Fanghänel, H.-H. Emons and K. Köhnke, *Z. Anorg. Allg. Chem.*, 1989, **576**, 99.
3. Th. Fanghänel and H.-H. Emons, *Abh. Sächs. Akad. Wiss. (Leipzig)*, 1992, 57, 1. Math.-Naturwiss.
4. J.H. van't Hoff, *Zur Bildung der ozeanischen Salzablagerungen*, Friedrich Vieweg & Sohn, Braunschweig, 1905.
5. J.H. van't Hoff, *Untersuchungen über die Bildungsverhältnisse der ozeanischen Salzablagerungen insbesondere des Stassfurter Salzlagers*, Akademische Verlagsges. m.b.H., Leipzig, 1912.
6. R. Beck, H.-H. Emons and H. Holldorf, *Freib. Forschh.*, 1981, **A628**, 7, 19.
7. H. Holldorf and N. Menzel, *Freib. Forschh.*, 1984, **A690**, 27.
8. N. Menzel and H. Holldorf, *Freib. Forschh.*, 1984, **A690**, 46.
9. E. Kropp and H. Holldorf, *Freib. Forschh.*, 1988, **A764**, 67.
10. W. Voigt, *Freib. Forschh.*, 1999, **A853**, 5.
11. R. Cohen-Adad, D.B. Hassen-Chehimi, L. Zayani, M.-Th. Cohen-Adad, M. Trabelsi-Ayedi and N. Kbir-Ariguib, *CALPHAD*, 1997, **21**, 521.
12. R. Cohen-Adad, M.-Th. Cohen-Adad, Ch. Balarew, St. Tepavitcharova, W. Voigt, L. Zayani, D.B. Hassen-Chehimi and S. Mancour-Billah, *Monatsh. Chem.*, 2000, **131**, 25.
13. I. Stahl, W. Beer, K. Wambach-Sommerhoff and R. Keidel, Produkte der Kaliindustrie, in *Chemische Technik: Prozesse und Produkte*, Vol 8, *Ernährung, Gesundheit, Konsumgüter*, Winnacker, K. (ed), Wiley-VCH Verlag, GmbH & Co. KGaA, Weinheim, 2005.

14. C.E. Harvie, N. Möller and J.H. Weare, *Geochim. Cosmochim. Acta*, 1984, **48**, 723.
15. R.T. Pabalan and K.S. Pitzer, *Geochim. Cosmochim. Acta*, 1987, **51**, 2429.
16. N. Möller, *Geochim. Cosmochim. Acta*, 1988, **52**, 821.
17. J.P. Greenberg and N. Möller, *Geochim. Cosmochim. Acta.*, 1989, **53**, 2503.
18. M.D. Yuan and A.C. Todd, *SPE Prod. Eng.*, 1991, 63.
19. G. Ziegenbalg, H.-H. Emons and Th. Fanghänel, *Kali & Steinsalz*, 1991, **10**, 366.
20. S.V. Petrenko, V.M. Valyashko and G. Ziegenbalg, *Zh. Neorg. Khim.*, 1992, **37**, 2111.
21. S.L. He and J.W. Morse, *Geochim. Cosmochim. Acta*, 1993, **57**, 297.
22. H. Voigt and W. Voigt, Development of a standard data file (bfs) for application in geochemical modelling (Ger.). Final report for BfS (Bundesamt für Strahlenschutz), Salzgitter, Germany, 1998, Vol 1–4.
23. G. Marion and R.E. Farren, *Geochim. Cosmochim. Acta*, 1999, **63**, 1305.
24. Ch. Monnin, *Chem. Geol.*, 1999, **153**, 187.
25. G. Marion, *Geochim. Cosmochim. Acta*, 2001, **65**, 1883.
26. Ch. Christov and N. Möller, *Geochim. Cosmochim. Acta*, 2004, **68**, 1309.
27. Ch. Christov and N. Möller, *Geochim. Cosmochim. Acta*, 2004, **68**, 3717.
28. D. Freyer, W. Voigt and V. Böttge, Kali & Steinsalz, 2006, 28.
29. F. Walkhoff and K. Ulrich, *Freib. Forschh.*, 1999, **A853**, 94.
30. D.E. Garrett, *Borates – Handbook of Deposits Processing, Properties and Use*, Academic Press, San Diego, London, NY, Sydney, Tokyo, 1998.
31. N.P. Nies, *Alkali metal borates: Physical and chemical properties, in Inorganic and Theoretical Chemistry*, 1980, **5**(A), 343.
32. Report of the Thaumasite Expert Group, DETR, London, 1999.
33. W. Voigt, *Pure Appl. Chem.*, 2001, **73**, 831.
34. F. Bellmann, *Zur Bildung des Minerals Thaumasit beim Sulfatangriff auf Beton*, Diss. Bauhaus-University, Weimar, 2005.
35. R.A. Edge and H.F.W. Taylor, *Nature*, 1969, **224**, 363.
36. A.E. Moore and H.F.W. Taylor, *Acta Cryst.*, 1970, **26**, 386.
37. S.J. Barnett, D.E. Macphee, E.E. Lachowski and N.J. Crammond, *Cem. Concr. Res.*, 2002, **32**, 719.
38. D.E. Macphee and S.J. Barnett, *Cem. Concr. Res.*, 2004, **34**, 1591.
39. F. Bellmann, *Adv. Cem. Res.*, 2004, **16**, 55.
40. D. Damidot, S.J. Barnett, F.P. Glasser and D.E. Macphee, *Adv. Cem. Res.*, 2004, **16**, 69.
41. F. Bellmann, *Adv. Cem. Res.*, 2004, **16**, 89.
42. E.F. Irassar, V.L. Bonavetti, M.A. Trezza and M.A. Gonzalez, *Cem. & Concr. Compos.*, 2005, **27**, 77.
43. G. Collett, N.J. Crammond, R.N. Swamy and J.H. Sharp, *Cem. Concr. Res.*, 2004, **34**, 1599.
44. M. Altmeier, V. Metz, V. Neck, R. Müller and Th. Fanghänel, *Geochim. Cosmochim. Acta*, 2003, **67**, 3595.
45. W. Voigt, *Expertise on thermodynamic stability of gel systems from* $MgCl_2$ *solutions*, unpublished.

46. E. Usdowski and M. Dietzel, *Atlas and Data of Solid-Solution Equilibria of Marine Evaporates*, Springer-Verlag, Berlin, Heidelberg, New York, 1998.
47. X. Liu, K. Cai and S. Yu, *Sci. in China*, 2004, **47**, 720.
48. Y. Yao, P.S. Song and J. Zhang, *Ocean. and Limin.*, 1999, **30**, 6.
49. B. Li, B. Sun and C.H. Fang, *Acta Chim. Sin.*, 1997, **50**, 545.
50. P. Song and Y. Yao, *Calphad*, 2001, **25**, 329.
51. P. Song, Y. Yao and J. Li, *Prog. Chem.*, 2000, **12**, 255.
52. Shi-hua Sang, Hui-an Yin, Ming-lin Tang and Ning-Fei Lei, *J. Chem. Eng. Data*, 2004, **49**, 1586.
53. Tianlong Deng, *J. Chem. Eng. Data*, 2004, **49**, 1295.
54. Shi-hua Sang, Hui-an Yin and Ming-lin Tang, *J. Chem. Eng. Data*, 2005, **50**, 1557.
55. Ying Zeng and Ming Shao, *J. Chem. Eng. Data.*, 2006, **51**, 219.
56. Tian-Long Deng, Hui-An Yin and Min-Ling Tang, *J. Chem. Eng. Data*, 2002, **47**, 26.
57. Ai-Yun Zhang, Yan Yao, Li-Juan Li and Peng-Sheng Song, *J. Chem. Thermodyn.*, 2005, **37**, 101.
58. Dewen Zeng and Jun Zhou, *J. Chem. Eng. Data.*, 2006, **51**, 315.
59. Ying Zeng, Hongmei Yang, Huian Yin and Minglin Tang, *J. Chem. Eng. Data*, 2004, **49**, 1648.
60. W. Voigt and D. Zeng, *Pure Appl. Chem.*, 2002, **74**, 1909.
61. G.A. Lane, *Solar Heat Storage: Latent Heat Material*, Vol I, CRC Press, Inc., Boca Raton, 1986.
62. D. Zeng and W. Voigt, *CALPHAD*, 2003, **27**, 243.
63. R.H. Stokes and R.A. Robinson, *J. Am. Chem. Soc.*, 1948, **70**, 1870.
64. M. Abraham, *J. Chim. Phys.*, 1981, **78**, 51.
65. W. Voigt, *Monatsh. Chem.*, 1993, **124**, 839.
66. M.R. Ally and J. Braunstein, *J. Chem. Thermodyn.*, 1998, **30**, 49.
67. D. Zeng, *Thermodynamische Modellierung von Salz – Wasser – Systemen von der Lösung bis zur Salzschmelze*, Dissertion, TU Bergakademie Freiberg, 2003.

Subject Index

Printed in the United Kingdom by
Lightning Source UK Ltd., Milton Keynes
138863UK00001B/50/P